T0297857

CAMBRIDGE BIOLOGICAL SERIES.

GENERAL EDITOR :—ARTHUR E. SHIPLEY, M.A.
FELLOW AND TUTOR OF CHRIST'S COLLEGE, CAMBRIDGE.

ZOOLOGY

ZOOLOGY

AN ELEMENTARY TEXT-BOOK

BY

A. E. SHIPLEY, M.A.,

FELLOW AND TUTOR OF CHRIST'S COLLEGE, CAMBRIDGE,
AND UNIVERSITY LECTURER IN ADVANCED MORPHOLOGY OF THE
INVERTEBRATA;

AND

E. W. MacBRIDE, M.A. (Cantab.), D.Sc. (Lond.),

SOMETIME FELLOW OF ST JOHN'S COLLEGE, CAMBRIDGE,
PROFESSOR OF ZOOLOGY IN McGILL UNIVERSITY, MONTREAL.

SECOND EDITION

CAMBRIDGE:
AT THE UNIVERSITY PRESS
1904

CAMBRIDGE
UNIVERSITY PRESS

University Printing House, Cambridge CB2 8BS, United Kingdom

Published in the United States of America by Cambridge University Press, New York

Cambridge University Press is part of the University of Cambridge.

It furthers the University's mission by disseminating knowledge in the pursuit of education, learning and research at the highest international levels of excellence.

www.cambridge.org
Information on this title: www.cambridge.org/9781107655508

© Cambridge University Press 1904

First edition 1901
Second edition 1904
First published 1904
First paperback edition 2014

A catalogue record for this publication is available from the British Library

ISBN 978-1-107-65550-8 Paperback

PREFACE.

WE have tried in the following book to write an elementary treatise on Zoology which could readily be understood by a student who had no previous knowledge of the subject. We have endeavoured to explain the technical terms as they occur, and since one of the difficulties proper to the science of Zoology is the enormous number and the prodigious length of these terms, we have in many cases given derivations which may help the beginner to fix them in his mind.

The attempt to construct the book on the plan that each section is built on what precedes it, has rendered it impossible to keep the treatment of the various groups at the same level, for much space is taken up in the earlier chapters in explaining processes a knowledge of which is assumed in the later. A book such as we have aimed at is bound to be progressive, and the later chapters will be intelligible to the beginner only if he has read the earlier. Thus the part of this book which deals with the Vertebrata is in many respects more advanced than those which deal with the several Invertebrate groups.

In order to give some account of the leading types of animal structure within a book of moderate compass, we have been compelled to make but scanty reference to Histology, Embryology and Palaeozoology; in fact the book in the main deals with the normal structure of the adult forms of recent

animals. Wherever possible we have endeavoured to exhibit
this structure as the outcome of function and habit. We have
tried to show that Zoology deals at least as much with living
as with dead organisms.

In tracing the relationship of the animals described to one
another we have at times put forward hypotheses which we fear
will not commend themselves to all zoologists, but we have
thought it better to run the risk of submitting views which
further research may compel us to abandon rather than leave
the student with the idea that the object of zoological study
is the mere collection of facts. We try everywhere to make
it clear that the ultimate end of the science is the discovery
of the laws underlying and binding together the facts.

At the end of the sections dealing with each phylum a
short table of classification is given. These tables do not
attempt to be complete but are intended to indicate the position of the animals mentioned in the text in the general scheme
of classification, and since the book appears in both an English
and an American edition, these examples are in most cases
drawn from the British and North American Faunas.

In one respect this book differs from many of the elementary treatises which have appeared within the last few
years. It has been drawn up with an eye to no examination
and does not claim to correspond with any of the numerous
syllabuses and schedules, issued from time to time by the
various Boards of Examiners scattered through the United
Kingdom and North America.

Many of the illustrations are new and whatever merit
they possess is due to the skill of Mr E. Wilson of Cambridge
and Mr F. M. Howlett of Christ's College. We owe our
grateful thanks to Mr S. H. Reynolds for permission to use
many of the illustrations of his book on "The Vertebrate

Skeleton," and to Messrs Macmillan & Co. and Messrs A. and C. Black and the Council of the Royal Agricultural Society for granting the use of certain other illustrations. We are also much indebted to many friends for help in special chapters. Mr E. S. Thompson and Mr J. Graham Kerr have read through the proof-sheets, a most tedious task, as the authors can abundantly testify; Dr Harmer, Dr Gadow, Dr Anderson, Dr Hopkins, Mr G. P. Bidder, Mr J. Stanley Gardiner, Mr R. Evans, Mr H. H. Brindley, and Mr C. Warburton, have most freely given us the help of their special knowledge. Whilst saving us from many mistakes they are by no means responsible for those that remain. We tender them all our sincere thanks.

A. E. S.
E. W. M.

PREFACE TO THE SECOND EDITION.

IN issuing a second edition of this Zoology, which has been for some time out of print, the authors desire to thank many critics who have pointed out errors in the book. They are again indebted for much help to the gentlemen mentioned in the former Preface, and above all they owe thanks to Mr H. H. Brindley of St John's College, who has kindly read the whole of the proofs and whose critical power has been most unreservedly placed at the use of the authors. In the subjects which they have made their own Dr Gadow and Mr L. A. Borradaile of Selwyn College have given the writers much valued help.

A. E. S.

E. W. M.

March, 1904.

TABLE OF CONTENTS.

b 2

LIST OF ILLUSTRATIONS.

CHAPTER I.

INTRODUCTION.

THE word Zoology (Gr. ζῶον, an animal; λόγος, an account)
denotes the science which concerns itself with animals,
Life. endeavouring to find out what they are and how they
came into being. It is a branch of the wider science of Biology
(Gr. βίος, life, λόγος, a discourse)[1], which deals with all living things,
plants as well as animals. Before any progress can be made with
the study of Zoology, it is necessary to get clear ideas on two points:
firstly, as to what is meant by life and living things; and secondly,
as to how an animal is to be distinguished from a plant.

The idea implied in calling a thing living, is that in some
respects its existence is similar to our own. Our own existence is
the only thing immediately known to us, the standard with which
we compare everything else. Every material object has certain
points of resemblance to our bodies, inasmuch as all are composed
of matter obeying the same laws of chemical affinity, gravitation,
and so forth; it is necessary therefore to define the amount of resemblance which constitutes life. Now everyone knows that human
beings grow, that is, increase in size at the expense of matter
different from themselves called food, and that further, they give
rise at intervals to fresh human beings. These two fundamental
characteristics—the power of growth and of multiplication—define
life; everything that can increase its bulk by building up foreign
matter into itself and that reproduces its like is said to be alive.

The idea originally underlying the word animal was a self-
moving object, as distinguished from a plant which was regarded
as motionless[2]. This distinction, however, will not stand close

[1] This term is too well established to admit of alteration but it implies a
mistranslation of βίος. This does not mean 'life' in the physiological sense but
a period of life, a career, a life-time or circumstances of life, environment.

[2] It is true that to all general statements of Zoology, as to this, exceptions
could be found. The rule followed in this book is to have regard only to the

examination. Plants as well as animals move, and although the motions of animals are conspicuous and such as to catch the eye, whilst those of plants are usually slow and imperceptible, yet there is no essential difference between the nature of the movements in the two cases.

The fundamental difference between animals and plants is to be found in the nature of their food. Animals can only live on complex substances, not very different in chemical composition from their own bodies, and further, they can live on solid food. Plants, on the other hand, build themselves up out of carbon di-oxide and other gases and water with a few simple salts in solution, and they only take in fluids or gases. There are, however, a certain number of living beings which combine the characters of animals and plants, and the question in which division they should be ranked is a matter to be determined only after a study of the special circumstances of each case.

Distinction between animals and plants.

It has been pointed out that our own existence is the original type from which the idea of life is derived. But we know ourselves not only as bodies in which growth and reproduction occur, but also as conscious, thinking beings, and we are naturally inclined to imagine that animals at least, which not only grow and multiply, but in many other respects also resemble us, are likewise conscious. How far this belief is well-founded is open to serious question, if by consciousness we mean anything at all resembling our own inner life— the only consciousness we know anything about. The movements of the higher animals suggest that they experience the feelings of fear, anger, desire, etc., and it would be foolish to deny all similarity between them and man in these respects, but the habit many people have of uncritically attributing purely human feelings to dogs, cats, horses, etc., is apt to lead us into serious error. Our forefathers went further than even we are inclined to do and supposed all natural objects, the sun, wind, trees, etc., to have spirits, that is, to be conscious. Since we can never learn much about the consciousness of beings with whom we cannot speak, zoologists content themselves with looking at animals entirely from the outside, without enquiring as to whether or no they are conscious ; animals are for them bodies in which certain changes take place, changes such as growth, reproduction, movement, and others.

vast bulk of normal cases which gave rise to the *idea*. The reasons for classifying abnormal cases in one category or another are not general but special, and have to be considered in each case.

A close study of animals reveals the fact that though the chemical constitution of no two is exactly alike, yet all contain certain

Protoplasm.

highly complex substances of very obscure chemical composition, known as proteids. These substances occur in the form of a thick, viscous fluid, in which are suspended not only numerous solid granules of most varied composition, but also minute drops of other fluids. Such a mixture is called by chemists an emulsion, and it is the emulsion just described which is the seat of all those processes which we call life. This emulsion is termed protoplasm (Gr. πρῶτος, first ; πλάσμα, a thing moulded).

Further, it has been found that, so long as any sign of life is visible, this protoplasm is in a continual state of slow combustion, absorbing oxygen from outside and decomposing with the liberation of energy, and whilst some of the products of decomposition are cast off, others· apparently reconstitute the original substance by combining with some of the materials of the food. The energy liberated is the cause of the movements which constitute the visible manifestation of life.

An animal then is only the more or less constant form of a flow of particles ; it may be compared to a flame, which has a constant form, although the particles which compose it vary from moment to moment ; unburned particles coming in at one end and the oxidised products escaping at the other.

The deepest insight which can be obtained into the nature of

Metabolism.

life viewed as a series of changes in the shape and position of bodies reveals to us this continual chemical change as the ultimate cause of all manifestations of life. It is known by the convenient name of metabolism (Gr. μεταβολή, change, changing). The ultimate object of Zoology is therefore to discover the nature, cause, and conditions of the metabolism in the case of every animal ; but the means of attaining this object are still to seek, and for the most part the zoologist has to be content with describing and comparing with one another the outer and visible effects of the metabolism in various cases.

The proteids, which form the essential basis of protoplasm, consist of carbon, nitrogen, hydrogen, oxygen, and sulphur ; besides these elements phosphorus, chlorine, potassium, sodium, magnesium, calcium and iron are constantly found in the bodies of animals, and some of them are doubtless chemically combined with the proteid. Phosphorus is a constituent of nucleic acid, a substance which in combination with proteid is characteristic of the nucleus (see

p. 16). Proteids have a percentage composition which varies somewhat, though not widely, in different cases.

Carbon	from	50	to	55	per	cent.
Hydrogen	„	6·5	to	7·3	„·	„
Nitrogen	„	15	to	17·6	„	„
Oxygen	„	19	to	24	„	„
Sulphur	„	·3	to	2·4	„	„

The size of the molecules of which proteids are composed is undoubtedly a large one. It is difficult if not impossible to determine exactly how many atoms are contained in a molecule of a particular proteid because it is difficult to obtain one such substance in a pure condition free from admixture with others. The best determinations which have been made show however that at least 1000 atoms must be contained in the molecule. But the proteids known to the chemist are of course taken from the dead bodies of animals and are themselves to be regarded as products of the decomposition of the molecules which existed during life. The proteid as the seat of life has probably a decidedly different composition from the dead substance. To avoid confusion, we may call the living molecules biogens.

The biogen molecule is continually absorbing oxygen from the outside. This process is called respiration or breathing. It decomposes and some of the products are no longer capable of being built up again into other biogen molecules and are therefore got rid of, since otherwise they would interfere with the chemical action, just as accumulating ashes will eventually put out a fire. The process of ejecting these waste products is called excretion, the waste substances themselves, excreta, and the chemical changes which lead to their production, katabolism (Gr. καταβολή, deposition). The commonest excreta are water, carbon dioxide, urea, and uric acid; the last two substances containing nitrogen. But it is not necessary that in all cases excreta should be ejected. They may remain within the bounds of a mass of protoplasm; if they are removed from the sphere of chemical action of the protoplasm this is sufficient. In some animals uric acid is stored up in this way. Many of the excreta, though injurious if they remain in the protoplasm, are indirectly useful to the animal after ejection. Such useful excreta are called secretions. Thus, all the hard skeletons of animals are really insoluble excreta. On the other hand, the gastric juice which digests the food in the human stomach, and the slime or mucus, which prevents a frog from drying up when taken out of water, are fluid excreta. A part of the body specially adapted to produce a secretion is termed a gland.

Other products of decomposition reconstitute, as we have seen, the original molecule by combining with the necessary elements from the food; this process is known as anabolism (Gr. ἀναβάλλειν, to put back or up) or assimilation. Inasmuch as, generally speaking, from the breaking up of one molecule more than one residue is produced capable of regeneration, there is an increase in the number of biogen molecules causing an increase in bulk of the protoplasm, or growth[1].

The regeneration of the biogen takes place at the expense of the food. Taking in food is called eating, or ingestion. Since however, the food must penetrate to every portion of the protoplasm it must be dissolved—a process effected by the chemical action of certain products of the decomposition of the biogens, known as ferments. The process is called digestion. The casting out of an insoluble remnant of the food is called defaecation, and inasmuch as such remnants have never formed part of the biogen molecule, this process is carefully to be distinguished from excretion. The accumulation of excreta soon stops metabolism, whereas the intermission of defaecation need only interfere very slightly with metabolism.

Of the numerous solid particles found in protoplasm some are secretions, others are solid deposits of partly assimilated food, which act as reserve stores, others are indigestible remains. The fluid drops consist largely of water—some have in solution excreta or secretions ; others contain the results of digestion.

Animals, as we have seen, possess the power of executing movements ; this power is exercised in order to seek their food and escape their enemies. However complicated these movements may be, they are all found to be dependent on the capacity of protoplasm to alter its shape, suddenly contracting and then slowly expanding. By contraction is meant such an alteration of shape of the moving part as will tend to diminish its surface but not its bulk; that is, the contracting part tends to assume a spherical shape ; by expansion, on the other hand, is meant an alteration of shape leading to increase of surface. A bird flies by contracting the muscles first on one side of the wing, then on the other ; a fish swims by alternate contractions of the two sides of the fleshy tail. Any part of an animal fitted to execute

Movement.

[1] See Verworn, *General Physiology* (Engl. Edition), 1899, p. 486.

movements more quickly in one direction than in another and so to bring about the movement of the whole animal, is called a locomotor organ. Protoplasm in which the power of contraction is highly developed is called muscle.

A contraction is the result of an explosive decomposition of the living substance; there have been a great many theories as to how the chemical change brings about the change of shape but, since all of them account for some of the facts and none of them for all, there is no need to mention any of them here.

The sudden chemical change which brings about contraction, although dependent on the unstable character of the biogen molecule, must be precipitated by some change occurring either in the living matter itself or in the surrounding medium, just as an explosion of gunpowder is not brought about without a spark. In either case the change causing the contraction is known as a stimulus, and the capacity of contracting under the influence of stimuli is known as irritability. Thus when a moth flies into a flame it is acting under the stimulus of light; when a hungry lion in the Zoological Gardens rises up and commences running violently round its cage it is obeying the stimulus of hunger. In the first case we have to deal with an external stimulus, in the second with an internal one. Of course since all internal changes are ultimately due to changes in the surrounding medium,—e.g. hunger to a disappearance by digestion of the food in contact with the stomach,— the distinction between external and internal stimuli, though convenient, cannot be sharply drawn. The power of protoplasm to originate movement through internal changes is called automatism. In the case of external stimuli we can often observe that the disturbance caused at the point of application of the stimulus is propagated to widely different parts of the animal. Nerves contain protoplasm in which this power of transmission is powerfully developed.

We have seen that at some period in the life of all animals when food is abundant, more living matter is formed than is broken down; in a word, that the animal increases in size, grows. But whereas volume increases proportionately to the cube of the length (or breadth), surface increases only proportionately to the square of the same dimension. Hence the amount of volume per unit of surface continually increases, and thus the chemical action between the internal portions of the protoplasm and the surrounding medium, which can only go on through the

Growth.

surface, is slowed down; in other words, the activity of growth is checked and when a certain size is reached waste becomes equal to repair.   At this stage there is a tendency for the protoplasm to divide into two or more pieces of smaller size.   This division into smaller pieces is called Reproduction, and it is a necessary result of growth.   When an animal divides into two equal portions, the process is called fission, but when one portion is very much smaller than the other, the process is known as gemmation; the smaller portion is called the germ, and the larger the parent, since the latter is—somewhat illogically—regarded as identical with the original animal before division.   A germ very rarely resembles the parent; usually it has to undergo a series of changes during growth by which it at last attains the shape of the animal which gave rise to it; this series of changes in shape and size is known as Development.

Reproduction in the higher animals is closely associated with another process called Conjugation or Sexual Union.   This process consists in the coalescence with one another of two portions of living matter.   Conjugation probably occurs in all animals, but the interesting thing about the higher animals is that they give rise to special germs of two kinds, called ova (eggs) and spermatozoa respectively, which cannot develope without first conjugating, one of the first kind uniting with one of the second.

Reproduction.

The ovum is devoid of the power of movement and has a larger or smaller amount of undigested or at any rate unassimilated food stored in it; this reserve material is called yolk.   The spermatozöon, on the other hand, has no such reserve and is in consequence very much smaller than the ovum, but it possesses in nearly every case the power of movement by which it is enabled to seek and find the ovum.   Reproduction, which thus requires conjugation before development can take place is called Sexual Reproduction.   In most cases ova and spermatozoa are developed in different individuals. The individual giving rise to ova is called the female, that giving rise to spermatozoa the male.   In this case the animals are said to be bisexual.   When both ova and spermatozoa are developed in the same individual it is spoken of as hermaphrodite.

It is obvious to the most casual observation that there is an amazing variety of animals in the world.   Closer observation reveals the fact that while no two animals are exactly alike, all can be nevertheless sorted into a

Species.

number of kinds called species, the individuals composing which—apart from the difference between males and females—resemble each other exceedingly closely. Where the observation has been made, it is always found that the members of a species conjugate freely with one another; and indeed this is assumed to be the case in every species; that is, we group a number of specimens into a species under the assumption that they can conjugate with one another, and that young like themselves will develop as the result. If this can be shown to be not the case, we conclude that a mistake has been made and that two or more species have been confounded with one another. It follows that the vast majority of species rest on provisional hypotheses; these hypotheses nevertheless possess a very high degree of probability, for by the use of them only can the great resemblance between the individuals grouped together in the same species be accounted for. When, as occasionally happens, members of different species are fertile *inter se*, the offspring is termed a hybrid, and hybrids may or may not be fertile.

It has been pointed out, that whereas germs are in most cases exceedingly different from their parents, they never-theless in process of growth come to resemble them. This tendency to reproduce the characters of the parent is called heredity. If the germ undergoes a large part of its development within a hard case, like a chick within the eggshell or in a cavity of the parent's body, it is called an embryo; if it moves freely about, it is termed a larva.

Heredity and Variation.

In the case of the development of an animal which has originated sexually, that is from the coalescence of two germs, the tendency is for it to assume characters intermediate between those of the two parents. Thus it is easy to see how sexual reproduction tends to annul the differences existing between members of the same species, by constantly producing means between them. When therefore a large number of individuals are found with very close resemblances, it is a reasonable supposition that the agent, which has caused this, is sexual reproduction; in other words, that they constitute a species. It is not however to be assumed that in every case conjugation results in the production of an animal exactly inter-mediate in character between the parents. Sometimes the child resembles closely the father or the mother, a result denoted by the term prepotency of the father or of the mother. Some-times in an unexplained way an exaggeration of a character found in one or both parents is produced. Sometimes even an apparently entirely new character arises. Such deviations of the offspring from the average of the parents constitute variation. If the difference is striking the individual exhibiting it is called a sport.

It is obvious that so vast a science as Zoology must be divided into various branches, since the different questions it seeks to solve require that special attention should be given to each side of the subject. Thus, the nature and conditions of the metabolism and the mechanism by which movements are effected, etc., constitute the subject-matter of Physiology; the investigation of the structure of individuals and of the differences in structure between the various species and the search for the causes of these differences is termed Morphology; whilst Bionomics is the name given to the study of the means whereby an animal obtains its food and orders its life, in other words, of its habits. But it must be remembered that all such divisions are purely arbitrary, and indeed no great progress can be made in any one department if the others be ignored. Bionomics, when followed to its sources, passes into Physiology, and in trying to explain the different structures studied in Morphology constant recourse must be had to both Physiology and Bionomics.

Of all divisions of the subject, that of Physiology has been most neglected; it has indeed only been studied systematically in the case of man and of a few of the higher animals. Hence this work will be mainly concerned with the questions of Morphology and Bionomics. Of these questions, by far the greatest is the problem how the distinctions between the various species are to be explained. The question of the "Origin of Species" involves nearly all others in Zoology.

The distinctions between species are of very different degrees, so that for convenience species closely resembling each other are collected into genera—genera into families—families into orders—orders into classes—and classes into phyla. These are the names in commonest use, but often the nature of the subject requires the introduction of further grades of difference, and the number of grades actually employed depends to a large extent on the point to which the analysis is pushed.

The only theory of the origin of species which has so far commanded any considerable agreement amongst naturalists is the famous theory of Charles Darwin. According to this theory, the resemblances between a number of living species are due to the fact that these species are descended from a common ancestral species which possessed the common features as characters of its own. Therefore, the degree of likeness between species is the expression of a nearer or remoter blood relationship, and it logically follows that, since no part of the animal kingdom is without resemblances to the rest, if we recede far enough in time we reach a period when all the animals in the world constituted one species.

To a certain extent Darwin's theory was only the expression of ideas that had first occurred to Greek philosophers, and had in one form or other been put forward by many naturalists before him. His special merit lies in that he pointed out various processes at present going on in nature which must lead to the modification of species.

He recalled attention to the well-known fact, that although the offspring in general resemble the parents, yet this resemblance is never exact, and further that the young of one brood often differ quite perceptibly from one another, and that these differences are often inherited by the offspring of the individuals showing them. Such differences, as has been mentioned above, are denoted by the term Variation.

Again, another fact well-known but usually ignored, was emphasised by Darwin: viz., that if the state of the animal population of the globe remains fairly constant, out of all the young produced by a pair of parents during their lifetime on an average only two will survive, since if more were to live the species would inevitably increase in numbers. Hence since each animal tends to multiply at a rate at which if unchecked it would soon overrun the globe, a competition must result between the members of each species both for food and in the escape from enemies, as a result of which the "fittest" will survive. So long as the surroundings of the species remain the same, this struggle for existence will only weed out those individuals least perfectly adapted to their environment, so that the species will be kept up to a high level of adaptation to its surroundings. This elimination of imperfect individuals which results in the survival of the fittest is known as Natural Selection. Thus we can well imagine that if white-haired individuals turned up amongst hares, they would be more conspicuous and hence more easily discovered by the animals which prey on hares. If however the circumstances of a species change, a different class of individuals will survive. For instance, if for the greater part of the year the country inhabited by the hares were covered by snow, as is the case in the North of Canada, the whitest-haired individuals would have the best chance, and from generation to generation would be selected until the colour of the hare was totally changed. The progressive modification of species by the agency of natural selection is called Evolution. If the modification tends towards simplification of structure it is called Degeneration, if on the contrary it tends towards great complexity it is spoken of as Differentiation.

So far the theory shows how a species will become slowly modified as its surroundings change. But it has been postulated that distinct species have arisen from the same ancestors. It is of course not difficult to see that if a species is distributed over a wide area the conditions in different portions may vary independently of one another, and hence the species may become modified in one place in one direction and in another situation in a different direction by the agency of natural selection. So long however as the species inhabits a continuous area this tendency to split up into divergent groups will be checked by inter-breeding between the sections of the species which are thus becoming modified in different directions. But if through geographical changes the species becomes divided into groups of individuals cut off from access to another, then no interbreeding can take place and in time two species will be formed. Thus when birds have been blown far out to sea and have colonised a distant island they have often given rise to a new species. The same result may be brought about by the sea overflowing a part of the area inhabited by the species, an event which we know from geology to have often occurred. The important fact to be borne in mind is that at bottom the evolution of several species out of one is due to the formation of colonies, and that the same causes which have led to the differences between the American and the Englishman have acted again and again in the world's history so as to produce the marvellous variety of species inhabiting the globe, the only difference between human and animal colonies being that, in the latter case, the divergence has become so great that animal colonists will no longer breed with the original race. Thus, accepting Darwin's theory, we find it possible to give a rational explanation of those resemblances between animals which are expressed in a system of classification[1]. If the theory be rejected these resemblances are pure figments of the human mind, and the species must be regarded

---

[1] Most of the names employed in classification were in use before Darwin's views were accepted. The word phylum (Gr. φῦλον, tribe or stock) is however an exception. This term expresses the central idea of the evolution theory, and its proper use is to denote the whole of a group of animals characterised

as just as independent of one another as are the chemical atoms. Hence since it is a choice between this explanation or none, the Darwinian theory is accepted by the overwhelming majority of naturalists.

One or two interesting consequences follow from the acceptance of this theory. The structural features of animals are to be regarded as adaptations to their surroundings, since they have been built up by natural selection. Hence an isolated resemblance in a particular feature between two species need not necessarily indicate that this feature was present in the common ancestral species, for similar surroundings may have evolved a similar modification in two animals only remotely related. Such similarities are called Homoplasy, whereas resemblances believed to indicate blood-relationships are grouped under the term Homology.

Again, the immature forms of some animals are found to exhibit strong resemblances to the adults of others, and the eggs of all the highest animals show the strongest general resemblance to the simplest animals—the so-called Protozoa (Gr. πρῶτος, first, ζῷον, animal). If these resemblances are to be interpreted in the same way as those prevailing between adults—and it is illogical to refuse to do so—then we are driven to conclude that most animals in their development pass through stages when they exhibit many characters once possessed by their ancestors, commencing at the stage of the Protozoa. These latter animals, since they are about as simply constructed as we can imagine living matter to be, may be looked on as slightly modified survivors of the first animals which appeared on the globe.

This method of interpreting the changes which occur during development is what is known as the Recapitulation Theory, because during Ontogeny (Gr. ὄν, ὄντος, being) or the development of the individual, nature recapitulates to some extent the development of the species in past time, Phylogeny (φῦλον, a stock, a race). There are, however, a great many other factors which have modified development, and the determination of these and their separation from the hereditary factor is a task requiring careful study and one which is as yet far from complete.

by having the same ground-plan of structure and believed to be the descendants of a common ancestor, from whom no other living animals are descended. The essential feature about a phylum is its isolation, in the present state of our knowledge, from other phyla. Of course it is believed that at bottom all living beings constituted one phylum, but there are enormous differences in structure which can only be bridged by imaginative hypotheses.

CHAPTER II.

PHYLUM PROTOZOA.

THE Protozoa are distinguished from all other animals (1) by the fact that they do not produce ova and spermatozoa but that the whole animal engages in the processes of conjugation and reproduction, and (2) by the fact that the protoplasm of the body is never differentiated into tissues nor exhibits cellular structure (see p. 27)[1]. The higher animals are often grouped under the name Metazoa (Gr. μετά, after; ζῶον, an animal) in order to contrast them with the Protozoa, but whereas the Protozoa, since they have a common structural ground-plan, constitute a phylum in the sense defined in the last chapter the same is by no means true of the Metazoa. Hence the name Metazoa does not denote a phylum but is a mere convenient collective term.

The term Invertebrata is also a mere collective name; it is employed to designate all animals which do not belong to the phylum Vertebrata. Like the name Metazoa its convenience in promoting terseness of expression is its only justification. The Protozoa are thus Invertebrata and the Vertebrata are Metazoa.

The phylum Protozoa includes the simplest and lowest members of the animal kingdom. With few exceptions the members of this phylum are too small to be seen by the naked eye, and yet many of them are of great importance in the economy of nature.

In order to fix our ideas we may select one of the simplest
Amoeba. Protozoa as a type for examination. *Amoeba*, sometimes called the Proteus animalcule, from its power of continually changing its shape, is found in the mud at the

---

[1] These statements are true of the vast majority of animals classed as Protozoa. The exceptions are for convenience classified as Protozoa, but it seems to the authors that in the light of a fuller knowledge they may turn out to be survivors of that great series of forms which must, if the evolution theory be true, have intervened between the Protozoa and the Metazoa.

bottom of ditches, ponds and pools of stagnant water. There are several species varying somewhat in size included under the generic name *Amoeba*, all of them, however, are so small as to necessitate the use of a microscope for their examination. When magnified an *Amoeba* appears like a small, almost transparent lump of jelly, in which we can distinguish a thin outer rind and inner

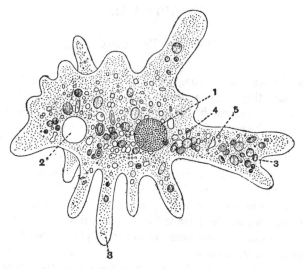

Fig. 1. *Amoeba proteus* × 330. From Gruber.

1. Nucleus.   2. Contractile vacuole.   3. Pseudopodia, the dotted line points to the clear ectoplasm.   4. Food vacuoles.   5. Grains of sand.

substance. The first, called the ectoplasm, is almost absolutely transparent, the second, called the endoplasm, has usually a grayish tinge, due to the presence of minute solid particles or granules, and is therefore described as granular. Often indeed, good sized objects of various shapes and generally of a green or yellow colour, can be seen in the endoplasm; these are the undigested remains of the microscopic plants which the animal has eaten and are surrounded by bubbles of water, termed vacuoles. *Amoeba* frequently engulfs particles of sand, though for what purpose is unknown; possibly to render itself less palatable to animals which might eat it. If the *Amoeba* is healthy we shall see it move. The transparent ectoplasm slowly sends out a projection, and then the granular endoplasm flows into it. As of course the size of the animal does not alter, when a process is thrust out in front, the rest of the animal must follow it by shrinking away behind; indeed it would no doubt be more correct to say, that it is the shrinking or contraction of the animal's body behind, which forces out the projection in front, for the movement of an *Amoeba*, like the movement of every other kind of animal, is brought about by a series of contractions.

These projections are called by the awkward name of pseudopodia (Gr. ψευδής, false , πόδιον, a little foot); the adjective pseudo- implies that they are not fixed organs like our own limbs, but are made at any part of the surface of the body. When *Amoeba* comes across anything it desires to eat, it throws out pseudopodia on each side of it; these then unite beyond the object, and so the latter becomes engulfed, so to speak, in the body of the animal, where it is digested. It may thus be said, that *Amoeba* flows round its prey. Once the prey is inside, it is surrounded by a drop of water poured out of the surrounding protoplasm or enclosed with the food. There is probably some substance secreted into this water which acts on the prey and dissolves it.

One of the most marked features in which *Amoeba* differs from other animals, from ourselves for instance, is, that it possesses no separate parts or organs, such as stomach, heart, lungs, etc., fitted to perform the separate vital actions, or functions as they are called. It breathes, that is, absorbs oxygen and gives off carbon dioxide all over the body; and it likewise excretes, that is, gets rid of the oxidized protoplasm, at all points of the surface. If, however, we are so fortunate as to come across a large *Amoeba*, which is at the same time comparatively clear of granules, and moving only sluggishly, we may be able to make out two definite objects in the endoplasm. The first of these is called the contractile vacuole (2, Fig. 1); this is a clear round space, which slowly enlarges and then suddenly vanishes, and then reappears in the same place and goes through the same series of changes. It is believed that the cause of this appearance is that at a certain point in the endoplasm a substance is produced by katabolism with a strong affinity for water; this substance attracts to itself from the surrounding protoplasm water, carrying in it the soluble waste products, in fact draining the protoplasm and forming a drop. This drop swells until it, so to speak, bursts the covering of protoplasm separating it from the outside water; the space it occupies then collapses, but as soon as the fluid has escaped the rent in the protoplasm joins up again, and as the excretory process continues the drop of fluid again accumulates.

The other object which we may perceive is the nucleus. This
Nucleus.   is a spherical body consisting apparently of the same kind of material as the endoplasm, only slightly denser (1, Fig. 1). If we, however, kill the animal by running in some iodine under the coverslip, the nucleus stands out at once in contrast to the rest of the protoplasm by its property of taking up more iodine and appearing stained a much deeper colour, and this happens in the case of any *Amoeba*, whether we have been able to see the nucleus whilst it was living or not. The material contained in the nucleus is an essential part of the body: when deprived of it metabolism within the protoplasm slackens and finally stops. Nearly all living things, animals or plants possess one or more nuclei, though in some rare cases the essential nuclear material is dispersed throughout the protoplasm. The bigger the plant or animal, the more nuclei it possesses. The so-called "Flowers of Tan" (Mycetozoa), which creep over the hides in tan-pits, are some of the few Protozoa which are distinctly visible to the naked eye; they may be compared to gigantic *Amoebae* with branching pseudopodia, and they have thousands of nuclei. In the case of certain Protozoa it has been proved that if the animal be broken in pieces, those bits which contain a nucleus can repair themselves and continue to live, eventually growing to form an animal like the one of which they are fragments; but those bits which contain no nucleus, though they continue to live for a short time, have no power of feeding themselves nor of growth. On the other hand if the nucleus be freed from protoplasm it dies; life depends on the mutual reactions of protoplasm and nucleus.

The reproduction of *Amoeba* ordinarily takes place by the simplest
Reproduction   conceivable process; the animal divides itself into
and Encyst-   two. This process is called fission: and it is found
ment.   that the nucleus always divides into two before the body as a whole shows any signs of the process. When *Amoebae* are exposed to unfavourable conditions, such as the drying up of their surroundings, they have the power of enclosing themselves in a cyst. They draw in all their pseudopodia and assume a spherical form, and the cyst appears as a membrane on the outside which then thickens. Once enclosed within its cyst, *Amoeba* can be blown about like a particle of dust, and in this way we can account for

the fact that we sometimes find *Amoeba* in infusions, that is, solutions made by allowing some animal or vegetable matter to stand in water exposed to the air. If we put some hay or meat into perfectly pure water and expose it to the air, it will putrefy; this is due to the development of minute microscopic plants called Bacteria, the spores of which are carried by the air: at a later stage, various Protozoa and sometimes *Amoebae* will appear. At one time it was supposed that both Bacteria and Protozoa were spontaneously developed out of the dead meat, but it has been shown that if the water and meat be boiled, so as to kill any spores which may be in them, and the mouth of the vessel plugged with cotton-wool whilst steam is issuing, so that the air penetrating from outside through the interstices of the wool has all the spores it may carry strained off before it comes in contact with the water, neither Bacteria nor Protozoa will appear. The cyst which invests the body of the *Amoeba* is the first instance we have met with of what is called a secretion. A secretion has already been defined as dead substance which is of use to the animal, and which is produced by the decomposition of protoplasm.

In one or two cases an *Amoeba* enclosed within its cyst has been seen to break up into a number of rounded germs which were eventually set free by the breaking of the cyst and each of which then took on the form of a minute *Amoeba*. This process is called sporulation and the germs to which it gives rise spores. It is unknown whether every species of *Amoeba* sporulates and if so under what conditions this occurs.

When we were describing the endoplasm of *Amoeba* above, we called it granular, owing to its containing solid particles. When the highest powers of the best microscopes are used, it appears that both endoplasm and ectoplasm have a structure comparable to that exhibited by a mass of soap-bubbles. The walls of the bubbles consist of the actual living substance which is probably composed of the biogen molecules; the cavities are filled with water which has in solution the products of digestion, from which the living framework repairs itself, and likewise the excretory products. This water also conveys in solution the oxygen necessary for life and removes the carbon dioxide. It is only by means of a structure like this that the complicated chemical changes which constitute life can be perfectly carried out by every particle of the living substance. The granules are temporary deposits in a solid form, either of matter resulting from katabolism, or of nutritious matter not yet assimilated.

We must now glance at some animals allied to *Amoeba*, in order to gain some idea of the group Protozoa as a whole.

Difflugia and *Arcella* are both found in the mud of pools and
Lobosa. ponds; they resemble *Amoeba* in general structure but differ from it in being provided with shells. In consequence of having these they are only able to put out pseudopodia at one spot, the mouth of the shell. The shell of *Difflugia* is composed simply of grains of sand stuck together with a secretion; it has the shape of a pointed egg with the thick end cut off (1, Fig. 2). *Arcella*, on the other

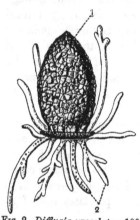

FIG. 2. *Difflugia urceolata* × 100.
After Leidy.

1. Shell composed of particles of sand containing body of the animal. 2. Pseudopodia.

hand, makes its shell entirely out of its own secretion; this is colourless when thin, but as the animal grows older the shell becomes thicker and acquires a characteristic brown colour, and we are enabled to recognize that it consists of chitin. This is really a name for a class of substances which are constantly met with in the animal kingdom and which are probably allied in composition to uric acid. Out of chitin, for instance, all insects construct their hard cases. It seems probable, that the self-destruction of protoplasm, which results from the ordinary vital functions, may in many cases give rise to chitin, so that perhaps in *Arcella*, the shell is at once a protection and the ordinary excretion. The shape of this shell is like a watch-glass with a flat lid resting on it, and in the middle of the lid there is a round hole through which the pseudopodia come out. Sometimes gas bubbles (7, Fig. 3) can be seen in the body of the animal, which tend no doubt to balance the weight of the shell. Owing to the blunt character of the pseudopodia, *Amoeba, Difflugia, Arcella* and similar forms are united into a class termed Lobosa (Gr. λοβός, a lobe).

The Protozöon *Gromia* possesses a thin membranous shell,
Foraminifera. shaped somewhat like that of *Difflugia*: but the animal shows two important differences; first, the

protoplasm of which the body is composed, besides filling the shell,
extends in a thin layer all over its outer surface (2, Fig. 4), and
secondly, the pseudopodia, which are given off from this layer, are
thin and delicate threads which join and interlace with each other
so as to form a network. *Gromia* seizes its prey by entangling it in
these fine pseudopodia; these then flow together and form a little
island of protoplasm surrounding the captive, which is thus digested
quite outside the main part of the body; the products of digestion

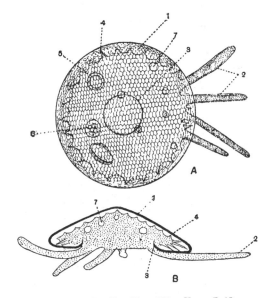

FIG. 3.   *Arcella discoides* × 500.   From Leidy.
A. Seen from above.   B. Seen from the side, optical section.   1. Shell.
2. Pseudopodia.   3. Edge of opening into shell.   4. Thread attaching
animal to inner surface of shell.   5. Nucleus.   6. Food vacuole.
7. Gas vacuole.

being carried along the pseudopodia into the protoplasm which is
inside the shell.

We may next consider a rather larger Protozöon, allied to
Gromia and like it possessing a shell, which however is composed
not of chitin but of calcium carbonate. The name of this animal
is *Polystomella* (Fig. 5). Like *Gromia* it possesses delicate inter-
woven pseudopodia which spring from the whole surface, since
there is a thin layer of protoplasm covering the outside of the
shell as well as the main mass inside it. Unlike *Gromia*, however,

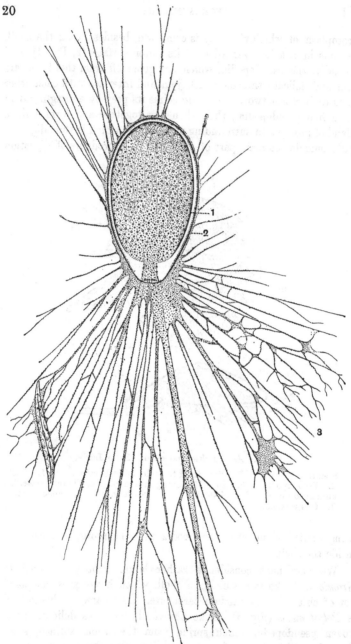

FIG. 4. *Gromia oviformis* × 250, but the pseudopodia are less than one-third their relative natural length. From M. S. Schultze.

1. Shell. 2. Protoplasm surrounding shell. 3. Pseudopodia, fusing together in places and surrounding food particles such as diatoms.

Polystomella has a shell which is perforated by a large number of
minute holes, through which pass cords of protoplasm, connecting
the inner and outer parts of the animal. *Polystomella* is therefore
a typical example of the Foraminifera (Lat. *foramen,* a hole;
fero, to carry), a class which includes countless varieties of microscopic
shells, generally composed of calcium carbonate, less frequently
of flint (silica). *Gromia* is included in the Foraminifera, though it
does not possess the peculiarity indicated by the name, because its
structure as a whole shows that it is really the same kind of animal.
There is another most instructive feature of the structure of
Polystomella wherein it differs from *Gromia.* If we examine the
shell with the low power of a microscope, we shall see that it is
shaped like a rather flat snail shell or the shell of the Pearly
Nautilus. If, however, we dissolve away the shell with dilute acid
so as to expose the proper body of the animal, it will be seen that
this is made up of separate parts, united to each other by two or
three little bridges of protoplasm, and arranged one behind the
other in a spiral series. This is the first example we have met with
of the repetition of similar parts in a definite order, but upon this
principle of the repetition of similar parts the bodies of the most
complicated animals are built up. It is no doubt fundamentally
the same thing as reproduction, only the various units which are
produced, instead of separating from each other and leading separate
existences, remain connected, and, as we say, are co-ordinated
to form an individual of a more complex kind. In *Polystomella*
the various parts are called chambers; a name which properly
belongs, and was first applied to, the segments of the shell enclosing
them. It is worthy of note that it is only the protoplasmic body of
Polystomella which shows this composition out of definite units
arranged in a definite order; there may be one large nucleus or
a considerable number of smaller nuclei, but they are not arranged
in correspondence with the chambers.

The group Foraminifera, of which *Gromia* and *Polystomella* are
examples, is of an enormous extent, and includes an immense variety
of forms, the variety being brought about by differences in the
number of chambers and the way they are arranged in series. The
Foraminifera almost all live in the sea; some, like the two we
have described, creep about amongst the sand and débris at the
bottom of pools or other places where the water is quiet; many
others float at the surface of the ocean, the protoplasm which clothes
the outside of the shell having numerous vacuoles filled with fluid

probably less dense than the sea-water, and thus serving as floats. In such inconceivable myriads do these floating Foraminifera exist,

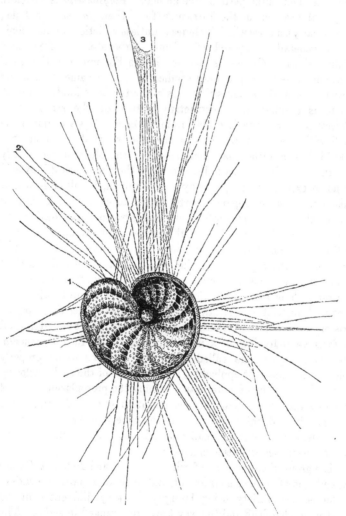

FIG. 5. *Polystomella crispa.* Highly magnified. After M. S. Schultze.
1. Shell.   2. Pseudopodia.   3. A mass of protoplasm formed by the fusion of pseudopodia.

that their empty shells form thick banks of impalpable white chalky mud at the bottom of the ocean, and the familiar white chalk of our English cliffs and hills is largely made up of the shells of Foraminifera.

The Foraminifera show the same two methods of reproduction, viz. fission and sporulation which were mentioned in the case of the Lobosa. Indeed the progress of research has rendered it probable that they are found in all classes of Protozoa. The presence of a hard skeleton has however produced modifications in the process of fission. When such a form as *Polystomella* for instance is about to reproduce by fission the protoplasm emerges from the old shell and divides repeatedly into a number of pieces each as large as the central chamber of an ordinary individual. These then secrete round themselves shells and begin to bud off new chambers and gradually acquire the size and shape of adults. When on the contrary *Polystomella* sporulates the protoplasm whilst still within the shell divides into a large number of small rounded pieces—these acquire hair-like projections called flagella which can be moved to and fro and by means of which they swim and then escape from the shell and move freely about. They coalesce with one another in pairs, and the resultant mass or zygote acquires a shell and begins to bud off new chambers. Since in spite of the fact that the zygote has resulted from the union of two spores it is much smaller than the product of fission—the adult resulting from the growth of a zygote is distinguished from that resulting from fission by the small size of the initial chamber. Hence we can distinguish a microspheric form resulting from sporulation from a megalospheric form resulting from fission. Lister has shown that in *Polystomella* there is an alternation of fission and sporulation, probably more than one generation of fission intervening between two periods of sporulation.

We may pass now to the consideration of some Protozoa, which show a good deal of resemblance in many points to the Foraminifera, though they have very marked peculiarities of their own. These are the Radiolaria; they have delicate threadlike interlacing pseudopodia, and their protoplasm is divided into two parts—an inner and an outer—by a membranous case having pores, through which the two parts communicate with each other. This case, the central capsule, may be compared to the Foraminiferan shell, but the interesting fact is that these Radiolaria have in addition to this another skeleton composed not of chalky—calcareous—but of flinty—siliceous—substance, as are also some of the shells of the Foraminifera. This

flinty skeleton may consist simply of isolated needles sticking out
on all sides from the centre; oftener, however, it consists of a
beautiful basketwork as in *Heliosphaera inermis* (Fig. 6), and some-

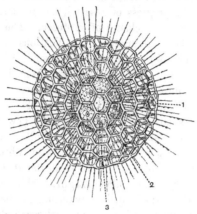

FIG. 6. *Heliosphaera inermis* × 350. From Bütschli.
1. Skeleton.   2. Central capsule.   3. Nucleus.

times we find several of these baskets one within the other, like the
Chinese ivory ball. The Radiolaria, like the free-swimming Foraminifera, have a bubbly outer protoplasm, and often drops of oil in
the inner protoplasm; these structures serve to sustain them and
they are found floating at the surface of the sea amongst the Foraminifera. At the bottom, in medium depths, their flinty skeletons,
though mixed with the calcareous shells of the Foraminifera, do not
affect the general character of the chalky mud (called the *Globigerina*
ooze, from the name of one of the commonest Foraminifera found
in it), but at greater depths, owing to the enormous pressure, the
quantity of carbonic acid dissolved in the water increases very
much—on the same principle that the pressure inside a soda-water
bottle keeps the gas dissolved—and all the shells composed of
chalky matter are dissolved, only the flinty skeletons being left.
The bottom mud here entirely changes its character and is called
Radiolarian ooze.

The next group of Protozoa to be considered is a very remarkable one, including the largest forms known. The
so-called "Flowers of Tan" (Mycetozoa) are brightly
coloured patches, which may be seen on the surface of the oak-bark
used in tan-pits. Similar patches may be seen on old tree stumps

Mycetozoa.

and on the surface of beanstalks which have been wet for a considerable time. These patches under the microscope are seen to resemble enormous *Amoebae* with thin branching pseudopodia,

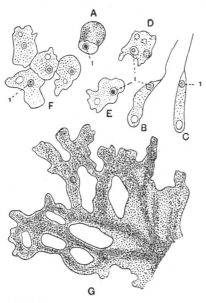

FIG. 7.   Various stages of *Chondrioderma difforme*.   From Strasburger.

A. Flagellula leaving cyst.   B & C. Flagellulae.   D young and E older
    amoebulae.   F. Amoebulae fusing to form plasmodium.   All × 540.
    G. Plasmodium × 90.   1. Nucleus.

which are apt to join one another to form networks, although these networks are much coarser than in the case of the Foraminifera. The fluid endoplasm is seen to have a regular flow alternately backwards and forwards in these pseudopodia ; the movement of the whole mass in any direction being due to the predominance of the forward flow over the backward, or vice versa.   When stained the protoplasm is seen to include thousands of very small nuclei. The name Mycetozoa literally means Fungus animals (from Gr. μύκης, a fungus, ζῷα, animals).   They are also often called Myxomycetes— literally Slime-Fungi, both names having been suggested because their special mode of reproduction leads some naturalists to consider them to be plants.   Their power of encystment is very marked, the slightest tendency to drought calls it into action, and then a mass will break up into numerous cysts, which will remain perfectly passive until wetted. Before reproduction, the same process occurs, but the contents of the cyst divide repeatedly so as to form a mass of small germs—spores—which acquire walls of cellulose, a constant product of plant life. Such spores are called chlamydospores. Some of the protoplasm not used in the formation of spores forms long threads of cellulose, called collectively a capillitium, which when wetted expands and so expels the spores. The appearance of cellulose was the only justification for regarding these animals as in any way allied to plants, and it is known that cellulose is quite a constant product in some groups of animals. The contents of the spore escape as a germ propelling itself by a vibratile thread called a flagellum (B and C, Fig. 7), the germ itself being termed a flagellula. This thread is soon withdrawn and the germ takes on the form of a small Amoeba and is then called an amoebula (D, E, and F, Fig. 7). Many of these amoebulae coalesce to form the adult form, which is called the plasmodium (G, Fig. 7), a name given to the result of the fusion of a number of originally separate animals.

The Sun animalcules, or Heliozoa (Gr. ἥλιος, the sun), which inhabit, with few exceptions, fresh water, were formerly confounded with the Radiolaria, but they are in reality very different from these. They are spherical in shape and have a large number of stiff pointed pseudopodia sticking straight out all round them, like the conventional rays in pictures of the sun. The common and scientific names are taken from this circumstance. Since these animals float about, it is not surprising to find much the same structure in the outer protoplasm as we found in Radiolaria and the floating Foraminifera. The pseudopodia are different in character from those of the Radiolaria, since they do not interlace, nor do they run together when they seize prey; the captured food is simply pressed in towards the body by the bending of the pseudopodia, and when it is brought quite close, a broad irregular pseudopodium, like one of those of *Amoeba*, shoots out and engulfs it. Pseudopodia were defined in the case of *Amoeba* as irregular projections shot out at intervals from the body and soon withdrawn, and the question arises how far we have any right to call by the same name these stiff projections of the Heliozoa. They are, however, true pseudopodia, for if the animal be subjected to strong irritation they are all withdrawn. These animals show a most interesting example of the repetition of parts. The species *Actinophrys sol* and *Actinosphaerium eichhornii* are both comparatively common inhabitants of our ditches. The first is, however, exceedingly minute, not more than $\frac{1}{1000}$th of an inch in diameter and possesses only a single nucleus, whereas the second is large enough to be just visible to the naked eye, and has about 200 nuclei. There is here repetition of the nuclei, but no division of the protoplasm, whereas in *Polystomella* there is segmentation of

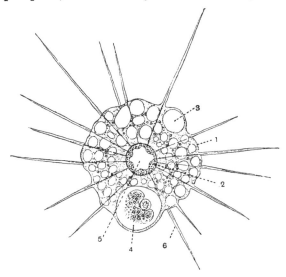

FIG. 8.   *Actinophrys sol* × about 800.   From Bronn.
1. Ectoplasm.   2. Endoplasm.   3. Contractile vacuole.   4. Food vacuole.
    5. Nucleus.   6. Axis of a pseudopodium, stiffer than the protoplasm which covers it.

the protoplasm, but no corresponding multiplication of the nuclei. If both were to occur simultaneously and to correspond so that the body were to consist of a number of segments of protoplasm, each with its nucleus, the animal would be said to be multicellular, each unit being spoken of as a cell; but it would no longer be a Protozöon.

So far we have been considering animals which, however much they may differ in details, are all essentially naked masses of protoplasm, and in them no very definite organs are set apart for the performance of special functions. We must now examine some Protozoa of a distinctly higher grade of structure, wherein definite organs exist; by the word organ being meant a part of the body definitely fitted to perform some special function for the general benefit of the whole. If we examine the

Ciliata.

roots of Duckweed with a lens, we shall probably find some coated
with a whitish scum ; if these be cut off and mounted in a drop of
water on a slide and examined with a low power, they will be seen

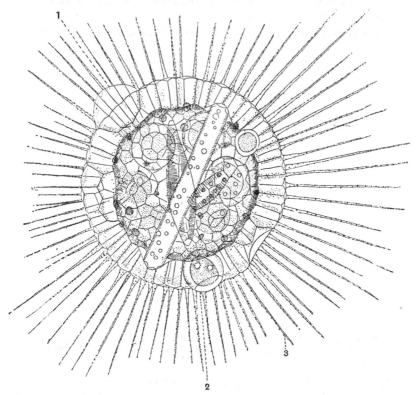

Fig. 9. *Actinosphaerium eichhornii* × 200. From Leidy. The endoplasm is
crowded with food vacuoles containing Diatoms, and nuclei are represented
in the figure by the dark areas.
1. Contractile vacuole. 2. Food vacuole which has just swallowed a Rotifer.
3. Pseudopodia.

to be covered by numerous specimens of a most beautiful animal
called *Vorticella*. *Vorticella* has something the shape of a blue-bell
flower ; it consists of a long delicate stalk and a bell-shaped body ;
by means of the stalk it is fixed to some definite support, such as,
for instance, the Duckweed root. The part of the body corresponding to the lip of the bell is broad and turned outwards ;
encircled by this lip, which is called the peristome, there is a

flattened projection called the disc. Between the peristome and disc there is a groove, and in this we can make out some short hair-like structures waving to and fro; there is a circle, or rather one twist of a spiral of these, as we can see when the animal turns the surface of the disc upwards. By the regular rhythmical bending of these cilia (Lat. *cilium*, an eyelash) as they are called, and possibly by the movement of a rolled membrane which projects into the mouth, a vortex is produced in the water, which draws particles of food to the *Vorticella*. The cilia and the stalk are definite permanent organs, the first of the kind we have met with. But the possession of these organs is not by any means the only difference between *Vorticella* and the lower Protozoa. The shape of the body, though it varies slightly with the state of expansion or contraction, is practically constant; no pseudopodia are given out. This is the

Fig. 10. *Vorticella microstoma* × about 200.
From Stein.

result partly of the possession of a firm membrane covering the whole of the body, called the cuticle, which is a protective secretion like the shell of *Arcella* only much thinner and so intimately connected with the protoplasm under it as to be inseparable; but the constancy of shape is also due to the fact that the outermost layer of the protoplasm itself is finely striated, constituting a specially contractile sheet surrounding the body and distinctly marked off from the inner protoplasm. This sheet is called the cortical layer and the striation is caused by the differentiation of the protoplasm into parallel strings called fibrils, embedded in different material acting as a cement. This arrangement of protoplasm makes its appearance whenever the contractile power is specially developed. The stalk, which is entirely composed of this layer, might almost be regarded as a muscular fibre. The stalk is slightly twisted and attached in a long spiral to the inner side of an elastic tube of cuticle. When contraction occurs the stalk is necessarily thrown into the most evident spiral curves, like a corkscrew; the restoration of the form after contraction is due to the elasticity of the tube of cuticle.

Vorticella possesses a contractile vacuole and a nucleus just as *Amoeba* does; in small nearly transparent specimens both are easily detected during life; in fact, if the specimen under observation only

keeps moderately still, we can follow the expansion and contraction
of the vacuole with the greatest ease. The nucleus is very large
and has more or less the shape of a horse-shoe, though the two ends
are generally at different levels, so that in reality it forms part of a
spiral. If we run in some iodine it at once absorbs the stain and
stands out very distinctly ; the *Vorticella*, however, frequently shows

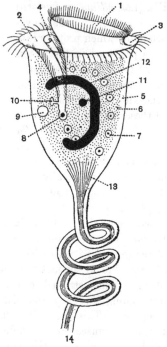

Fɪɢ. 11. Diagram of *Vorticella*. The cilia at the side of the mouth have
been omitted.

1. Disc.   2. Mouth.   3. Peristomial groove.   4. Vibratile membrane in
mouth.   5. Cortical layer.   6. Endoplasm.   7. Food-vacuoles. The
last of the food-vacuoles is nearing the position of the anus.   8. Pharynx
showing formation of food-vacuoles.   9. Contractile vacuole.   10. Permanent receptacle into which the contractile vacuole opens.   11. Micronucleus.   12. Nucleus.   13. Contractile fibrils running into muscle
in stalk.   14. Stalk contracted (the axial fibre should touch the cuticle
in places).

its dislike to the operation by contracting its body into the shape of
a ball and snapping itself off from the stalk : it is then apt to get
washed away from its position by the inflowing iodine and we may
have to search over the slide to find it. When a *Vorticella* is
irritated, the peristomial lip is turned in so as to lie against the disc,

and thus the groove in which the cilia lie becomes converted into a tube and so they are efficiently protected.

We have seen above that *Vorticella* uses its cilia in order to produce a miniature whirlpool in the water by means of which particles of edible matter—whether living or not—are drawn towards it. Since, however, it possesses a firm cuticle and in addition a specialized outer layer of protoplasm, the question arises how the food is taken into the interior of the body. If we run some Indian ink under the cover-glass we shall have a demonstration of how this is managed. The black particles are caught in the whirlpool made by the cilia, they course round and round and finally accumulate in a pit which opens into the ciliated groove and from the bottom of this they pass one by one into the internal protoplasm of the body. This pit which obviously passes through both the cuticle and the outer protoplasm—the cortical layer as it is called—is termed the pharynx and its opening the mouth. The particles of Indian ink which have passed into the body are surrounded by little drops of fluid which are partly swallowed with it and partly secreted by the protoplasm, in order to effect the solution of the particle, that is, its digestion. Such a drop is called a food-vacuole to distinguish it from the contractile vacuole, which, as we have seen, has probably an excretory office to perform. As there is nothing nutritious in Indian ink, the *Vorticella* soon gets tired of trying to digest it, and the particles after having travelled through the body in a more or less definite tract are thrust out into the ciliated groove. Since this takes place at only one spot, there must be a permanent hole in the cuticle here though we cannot discern it, and this opening may be called the anus. Therefore in contradistinction to *Amoeba*, where food can be taken in and undigested remnants cast out at any spot on the surface, in *Vorticella* it is only at one particular spot that either action can take place.

The reproduction of *Vorticella* is a most interesting process. It takes place by longitudinal splitting, or, as it is technically called, fission. The disc splits into two, and the cleft soon reaches right down to the beginning of the stalk, so that for a time we have two bodies attached to the same stalk. One of these acquires a new row of cilia round its base; soon after the original circle of cilia and the peristomial groove disappear; the animal then breaks loose from the stalk and swimming by means of its new circle of cilia seeks a new place of rest. The other body remains on the original

stalk and resumes its ordinary life. This simple mode of reproduction can go on for a long time unchecked, but experiments made on
other Protozoa more or less allied to *Vorticella* show that it has a
limit. When we were speaking of reproduction in the introductory
chapter, we mentioned that in sexual reproduction two germs had
to fuse together before they could give rise to a new individual.
Something like this has to happen at intervals in the case of
Vorticella in order that the reproduction by division may go on in
a healthy manner. When this process (conjugation) is about to
take place, one individual divides repeatedly by longitudinal division
without any of the new individuals produced breaking away from
the stalk, so that we have a bunch of minute *Vorticellae* attached to
the same stalk. This rapid division is clearly comparable to the
process termed "sporulation" which has been already described in
the case of the Lobosa and the Mycetozoa. The "spores," as we
may term the small *Vorticellae,* become free and swimming away
attach themselves to the sides of the large stalked individuals. There
then ensues an interchange of substance between the two individuals
which thus adhere to one another; the large nucleus of each breaks
up and disappears and a small subsidiary nucleus, the micronucleus
(11, Fig. 11), exceedingly difficult to detect at ordinary times, now
comes into view. It also breaks up and many of the portions
disappear, but a part of the micronucleus of the large one passes into
the little one and vice versa. In the case of the allies of *Vorticella,*
when the two individuals which thus conjugate are of the same
size, they separate after the operation, and each goes on dividing on
its own account. In *Vorticella,* however, the small individual
seems to be exhausted by the process and is absorbed into the
body of the other. It appears that this process of conjugation is
only effective when it takes place between individuals of different
parentages; if care is taken to exclude all *Vorticellae* of foreign
stock from a collection consisting of the descendants of a single one,
either no conjugation takes place, or, if it does take place, it fails
to produce the results which normally follow, viz., increased vigour
of reproduction and other vital processes. When conjugation with
individuals of different parentage is prevented the individuals which
are produced by fission, after a certain number of generations, are
said to be badly formed, unable to feed themselves and die, and
this is the only instance of natural death which is met with amongst
the Protozoa.

Like *Amoeba, Vorticella* encysts and it appears still more

frequently than *Amoeba* in infusions. The *Vorticellae* which are found under these circumstances are usually small and transparent and more favourable for observation than those occurring in ditches. The genus occurs both in fresh and salt water.

Vorticella is but one example of a large class of Protozoa termed Ciliata, which agree with it in all the essential points of structure, but differ in the arrangement of cilia, the absence of a stalk, and more rarely in the absence of a mouth and pharynx. This last feature is found only in those species which live in places where the surrounding fluid contains dissolved nutriment which can percolate in at any spot. An example of such a Ciliate is *Opalina*, found in the intestine of the Frog. This animal is thin and plate-like and covered all over with cilia of the same size arranged in regular lines; this arrangement of cilia, which is called the holotrichous arrangement (ὅλος, entire ; θρίξ, hair), is always associated with the absence of a stalk and with a free-swimming life. *Vorticella*, on the other hand, is said to be peritrichous (περί = around). *Opalina* is further remarkable for possessing a large number of nuclei (1, Fig. 12), which is a rare occurrence amongst the Ciliata. When, however, division commences it continues until the resulting pieces have only one nucleus each; they then grow and do not divide again till they acquire the size they had before division took place and also the same number of nuclei. Hence we might regard the multiplication of the nuclei as the real reproduction of this form, the division of the protoplasmic body being of lesser importance and setting in later.

FIG. 12. *Opalina ranarum.*
Highly magnified.
From Bronn.

1. Nuclei.   2. Ectoplasm.

Paramecium is one of the commonest free-swimming Ciliata. It is of an elongated oval outline; seen sideways it has a thin scoop-like anterior end and a thick posterior part, so that it is usually described as slipper-shaped. It is holotrichous like *Opalina*, but like *Vorticella* it possesses only two nuclei, one large and easily visible and one, the micronucleus, small and difficult to detect. It has a well-developed mouth and deep pharynx situated on one side and lined with specially long cilia. *Paramecium* is a beautiful form in which to study the contractile vacuole; there are two

of these present, one in the anterior and another in the posterior
portion of the animal. If one of these vacuoles be watched it can
be seen to contract and then slowly to re-appear. In the process
of reappearance five or six isolated drops are seen which elongate
into streaks arranged like the rays of a star. These streaks coalesce
with one another and soon form a perfectly spherical drop.

   Paramecium possesses peculiar organs named trichocysts
(5, Fig. 13) embedded in the outer layer of the protoplasm. These

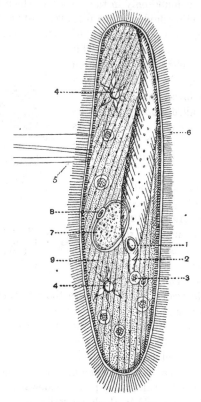

FIG. 13.   *Paramecium caudatum* × about 250.   After Bütschli.

1. Mouth at bottom of groove.   2. Oesophagus.   3. Food vacuole just
   being formed.   4. Contractile vacuoles.   5. Trichocysts which have
   exploded: the unexploded ones line the cuticle.   6. Cilia.   7. Nucleus.
   8. Micronucleus.   9. Contractile fibrils.

look like minute rods. When the *Paramecium* is irritated—as
for instance if it is deluged with dilute iodine—or approaches

prey it wishes to seize, these are suddenly shot out, assuming the form of long threads, and they appear to exercise a stunning effect on any small animal with which they come in contact.

The process of conjugation has been carefully worked out in the case of *Paramecium*, so that some more details may here be given of what happens in that case. Two individuals about to conjugate become attached to one another, the larger nucleus in each breaks up into a number of minute pieces which are apparently used as food by the rest of the protoplasm—at any rate they gradually disappear. The smaller nucleus breaks up in eight pieces in each case, and of these seven are cast out, the eighth divides into two and of these two one passes over into the body of the other *Paramecium* and fuses there with the remaining piece of the corresponding smaller nucleus. In this way each *Paramecium* becomes a zygote. The two individuals separate, the single nucleus in each divides twice so that four nuclei are produced, the *Paramecium* then divides transversely so that each half has two nuclei, one of which becomes the large and the other the small nucleus of the new individual.

Passing from the Ciliata, we next come to a small group called the Suctoria, which are allied to the Ciliata, for their buds commence life as holotrichous forms. When they grow up they frequently become stalked like *Vorticella*, lose their cilia, and acquire instead a number of stiff rod-like outgrowths ending in knobs ; these structures are termed tentacles. These are able in some way we do not understand to seize small animals and suck out their contents. Some secretion must be produced which eats its way through the cuticle and dissolves the contents of the prey.

Suctoria.

The next group of the Protozoa we shall consider is a very large and important one. It is called the Flagellata. The members of it agree with the Ciliata in having a fixed shape and a firm cuticle and probably (though this is difficult to make out in the smaller ones) a specialized cortical layer of protoplasm; they differ, however, in not having cilia, but in possessing instead, one or two—rarely more—whip-like organs called flagella, which lash about in the water, and drag the animal after them by a spiral screw-like motion, just as a steamship is dragged by the screw when the engines are reversed. We may take as a type *Euglena viridis*, one of the numerous inhabitants of ditches. This animal has a narrow elongated shape, pointed at one end, and at the other—which is its front end—it possesses the vestige of

Flagellata.

a pharynx, which is exceedingly narrow.   It has a flagellum which
arises from the pharyngeal wall near its inner end.   Slightly behind
the pharynx there is a small contractile vacuole, and at the one side
of this a small red spot, which may very possibly be associated with
a sensitiveness to light.   About the middle of the body is a nucleus
which can sometimes be made out as a clear spot in the living
animal, but which is most satisfactorily observed when the animal is
killed with osmic acid and stained with picrocarmine.

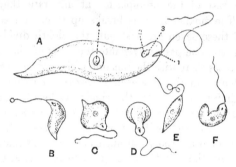

FIG. 14.   *Euglena viridis.*

A × 100, B, C, D, E, F × 200 showing the different shapes assumed by the
animal during the euglenoid movements.   1. Pharynx.   2. Contractile
vacuole.   3. Pigment spot.   4. Nucleus.

Two features in *Euglena*, however, will strike us as very peculiar.
One is, that in spite of possessing a cuticle and a cortical layer of
protoplasm, it is able to change its shape.   It does not possess the
power of throwing out pseudopodia, but it bends its body in the
most extraordinary way, and contracts it till it is almost spherical.
The peculiar wriggling movements which it thus executes are so
unlike anything else that they have been called euglenoid.   The
reason for their possibility is no doubt that the cuticle is flexible and
the cortical layer powerfully contractile.   The other peculiarity is
still more striking, and it is that the protoplasm is coloured bright
green and that it contains particles of a substance very like starch.
Now these things indicate that *Euglena* feeds itself like a plant, and
that it constructs its protoplasm out of carbon dioxide and mineral
salts dissolved in water in the presence of sunlight.   The only points
therefore that can be suggested in which it differs from plants are
that it has a flagellum and moves, and that it does not possess a
covering of cellulose.   These supposed differences, however, will not
stand examination ; the germs of many undoubted plants, such as,

for instance, the sea-weeds, have no cell wall and propel themselves by means of flagella. What justification then, it may be asked, have we for reckoning *Euglena* as an animal? What do we mean by so classing it? It must indeed be admitted that when we come to deal with the simple Flagellata, the animal and plant kingdoms merge into one another, and the only valid line of division we can draw is between forms which feed on solid food, and those which absorb dissolved nutriment; and amongst the latter we call those forms animals which we believe to have been derived from ancestors which fed on solid food. Now the pharynx in *Euglena* takes in solid particles from time to time and these passing into the protoplasm are apparently digested. We might therefore imagine either that *Euglena* is a plant which has acquired a pharynx and is commencing to live like an animal or else that it is an animal that has acquired chlorophyll and has commenced to live like a plant. The fact that the pharynx is small and of little use is against the idea that it is an organ which has been newly acquired—as all organs are acquired—on account of its usefulness. It has the appearance of being the *vestige* of a once useful organ and therefore we conclude that *Euglena* is an animal which has begun to live like a plant.

The reproduction of *Euglena* and of the Flagellata in general is quite similar to that of the Ciliata; they increase by longitudinal division, but they also divide when in an encysted condition into two or four, or a larger number of germs; these germs are not killed by drought. When dry they are blown about, and so appear in infusions. In infusions Bacteria appear first, then Flagellata, and finally Ciliata.

Many Flagellata are devoid of a pharynx altogether, but these rarely have chlorophyll and subsist on the nutritive substances which are dissolved in the fluid in which putrefying matter is soaking. These are reckoned as animals on no very good grounds; for it is well known that plants can lose their green matter when they can get the materials of protoplasm without building it up from carbon dioxide. How entirely arbitrary the decision is is best shown by the fact that many forms are claimed by both botanists and zoologists. For this reason it is convenient to have a name which denotes simply a living thing without prejudging the question as to whether it is an animal or a plant. Such a name is supplied by the word organism, which is frequently used.

The last group of Protozoa, the Sporozoa, agree with the
Ciliata in possessing a firm cuticle and a highly
developed contractile cortical layer, but differ in
never having any organs such as cilia or flagella; their movements,
which are very sluggish, are carried out entirely by contractions

Sporozoa.

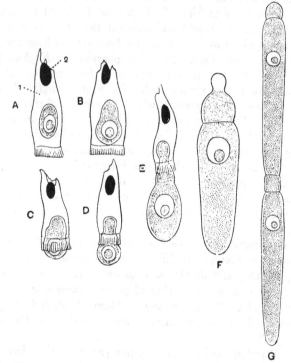

Fig. 15. *Clepsidrina longa*, from larva of *Tipula*, the Daddy-long-legs.
Highly magnified. From Léger.

A, B, C, D, E. Stages of the development of *C. longa*, at first within and then
pushing its way out of one of the cells of the intestine of the *Tipula* larva.
F. Mature form.   G. Two forms conjugating.   1. Cell of intestine
of host.   2. Its nucleus.

of the cortex, which give rise to worm-like wrigglings. The name
Sporozoa which was suggested by the frequently recurring *sporulation* which is a marked feature in the life-history is inappropriate,
as we have seen reason to believe that something analogous occurs
in all Protozoa. All the Sporozoa are parasitic; that is to say all
live at the expense of some other animal which is termed the host.
All as a matter of fact pass the first period of their existence

embedded in the protoplasm of some animal. Some—the Coccidea —remain throughout life in this position, but others when fully grown become at any rate partly free, adhering to their hosts only by one end. Only fluid nourishment is absorbed and consequently there is neither mouth nor anus. There is never more than one nucleus, although the body may be divided by partitions running across the protoplasm into two or even three portions, placed one behind the other. Reproduction takes place after encystment, and this encystment is in most cases preceded by conjugation, so that two individuals are enclosed by a common cyst. The contents of the cyst break up into spores, which surround themselves with flinty cases and hence are called chlamydospores (χλαμύς = a cloak). The protoplasm inside these spores is sometimes liberated as a small amoeba-like creature which usually divides into two worm-like forms, which wander into suitable positions and become metamorphosed into the adult form. In very many cases, however, from two to eight worm-like forms are formed by the division of the contents of the spore before the case breaks. The name falciform embryo has been given to these germs.

The best known Sporozoa are *Monocystis* found in the vesicula seminalis of the earthworm, with a long worm-like undivided body, *Clepsidrina blattarum* found in the intestine of the Cockroach and *C. longa* in the intestine of the larval grub of the Daddy-long-legs, which lives in damp soil. This last form is quite free when adult and is distinctly divided into two portions.

The *Haemamoeba* and *Haemomenas* which cause Malaria in man and allied forms which infest other vertebrates are classed with the Sporozoa.

The Coccidea is the name of one of the sub-divisions whose members remain entirely enclosed in the protoplasm of the infected animals throughout life, and in some cases cause disease.

The larger Sporozoa are often easily detected by the intense opaque white colour of the protoplasm, due to the inclusion of an immense number of granules. This is in marked contrast with the translucent protoplasm in which they are embedded.

Phylum PROTOZOA.

The Protozoa are classified as follows :—

Class GYMNOMYXA.

Naked forms without distinct cortical layer and capable of emitting pseudopodia.

Order 1. Lobosa.

Simple forms with blunt pseudopodia which do not form networks ; with or without a shell.

Ex. *Amoeba, Difflugia, Arcella.*

Order 2. Reticularia.

Protozoa with thread-like pseudopodia which form networks ; a shell is formed and protoplasm covers the outside as well as the inside.

Suborder (*a*) **Foraminifera.** The shell consists of one or a series of chambers composed of lime or flint.

Ex. *Gromia, Polystomella, Globigerina.*

Suborder (*b*) **Radiolaria.** The shell is a single sac of membrane. In almost every case there is an additional skeleton of flinty needles often joined so as to form complicated basket-works.

Ex. *Thalassicola, Heliosphaera.*

Order 3. Mycetozoa.

Protozoa with branching pseudopodia forming coarse networks and devoid of skeleton. Reproduction by means of spores coated with cellulose. The contents of many spores coalesce to form one individual.

Ex. *Chondrioderma.*

Order 4. Heliozoa.

Protozoa with stiff radiating pseudopodia. Skeleton when present only in the form of isolated needles.

Ex. *Actinosphaerium, Actinophrys.*

Class CORTICATA.

Protozoa with a distinct cuticle and almost always a distinct cortical layer.

Order 1. **Ciliata.**

　Forms provided with cilia.
　Ex.　*Vorticella, Paramoecium, Opalina.*

Order 2. **Suctoria.**

　Forms provided with sucking tentacles.
　Ex.　*Acineta.*

Order 3. **Flagellata.**

　Forms provided with flagella.
　Ex.　*Euglena.*

Order 4. **Sporozoa.**

Parasitic forms, devoid of mouth, cilia, flagella or tentacles.
The younger stages at least are cell-parasites.
　Ex.　*Monocystis, Clepsidrina, Coccidium, Haemamoeba.*

CHAPTER III.

PHYLUM COELENTERATA.

IT is difficult to say what idea the originator of the name Coelenterata meant to convey. Most animals have hollow insides (Gr. κοῖλος, hollow; ἔντερον, inside); the Coelenterata however are distinguished from all the more highly organized groups in the animal kingdom by containing inside only one set of spaces, which all communicate with each other and with the exterior through the mouth.

The Coelenterate of simplest structure is undoubtedly the common fresh-water Polyp (*Hydra*), (Fig. 16). If a *Hydra.* mass of weed and other débris from a ditch or even the edge of a river be placed in a glass vessel along with some of the water in which it was grown and allowed to settle, a number of these small animals frequently termed polyps will usually be found collected on the side of the vessel nearest the light. Several distinct species are collected under the name Hydra. There are three species recognized in Great Britain; *Hydra fusca*, about a third of an inch long when expanded and of whitish yellow colour, *Hydra viridis*, a quarter of an inch long, of a green colour, and *Hydra vulgaris*, which is almost colourless. Similar species to the first two, if indeed they are not identical, are common in Lower Canada. *Hydra fusca* may be selected as a type.

The shape of this animal is that of a minute cylinder. The base or foot is attached to the surface of the glass by an adhesive disc, whilst the other extremity carries a circle of delicate thread-like appendages called tentacles. In the centre of these, near their point of origin, we can with a lens detect a minute conical elevation, the oral cone (2, Fig. 16), at the end of which is the mouth. The mouth is the only opening in the body and it leads into a space which occupies the whole extent of the animal, so that we might with justice say that the polyp is really simply a tube closed at one end and open at the other: further, the tentacles can be seen with the microscope to be nothing but thin hollow tubes, opening into the central one (Fig. 17). The central space is often termed a stomach, and in the case of *Hydra* the idea suggested by this

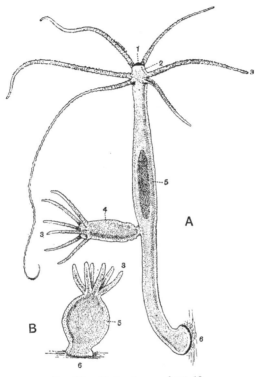

FIG. 16.   *Hydra fusca* × about 12.

A. Expanded condition. This specimen is budding off a young *Hydra*. It contains a large food mass in its coelenteron, probably a Daphnia or some other fresh-water Crustacean.   B. Retracted condition.   1. Mouth. 2. Oral cone.   3. Tentacles.   4. Bud.   5. Endoderm.   6. Foot.

term is correct. In other Coelenterata the space performs other functions besides those of the human stomach, and the term coelenteron, which does not imply any function, is preferable.

With the microscope, however, we can make out a number of further points. If the edge of the animal be carefully focussed it can be seen that the body-wall consists of two layers, an outer

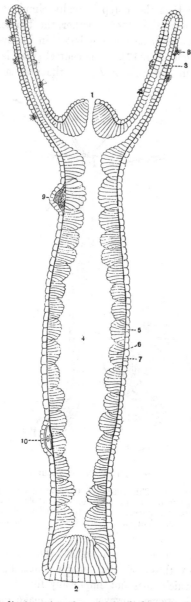

Fig. 17. A longitudinal section through the body of a *Hydra*: somewhat
diagrammatic, the details of the cells being omitted. Magnified.

1. Mouth. 2. Foot. 3. Tentacle. 4. Coelenteron or digestive cavity.
5. Ectoderm. 6. Endoderm. 7. Mesogloea or structureless lamella.
8. Batteries of thread cells. 9. Testis. 10. Ovary with single ovum.

clear one, termed the ectoderm (Gr. ἐκτός, external ; δέρμα, skin),
and an inner one called the endoderm (Gr. ἔνδον, inside), which
is green in *Hydra viridis* and brownish in *Hydra fusca* ; so
that we may speak of a skin as distinct from the lining of the
coelenteron (Figs. 17 and 18). It is further possible to make out
under the microscope that at any rate the outer layer is not
homogeneous, but is composed of separate small pieces. It is
necessary, however, to examine thin sections of specimens which
have been hardened by being soaked in corrosive sublimate or some
similar reagent, before one can really get a good idea of the
structure of the "skin" and of the "inner lining" of the polyp.
Then it is seen that both are made up of the repetition of similar
parts, and that in each of these parts there is a single nucleus.
Such a portion of protoplasm, marked off from the surrounding
parts by a definite boundary, is called by Zoologists a cell. The
wall or boundary of the cell probably consists of a thin layer of
some secretion, in many cases, if not in most, traversed by bars or
sheets of protoplasm connecting the cell with its neighbours.
Around this term "cell" many battles have been waged and its
indiscriminate use has led to much misconception.

Cell.

It used to be said, for instance, that the Coelenterata
were multicellular animals, as opposed to the Protozoa, which
were unicellular. Now it has already been pointed out that the
centre of the vital processes is the nucleus, which controls the
processes going on in the protoplasm, and that in some of the
Protozoa, such as *Actinosphaerium* and *Opalina*, this essential organ
is repeated several hundred times. But an *Actinosphaerium* or an
Opalina certainly does not correspond to a so-called cell of *Hydra*,
with its single nucleus ; the relation between them may rather be
defined by saying that, whereas in *Actinosphaerium* the areas of
control of the various nuclei are not visibly delimited from each
other, in *Hydra*, on the other hand, this delimitation has to some
extent taken place, leading to the appearance of cell-structure.
But not only are cells to be detected in *Hydra* ; the cells are not
all of the same kind. Those forming the endoderm are very
big and often have great watery vacuoles near their inner ends ;
they also contain the coloured granules to which the colour of the
animal is due. The cells of the ectoderm or outer skin, on the
contrary, are much shorter than those of the endoderm and are
more or less pear-shaped, the broader end being turned out. Between their narrower bases we find groups of very small round cells

(2, Fig. 18). These so-called interstitial cells are young cells, which partly, no doubt, become developed into ordinary ectoderm cells as the older cells die and drop off, and in certain seasons of the year they increase very much in number at certain spots and form the reproductive organs (9 and 10, Fig. 17). The two kinds of organs, male and female, are borne by the same individual; in the male organ or testis all the cells remain small and become converted into the small spermatozoa; in the female organ or ovary one cell increases very much in size at the expense of the rest and becomes the egg-cell or ovum (10, Fig. 17). There is,

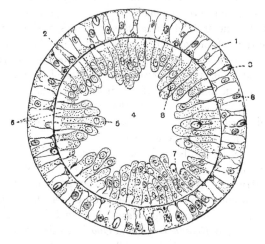

Fig. 18. Transverse section of *Hydra fusca*.

1. Ectoderm cells (myo-epithelial). 2. Interstitial cells. 3. Nematocysts.
4. Coelenteron. 5. Endoderm cells. 6. Vacuoles. 7. Food granules. 8. Nuclei.

however, a third change which these interstitial cells may undergo, which is of the utmost importance to the animal. Some of them move outwards and become wedged between and even embedded in the large ectoderm cells near the surface, each developing in its interior an oval bag filled with fluid. One end of this bag is turned into the interior of it, forming a long hollow thread. The whole bag is called a thread-capsule or nematocyst (Gr. νῆμα, a thread; κύστις, a bladder) (Fig. 19). If now the cell in which the thread-capsule is situated contracts, since the fluid in the capsule is incompressible, the hollow thread must be quickly turned

inside out and thus thrust out of the capsule. If the irritation
of skin continues the whole capsule will be pressed out by the
animal. These thread-capsules are most abundantly developed in
the tentacles, and a small amount of observation of the habits
of *Hydra* will show how they are used. If a small Crustacean,
or other animal, approaches too near a *Hydra*, the latter makes

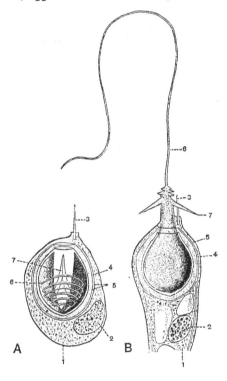

Fig. 19.   Cnidoblast with large Nematocyst from the body-wall of *Hydra fusca*.
                    Highly magnified.   From Schneider.

A. Unexploded.   B. Exploded.   1. Cnidoblast.   2. Nucleus of cnidoblast.   3. Cnidocil.   4. Muscular sheath.   5.   Wall of
nematocyst.   6. Thread.   7. Reflexed processes.

one swift lash with its tentacles and the luckless water-flea is seized
and at the same time paralysed. If we now remove and examine
the prey, we shall find it covered with exploded thread-capsules,
the threads of which have entered its body, and exerted a poisonous
action on it. It is possible to induce a *Hydra* which is being
observed under the microscope to eject its thread-capsules: we

have only to irrigate it with a little ten per cent. solution of common salt, and from all parts of the skin we shall see first the threads shot out, and then the capsules follow.

In the case of a fluid like salt solution, the stimulating action is no doubt exerted over the whole surface of the animal, but an examination of the tentacle when it is extended reveals an arrangement for bringing about the explosion of the thread-capsule. The surface of the tentacle is seen to be covered with little swellings, in which are collections—one might say, batteries—of thread-capsules (8, Fig. 17); and from the surface of the ectoderm, in which they are embedded, delicate hair-like rods project out into the water (3, Fig. 19). These rods are called cnidocils (Gr. κνίδη, a nettle; Lat. cilium, an eyelash) and are the simplest form of sense-hairs met with in the animal kingdom. If one of these be touched, it transmits a stimulus to the cell containing the thread-capsule, the cnidoblast (Gr. κνίδη, a nettle; βλαστός, a sprout), as it is termed; in response to this stimulus the cell contracts, presses on and explodes the capsule.

In the first chapter it was pointed out that protoplasm, when it effects movements, always does so by contracting. We saw, for instance, that the extrusion of the pseudopodia of *Amoeba* could be accounted for by supposing that part of the outside protoplasm contracted and, so to speak, squeezed out part of the more fluid interior. In the life of *Hydra* the principal movements which occur are the shortening and lengthening of the body and the tentacles (B, Fig. 16). Now it has been found that in these movements, the shortening is effected by the contraction of the ectoderm in a longitudinal direction, and the lengthening by the contraction of the endoderm in a transverse direction, in consequence of which the animal is rendered thinner and longer. It has been further ascertained, by the examination of very carefully prepared longitudinal sections, that each ectoderm cell possesses at its base a tail running vertically, which is embedded in the thin layer of jelly sometimes called the structureless lamella or mesogloea (Gr. μέσος, intermediate; γλοία, glue), which separates ectoderm and endoderm (Fig. 20). The endoderm cells similarly possess short tails, embedded in the jelly, but these run transversely. These tails then are instances of the tendency of protoplasm, which contracts regularly in one direction, to be drawn out into fibres in that direction, or, in other words, we have before us the first step in the conversion of an ordinary cell into a muscle cell. Cells showing this modification are termed myo-epithelial (Gr. μῦς = muscle): the word epithelial is used to signify the arrangement of cells in a layer to form a pavement or mosaic.

The most important function of the endoderm cells is to digest the prey which is captured by the tentacles and thrust into the coelenteron. For this purpose they secrete a fluid which has a great power of dissolving protoplasm. This fluid, termed digestive juice, is poured forth into the coelenteron and a large portion of the prey is dissolved by it and passes by diffusion into the endoderm cells, from which part is transferred in a similar manner to

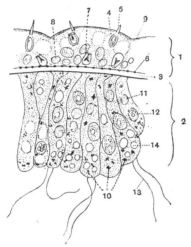

Fig. 20. Section through the body-wall of *Hydra fusca*. Highly magnified. After Schneider.

1. Ectoderm. 2. Endoderm. 3. Mesogloea or structureless lamella. 4. Nematocyst. 5. Cnidocil. 6. Muscle-fibres of ectoderm cells cut across. 7. Nucleus of ectoderm cell. 8. Interstitial cells. 9. Cuticle. 10. Pigment granule. 11. Food granule. 12. Nucleus of endoderm cell. 13. Flagellum. 14. Water vacuole.

the ectoderm. Certain portions of the prey, consisting of some of the proteids, resist the action of this juice. These are seized by pseudopodia emitted by the endoderm cells and bodily engulfed, to be subsequently slowly digested in food vacuoles. Any insoluble parts of the prey, such as cuticle, skeleton, etc., are ejected by the mouth. Some of the endoderm cells also bear flagella, whose movement doubtless aids the circulation of the fluid in the coelenteron.

We have already seen that *Hydra* at certain seasons of the year, viz., the late autumn, produces egg-cells (ova) and male germs (spermatozoa). The latter are shed out into the water, and eventually some of them reach the egg-cells and unite with them. This process is called fertilization, and the fertilized egg-cells cover themselves with spiny coats and drop off into the mud. Here they remain through the winter; in the spring the hard coat cracks and out issues a minute *Hydra*.

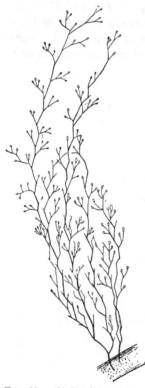

FIG. 21. *Obelia helgolandica* × 1. From Hartlaub. This is the hydroid generation, natural size as it appears to the naked eye.

But *Hydra* is by no means limited to this method of sexual reproduction in its power of multiplying itself. All through the spring and summer, if it be well fed, it buds or reproduces itself by Gemmation. A small swelling makes its appearance on the side of the body; this is really a hollow pouch containing a cavity in communication with the coelenteron (4, Fig. 16). The walls of the pouch are merely continuations of the body-wall of the *Hydra*, and hence consist of the same two layers. The pouch rapidly lengthens, and after a while a circle of tentacles sprouts out from its free end, and a mouth is formed in the centre. We thus have a daughter *Hydra* still in close connexion with the parent, the coelentera of the two being in open communication; later, however, this communication becomes closed and the offspring separates from the parent and leads a free existence. A third method of reproduction, which probably rarely occurs except artificially, is Fission. If a *Hydra* be divided into two halves, each half will grow up into a new individual.

A large number of the Coelenterata, called the Hydromedusae, agree with *Hydra* in all essential points of structure; the most important point of difference is that in them the buds do not become separated, but remain permanently in connection with the parent, and thus complicated colonies are built up (Fig. 21). Other differences of less

importance are that there is a horny shell, the perisarc (Gr. περί,
around ; σάρξ, flesh) (Fig. 22), secreted by the ectoderm at any rate
on the lower portion of the body, also that the tentacles are nearly

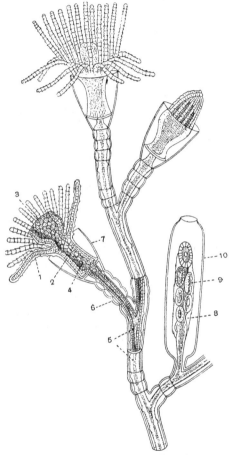

FIG. 22.   Part of a branch of *Obelia* sp.   To the left a portion is shown
               in section.   After Parker and Haswell.

1. Ectoderm.   2. Endoderm.   3. Mouth.   4. Coelenteron.
    5. Coenosarc.   6. Perisarc.   7. Hydrotheca, prolonged at base of
    Hydroid as a shelf.   8. Blastostyle, a mouthless hydroid bearing medusa-
    buds.   9. Medusa-bud.   10. Gonotheca, part of perisarc which protects
    the medusa-buds.

always solid, containing, instead of tubular outgrowths of the endoderm, a solid cord of cells (Fig. 22) with firm outer membranes and

partially fluid contents, so that the cells have the same kind of stiffness as a well-filled water-pillow. These cords likewise bud out from the endoderm, but, as apparently the animal does not need the tentacle cavity which exists in the *Hydra*, it has disappeared, and the solid axis is essentially a strengthening or skeletal structure. As in *Hydra*, there is an oral cone; and in some species of Hydromedusae, at any rate, there is an additional row of short tentacles at the tip of this. It has been stated above that the buds do not become detached, but there is one kind of bud differing much in shape from the rest which does become detached. In such a bud, the whole body becomes very much shorter and at the same time much flattened out in its lower portion, so that the main circle of tentacles is widely separated from the oral cone; at the apex of the latter there is sometimes a second circle of small tentacles. The flattened part of the body becomes concave on the side towards the mouth so as to assume the form of a bell or umbrella, and, owing apparently to this circumstance, the part of the coelenteron which it contains becomes so pressed together, that by the adhesion of its upper and lower walls, its cavity for the most part disappearing, it becomes converted into a concave layer, called the endodermal lamella. Along four lines, however, the cavity does not disappear (4, Fig. 23), and it also remains open just beneath the circle of larger tentacles at the edge of the bell, so that in this way we have a circular or marginal canal established, communicating by four radial canals with the part of the coelenteron that still persists in the oral cone, and opening to the exterior by the mouth (1, Fig. 23). The upper surface of the bell is styled the exumbrella or aboral surface (Lat. *ab*, away from; *os, oris*, the mouth) the lower the subumbrella or oral surface.

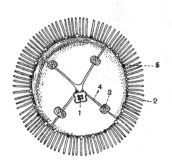

FIG. 23.　Free-swimming Medusa of *Obelia* sp.

1. Mouth at end of manubrium.
2. Tentacles.
3. Reproductive organs.
4. Radial canals.
5. Auditory organ.

The great mass of the bell is composed of the jelly intervening between the outer ectoderm on the convex side and the endoderm. In this jelly solid strings sometimes appear which give it a firmer consistence. The modification of the

Fig. 24. *Bougainvillia fructuosa*, × about 12.   From Allman.

A. The fixed hydroid form with numerous hydroid polypes and medusae in
various stages of development.   B.   The free swimming sexual Medusa
which has broken away from A.

base of the animal into the shape of an umbrella causes the oral
cone to resemble the handle, hence the name manubrium (Lat.
a handle), by which it is usually designated in a bud of this kind
(1, Fig. 23). Just above the circular canal in most Medusae a fold of
the outer skin grows in towards the oral cone, so as to form a broad
circular shelf: this structure is called the velum (Lat. an awning)
(B, Fig. 24; 1, Fig. 25). The bud now breaks loose and swims by
contractions of the bell, aided by vibrations of the velum. Anyone
would now recognise it as a minute jelly-fish, though it really is
quite different in many points from the larger and better known
animals denoted by that term. Zoologists speak of it as a Medusa,
and speak of the stock from which it was budded as a colony
consisting of medusoid and hydroid persons, the latter term
denoting the ordinary buds which resemble *Hydra*. The terms polyp
and hydranth are also often used to denote a hydroid person.
A medusoid is in many respects more highly developed than
the hydroid person. The ectoderm cells composing the velum
and those forming the lining of the under side of the bell or
sub-umbrella are strongly drawn out into processes which are
muscular. In the velum these are arranged so as to form two
bands running round the edge of the bell or umbrella, one band
being in connection with the' upper and another with the lower
layer of cells composing the fold of ectoderm of which the velum
consists. Just, however, where the velum is attached to the bell,
its cells—upper and lower—undergo another and more interesting
modification (4 and 5, Fig. 25). At their bases a tangle of delicate
threads of almost inconceivable fineness appear; these threads are
outgrowths of the cells, but far more delicate than those which
already in *Hydra* we recognised as the forerunners of muscles; the
threads we are now considering are, in fact, nervous in nature,
and the tangles of them connected with the upper and lower layers,
respectively, of the velum, constitute an upper and a lower nerve
ring. Each thread is to be regarded as the tail of an excessively
small ectoderm cell.

In *Hydra* we found the earliest appearance of sense hairs; and
the cells of which they are processes, viz., the cnidoblasts, may be
called sense cells. In the Medusa we meet with definite collections
of sense cells aggregated so as to form sense organs. These are
found close to the position of the nerve ring, either on the velum
itself or immediately outside it at the bases of the tentacles, so
that the stimuli which they receive are easily transmitted to the

nerve ring. Two main kinds of sense organs are frequently found, which may be roughly called eyes and ears; never, however, both kinds in one Medusa. The 'eyes' are little coloured patches of skin; some of the cells of which end in clear rods while others secrete a coloured substance or pigment. Both pigment and rods are necessary if there is to be vision, though we do not understand why. The ears are little pits in the base of the velum; they may be open or their edges may come together, so that the ectoderm lining them is entirely shut off from the outer skin. In either case, some of the

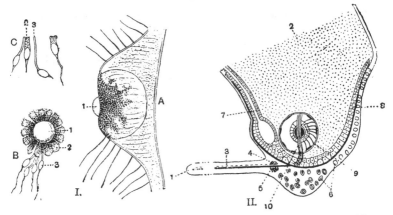

FIG. 25. I. A. Eye of *Lizzia koellikeri* seen from the side, magnified. B. The same seen from in front. C. Isolated cells of the same. From O. & R. Hertwig.

1. Lens.   2. Pigment cells.   3. Percipient cells.

II. Radial section through the edge of the umbrella of *Carmarina hastata* showing sense-organ and velum.

1. Velum.   2. Jelly.   3. Circular muscles of velum.   4. Upper nerve ring.   5. Lower nerve ring.   6. Nematocysts.   7. Radial vessel running into circular vessel, both lined by endoderm.   8. Continuation of endoderm along aboral surface.   9. Sense organ or tentaculocyst.   10. Auditory nerve.

cells forming the walls of the pits secrete particles of lime, others close to them develope delicate sense hairs. The result is that vibrations in the water, if they come with a certain frequency, will affect the heavy particles, and their vibrations in turn will affect the sense hairs. There is another kind of information, however, which organs like these give their possessor, and this is probably still more important to the floating Medusa, namely, information as to the position of the animal with regard to the vertical. In other

words, the Medusa learns from them whether it is moving upwards
or downwards or sideways : for when the animal shifts its position,
the heavy particles in the ear-sacs are shifted conformably and affect
different sense cells.

Through these different sense organs stimuli are continually
pouring in from the external world. If the stimuli only affected
the contractile cells nearest them irregular movements would result.
The function of the nerve ring, as of all nervous systems, is to coordinate the stimuli, that is to collect and rearrange and rapidly
distribute them to the whole animal so that a definite reaction of
the whole contractile tissue results, not a series of local reactions
interfering with one another.

The Medusa is very voracious and rapidly increases in size. It
feeds on the small organisms of all kinds, both plants and animals,
which are found at the surface of the sea. After some time it com-

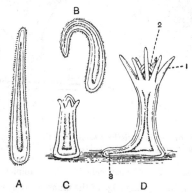

Fig. 26.   The ciliated larva or Planula of a Hydromedusan, *Clava squamata.*
Magnified.   From Allman.

A & B.   Swimming about in the sea.   C. Coming to rest on a rock.
  D. Developing tentacles, oral cone and stolon.   1. Tentacles.   2. Oral
  cone.   3. Stolon.

mences to give rise either to eggs or to spermatozoa, which usually
develope in exactly the same way in which they developed in *Hydra,*
i.e., from the interstitial cells of the ectoderm.   The accumulations
of these cells, called gonads or generative organs, are borne
either on the under side of the bell (3, Fig. 23), or on the sides of
the manubrium, and it is a curious fact that those Medusae which
have them in the former position usually possess ear-sacs, whereas
when the gonad is situated on the oral cone, ear-sacs are never
present, but eyes may be.   The eggs and spermatozoa are both shed

out into the water and coalesce there, and the fertilized egg developes
into a little oval larva, termed a Planula (Fig. 26), without
tentacles or mouth, and covered all over with cilia. It consists at
first of a hollow vesicle of ectoderm cells, later becoming filled with
a solid plug of endoderm. This little creature swims about for a
while and then attaches itself by one end to a stone or a piece of
sea-weed. The attached end flattens out (C and D, Fig. 26), but
the rest of the animal lengthens and a mouth and tentacles appear
at the free end and the endoderm becomes hollowed out, so that
the creature takes the form of an unmistakeable hydra-like organism.
It then begins to bud out a branch called a stolon which creeps along
the substratum. From this other polyps will arise, each of which
has only to bud in order to reproduce the colonial stock from which
its parent, the Medusa, was separated. The free-swimming young
or planulae furnish good examples of what is meant by the term
larva. This name is given to the young form of any animal when
it is very different to the fully-grown animal and leads a free life.

We have thus learnt that a Medusa gives rise to an egg which
develops into a Hydroid person, which after a time in turn buds off a
Medusa; such an alternation of generations is very characteristic
of a large number of Coelenterata. The Medusa represents a sexual generation, the Hydroid an asexual
generation, and inasmuch as the Medusoid is often
only produced as a bud of the third or fourth order (i.e. is budded
from a Hydroid person which has produced similarly from another
Hydroid person), it will be seen that several asexual generations
intervene between two sexual ones. One explanation of this life-
history is that the Medusa is only a specially modified Hydroid, which
has acquired the power of locomotion, in order to disperse the eggs
over a large area, and thus avoid the overcrowding of a limited area
with one species. The swimming bell and velum are contrivances
to enable the bud which bears the eggs to move about. If, however,
this explanation be adopted, it is a most remarkable fact that in
many species the Medusae are very imperfectly developed and
never become free. Such Medusae are usually more or less degenerate and are termed gonophores. Since the gonophore fails
to fulfil the purpose for which we believe the Medusa to have been
developed we must assume that conditions have now so far changed
that the same wide scattering of the eggs is not now so necessary as
formerly, possibly because the species in question are restricted to
particular strips of the shore. *Tubularia larynx* found growing on

seaweed is a good example of a form with degenerate Medusae, *Bougainvillia* or *Obelia* of forms with free Medusae.

The Hydromedusae include a large number of families, most of which are represented by small plant-like forms resembling the genera just mentioned, but there are several groups which show marked peculiarities and have been regarded by many zoologists as of co-equal rank with the order although they have probably been derived from ordinary Hydromedusae. Of these we may name (i) the Trachymedusae, (ii) the Narcomedusae, (iii) the Siphonophora and (iv) the Hydrocorallinae. In the first two groups the eggs develope from the planula stage directly into Medusae, missing out the hydroid stage completely. In both cases also the sense-organs are specially modified tentacles which are suspended like minute clubs round the edge of the bell. In the Narcomedusae these clubs are freely exposed and the wide baggy stomach occupies the whole under-surface of the umbrella, whereas in the Trachymedusae the sensory clubs are enclosed in pits (Fig. 25) and the stomach is small and suspended from the umbrella by a stalk traversed by the radial canals. The name Trachymedusae (Gr. τράχυς, rough) is derived from the circumstance that the umbrella is stiffened by numerous ribs of endoderm cells and the edge has a thick rim of ectoderm. The members of the third group are stocks consisting both of medusoid and hydroid persons which are not attached to any support but which freely swim or float in the sea. Some of the medusoid persons known as nectocalyces have taken on the function of locomotor organs and by their rapid pulsations not only drive themselves through the sea, but draw after them the rest of the stock much as an engine draws a train of carriages. A few forms, however, like the Portuguese Man-of-war, *Physalia*, have no nectocalyces and float passively about. The popular name of this genus is derived from the shape of the huge air-containing float from which the persons of the colony are suspended. It has been plausibly suggested that the Siphonophora have been derived from planulae which attached themselves to the surface-film of the water instead of to a solid support. The surface-film in consequence of its physical properties acts like an elastic membrane, and in artificial cultures it can often be seen that some planulae do attach themselves to this, and in consequence perish. But if by favourable variations, such as a tendency to cupping of the base and an inclusion of air-bubbles in the cavity, the stock were enabled to remain suspended, then it would be placed in a very favourable position for getting food, and thus it has been suggested the simply floating Siphonophora have been evolved from Hydro-medusae. If this view be taken, the three chief divisions of Siphono-phora represent three successive stages in the adaptation of the group to a pelagic life. Thus the Physaliidae simply float, the Physo-phoridae float and swim by nectocalyces, whilst the Calycophoridae have lost the float and trust entirely to their powerful nectocalyces.

The Siphonophora are remarkable for the varieties of person which compose their colonies. As varieties of the hydroid person may be named the palpons or tactile persons devoid of a mouth, but showing their equal rank with the nutritive person by the possession of similar tentacles. To the category of medusoid persons belong not only the nectocalyces but the bracts—transparent sheath-like structures sometimes present, which shelter groups of persons. This extreme variety of persons is foreshadowed in the ordinary Hydromedusae. *Hydractinia* for instance, which grows at the mouth of whelk shells inhabited by hermit crabs, has palpons amongst its hydroid persons, but in no case is such extreme diversity attained as among the Siphonophora. The Hydrocorallinæ hydroids form large colonies and are divisible into nutritive polyps or gastro-zooids and tentacle-like polyps, the dactylozooids. The skeleton is massive and they form encrusting growths. The medusoid persons attain varying degrees of perfection.

The Sea-Anemones are representatives of a second division of

Actinozoa.

the Coelenterata, which show a decidedly more com-plicated structure than the animals just considered. Unfortunately it is very difficult to obtain the ordinary sea-anemones in a sufficiently expanded condition to make out their structure, since when irritated they contract so much as to throw their internal structures into great confusion. Another animal belonging to the same group is the 'colonial' species *Alcyonium digitatum,* sometimes called "Dead men's fingers." It is comparatively easy to paralyse the members of the colony or polyps by adding cocaine, or some similar reagent, to the water in which the colony is living (Fig. 27). If then an expanded polyp be cut off and examined with a lens, we shall be able to make out most of its structure. We notice to begin with that there is a single circle of eight tentacles, each of which has a double row of short branches, so that it looks like a miniature feather; within the circle of tentacles there is, however, no trace of an oral cone; there is instead a flat disc, slightly sunken in the centre, where we find the slit-like mouth. If we look in at the lower cut end of the polyp we shall see that the internal cavity or coelenteron, instead of being a simple cylindrical space like that of *Hydra*, is partially divided into compartments by folds stretching in towards the centre, but not meeting. These folds are called mesenteries, and there are eight of them, corresponding in number (but not in position) with the

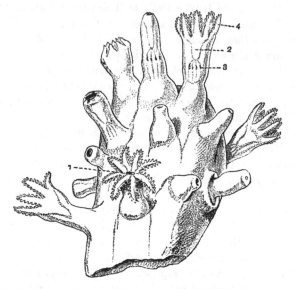

Fig. 27.   Part of a colony of *Alcyonium digitatum* × 8, showing thirteen polypes in various stages of retraction and expansion.

1. Mouth.	3. Mesenteries with reproductive cells.
2. Oesophagus.	4. Feathered tentacles.

tentacles (Fig. 28). We shall further see that the mouth does not, as in *Hydra*, open directly into the coelenteron, but leads into a flattened tube which projects into the interior of the body. This tube, the so-called oesophagus or gullet, is really lined by the ectoderm, which is merely tucked in at the mouth. Such a tube is known as a stomodaeum[1]. The mesenteries, although they end freely below, are attached to the sides of the stomodaeum above, so that in this region the coelenteron is divided into a number of compartments, each of which is prolonged into one of the hollow tentacles (Fig. 29).

[1] "I have proposed to designate this ingrowth...the stomodaeum (στόμοδαιον, like πυλόδαιον, the road connected with a gateway) and similarly to call another ingrowth which accompanies the formation of the second orifice (the anus) of the enteron, the proctodaeum" (πρωκτός, the anus). Ray Lankester.

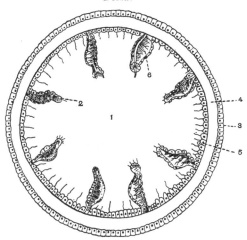

Ventral

Fig. 28. Transverse section through a polyp of *Alcyonium digitatum* below the level of the oesophagus × about 120. From Hickson.

1. Coelenteron.   2. Mesentery with free edge.   3. Ectoderm.   4. Mesogloea or jelly.   5. Endoderm.   6. Muscles in mesentery.

Dorsal

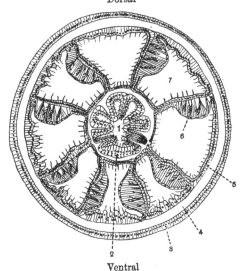

Ventral

Fig. 29. Transverse section through a polyp of *Alcyonium digitatum*, through the region of the oesophagus × about 120. From Hickson.

1. Cavity of oesophagus.   2. Siphonoglyph.   3. Ectoderm.   4. Mesogloea or jelly.   5. Endoderm.   6. Muscles in mesenteries.   7. Intermesenteric cavity.

A microscopic section of such a polyp shows us several other interesting points. We see that we have to deal with the same layers which we met with in *Hydra*, skin (or ectoderm) and coelenteron lining (or endoderm). Between them, however, there is the jelly, which was present as an exceedingly fine membrane in *Hydra*, and which, greatly thickened, formed the substance of the bell of the Medusa. This jelly is fairly thick in the minute sea-anemone we are examining, and here contains cells which have wandered into it from the ectoderm. Some of these cells have the power of secreting thorny rods of lime, termed spicules. These spicules are very abundant where the polyp merges into the general surface of the colony, so that they form a kind of stiff protecting crust round the base of the polyp and over the surface of the colony from which the polyps rise. In the organ-pipe coral, *Tubipora*, the spicules in the lower parts of the polyps are so felted together that they form a set of parallel tubes, suggesting the pipes of an organ; only the upper part of the polyp, where the spicules are not yet closely aggregated, being capable of movement. We have spoken above of the colony as distinct from the polyps, and this use of the word demands some justification. When we were dealing with the Hydromedusae, we used the word colony in the sense of the whole mass of the polyps which cohered together, and which had arisen by the growth of one original polyp. Now in *Alcyonium* and its allies, budding does not take place in quite the simple manner in which it occurs in *Hydra* and its allies. Instead of one polyp growing directly out of another, the coelenteron of the parent sends out a tube lined only by endoderm. This tube grows, pushing the ectoderm before it; but, as between the ectoderm and endoderm there is a thick jelly interposed, the endodermal tube can branch without the ectoderm becoming indented. Where the free ends of these tubes reach the surface, there fresh polyps are developed, mesenteries and oesophagus making their appearance. Something like these tubes does, in fact, occur amongst Hydromedusae: a complete colony is found to consist of a number of upright branches ending in polyps but connected at their bases by tubes called stolons which creep along the sea floor: the endodermal tubes of *Alcyonium* may be compared to these stolons, the great difference being that in their case, owing to the thickness of the jelly, the ectoderm is stretched uniformly over a mass of tubes, instead of each tube having its own ectodermal covering as in the Hydromedusae.

Still examining a section of the polyp the next point we notice is the structure of the mesenteries. These end in a free edge below which is much thickened and folded, and since it stands out in contrast with the rest of the mesentery as if it were an independent structure it has been called a mesenteric filament (Figs. 28 and 30). The cells composing six of these filaments are very tall and secrete a juice which digests the prey: the remaining two filaments are composed of ciliated cells of moderate height which maintain a constant outward current of water. The surface of the mesenteries is covered by cells which become very much folded so as to produce a marked projection from the face of the mesentery. The cells of the folded area are all produced into vertical muscle-tails so that together they give rise to one of the powerful longitudinal muscles (Figs. 28 and 29), by which the sudden retraction of the polyp is brought about. The slow expansion is effected by the reaction of the elastic jelly or mesogloea and perhaps also by the pressure exerted on the fluid contained in the body by a layer of circular muscles developed as outgrowths from the endoderm cells of the intermesenteric chambers.

A second difference is found in the position of the eggs and sperm cells. These are developed from the endoderm on the face of the mesenteries, very low down in the base of the polyp and nearer the free edge than the longitudinal muscle. The eggs when ripe are cast out into the coelenteron and so out by the mouth, though in many species they come in contact with the male cells whilst in the coelenteron of the parent.

The gullet has at one side a deep indentation or groove which is lined by powerful cilia (2, Fig. 29). The groove is termed a siphonoglyph (Gr. σίφων, a tube; γλύφω, to hollow out) and its cilia keep up an inward current of water whilst the rest of the gullet is choked with prey, and so fresh supplies of water charged with oxygen are brought in contact with the lining of the coelenteron and enable it to respire. The two mesenteries with which the lower end is connected are called the directive mesenteries, they are situated opposite to the two ciliated mesenteric filaments. By the cooperation of the latter with the siphonoglyph complete circulation of the water in the coelenteron is maintained. The ectoderm of course gets its oxygen directly from the surrounding water.

The ordinary sea-anemones or Zoantharia differ from *Alcyonium* in very many points. The tentacles, hollow as before, are never feather-like but always perfectly simple and round, and there is usually a large number of them arranged in several concentric circles. The mesenteries also are numerous, and extend inwards to different lengths, so that we can distinguish primary mesenteries which join the gullet from secondary ones which do not. Both primary and secondary are usually arranged in pairs, but there is much variety and all that can be universally asserted is that they never exhibit exactly the arrangement shown in Alcyonaria. A very common arrangement is·to have six pairs of primary mesenteries and two siphonoglyphs, one at each end. Spicules are never developed and in the ordinary anemones of our coasts there is no skeleton whatever. These commoner forms sometimes, though rarely, bud, but there is another large class of anemones which do form colonies, the buds occasionally arising as in the Hydromedusae from the body of the parent directly. These colonial anemones form the hard stony masses called coral (Fig. 30). If we look at a piece of coral we can see in it cups

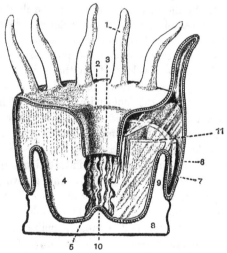

Fig. 30. Semi-diagrammatic view of half a simple Coral, partly after G. C. Bourne. On the right side the tissues are represented as transparent to show the arrangement of the theca and septa ; on the left side a mesentery is seen.

1. Tentacle. 2. Mouth. 3. Oesophagus. 4. Mesentery. 5. Mesenteric filaments, free edge of mesentery. 6. Ectoderm. 7. Endoderm. 8. Basal plate. 9. Theca. 10. Columella. 11. Septum.

with partitions radiating inwards, the whole reminding one of the structure of a sea-anemone: and it was a natural mistake to suppose, as the earlier naturalists did, that the hard skeleton was

formed inside the body of the polyps, the partitions representing
the mesenteries. Of course it is difficult to imagine how the animal
could move if it had all that mass of stone inside it. How the
corallum or stony skeleton is formed is a matter of dispute. It
is certainly situated outside the ectoderm, but whether it is secreted
by the ectoderm as a kind of sweat which hardens, or whether the
ectoderm cells are calcified and thrown off, or the ammonium carbonate, secreted by all animals, precipitates the calcium carbonate
of the sea water and so forms the skeleton, is not finally decided.
At any rate a calcareous cup is formed in which the polyp sits and
the partitions of the cup indent the base of the animal, pushing
before them folds of the body wall, which project into the coelenteron between mesenteries, so that the action of the longitudinal
muscles is not interfered with.

Under the name *Coral* the skeletons of quite a number of
different kinds of Coelenterata were included besides
Coral.   Zoantharia.

The so-called Millepore Corals belong to the first division of the
Coelenterata, the Hydrozoa, for Millepora itself gives off quite
typical Anthomedusae and the other genera have gonophores.

The Hydrocorallinae are really distinguished by the fact that the
perisarc only which covers the basal stolons is thick and calcareous.
After a while the stolons enclosed in the skeleton die, but fresh
stolons are thrown out at higher levels, so the skeleton grows in
thickness. The hydroid persons are of two kinds, nutritive persons,
gastrozooids, short and with wide mouths, and tactile persons,
dactylozooids, which surround each gastrozooid in a circle and
which are long and mouthless. Both kinds have short rudimentary
tentacles looking like knobs.

The so-called Organ-pipe coral is, as has been already explained,
an Alcyonarian in which the spicules cohere. Various fossil so-
called corals, e.g., *Syringopora*, belong to the same category. The
red Neapolitan coral of which ornaments are made is also an
Alcyonarian, the spicules of which are of a bright pink or red
colour and cohere to form a rod in the axis of the colony. In some
spots off the coast of Australia the Alcyonaria with coherent spicules
are so numerous that they form reefs.

Coral-forming anemones are found all over the world,—one
genus, *Caryophyllia*, being actually found at low spring tides on
the south-west coasts of England : but it is only in the tropics that
those species are found which keep on budding and growing with

sufficient persistence to build up the great reefs which form the famous coral islands of warmer seas. Of course as soon as the reef is built up to the surface the polyps cease to grow, and then the breakers soon pile up broken off pieces in sufficient quantity to raise the reef above the tide-marks.

A third group of the Coelenterata is constituted by the Acale-
Acalephae.   phae (Gr. ἀκαλήφη, a nettle). These animals are the larger and better known jelly-fish. They are to some extent intermediate in character between the Hydrozoa and the Actinozoa. Like the latter their genital cells are developed from endoderm, and in the larval condition there are mesenteries, but they do not possess a stomodaeum.

A common British species, *Aurelia aurita*, is in summer often cast by thousands on the southern shores of Great Britain. Viewed from the outside it very much resembles the medusoid persons of the Hydromedusae. Like them it possesses a swimming bell with a circle of tentacles at the margin. There is also a prominent oral cone or manubrium. This however does not bear real tentacles, but the four corners of the rectangular mouth are drawn out into long frilled lips (2, Fig. 31), along the inner sides of which are open grooves leading into the gullet. Perhaps the most marked difference is that the reproductive organs are here, as in the anemones, swellings of the stomach lining: the eggs and spermatozoa are shed into the coelenteron and escape by the mouth. The generative organs have the shape of four semicircular ridges, and along the inner side of each of these there is a row of filaments composed of cells somewhat similar to the cells on the edges of the mesenteries in the anemones (11, Fig. 31). Nothing like these gastral fila-ments, as they are called, are found in the Hydromedusae. There is, further, no velum in the Acalephae, and there is also no nerve ring. Sense organs however of an exceedingly interesting kind are present.

In *Aurelia*, for instance, there are eight minute tentacles which stick out from the edge of the bell and are covered by special little hoodlike outgrowths of the same (9, Fig. 31). Each of these tentacles contains a hollow outgrowth from the circular canal lined like it by endoderm. The endoderm cells at the tip secrete a mass of calcareous particles: the skin cells at the base of the tentacle have produced nervous fibrillae from their bases and so the tentacle, as it is caused to sway in one direction or another by the weight of its heavy end, affects now some of the nerve fibrillae and now

others, and so produces the same effect as the stones in the ear
sac of a medusoid, though the construction of the Acalephan organ
is quite different.   In the Trachymedusae and Narcomedusae, however, sense tentacles similar to those of the Acalephae are found.
Here the edges of the hood often join so as to form a sac enclosing
the organ, whence the name tentaculocyst (9, Fig. 25).

It has been proved experimentally that the ordinary stimuli
which cause the rhythmical pulses of the bell proceed from these

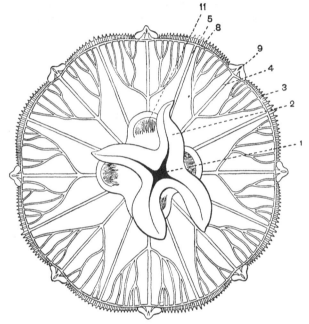

Fig. 31. *Aurelia aurita.* Somewhat reduced.

1.   Mouth.   2. Circumoral processes.   3.   Tentacles on the edge of the
umbrella.   4. One of the branching perradial canals.   There are four of
these, and four similar interradial canals; the perradial canals correspond
to the primary stomach pouches of the Hydra-tuba, the interradial
alternate with these.   5. One of the unbranched adradial canals.
8. The circular canal.   9. Marginal lappets hiding tentaculocysts.
11. Gastral filaments, just outside these are the genital ridges.

tentaculocysts, so that they act like minute brains. How the
co-ordination of the stimuli proceeding from the eight centres is
brought about we do not know, but it is probably due to the
presence of a very thin diffuse sheet of nerve fibrillae on the under
surface of the bell.

It has been mentioned above that the reproductive organs are swellings of the endoderm. The central space or "stomach" is a wide sac occupying the centre part of the bell and not, as in the Hydromedusae, confined to a large extent to the oral cone. In *Aurelia* this space is produced into four lobes, and in the floor of each lobe is one of the reproductive organs. From the edges of the stomach a number of branching canals lead out into the circular canal (4, Fig. 31), all these tubes being, as it were, burrows in a continuous sheet of endoderm cells, which stretches out to the edge of the disc and really represents a part of the coelenteron, the cavity of which has been obliterated. It thus corresponds exactly to the endodermal lamella of the Hydromedusae.

When the eggs fall out of the mouth they are caught in little pockets and there develope into little Planulae. These, as usual,

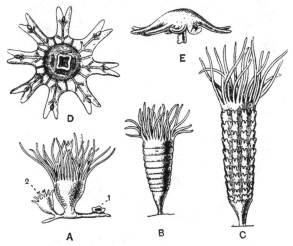

FIG. 32.   Strobilization of *Aurelia aurita*.   From Sars.

A.   Hydra-tuba on stolon which is creeping on a Laminaria. The stolon is forming new buds at 1 and 2.   B.   Later stage or Scyphistoma × 4.   The strobilization has begun.   C.   Strobilization further advanced × 6. D.   Free swimming Ephyra stage × 7·5, seen from below.   E.   The same seen in profile × 7·5.

become free and swim about, and finally each fixes itself and developes into a little polyp, called a Hydra-tuba, not unlike a *Hydra* in appearance (A, Fig. 32), but there are nevertheless important points of difference. Thus there is no oral cone but a flat oral disc in the centre of which the mouth opens into the coelenteron. The latter has four ridges projecting into it, the lower edges of which are free while the upper ones are joined to the gullet. These ridges being produced by the folding of the endoderm layer they are double and contain between their two limbs a space filled with jelly. Into this space a prolongation of the ectoderm of the mouth disc grows down so as to form a "septal funnel." The cells composing the septal funnel secrete longitudinal muscular fibrils, and thus four powerful septal muscles are formed which serve to shorten the Hydra-tuba. The hydroid persons of the Hydromedusae have also longitudinal muscles but these are disposed in a uniform sheet round the polyps in question and belong to the ectoderm cells forming the sides. During a large part of the year the Hydra-tuba multiplies by budding, just as a *Hydra* does, but at certain seasons it undergoes a very remarkable change (B and C, Fig. 32). The oral disc flattens out very much and its edges become drawn out into lobes, the tentacles at the same time dropping off. A short oral cone is developed from the centre of the disc, the mesenteries become perforated and finally the whole flattened-out top of the Hydra-tuba breaks off and swims away. This is known as an Ephyra larva (D and E, Fig. 32). It leads a free life and gradually develops into a large jelly-fish. But long before the primary oral disc has become free, the part of the Hydra-tuba next below has been growing out so as to produce a similar disc. This process, called Strobilization (Gr. στροβιλός, a whorl), is repeated until the Hydra-tuba resembles a pile of saucers, in which state it is called a Scyphistoma (Gr. σκύφος, a saucer).

We can get some idea as to how this extraordinary development may have arisen on the following hypothesis:—The original Acalephan was probably an organism like an anemone with a wide top and narrow base. In this top the generative organs were developed, and when the eggs became ripe it broke off and wandered away in order to disperse the species. The lower part of the polyp regenerated the head, exactly as a *Hydra* can do if the head with its ring of tentacles be cut off. Later this process of renewal became hurried on until it commenced before the separation of the head was complete and thus we have the Scyphistoma stage. It is a strong support of this theory that there exists a large coral-forming anemone, *Fungia*, in which there is a flat top and a stalk, and the flat top periodically falls off and is renewed.

A third great division of the Coelenterata is constituted by the

Ctenophora. animals called Ctenophora (Gr. κτείς, κτενός, a comb). These are widely different from both Hydromedusae or Actinozoa, even including Acalephae. They never bud and with one doubtful exception have no thread cells. In place of these there are **gripping** or **adhesive cells** covered with a secretion by which they adhere to the prey. They are often ripped off in the struggles of the prey. They are each provided with an elastic tail embedded in the jelly which pulls in the object to which they have adhered. The Ctenophora are free swimming but their locomotion is performed not by the agency of muscular bands but by eight bands of cilia which run like meridians of longitude over the

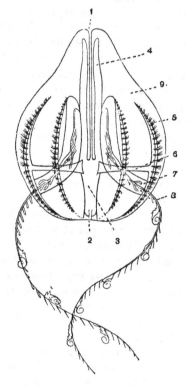

FIG. 33. *Hormiphora plumosa.* After Chun. Side view.

1. Mouth leading into stomach.   2. Aboral pole with sense organ.
3. Funnel.   4. Paragastric canal running back towards oral pole.
5. One of the eight bands of fused cilia.   6. One of the eight canals running towards 5.   7. A tentacular pouch.   8. A tentacle.
9. Gelatinous tissue.

generally oval body from the mouth to the opposite pole (5, Fig. 33). The cilia in each band are arranged in short transverse rows, and the cilia in each row are joined at the base and free at the tip. So each row has the form of a comb, and thus the name Ctenophora, comb-bearer, is seen to be appropriate. Further the principal sense organ is situated in the centre of the end of the animal opposite to the mouth at the spot where the bands of thickened ectoderm which carry the combs converge. These thickened ridges of cells are often termed "ribs." If we compare the animal to a globe, the end at which the mouth is may be called the oral pole, the opposite end the aboral pole (2, Fig. 33). The sense organ at the aboral pole is a plate of thickened ectoderm the cells of which have developed nerve tails. Similar nerve tails are developed by the bands of ectoderm which carry the combs. The cells at the edge of the plate carry cilia fused with one another which arch over the plate and cover it like a tent. Inside is a calcareous ball supported on four curved bars, each made of conjoined cilia, borne by some of the inner cells. This ball acts as a balancing sense-organ. If the animal inclines to one side the ball will bear heavily on the support on that side, and stimulate thus the corresponding ribs, which will thus act more vigorously than the rest and tend to restore the vertical position.

Like Actinozoa, Ctenophora have a well-marked stomodaeum, and the true coelenteron is represented by a series of branching canals, the central one being termed the funnel (3, Fig. 33). The funnel and stomodaeum are both flattened but in planes at right angles to one another. The funnel gives off (1) two canals, the so-called excretory canals, which open at the sides of the sense-organ; (2) two canals, paragastric, running back towards the mouth parallel with the stomodaeum (4, Fig. 33); (3) two canals running each to a branched tentacle, which can be retracted within a pouch (7, Fig. 33). This branched tentacle is covered with adhesive cells, there is one on each side of the animal. Each tentacle canal gives off four branches (6, Fig. 33) which lead into the meridional canals running under the ribs, from the cells lining which both ova and spermatozoa are produced, Ctenophora being hermaphrodite. The commonest British form is *Hormiphora plumosa*, which sometimes appears in shoals in the seas washing the Atlantic coast of Britain on the one hand and America on the other. The Ctenophora are good examples of what are called pelagic organisms, that is to say, organisms which pass their whole life from the egg to the adult condition floating at or near the surface of the sea. Such organisms are the only ones which are found in mid-ocean. Nearer the shore the waters are filled by a profusion of other animals, but these turn out on examination to be largely composed of forms which in some period of their existence are adherent to or creeping on the bottom. Other purely pelagic groups are the Siphonophora, the Trachymedusae and the Narcomedusae.

The Ctenophora contain many forms which differ widely in appearance from *Hormiphora*—for instance the *Cestum veneris*, or Venus's girdle, a beautiful transparent ribbon-like creature, a foot or so in length and two or three inches wide. On close examination the reason of this diversity of shape is found to be that the Ctenophora are not really radially symmetrical, but doubly bilaterally symmetrical. That is to say, not only right and left sides are like one another but also the back and belly are alike, but at the same time different from the sides. The difference is slight in *Hormiphora* but very strongly marked in *Cestum veneris*.

Phylum COELENTERATA.

The classification of the Coelenterata is as follows :—

Class I. HYDROZOA.

Coelenterata without mesenteries or gullet lined by ectoderm: genital cells derived from ectoderm.

Order 1. **Hydrida.**

Only hydroid persons present, not permanently attached but capable of locomotion : the buds become free.

Order 2. **Hydromedusae.**

Composite fixed colonies of hydroid persons from which medusoid persons are budded off.

Suborder (1) **Gymnoblastea.** Perisarc confined to the base of the hydroids: medusoids have eyes and bear gonads on the manubrium.

Suborder (2) **Calyptoblastea.** Perisarc expanded to form cups called hydrothecae, into which heads of hydroid persons can be retracted : medusoids have ears and bear gonads on under side of umbrella.

Order 3. **Narcomedusae.**

Only medusoid persons present. The manubrium poorly developed, the wide stomach occupying the under side of the bell. The sense-organs are reduced tentacles projecting at the edge of the bell.

Order 4. **Trachymedusae.**

Forms in which like the foregoing group only medusoid persons are present, but there is a long manubrium traversed by the radial canals, and the stomach is only at the bottom of it.

Order 5. **Siphonophora.**

Free-swimming colonies consisting of hydroid and medusoid persons in which the base is modified into a float or some of the medusoids are transformed into swimming organs, or both arrangements are combined.

Order 6. **Hydrocorallinae.**

Composite fixed colonies of hydroid persons: medusoid persons budded off in only one or two genera.

Perisarc thick and calcareous, surrounding chiefly the stolons which are given off at various levels and form a thick mass.

Class II. ACTINOZOA.

Solitary or colonial Coelenterata with gullet lined by ectoderm: coelenteron provided with radiating mesenteries: genital cells derived from endoderm.

Order 1. **Alcyonaria.**

Eight mesenteries and eight fringed tentacles: spicules in the jelly.

Order 2. **Zoantharia.**

Mesenteries usually in pairs, either six pairs or some multiple of six: tentacles conical: no spicules but often an external calcareous skeleton formed by the ectoderm.

Class III. ACALEPHAE.

Coelenterata with alternation of generations: mesenteries present in young, but later becoming absorbed: oral part breaks loose and becomes developed into a free-swimming organism externally resembling a medusoid: the stalk of the original polyp reproduces the lost parts.

Class IV. CTENOPHORA.

Very widely different from the two preceding divisions: free-swimming animals with a sense organ and nervous disc of skin at the pole opposite the mouth: swim not by muscular contractions but by vibrations of eight longitudinal bands of cilia radiating from nervous disc, which bands consist of successive transverse rows of cilia, the cilia of each row fused at base so as to form a comb-like structure: only two tentacles, a gullet lined by ectoderm: stomach represented by a system of branching tubes.

CHAPTER IV.

PHYLUM PORIFERA.

THE group of the sponges or Porifera occupies an almost isolated
position in the animal kingdom. Sponges agree, it is
General
description.
true, with Coelenterata in exhibiting cellular structure
and having their protoplasm arranged in tissues ; and
further in the fact that all the internal cavities of the body are in
communication with one another, so that both Coelenterata and
Porifera might be described as systems of branched tubes. A
closer inspection however reveals the fact that the tissues of the
Porifera are very different from those of Coelenterata and originate
in a different way from the larva, so that the opinion is gaining
ground that whereas most, if not all, of the higher groups of
animals have descended from ancestors which had we seen them
we should have classed as Coelenterata, Porifera on the other hand
have been independently derived from Protozoa.

In Coelenterata the colonies can be analysed into persons
(medusoids or hydroids) and stolons, and many of the Porifera show
a like aggregation of persons. But in many forms it is impossible
to suggest how many individuals are contained in the branch
system of a single aggregate since all distinctness of individuals is
lost. Further analysis shows that the apparent persons or units,
even when most clearly demarcated, are of very varying morpho-
logical value.

The salient peculiarities of sponges will be best appreciated by
a short description of one of the simplest types
Leucosolenia.
known, a sponge called *Leucosolenia*, which is
common on most clean rocky shores.

In this animal we can recognise a foundation consisting of a
network of horizontal stolons, adherent to some foreign object,
from which a number of upright tubes spring. Each upright tube

ends in a large opening, the osculum (1, Fig. 34), which can be closed if the animal be irritated and which in *Leucosolenia* is partly closed by a perforated membrane. This opening, which at first sight recalls the mouth of *Hydra,* is really used for a quite different purpose. It is an e f f e r e n t opening (Lat. *effero,* to carry out) and from it the water which has passed through the animal is expelled. Water enters the internal cavity through a multitude of very fine p o r e s in the walls of the tube (Fig. 35, and 1, Fig. 37): it is the universal presence of these pores which gives the name P o r i f e r a to the group.

FIG. 34. View of a branch of *Leucosolenia* sp., showing the sieve-like membrane which stretches across the osculum. The lower part of the sponge shows spicules only × 10. From Minchin.

1. Sieve-like membrane.

The wall of the tube is made up of two layers, but we must guard ourselves against rashly comparing these with the layers of the body wall of *Hydra,* and hence it is better to avoid the names ectoderm and endoderm and adopt the terms d e r m a l and g a s t r a l layers.

The dermal layer consists of flat cells which cover the external surface and extend for a short distance inside the osculum, and of cells termed amoebocytes from the resemblance of their movements to those of an *Amoeba*: the whole of the rest of the tube and the stolons are lined by a tissue consisting of peculiar cells called c h o a n o c y t e s (Gr. χόανος, a funnel; κύτος, a hollow vessel), or collar cells, which alone constitute the gastral layer (3, Fig. 35 and Fig. 36). Each of these is cylindrical and provided with a funnel-shaped transparent rim called the collar, turned towards the cavity of the tube. The collars of adjacent cells are not

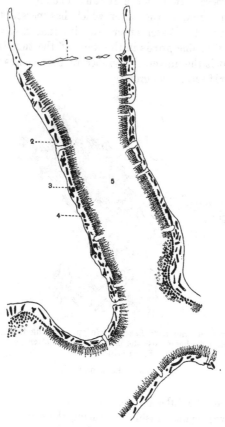

FIG. 35. Vertical section through an osculum with sieve-like membrane, and a tube of *Leucosolenia* sp. Highly magnified. From Minchin.

1. Sieve-like membrane. 2. Outer layer. 3. Flagellated or collar cells (choanocytes). The pointer should have been continued to indicate the cells lining 5. 4. Spicules. 5. Internal cavity

normally in contact, and the outer part of the cell bodies are widely separate, so that here the distinctness of the elements of a cellular tissue is carried to an extreme. From the centre of each collar a long flagellum arises, and it is by the action of these flagella that water is drawn in through the pores. The sponge lives on the organisms carried in by the current; these appear to be carried within the collars by the minute whirlpools produced by the individual flagella: they adhere to the collars and are swallowed by them and digested by the cells. It will be seen that the collar is a real living structure, not a cuticular tube, such as the hydrotheca of a calyptoblastic hydroid, and this is further illustrated by the fact that it is withdrawn by the collar-cell under certain conditions.

The water after being exhausted of its food is expelled through the osculum, carrying with it all excreta.

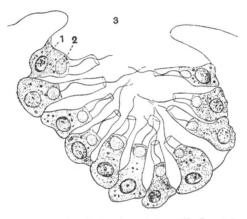

FIG. 36. Section of flagellated chamber of *Spongilla lacustris*, showing the flagella and collared cells × 1500. From Vosmaer.

1. Nucleus.   2. Vacuole.   3. Opening into the inner space of the sponge.

The outer layer of the body-wall consists, in the ordinary condition of the sponge, of flattened cells. These however, especially in the region of the osculum, have the power of changing their shape so as to become shorter and thicker; in a word they can contract, although they show no trace of the fibrillae found in all muscle and in the muscle tails of the contractile cells of Coelenterata. The contraction is slow, not quick, as in true muscle.

It has been proved that the pores are formed by specially large cells, the porocytes, which extend from the outer layer and push aside the choanocytes, and then become hollowed out.

Between the two layers is found a certain amount of secretion which may be termed jelly, in which in many sponges a large number of cells is found. These form a portion of the dermal layer, and are, for the most part, amoeboid. Some of these probably act as carriers of food and possibly of excreta from one layer to the other. Others at first very similar give rise to ova and spermatozoa. A third class called scleroblasts—derived in *Leucosolenia* from the flat cells which cover the surface, but not so derived in all sponges—secrete the rods which form the skeleton and which are termed spicules (Fig. 34, and 6, Fig. 37). In *Leucosolenia* these are calcareous and have three rays, more or less in one plane—a

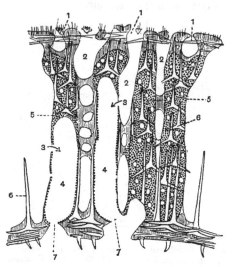

Fig. 37. Section of a portion of *Grantia extusarticulata*. Highly magnified. From Dendy.

1. Openings of the inhalant canals.   2. Inhalant canal.   3. Openings of inhalant canals into flagellated chamber (prosopyle).   4. Flagellated chamber.   5. Flagellated or collar-cells (choanocytes).   6. Spicules. 7. Exhalant opening of flagellated chamber.

shape technically named triradiate. One limb is usually directed parallel with the long axis of the tube, and often bears a fourth ray or spine making a quadriradiate spicule. The spicules although remaining unconnected are numerous enough to form a loose meshwork.

The most important points in which the higher sponges differ from *Leucosolenia* are the folding of the outer and inner layer, the restriction of the choanocytes to small portions of the latter, and the differentiation of the body into distinct regions.

Complex Sponges.

A common sponge on the British coast, *Sycon (Grantia) compressum*, will illustrate the first step in this complication. This animal has the form of a series of flattened thick-walled upright tubes. The layer lining the central cavity consists of flattened cells, but from this cavity pouches lined by choanocytes extend out into the substance of the wall. These flagellated chambers, as they are often called, communicate with the exterior by a series of inhalant canals which intervene between them and into which the pores open (Fig. 37).

When a sponge becomes still more complicated the central cavity becomes broken up into a series of branching canals, which are termed exhalant or efferent, and the ciliated chambers become small and rounded (Fig. 36), each often connected only by a single opening or prosopyle (Gr. πρόσω, forwards ; πύλη, a gate) with the afferent system of canals. Numerous oscula are found in one sponge mass, so that no pretence of discriminating the individual can be made.

A still further complication arises from the presence of subdermal spaces. These are wide cavities immediately beneath the surface of the sponge into which the inhalant pores open and from which the inhalant canals take their origin. In this way a rind or crust of the sponge can be separated from a deeper part containing the flagellated chambers. Sponges are by some of the best authorities divided into two main classes, viz.:

Class I.   CALCAREA.

This group includes all those sponges with calcareous spicules and comparatively large flagellated chambers.

It is divided into two main orders :

Order 1.   **Homocoela.**

Sponges consisting of tubes lined throughout with choanocytes.

Order 2.   **Heterocoela.**

Sponges in which the choanocytes are restricted to special chambers—which may be cylindrical as in *Grantia* or spherical as in *Leucandra*.

Class II.   HEXACTINELLIDAE.

Sponges in which the skeleton consists of a coherent network of siliceous spicules each consisting of three axes placed at right angles to one another. The flagellated chambers are large and cylindrical but are separated from the central space by a system of canals. The central space may be deep and narrow and covered with a plate pierced by numerous oscula, or short, open and shallow.

These sponges inhabit as a rule very deep water and most species are provided with a tuft of long needle-like spicules which root them in the soft mud which forms the bottom of the sea at these depths.

Class III. Demospongiae.

These sponges derive their name from the fact that their spicules, which are always siliceous, are arranged in cords so as to form a network traversing the substance of the sponge. The spicules composing these cords are nearly always cemented together by a horny elastic material called *spongin*. The flagellated chambers are always extremely small and there is never a central chamber. Besides the skeletal spicules, as those composing the cords are called, smaller ones called *flesh spicules* are scattered singly in the intervals of the network.

There are several exceptional genera in which interesting modifications occur.

Oscarella is totally devoid of any skeleton and has the appearance of a whitish yellow scum on the rocks to which it adheres. *Euspongia* possess spongin cords but no spicules in them, and for this reason it can be employed for domestic purposes.

Two fresh-water species, namely, *Spongilla lacustris* with a bush like appearance and *Ephydatia fluviatilis* with an encrusting form, are often found growing on the side of canals and on the timbers of river-locks or weirs in Great Britain. The two species are bright green when they grow in the light, but they are pale flesh-colour when they grow in the shade. In Canada similar species adhere to stones in the river St Lawrence.

Larva.

The larvae of sponges are best understood by a short description of the simplest form, viz. the larva of *Oscarella*. This has the form of a simple hollow sphere of ciliated cells like the planula of Coelenterata in its first stage. The cells at one pole lose their cilia, become pigmented and granular and then the larva fixes itself by the ciliated pole. The whole animal flattens and the granular cells extend over the ciliated cells which become tucked into the interior and there arranged as an inner lining to a cavity. The flagellated chambers of the adult arise as small pocket-shaped outgrowths from this cavity and the osculum is a later perforation. The ciliated cells are eventually restricted to these chambers where they form the choanocytes and all the rest of the sponge is formed from the granular cells. Other larvae differ from that of *Oscarella* in the early multiplication of granular cells which form a solid mass at one end of the larva and often—indeed generally—of such extent as to project into the interior.

To compensate for greater dead-weight, so to speak, the ciliated layer—the locomotor organ of the larva—becomes extended to surround the granular material, so that we are presented with the remarkable phenomenon of the internal layer of the larva bursting forth and becoming the outer layer of the adult. This is the case in the larva of *Leucosolenia*. In the larvae of other calcareous sponges, the ciliated cells at first surround the granular cells, but the latter are afterwards exposed and the larva in this form has been called an amphiblastula. In the case of most of the Demospongiae the ciliated cells nearly, but not quite, surround the granular cells, and these last often contain a number of spicules ready formed in a central bundle which are scattered in all directions when the sponge flattens on fixation. Comparing the development of a sponge with that of the planula of a Coelenterate we see that in the first the ciliated cells form the internal layer, in the second the external layer of the adult; in the first the animal fixes itself by the pole at which the invagination or intucking of the cells destined to form the inner layer takes place—in the Coelenterata at the opposite pole—so that if Coelenterata and Porifera had an ancestor in common it could only have been an animal like the organism *Volvox*, consisting of a single sphere of cells—in a word it would have been classed were it living now as a Protozoon.

The study of the development of sponges like *Sycon* shows that at first, after the metamorphosis, the sponge has the form of *Leucosolenia* by a simple cylinder lined by choanocytes. The flagellated chambers arise as horizontal cylindrical branches on the primitive chamber and soon become so numerous that their walls come into contact and the afferent or inhalant canals are simply the crevices left between these chambers. As the chambers develop, flattened cells come inwards from the pores and displace the choanocytes except in the chambers.

Porifera then may be defined as animals consisting of branch-systems of tubes, the principal openings of which are exhalant, whereas the inhalant openings are minute perforations of the walls. The wall consists of two layers; some cells of the inner layer have the form of choanocytes, whilst the skeleton consists of siliceous or calcareous needles formed by cells of the outer layer which wander in, or of spongin. There are never any thread-cells or differentiated muscle or well-marked nerve-cells, nor any such organs as tentacles.

CHAPTER V.

INTRODUCTION TO THE COELOMATA.

THE last two groups of animals studied, although very different from one another in most respects, yet agree in this that the groundwork of their structure is a set of tubes branched in different ways and with walls of varying thickness but consisting always of two layers with an intervening jelly. It would not be straining the truth to assert that in the Coelenterata and the Porifera we find but two tissues, an outer more or less differentiated skin—the ectoderm with the underlying jelly,—and an inner layer mainly digestive in function.

The phyla which are next to be considered, and which may be grouped together under the name Coelomata, differ

Coelom.

from the two mentioned above in the possession of an important organ termed the coelom (Gr. κοίλωμα, a thing hollowed out). This is often described as a space intervening between the ectoderm and endoderm, and the term coelomic cavity or body cavity has been used to describe it. In spite of the etymological difficulty we propose in the following pages to deal with this organ under the term *coelom*, and its cavity under the term *coelomic cavity*. In reality it consists of one or more pairs of sacs with perfectly defined walls lying at the sides of the endodermic tube. In the adult these sacs join each other above and below the endoderm, and the adjacent walls entirely or partly break down, and thus one continuous cavity results. The wall of the coelom and the tissues derived from it are known as the mesoderm. To describe the coelom as a split or space is to describe it negatively: with as much justice the endodermic tube might be described as a split. In each case the real object of consideration is the wall.

If we leave out of account cases in which the facts of development have not been fully elucidated and confine our attention to

6—2

those instances where the whole history of the coelom has been exhaustively worked out, we find that this important organ arises in one of two ways, either (1) by the formation of pouches of the endodermic tube, which become nipped off (Fig. 38); or (2) by the budding of two large cells, formed themselves by budding from the endoderm (Fig. 39), these cells subsequently growing rapidly and dividing so as to form bands, the so-called germinal bands, which subsequently become hollowed out. These initial cells are termed pole-cells.

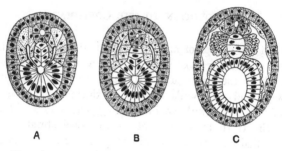

Fig. 38.   Three transverse sections through a developing Amphioxus to show origin of mesoblast from endodermal pouches × 435.   From Hatschek.

The ectoderm is deeply shaded, the mesoderm is lightly shaded, the endoderm —alimentary canal and notochord—is unshaded. A. shows the origin of the paired mesodermal pouches from the archenteron ; the cavity— coelomic—of the former is still in communication with the cavity of the alimentary canal. The notochord is arising in the middle line from the endoderm, and the tubular nervous system above it is already separated from the ectoderm. B. shows the mesodermal pouches completely shut off ; they each enclose a cavity, the coelom, and each consists of an outer wall next the ectoderm, the 'somatopleur,' and an inner wall next the endoderm, the 'splanchnopleur.' C. shows the mesodermal pouches extending ventrally beneath the notochord, now completely separated from the wall of the alimentary canal and also round the alimentary canal. The coelomic space is larger, and the splanchnopleur is beginning to form muscle-cells.

A sharp controversy has raged round the question which of these two processes gives us the best representation of what occurred in the evolution of Coelomata from simpler Coelenterata-like ancestors.

If however we recall the fact that in the Actinozoa the endo-dermic sac has the form of a series of pouches ranged round a central cavity, and that the walls of these pouches become con-verted into muscles and generative cells exactly as in the case of the coelom, and that pores exist in many cases placing the cavity of these pouches in communication with the outside world, we shall be induced to conclude that the coelom was probably evolved from lateral pouches of the gut and that the mesoderm is therefore derived from the primitive endoderm. Where pole-cells occur the cavity of the alimentary canal is small in proportion to the thickness of its wall, and the pole-cell might be looked on as a solid pouch.

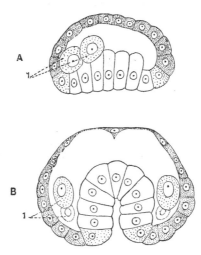

FIG. 39. Two stages in the early development of a common fresh-water mollusc, *Planorbis*, to show the origin of the mesoderm cells × 320. From Rabl.

The ectoderm cells are deeply shaded, the endoderm cells are unshaded. A. Young stage in which the endoderm has not begun to be invaginated; it is a lateral optical section. B. Older stage, optical section seen in front view; the endoderm cells are invaginating, and the two mesoderm cells are seen on each side. 1. Mesoderm or pole-cells; in B, each has budded off another mesoderm cell.

In most Coelomata the mesoderm forms by far the greatest portion of the body, and it may be roughly stated that the mesoderm gives rise in the fully developed animal to " the muscles, the bones, the connective tissue, both arteries and veins, capillaries and lymphatics, with their appropriate epithelium," and to the excretory and generative organs.

In the Porifera and in the more complex Coelenterata, where the thin structureless lamella of *Hydra* has swollen into a bulky mass of jelly such as we find in a Medusa, cells begin to wander into it from the ectoderm, and thus a kind of tissue is formed to which the name mesoderm has unfortunately been applied. We see that its origin and nature are quite different, although both in Sponges and Coelenterata it may give rise to the skeleton, and in this respect at least it may be regarded as a forerunner of the true mesoderms.

The endoderm, after the separation from it of the mesoderm, forms the lining epithelium of the digestive tube and of its appendages, which in the higher Vertebrata are the organs known as lungs, liver, pancreas, and urinary bladder. The basis of the skeleton of Vertebrata, the gelatinous rod called the notochord, also arises from it.

In Coelenterata there is one opening only to the digestive sac, which is used both as a mouth to take in food and as an anus to cast out indigestible material. In the overwhelming majority of the Coelomata there is a second opening, the anus, the mouth being restricted to the function of taking in food. As a consequence the digestive sac takes the form of a tube open at both ends, and is known in the higher groups as the alimentary canal. Often this endodermic tube is much longer than the ectoderm, and so in order to be contained in the bounds of the ectoderm it has to be bent and looped on itself.

Alimentary Canal.

Round both mouth and anus the ectoderm is generally tucked in, so as to form as it were vestibules to the true alimentary canal. Of these, the ectodermic vestibule to the mouth is called the stomodaeum, and is found amongst the Actinozoa, where it has been already described. The proctodaeum is the term applied to the vestibule around the anus. Although not strictly parts of the alimentary canal stomodaeum and proctodaeum are usually included in descriptions of it, and indeed in some cases—Crustacea—they form by far the greatest portion of the apparent digestive tube.

The internal anatomy of the lower animals was first studied by physicians and others who were primarily interested in human anatomy. An unfortunate consequence is that a large number of names are used in the description of simpler animals which are based on fanciful resemblances between their organs and those of man. As a consequence many of these names are quite misleading. To give some instances: the word stomach in the Lobster denotes part of the stomodaeum, in the Vertebrata it signifies part of the endodermic tube. The pharynx of an earthworm is the stomodaeum, in a fish it includes both stomodaeum and the first part of the endodermic tube. The term liver has also been much abused.

The names taken from the anatomy of the higher animals which are customarily used in the description of the alimentary canal are as follows: mouth- or buccal-cavity, pharynx, oesophagus, stomach or crop, gizzard, intestine, and rectum. They are applied generally to parts of it succeeding one another in the order above given. The significance of these will be explained in each case: it would perhaps be more logical to sweep away altogether these and a host of similar terms employed to designate other parts of the body, but so deeply are they engrained in zoological literature that such a course would render unintelligible most anatomical descriptions of species that we possess.

Besides forming the outer layer of the skin or epidermis of the animal and the stomodaeum and proctodaeum, the ectoderm gives rise to the brain and nervous system and to the essential cells of the sensory organs.

In the group Coelenterata a general circular outline of the body predominates, the principal external organs being arranged like the spokes of a wheel around the mouth as a centre. Such an arrangement is spoken of as a radial symmetry, and is in all probability connected with the fixed life so common amongst Coelenterates, a condition of affairs which renders it advantageous to have organs developed so as, to use a familiar phrase, to be on the look-out all round.

General shape of the Body.

The free-swimming Medusae, it is true, move always with the apex of the bell directed forwards, but as has been pointed out (see p. 55) they are to be regarded as specially modified hydroids, and they have not modified the radial symmetry so deeply impressed on them by the habits of ancestral Coelenterata.

In the higher groups of animals there is usually one fixed part of the body which moves first when the animal changes its position; and when such a portion is definitely set aside to move first we can distinguish the front end of an animal from the hind end. It is usual to term the former the head or anterior end and the latter the posterior end, and when an organ such as a man's arm lies nearer the anterior end (head) than another, as for instance his leg, we say the former is anterior to the former, and that the latter, i.e. the leg, is posterior to the former, i.e. the arm. In describing the parts of an appendage such as the arm it is usual to speak of the part nearest the base of attachment as proximal and the part further away as distal.

Corresponding with the appearance of the head as distinct from the rest of the body, which is in contradistinction termed the trunk, we find a difference arising between the surfaces of the body. With few exceptions animals, whether creeping, swimming, flying or walking, keep the same surface turned towards the earth. This lower surface is termed the ventral (Lat. *venter*, belly), whilst the upper surface or back is termed the dorsal (Lat. *dorsum*, back). As a rule the difference in their relationship to their surroundings induces a difference in the aspect of these two surfaces, and it is seldom difficult to determine which is the ventral and which the dorsal surface of the body. As a general rule the ventral surface of an animal is much lighter in colour than the dorsal. In most Coelomata the nervous system is for the most part ventral and the chief blood-vessel dorsal, but in the Vertebrata the reverse is the case. In both the alimentary canal lies between the chief blood-vessel and the main nervous system.

The two sides of an animal, the right and left, are however exposed to much the same conditions and as a rule resemble each other very closely. When this is the case an animal is termed bilaterally symmetrical, and it may then be divided in one plane—and in *one* plane only—in such a way that each half forms a reflected image of the other, such as we should see if we held half the animal up to a looking-glass. This bilateral symmetry may extend to all the internal organs, as it does in an earthworm or a crayfish, or it may be confined to the external features and some of the internal organs only, as in insects or in most vertebrates, where the coiling of the alimentary canal, etc., interferes with the bilateral symmetry of the internal organs.

In some animals, and these are for the most part such as move sluggishly or have become permanently attached to some substratum and do not move at all, this bilateral symmetry has been lost and the two sides do not resemble one another. Such animals are called asymmetrical. The Snail is a familiar example of such asymmetry. Amongst the Echinodermata (Star-fishes and Sea-urchins) this asymmetry is replaced by a radial symmetry.

CHAPTER VL

PHYLUM ANNELIDA.

THE name Annelida (Lat. *annulus*, a ring) means ringed, and refers to the fact that the bodies of the creatures grouped under this name are built up of a series of parts more or less resembling each other placed one behind another. This division of the body into more or less similar parts is called segmentation; each part is called a segment (or somite), and the animal is said to be segmented. Like the symmetry, the segmentation may be merely external or may affect both the exterior and

Segmentation. a greater or less number of the internal organs. Sometimes, however, as in the case of the longer half of an earthworm's body, the segmentation affects all the organs, and the likeness of one segment to another is so great that it would be impossible to say what part of the body any given isolated segment was taken from. More often, however, one or another of the organs of the body differs in shape or size in successive segments, and this is the case with the internal organs of the first twenty segments of the earthworm's body, so that if these segments were all separated it would not be very difficult to place them together in their natural order.

If we take an earthworm and kill it by placing it in alcohol for a few minutes and examine it carefully, we shall see

The Earthworm. External features. that the body is composed of some 150 rings, each of which corresponds with a segment. The rings are separated from one another by slight grooves. At each end of the body there is an opening, the mouth (2, Fig. 40) in front and the anus (3, Fig. 40) behind. Besides these, two slit-like pores with rather swollen lips, situated on the under surface of the fifteenth segment (5, Fig. 40), may be seen. These are the pores through which spermatozoa are discharged, and are consequently known as the male genital openings. The other openings into the body are minute and require the aid of a lens to make them out. There are paired openings on each segment, except the first three and the last, situated latero-ventrally; these are the openings of the tubes known as nephridia (Gr. νεφρίδιον, a little kidney), which act as kidneys; in addition to these a median dorsal pore opening into the body-cavity is situated in each groove behind the tenth segment (11, Fig. 44). The earthworm is hermaphrodite, that is, it contains both male and female organs in its body. Through two slit-like openings in the ventral surface of the fourteenth segment the eggs are discharged: these are called the female generative openings. Two pairs of pouches called spermathecae, which are reservoirs for spermatozoa received from another worm (v. p. 107), open, one pair between the ninth and tenth, the other between the tenth and eleventh segments, all on the ventral surface.

If a worm killed in alcohol be drawn through the fingers a certain roughness may be felt along the sides and lower surface. This roughness is due to the presence of a number of small bristles, called chaetae (Gr. χαίτη, hair), which project from the body (7, Fig. 40, and Fig. 43). Each segment bears eight of these chaetae arranged in four pairs, one pair on each side being lateral and the other nearer the ventral middle line. It is by means of the chaetae that the worm crawls about; since by protruding the chaetae and implanting them in the soil a fixed point is obtained from which the anterior end of the

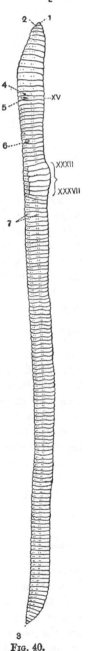

FIG. 40. Latero-ventral view of *Lumbricus terrestris*, slightly smaller than life-size. From Hatschek and Cori.

1. Prostomium. 2. Mouth. 3. Anus. 4. Opening of oviduct. 5. Opening of vas deferens. 6. Genital chaetae. 7. Lateral and ventral pairs of chaetae.

xv. xxxii. and xxxvii. are the 15th, 32nd, and 37th segments. The 32nd to the 37th form the Clitellum.

FIG. 40.

body can be pushed forward and to which the hinder end of the body can be drawn up.

The colour and thickness of the body from the thirty-second to the thirty-seventh segment differ in adult worms from those of the segments which lie before and behind this band. This is due to the presence in this region of certain ectodermal glands whose secretion forms the cocoons in which the eggs are laid. This region of the body is called the Clitellum (XXXII—XXXVII, Fig. 40).

The surface of the body of an earthworm is glistening and somewhat slippery. This is due to the cuticle, which is a thin membrane secreted by the ectoderm cells of the skin; if a dead earthworm be soaked in water for a few hours the cuticle can be easily stripped off the body. In the crayfish, insects, etc., a similar cuticle is present, but it is much harder and forms an external protective skeleton; even in the earthworm, where it is soft, it acts as a protection to the underlying cells, and its smooth surface enables the worm to creep into narrow holes without hindrance. The chaetae are simply large local thickenings of the cuticle: they protrude from pockets called chaeta-sacs, each of which is a portion of the ectoderm tucked in. In the bottom of each sac is a specially large cell which rapidly secretes a column of cuticle and builds up the chaeta.

If we cut through the skin of an earthworm we do not make our way into the cavity of the alimentary canal but *Internal Anatomy.* into the coelomic cavity, in which not only the alimentary canal but the blood-vessels, kidneys, reproductive organs, apparently lie. The relation of the alimentary canal to the body-cavity might be roughly represented by introducing a piece of glass tubing loosely into an india-rubber pipe. The alimentary canal would be represented by the glass tube and the body-cavity by the space between the glass and the india-rubber.

The coelomic cavity is a very important feature in all the higher animals; it may become very reduced, as in the Arthropods, but it is always present, although it may not at first sight be easy to recognise. There are, however, certain features which it always presents: (i) it always possesses a proper wall, never being a mere slit intervening between various organs, and it is always surrounded by mesoderm; (ii) its walls give rise to the cells which form the reproductive cells; (iii) the kidneys, which are primitively tubes with open ends, open into it.

There is no difficulty in recognising the body-cavity of an

earthworm. It is comparatively spacious and is divided by a series of partitions into a number of chambers which correspond in number and position with the segments of the body. These partitions or septa (8, Fig. 41) are pierced by the alimentary canal, the nervous system and blood-vessels; they are not complete but are provided with holes so that the space in one segment is not shut off from the spaces in the neighbouring segments. Fundamentally in Annelida the body-cavity consists of a series of pairs of sacs interposed between the skin (ectoderm) and the gut-wall (endoderm); there is in the embryo a pair in each segment, but the walls of these come into contact above and below the alimentary canal and then break down, so that the cavities of the right and left sacs open into one another and a ring-shaped space results. This space has distinct inner and outer walls of its own which are known collectively as the peritoneum (Gr. περί, around; τόνος, a stretched band).

The septa are formed where the adjacent walls of two sacs, placed one behind the other, come in contact. If this description of the relations has been followed it will be seen that the coelom in the adult consists of a series of ring-shaped spaces, and that the alimentary canal is not truly in the coelom nor, it may be added, is the nervous system or the blood-system.

Like all similar spaces in animals the body-cavity of an earthworm contains a fluid, and in this fluid certain cells float which change their shape as an *Amoeba* does, and hence are called amoebocytes. As a rule the body-cavity is completely shut off from the outside world, but in the earthworm it opens to the exterior by means of the dorsal pores (11, Fig. 44), and at times the fluid which it contains escapes through these holes and pours over the cuticle. This fluid has a certain poisonous action on bacteria, and helps to keep the outside of the body clean and free from parasites. Somewhat similar pores leading from the exterior to the body-cavity are found in certain fishes.

The first segment is divided into two parts, a lobed lip or prostomium (1, Fig. 43), overhanging the somewhat crescent-shaped mouth, and a peristomium containing the mouth which leads into an oral cavity extending through three segments (Fig. 41). There are no teeth in this cavity and the food is probably sucked in by the action of the muscular stomodaeum, called the pharynx, which succeeds it and reaches back to the sixth or seventh segment. This is followed by the true endodermic tube. The first part is narrow and is called the oesophagus; it reaches to the twelfth segment and has three pairs of lateral pouches developed on its walls. These pouches secrete calcareous particles, and hence are termed calciferous glands. The oesophagus dilates behind into a thin-walled sac, called the crop, situated in the region of segments thirteen to sixteen, and this is separated by a groove from a thick-walled sac, with hard, horny walls, termed the gizzard, which extends to about the twentieth segment. The exact segment in which the above-mentioned parts of the alimentary canal lie varies with the amount of food they contain, the septa which are pierced by them being stretched forward or backward according to their state of fulness or emptiness.

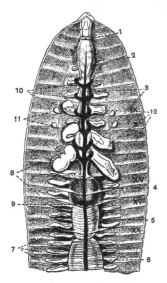

Behind the twentieth segment the intestine stretches without change to the anus. It is a thin-walled tube, supported by the septa between each segment and swelling out slightly in each segment, so that it presents an outline like a string of beads. A deep fold, called the typhlosole (Gr. τυφλός, blind; σωλήν, a gutter), runs along the upper surface of the intestine, projecting into its cavity. Its presence causes the wall of the intestine to be pushed in, and thus the internal absorbing portion is increased (7, Fig. 42). The intestine is covered everywhere by a number of cells of a yellow colour. These form the inner wall of the coelomic sac and are actively engaged in excretion.

FIG. 41. Anterior view of the Internal organs of an Earthworm, *Lumbricus terrestris*. Slightly magnified. From Hatschek and Cori.

1. Central ganglion or brain. 2. Muscular pharynx. 3. Oesophagus. 4. Crop. 5. Muscular gizzard. 6. Intestine. 7. Nephridia (the reference lines do not quite reach the nephridia). 8. Septa. 9. Dorsal blood-vessel. 10. Hearts. 11. Spermathecae. 12. Vesiculae seminales.

The Roman figures refer to the number of the segments.

The exact part that each of the above-mentioned parts of the alimentary canal plays in digestion is not thoroughly understood. The pharynx helps to take food in by a sucking action which is caused by the contraction of the

muscles running from it to the body-wall, resulting in an enlargement of the cavity of the pharynx so that food may pass in by
atmospheric pressure.   The food passes down the oesophagus,
being propelled by a series of contractions of the walls of the
alimentary canal which push it along; on its passage it is mixed
with the secretions of the calciferous glands.   The crop serves
as a resting-place in which the food accumulates before passing into
the gizzard.   The hard, horny walls of the last-named chamber help
to grind up the food and render it fit for the action of the juices
which digest it.   The process of digestion, or the rendering of the
food soluble, probably takes place in the intestine, and through the
walls of this portion of the alimentary canal the soluble products of
digestion soak, and are taken all over the body by the blood-vessels
and probably also to some extent by the fluid in the coelom.

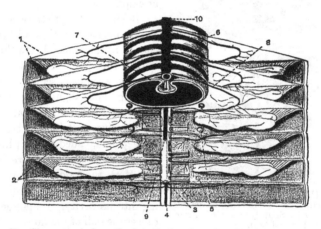

Fig. 42.   Six segments from the intestinal region of an Earthworm, *Lumbricus
terrestris*, dissected to show arrangement of parts.   Magnified.   From
Hatschek and Cori.

1. Septa.   2. Nephridia.   3. Ventral nerve-cord.   4. Sub-neural
blood-vessel.   5. Nephrostomes, internal funnel-shaped openings of
nephridia.   6. Intestine.   7. Typhlosole.   8. Circular blood-vessels.
9. Ventral or sub-intestinal blood-vessel.   10. Dorsal blood-vessel.

The series of contractions which squeeze the food onwards
towards the anus are known as peristalsis; they constitute the
sole movements of which the alimentary canal is capable and are
carried out by muscles developed from the cells of the inner wall
of the coelom, which pass round the canal like a series of rings
or tight india-rubber bands.

The earthworm eats earth and manages to find sufficient nourishment for its needs in the small amount of organic matter, broken-down débris of leaves, etc., which is contained in the earth. The actual minerals of the earth are not digested but are passed out of the body in the form of those coiled and thread-like castings which are so commonly seen on a lawn in the early morning. Earthworms also eat fallen leaves and to this end they drag the leaf-stalks into their burrows, and on autumn mornings it is a common sight to see lawns studded with the stalks of horse-chestnut leaves or the needles of fir trees, the stalks having been dragged a little way into the burrows by the worms. The burrows that they make admit both air and rain to the deeper layers of the soil, and the earth which they swallow in their burrows is brought to the surface and spread about in the form of castings. This is carried on to such an extent that the whole surface of the soil soon becomes covered by a layer of earth brought up from below. It is thus clear that the earthworm is of great use as an agricultural agent.

All the blood-vessels are for the most part merely crevices between the coelomic wall on the one hand and the ectoderm and endoderm on the other. Those described are merely the larger channels in a continuous network of spaces. The contractile power which some, like the hearts, dorsal vessel, and sub-intestinal vessel, possess is due to the presence of a special wall of muscular cells derived from that part of the coelomic wall which lies next them.

The earthworm is the first animal that we have studied possessing a distinct and well-marked body-cavity; it is also the first in which we find a distinct blood-system. In the Coelenterata the cavity in which digestion is carried on permeates the body in all directions, and the soluble products of digestion are never far from the tissues or cells which may need them. But in the earthworms the alimentary canal is a straight tube separated from a number of the other systems of organs by a space or coelomic cavity, and hence a vascular system is of great use in conveying the digested products to where they are most needed. Thus the blood serves to take up the nutriment from the intestine and distribute it to all the active cells in the body. The blood is also the medium by which the waste products resulting from katabolism are collected and taken to the appropriate organs whose duty it is to separate them from the blood and cast them out of the body of the animal. Amongst

the products which do not contain nitrogen the most important is
carbon dioxide, which is carried by the blood to the skin and got
rid of through the ectoderm, at the same time as the oxygen
needed for respiration is absorbed.

The dark streak which runs along the body of the worm from
head to tail in the middle line is caused by the dorsal blood-
vessel (10, Fig. 42), in which the blood flows forward. A parallel
sub-intestinal vessel in which the blood flows backwards underlies the intestine, and a third but smaller vessel, the sub-neural, lies
still more ventrally under the nerve-cord. The dorsal vessel receives
blood from the yellow cells covering the intestine by two pairs of
minute vessels in each segment, and anteriorly it breaks up into a
network of small vessels which branch over the pharynx. But by
far the larger part of the blood from this vessel passes into the sub-
intestinal vessel by means of five pairs of loops, called hearts,
situated in the seventh, eighth, ninth, tenth, and eleventh segments
(10, Fig. 41). Each pair of these hearts encircles the oesophagus
and contracts at regular intervals from above downwards. Their
contractility has suggested the name heart. As they pass from
the dorsal vessel into the sub-intestinal the effect of their contractions is to drive the blood which is passed forward on the dorsal
side of the animal into the ventral system, whence it passes toward
the tail. These contractile hearts thus take a large share in main-
taining the circulation of the blood. The sub-intestinal vessel
gives off a special vessel in each segment to the nephridia, and the
blood which is purified in these organs is returned to the dorsal
vessels by another series of vessels. The dorsal vessel and the
sub-neural vessel are put into communication in each segment by
two lateral vessels which lie on the outer wall of the coelom
and which receive numerous small vessels from its substance.

The earthworm breathes through its skin. The blood-system
sends up into the skin innumerable minute vessels or capillaries
which come so near the outer surface of the worm that the oxygen
can pass in from the air into the blood. The name capillary (Lat.
capillus, a hair) was suggested by a comparison of the exceedingly
small calibre of these vessels with the diameter of a human hair.

The blood is red, and the red colour is due to the same substance
which colours our blood, haemoglobin, but there is this difference,
that whereas in Vertebrates the haemoglobin is contained in certain
cells which float in an almost colourless fluid, in the earthworm it
is dissolved in the fluid itself. This substance has a strong attrac-

tion for oxygen which it takes up from the air that comes into the neighbourhood of the skin-capillaries, forming a bright red compound called oxy-haemoglobin. This compound is unstable, and when the blood in its course round the body encounters a cell hungry for oxygen, the oxy-haemoglobin is decomposed: the reduced haemoglobin is purplish in colour. At the same time the cell gives up carbon dioxide to the blood. The relations of this gas in the blood are less understood than those of the oxygen, but like the latter it is in loose chemical union, though not with the haemoglobin. In Vertebrate animals the sodium of the blood provides the means of conveying the carbon dioxide to the respiratory organs. When the blood again approaches the skin carbon dioxide is got rid of, oxy-haemoglobin being again formed by fresh oxygen taken in.

In the Vertebrata the excretion of the waste nitrogenous material is performed by a pair of compact organs, the kidneys. In the earthworm this function is carried out by the nephridia, which fundamentally resemble the tubules composing the kidney of Vertebrates, but are not compacted into a solid organ. They are distributed throughout the body, one pair being situated in each segment, except the last segment and the first three, which have no nephridia (7, Fig. 41, and 2, Fig. 42). Each nephridium is a minute tube, opening at one end on to the surface of the worm near the outer chaeta of the more ventral pair, and at the other end into the body-cavity. This inner opening or nephrostome has cilia on its funnel-shaped rim, and these flicker with an untiring movement. The nephrostome does not lie in the same segment as the rest of the tube but pierces the anterior septum, and projects into the cavity of the segment in front, somewhere near the sub-intestinal vessel. Thus each segment contains a funnel-shaped opening and a tube which opens externally, but they do not belong to the same nephridium. The tube is not straight but is coiled and lies as a white glistening tangle close to the muscular body-wall. Each nephridium is to be regarded as a portion of the coelomic sac into which it opens internally. It is, so to speak, a tail of this sac which projects backwards into the next one—not, of course, piercing it, but indenting, so to speak, its anterior wall.

When we examine a nephridium through a microscope we see that the walls of the tube are very richly supplied with minute blood vessels. The tube is really a cord of glandular cells placed end to end and traversed by a minute cavity. It is these cells which take up the waste nitrogenous matter from the blood and convey it out of the body. The part of the nephridium nearest the external opening is swollen so as to form a bladder. The cavity is here intercellular instead of piercing the cells themselves, and surrounding it is a muscular wall by the contraction of which the contents are from time to time expelled.

The blood thus takes digested food to the living cells all over the body and brings from them certain nitrogenous excreta to the nephridia, which cast them out of the body. But the nephridia also exert some action on the other great fluid of the body—the coelomic fluid—which bathes all the organs of the body. It has been mentioned above that the funnel-shaped ciliated openings of the nephridia open into the coelom, so that the fluid of this cavity can pass out of the body not only by the dorsal pores but by the tubular nephridia. This fluid has suspended it in numerous amoebocytes (v. p. 92), and these corpuscles act as scavengers, taking up into themselves any foreign bodies, such as bacteria, which have made their way into the coelom, and breaking them up.

The yellow cells (7, Fig. 44), which surround the gut and form the inner wall of the coelom, are also actively engaged in extracting nitrogenous waste from the endoderm cells and the blood-vessels which pass near them. When the excreta have accumulated to a certain extent in a yellow cell it dies, and its remains fall out into the coelomic fluid, where they are eaten by the amoebocytes. These latter then wander to the nephridium and become pressed close against its wall, the cells of which extract the excreta from the amoebocytes and pass them into the cavity of the nephridial tube. The funnel of the nephridium is too small to admit the amoebocytes—it serves as a flushing apparatus, since its cilia draw in water from the coelom which is swept down the tube and carries the excreta into the terminal bladder whence they are from time to time expelled.

It is probable that the yellow cells represent a primitive mode of excretion and that originally the whole coelomic wall undertook this function, the products escaping either by simple pores or by being taken up by amoebocytes which forced their way out through the skin, as in Echinodermata. The yellow cells and the nephridia are then to be regarded as portions of the coelom in which the power of storing up excreta is specially developed, and in this limitation of this power to a special area we have the first type of an excretory organ. A localized excretory organ requires some means of bringing to it the products of katabolism of all portions of the body—since poisonous excreta are produced by all living protoplasm—and this means is supplied in the earthworm by the blood-system and the amoebocytes.

The earthworm, although it lives in earth, has a clean, glistening look, and this is partly due to the fact that the coelomic fluid is poured out from the dorsal pores (11, Fig. 44) and keeps the skin moist and lubricated. This fluid is also antiseptic in its action and thus its presence prevents foreign organisms, such as bacteria, which swarm in the mould in which the worm lives, settling upon the skin and growing there. Numerous glandular cells belonging to the ectoderm also pour forth a secretion through minute pores in the cuticle.

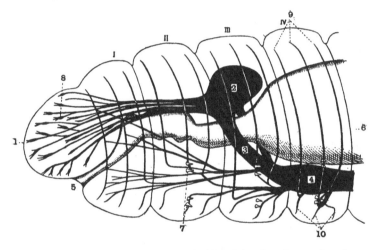

Fig. 43. Diagram of the anterior end of *Lumbricus herculeus* to show the arrangement of the nervous system. After Hesse.

I. II. III. IV. The first, second, third, and fourth segments.

1. The prostomium.　2. The cerebral ganglia.　3. The circumoral commissure.　4. The first ventral ganglion.　5. The mouth.　6. The pharynx.　7. The dorsal and ventral pair of chaetae.　8. The tactile nerves to the prostomium.　9. The anterior, middle and posterior dorsal nerves.　10. The anterior, middle and posterior ventral nerves.

If we cut open an earthworm by a median dorsal incision and attentively examine the upper surface of the pharynx

The Nervous System.

we shall find at its anterior end, tucked away between it and the skin, two little whitish knobs lying close to one another. These are the cerebral or supra-pharyngeal

7—2

ganglia (1, Fig. 41 ; 2, Fig. 43). At their outer ends the supra-pharyngeal ganglia pass into two cords (3, Fig. 43). If we now cut away the pharynx and remove the alimentary canal we can trace these two cords towards the ventral middle line where they unite and form the first sub-pharyngeal ganglion (4, Fig. 43) : from this a long white cord—the ventral nerve-cord—runs back to the extreme posterior end of the animal. If we examine these structures with a lens we shall be able to see that the supra-pharyngeal ganglion gives off small nerves to the sensitive pro-stomium, and that the ventral nerve-cord swells out between each pair of septa, that is, in each segment, into a thicker portion which gives off both dorsally and ventrally and on each side three pairs of nerves to the surrounding parts. Each of these swellings is termed a ganglion (Gr. γάγγλιον, a knot[1]) (4, Fig. 43).

The nervous system is made up of a number of cells termed neurons. These, as proved by a study of the development are ectoderm cells which have become pushed inwards from amongst the others. Each neuron consists of a body with a comparatively large nucleus difficult to stain. From the body in one direction is given off a tuft of root-like processes (which some suppose to be actual retractile pseudopodia) called receptive dendrites, by means of which stimuli are received into the cell. In the opposite direction is given off a long straight process called an axon which may branch once or twice, the branches being called collaterals. The axon itself and its branches end finally in tufts of root-like processes which are in close contact but apparently not in continuity with either a muscle-fibre or the receptive dendrites of another neuron and are called terminal dendrites.

Through the axon and its branches stimuli are transmitted to other neurons and to the muscles.

A bundle of collateral branches of axons bending outwards to convey stimuli to a group of muscles is known as a motor peri-pheral nerve.

The nervous system of an earthworm thus consists of two supra-pharyngeal ganglia situated in the third segment, a pair of connecting cords called commissures which form a ring round the pharynx, and a ventral cord which swells out into a ganglion in every segment behind the third. The ring round the mouth and the solid nature of the nervous system is common

---

[1] Γάγγλιον was used by the old medical writers to indicate the swelling or "knot" in a muscle caused by cramp.

to nearly all the Invertebrata, and in those which have a bilateral symmetry and are segmented there are supra-pharyngeal ganglia and a ventral nerve-cord bearing segmentally repeated ganglia.

The earthworm has no specialized sense-organs, it has neither eyes to see, nor nose to smell, nor ears to hear with. Still, although it is apparently deaf, it is not devoid of the power of appreciating those stimuli which in us excite the sensation of sight or smell. A strong light suddenly turned on the anterior end of the body will cause the worm instantaneously to withdraw into its burrow, and worms readily recognise the presence of such favourite food as onions and raw meat. Their sense of touch is well developed and they are very sensitive to vibrations; for instance, a stamp of the foot on the ground will cause all those in a certain radius to disappear into their burrows. It is further possible that earthworms possess other senses with which we are totally unacquainted.

In each segment of the worm scattered here and there amongst the ectoderm cells are a number of sense-cells. Each of these has a minute sense-hair which projects upwards through a hole in the cuticle, and by means of this hair stimuli of various kinds are received by the outer world. The body of the cell is small—just large enough to contain the nucleus—and from the base proceeds an axon which runs inwards and terminates inside the central nerve-cord in a brush of terminal dendrites in close contact with the receptive dendrites of a neuron. In this way the neurons receive impressions from the outside world. A bundle of the axons of sense-cells proceeding inwards is known as a sensory peripheral nerve.

The swelling called a ganglion is due to an aggregation of a number of the bodies of neurons, so that in this region the nerve-cord is broader than at other places, though everywhere some bodies can be seen in transverse section of the cord.

The nervous system is one of the most important organs of the body. It governs and controls the action of every tissue and cell. It receives and registers impressions from the outside world and co-ordinates the movements and activities of every part of the body. It further serves to put each organ and each part of each organ in communication with all the others, and thus this vast accumulation of tissues and cells acts in an orderly way and towards a set end.

A transverse section of an earthworm, such as can be cut by a microtome from a specimen embedded in paraffin wax, is most instructive, in exhibiting the relation to one another of the various tissues which make up the body of the earthworm. The outermost boundary is constituted by the cuticle. (1, Fig. 44), a hardened secretion poured out by the ectoderm (2, Fig. 44). The ectoderm

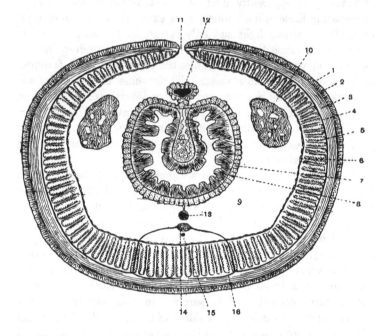

Fig. 44. Transverse section through *Lumbricus terrestris* in the region of the intestine and of a dorsal pore. Magnified.

1. Cuticle.   2. Ectoderm or epidermis.   3. Circular muscles.   4. Dorsal nerve.   5. Longitudinal muscles.   6. Somatic epithelium. 7. Splanchnic epithelium or yellow cells.   8. Endoderm or epithelium lining the intestine.   9. Coelom.   10. Nephridium cut in section. 11. Dorsal pore.   12. Dorsal blood-vessel lying along the typhlosole or groove in the wall of intestine.   13. Sub-intestinal blood-vessel. 14. Ventral nerve-cord.   15. Sub-neural blood-vessel.   16. Ventral nerve.

The dorsal and ventral nerves are added diagrammatically. The other structures are drawn from nature.

is composed of tall cylindrical cells, amongst which are isolated "goblet cells"—that is, cells with a round body situated beneath the level of the rest and with a long neck. The name is suggested by their shape. In the body of these cells mucus is secreted, which is poured forth through a hole in the cuticle opposite the end of the cell-neck and helps to keep the surface of the worm moist.

Beneath the ectoderm is a thin and hardly perceptible layer of jelly forming a bed on which the ectoderm cells rest. This foundation is called the dermis, and is included with the ectoderm in the ordinary conception of the "skin." In contradistinction to the dermis the ectoderm is often spoken of as the epidermis (Gr. ἐπί, upon).

Beneath the dermis comes a layer of circular muscles (3, Fig. 44), and beneath these again a much thicker layer of longitudinal muscles. The circular muscles consist of a few layers arranged to form rings round the section. The longitudinal muscles are arranged very regularly, and in the section they have the form of a series of feathers (5, Fig. 44), since the individual fibres appear arranged in oblique rows between which tongues of jelly extend, giving off lateral branches on which the fibres rest.

Both sets of muscles are composed of muscle-cells. These are long fibre-like structures pointed at both ends. Most of the protoplasm is differentiated into fine fibrillae, which indicate (see p. 29) contractile power. In the centre of the cell is a patch of unmodified protoplasm with a nucleus. The whole cell may be compared to a myo-epithelial cell of *Hydra* in which the epithelial part has diminished in size and the tail increased. Nor is this a fanciful comparison, for the study of development teaches us that the cell is actually derived in this way from the originally simple cells of the wall of the coelomic sac or in the case of the circular muscles from an ectoderm cell.

The movements of the earthworm can be more easily understood when the arrangement of the muscles is known. The longitudinal muscles serve to shorten the body, and as the coelomic fluid, like water, is practically incompressible, the diameter of the animal must be increased, and thus the chaetae can be driven into the sides of the burrow. On the other hand, the circular muscles diminish the diameter of the coelom, and the contained fluid being forced to move in a longitudinal direction stretches the body out. The holes in the septa equalize the pressure in the various segments by permitting the fluid to escape from one into the next.

Mention has been made above of "jelly" as forming a support for the ectoderm and the longitudinal muscles. It forms also the main part of the substance of the septa. In the worm and higher animals generally jelly fulfilling this function is known as con-

nective tissue.   Its nature will be more fully dealt with in
the section relating to Arthropoda.

Within the longitudinal muscles there is a layer of cells called the
somatic peritoneum (6, Fig. 44) which forms the immediate wall
of the coelom.   As the coelom in a segment of the worm has a ring-
shaped form there is an inner as well as an outer wall of the coelom ;
the former, since it closely invests the alimentary canal, is called the
visceral or splanchnic (Lat. *viscus* ; Gr. σπλάγχνον, entrail)
peritoneum (7, Fig. 44)—the latter the parietal or somatic
(Lat. *paries*, an outer wall ; Gr. σῶμα, body) peritoneum.   The
parietal peritoneum is composed of flattened cells ; the visceral
peritoneum, on the other hand, consists of large cubical cells, the
yellow cells already described.

Beneath the visceral peritoneum there is a thin layer of circular
muscles, the splanchnic muscles derived from the peritoneal
layer and forming the agency by which the peristalsis (*v.* p. 94) of
the gut is carried out.

The endoderm (8, Fig. 44) consists of a single layer of long
cylindrical cells bent in dorsally to form the typhlosole.   Within
the limbs of this fold the splanchnic peritoneum is very much
thickened.

The dorsal blood-vessel can be seen embedded in the yellow
cells lying in the typhlosole (12, Fig. 44), whereas the ventral vessel
is attached by a membrane to the ventral side of the intestine.
This membrane is really a part of the partition which separated
the two coelomic sacs which originally existed in the segment.

The nerve-cord, apparently lying loosely in the coelom, is surrounded by a layer of cells similar to those forming the somatic
peritoneum of which they once formed a part (14, Fig. 44).   Hence
the coelom has extended in a ring-shaped manner round the nerve-
cord exactly as it has surrounded the gut.   At the sides of and below
the nerve-cord may be seen sections of vessels, the sub-neural and
latero-neural vessels.   The mass of the nerve-cord is made up of the
sections of axons, whilst the nuclei of neurons can be seen forming
a sheath on the outer border of the cord.   The fibres are divided
into two bundles by a septum of connective tissue.   On the dorsal
surface of the cord there are seen three apparent tubes, these are
sections of the so-called " giant " fibres—colossal axons which are
outgrowths of correspondingly large neurons.

Sections of chaeta-sacs and nephridia may be seen in favourable
sections.

It has been mentioned above that it is one of the characteristics
of the coelom that the cells lining it should produce
the reproductive cells.   This does not mean that
any cell lining the coelom can become an ovum or
a spermatozöon, but that at certain spots the cells forming part
of the coelomic wall turn into either female or male generative cells.
In the earthworm the paired ovaries (5, Fig. 45) are situated in
the thirteenth segment and may be seen by cutting through the
intestine about the region of the gizzard and gradually lifting it up
from behind forwards ; when it is freed up to the twelfth segment

Reproductive
Organs.

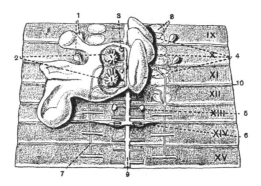

Fig. 45.   View of Reproductive Organs of the Earthworm, *Lumbricus terrestris.*
Part of the vesicula seminalis is cut away on the left side to expose the
testis and the inner opening of the vas deferens. Slightly magnified.
From Hatschek and Cori.

1.   Spermathecae.   2.   Funnel-shaped internal openings of the vas deferens.
3.   Anterior testis.   4.   Vesiculae seminales.   5.   Ovary attached to
posterior wall of septum separating xii and xiii.   6.   Oviduct traversing
septum separating xiii and xiv.   7.   Vas deferens.   8.   Glands in the
skin.   9.   Ventral nerve-cord.   10.   Septum.

The Roman figures indicate the number of the segments.

the ovaries may be seen as minute white pear-shaped bodies lying
one on each side of the nerve-cord.   They are attached by their
broad end to the posterior wall of the septum separating segment
twelve from segment thirteen, and they are formed by the accumulation and growth of some of the cells which cover this septum,
that is, from cells lining this portion of the coelom.

If one of the ovaries be removed and examined under a microscope it will be seen that many of the cells composing it are large,
spherical and crowded with granules.   The largest lie in the
narrow end of the ovary which waves about in the coelomic fluid.

These cells are the full-grown eggs or ova and when ripe they drop off from the ovary into the coelom, but are probably at once taken up by the wide funnel-shaped openings of the oviducts, one of which is situated opposite each ovary. Like the nephridia, the two oviducts pierce a septum, the one between the thirteenth and the fourteenth segments. They are short tubes which open into the coelom by the above-mentioned funnel-shaped opening in the thirteenth segment and to the exterior by a small pore just outside the inner pair of setae on the fourteenth (6, Fig. 45). They bear on their course a diverticulum or sac which is called the receptaculum ovorum, in which the ova collect until the earthworm is ready to make a cocoon to receive them.

The male reproductive cells are formed in the testes, of which there are two pairs situated in a similar position to the ovaries but in the tenth and eleventh segments (3, Fig. 45). They are in many respects similar to the ovaries but are hand-shaped, the broad end of the hand being attached and the fingers free. Their ducts which convey away the spermatozoa are called the vasa deferentia (Lat. vas, vessel; deferens, carrying away). They have similar funnel-shaped openings to those of the oviducts and they traverse the septum behind the segment in which these openings lie, but they do not at once open to the exterior. The two ducts of each side unite in the twelfth segment, and the common duct thus formed runs back to open by a pore with swollen lips on the fifteenth segment, the one behind that on which the oviducts open (7, Fig. 45).

There is, however, one great difference between the male and female organs. Whereas the ovaries lie freely in the body-cavity and can be seen readily if the intestine be removed, each pair of testes and the corresponding inner funnel-shaped openings of the vasa deferentia are concealed by a certain sac or bag called the vesicula seminalis, and it is only by cutting away the wall of this sac that these structures come into view (4, Fig. 45). Each vesicula seminalis is a flat, oblong bag extending backwards from the front wall of the segment in which it lies and situated beneath the alimentary canal. The angles of the front vesicula seminalis are produced into two long pouches which project upwards at the sides of the alimentary canal, and are often called lateral vesiculae seminales, though they ought to be termed lateral horns of the anterior vesicula seminalis. A similar projection is produced from the hinder angles of the posterior vesicula seminalis, so that on opening a worm three pairs of greyish white sacs are seen at the

sides of the gut. The study of the way in which the vesicula
seminalis is formed shows that the space it contains is really part
of the coelom which has become cut off from the rest by the outgrowth of folds from the septa, so that, although at first sight the
testes seem to differ from the ovaries and to be exceptions to the
general rule that reproductive cells have their origin from the walls
of the coelomic cavity, a closer examination shows that this apparent
divergence is not a true one.

Every earthworm has grown up from an egg which has been
fertilized by a spermatozoon. As the earthworm is hermaphrodite,
that is to say, contains both male and female organs, it might be
thought that the spermatozoa of an individual would fertilize its
own ova, but this is not the case. Cross fertilization or the
fertilization of the ova of one individual by the spermatozoa of
another is the rule in Nature, and the earthworm is no exception
to the rule. The method by which the spermatozoa reach the ova
is not clear in all its details, but it is something like this. The
cells which are to form the spermatozoa break off from the testes
and whilst lying in the fluid contents of the vesicula seminalis
they divide and the products of the division or spermatozoa develope each a long vibratile tail by whose aid they swim actively
about. Two earthworms then approach each other and the
spermatozoa pass down the funnel-shaped opening and vasa deferentia of each and into the spermathecae of the other. The
earthworms then separate, each carrying away the spermatozoa of
the other.

The spermathecae in which the earthworm stores up the
spermatozoa received from another individual are pockets of the
skin (1, Fig. 45). They belong, strictly speaking, to the female
reproductive system. Seen from the interior of the animal, they
appear as four small white spherical bodies, lying one pair near the
hind end of segment nine, and the other pair near the hind end of
segment ten, and each pair opens by a very short neck or duct on
the grooves between segments nine and ten and ten and eleven, just
inside the outer pair of chaetae. It is through these ducts that
the spermatozoa from another worm enter.

Earthworms lay their eggs in cocoons, which at one time were
mistaken for the eggs themselves. These cocoons are usually
brown and horny and vary in size in different species of earthworm ;
some are about as large as rape seed, others almost equal in bulk to
a small grain of wheat. They are formed from the secretions of the

peculiar ectoderm cells found in the clitellum and at first have
a ring-like shape. The secretions harden when in contact with the
air. The animal begins to wriggle out of the band, which at first
surrounds its body in the neighbourhood of the thirty-second to the
thirty-seventh segment. As the band passes over the openings of
the oviducts in the fourteenth segment it carries away with it a
certain number of ova, and as it passes the orifices of the spermathecae between the eleventh and tenth and tenth and ninth
segments, some of the spermatozoa which have been received from
another individual are squeezed out. Besides ova and spermatozoa
the cocoon contains a certain amount of a milky and nutritive fluid
in which these cells float; this is probably supplied by certain other
glands in the skin of the earthworm. At the moment the last
segment, that is, number one, is withdrawn, the anterior end of the
cocoon contracts and closes, and as the posterior end of the band-
like ring passes over the head it also closes, so that the cocoon lies
in the earth as a closed vesicle containing eggs, spermatozoa and a
nutritive fluid. The spermatozoa fuse with the ova and from the
fertilized ova, by division into a number of cells and by the
differentiation of the cells into muscle cells, epithelial cells,
digestive cells, nerve cells, etc., a young earthworm is built up.
Before being hatched out of the cocoons the young embryos are
nourished by the milky nutritive fluid in which they float.

In Great Britain there are several species of earthworm, which
are grouped into two genera, viz. *Allolobophora*, with
fourteen species, which, with one exception, have the
prostomium not dove-tailed into the peristomium;
and *Lumbricus*, with five species, in which the prostomium is completely dove-tailed into the peristomium. The above account has
been taken from the anatomy of *L. herculeus*, the largest of our
indigenous species, but with the exception of a few minor details
the account applies to most British earthworms.

Species of Earthworm.

Order I. **Oligochaeta.**

The sub-order to which earthworms belong, the Terricolae,
are for the most part inhabitants of the land, and occur widely
distributed over the Earth, being, as a rule, only absent from
sandy and desert soils. Some of them are aquatic but not
many. On the other hand the allied sub-order the Limicolae
are for the most part denizens of fresh water. A few Limicolae

possess gills or finger-like processes well supplied with blood-vessels which take up oxygen from the surrounding water. Both sub-orders contain numerous genera and families; together they form the order Oligochaeta, which is characterised by being hermaphrodite, by having the reproductive organs few in number and definite in position, by developing directly from the egg without the intervention of any larval stage, and lastly by the absence of certain structures which are very characteristic of the other great division of the true worms or Chaetopoda.

Order II. Polychaeta.

The Polychaeta differ from the Oligochaeta, as their name implies, by possessing a large number of chaetae on each segment. The sides of each segment are further as a rule drawn out into hollow flaps or lobes called parapodia, which bear the chaetae. Each parapodium may be divided into a dorsal half, the Notopodium, and a ventral half, the Neuropodium (15 and 16, Fig. 47). Both notopodium and neuropodium carry bunches of chaetae, and each has as a rule one particularly large chaeta, the aciculum, completely concealed in a very deep chaeta-sac, which is moved by muscles attached to its base and serves as a kind of skeleton for the parapodium. There is usually above the notopodium and beneath the neuropodium a process called a cirrus. The dorsal cirrus may be modified into a gill, and both dorsal and ventral cirri are absent in some cases.

The coelom is often divided into three longitudinal compartments by two muscular partitions (5, Fig. 47) which run from the dorso-lateral line towards the median ventral line near the nerve-cord. The septa which divide the coelom in one segment from that in the next are in many forms incomplete or absent.

Fig. 46. *Nereis pelagica*, L. After Oersted.

As a rule Polychaets have a certain number of the anterior segments modified to form a head, which usually carries tentacles and organs for absorbing oxygen from the water, called branchiae or gills. They are generally of separate sexes, and the eggs develop into a larva which swims in the sea and gradually changes and grows up into a worm. This group includes a very great variety of forms, almost all of which are marine. With few exceptions they form burrows for themselves, which most

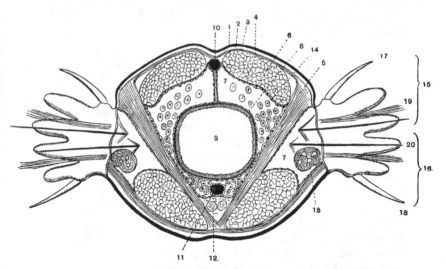

Fig. 47. Transverse section through *Nereis cultrifera*, slightly simplified. The parapodia are shown in perspective. Magnified.

1. Cuticle.   2. Epidermis.   3. Circular muscles.   4. Longitudinal muscles.   5. Oblique muscles forming a partition.   6. Somatic layer of epithelium.   7. Coelom.   8. Splanchnic layer of peritoneum. 9. Cavity of intestine.   10. Dorsal blood-vessel.   11. Ventral blood-vessel.   12. Ventral nerve-cord.   13. Nephridium cut in section. 14. Ova.   15. Notopodium.   16. Neuropodium.   17. Dorsal cirrus.   18. Ventral cirrus.   19. Chaetae.   20. Aciculum with muscles at inner end.

of them occasionally desert in order to seek prey and to discharge the reproductive cells. Some however never leave the burrows, which in this case often take the form of tubes composed of a secretion of the ectoderm.

Order III. Hirudinea.

Besides the Oligochaeta and Polychaeta the order Hirudinea, the members of which are popularly known as leeches, is included amongst the Chaetopoda. They were for some time regarded as a distinct order of Annelida, since the great majority of species possess no chaetae and have other peculiarities ; but the recent discovery of species possessing chaetae, and the close resemblance between the development of all Hirudinea and that of Oligochaeta, renders it evident that they are true Chaetopoda and that the absence of chaetae is a secondary characteristic.

There is little doubt that the Hirudinea are closely allied to the Oligochaeta ; indeed there are certain families which it is not easy to assign definitely to either group ; but the more typical forms are easily distinguished. Externally leeches may be recognized by the possession of a sucker at each end of the body, the anterior one being formed by the mouth, whilst the posterior one is a special organ. By alternately attaching and releasing these suckers and bending the body the animal crawls along.

Leeches. External features.

With the exception of *Branchellion,* which bears tufted gills, the bodies of leeches are without external processes. There are no parapodia, as in the Polychaeta, and no branchiae or tentacles, and only one genus of the family has any chaetae. The body is segmented, and recently it has been

Fig. 48. *Hirudo medicinalis,* about life size.

1. Mouth. 2. Posterior sucker. 3. Sensory papillae on the anterior annulus of each segment. The remaining four annuli which make up each true segment are indicated by the markings on the dorsal surface.

shown that the number of segments is always thirty-three. Some however of the segments are fused together ; thus for example the

posterior sucker contains traces of six or seven true segments. The
best test of the number is to count the ganglia on the ventral nerve-
cord. But even this is not decisive, because although there are
twenty-one free ganglia in the centre of the body a certain number,
some say five and some six, are fused into the sub-pharyngeal
ganglion, and a certain number, some say seven and some say six,
coalesce to form the ganglion of the posterior sucker. Whichever
view is taken the total number of segments is thirty-three.

The body of the leech is ringed or divided into a number of
annuli. These do not, however, represent the segments, but a
number, varying in the different genera, make up a segment. In
Hirudo, the medicinal leech, there are five annuli to a true
segment; in *Clepsine*, a common fresh-water leech, the number is
three. The real segmentation is, however, to some extent indicated
by markings on the skin.

The animal is covered like the earthworm by a thin cuticle
secreted by the outermost cells, and the ectoderm contains numerous
goblet cells which are especially well-developed over the segments
abutting on the generative orifices. Here they form a clitellum,
and the secretion the cells pour out forms a cocoon in which the
eggs are laid.

The nervous system of a leech does not differ in essentials from
that of the earthworm, but the nephridia, of which
there are in *Hirudo* seventeen pairs, are peculiar.
They are no doubt a modified form of the same
organ as the nephridium of the earthworm, and they consist of
coiled cellular tubes. The outer end communicates with the exterior
through a muscular vesicle. The inner end, or so-called testis lobe,
lies near the testis in the genital segments. The whole is traversed
by a ramifying network of chambers opening by minute pores on
the testis lobe.

The other systems of organs are still more unlike what has been
described in the case of the earthworm and deserve a short account.
Leeches live by sucking the blood or juices of other animals,
usually of Vertebrates. They are divided into two large groups—
(a) the *RHYNCHOBDELLIDAE*, which pierce the tissues of their hosts
by means of a fine protrusible stomodaeum, the so-called proboscis,
and (b) the *GNATHOBDELLIDAE*, which bite their prey by means of
horny jaws. The medicinal leech is one that bites, and the triradiate little scar which its three teeth make in the skin was well-
known to our forefathers in the times of bleeding and cupping.

The three teeth, which are notched like a saw, are really only thickenings of the cuticle borne by the wall of the pharynx, which contains many unicellular glands whose secretion prevents blood from coagulating. Thus the leech when fixed on to its victim by the oral sucker readily obtains a full meal.

From the pharynx a short narrow tube, the oesophagus, leads into an enormous dilatation, the crop. This extends to the fourteenth segment and gives off on each side a series of eleven pouches or caeca (Lat. *caecum*, blind) which increase in size from before backward. The posterior caeca are very large and reach back to the level of the anus, lying one on each side of the intestine. The leech has the habits of a boa-constrictor. It makes a hearty meal, absorbing as much as three times its own weight of blood, and the blood it absorbs is stored up for many months in this enormous crop. It slowly digests the food in a small globular stomach situated just behind where the posterior caeca leave the crop. The stomach opens into a short intestine which ends in the anus, a minute pore situated dorsally between the posterior sucker and the body (Fig. 49).

In one genus at least, *Acanthobdella*, the coelomic cavity is almost as well-developed as in an earthworm, and is divided up by septa as in that animal. In other leeches the cavity tends to disappear, becoming in fact filled up by a great growth of tissue, and thus reduced to a few narrow channels. In many leeches it contains a fluid closely resembling the true blood, so that unless very careful microscopic examination be made these channels may be mistaken for true blood-vessels. The capsules in which the ovaries and testes lie are also parts of the coelom.

Coelom.

The medicinal leech, owing to a great growth of this above-mentioned tissue, is almost without a coelomic cavity. When the body is opened a narrow vessel full of a red fluid is seen running along the middle dorsal line above the alimentary canal. This is the dorsal sinus, a remnant of the true coelomic cavity; a similar sinus runs along the ventral surface underneath the alimentary canal, which is called the ventral sinus. It communicates with the dorsal sinus by lateral channels which run between the intestine and the posterior caeca of the crop. It surrounds the ventral nerve-cord, which thus seems to float in blood but really lies in the red coelomic fluid, and it gives off lateral sinuses which surround the inner openings of the nephridia. The true blood-vessels

comprise a vessel running on each side of the body and connected together by transverse branches which run from side to
side below the ventral sinus. The
lateral vessels further supply capillaries
to the nephridia, alimentary canal,
reproductive organs, etc., and a very
extensive system to the skin where the
haemoglobin of the blood takes up
oxygen. Except in *Branchellion*, which
has special gills, the respiration of
leeches is carried on by the skin.

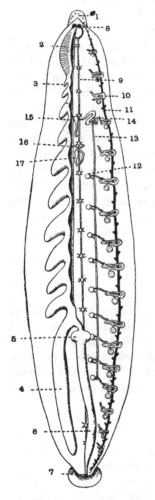

Leeches are, like the earthworm,
hermaphrodite, but their
reproductive organs differ
in some respects from those of that
animal.

Reproduc-
tion.

In *Lumbricus* the testes are repeated in two segments only, but in
Hirudo there are usually nine pairs of
testes. The cavities of both the testis
and of the ovary are to be regarded as
part of the original coelom; in strictness
the testes probably correspond to the
vesiculae seminales in an earthworm,
which are part of the coelom, and
enclose the true testis and the sperm-
funnel. Each testicular sac produces
spermatozoa on one side and on the
other side is ciliated. The ciliated
tract is the sperm-funnel and leads into
a short transverse duct which passes
into a longitudinal canal termed the
vas deferens, there being one such
canal on each side of the body. At its

FIG. 49. View of the internal organs of *Hirudo medicinalis*. On the left side
the alimentary canal is shown, but the right half of this organ has been
removed to show the excretory and reproductive organs.

1. Head with eye spots.   2. Muscular pharynx.   3. 1st diverticulum of
the crop.   4. 11th diverticulum of the crop.   5. Stomach.
6. Rectum.   7. Anus.   8. Cerebral ganglia.   9. Ventral nerve-
cord.   10. Nephridium.   11. Lateral blood-vessel.   12. Testis.
13. Vas deferens.   14. Prostate gland.   15. Penis.   16. Ovary.
17. Uterus—a dilatation formed by the conjoined oviducts.

anterior end each vas deferens passes into a convoluted mass of tubes—the so-called epididymis—whose walls secrete a substance which binds the spermatozoa together into packets called spermatophores. It is to be remembered that the names epididymis, prostate, etc., are given from fanciful resemblances to parts in the anatomy of man by no means homologous with the organs bearing the same name in the leech. From each epididymis a short duct passes towards the middle line, and these two ducts fuse and enter the base of the penis, which is protruded from the segment which contains the sixth distinct post-oral ganglion.

The penis is simply the muscular end of the conjoined male ducts or vasa deferentia; it is the organ by which the spermatophore is deposited in the body of another leech. The spermatozoa in *Clepsine* seem to penetrate the skin at any point and make their way to the ovaries, where they fertilize the eggs. In other species the spermatozoa enter in the usual way by the female genital pore.

As in the earthworm, there is but one pair of ovaries. These are minute filamentous bodies each enclosed in a small coelomic sac. From each sac a short oviduct proceeds and uniting with its fellow forms a twisted tube surrounded by many glands. This finally opens by a median pore on the segment behind the one bearing the male opening.

Thus in leeches, unlike the condition in the earthworm, the genital pores are single and median. The medicinal leech lays its eggs in a cocoon and buries them in holes in the banks of the ponds it inhabits. *Clepsine,* one of the Rhynchobdellidae which is very common in Britain, attaches its eggs to some stone or water-plant, or in some species carries them about on its ventral surface. It has developed a quite maternal habit of brooding over the eggs, and when the young are hatched it carries them about and they feed on some secretion from its body.

Of the Gnathobdellidae, *Hirudo medicinalis* is found in Great Britain, but is commoner in some parts of the Con-
Habits, etc. tinent. It is cultivated in some districts, but the demand for it is decreasing with the disappearance of blood-letting. It becomes mature in three years. In the young stages it sucks the juices of insects. Another common but small Gnathobdellid leech is the brownish *Nephelis*, which frequents our ponds and pools ; it feeds on snails and planarians. A large species of the same genus is common in the shallows of the St Lawrence, in Canada. In warmer climates many leeches take to living on land,

and are a source of great annoyance to travellers whose blood they suck. Even water-forms do much damage unless carefully guarded against. Certain species make their way with drinking water into the throat and back of the mouth, on which they fasten, and so cause great suffering both to man and cattle.

Phylum ANNELID'A.

This phylum includes segmented animals with, as a rule, a well-developed coelom and metamerically repeated nephridia. The cuticle is always thin and flexible, and the nervous system consists of a pair of supra-oesophageal ganglia, a nerve collar and a ventral nerve-cord which has a ganglionic swelling in each segment.

Class I. CHAETOPODA.

Annelida which possess bristles (chaetae) embedded in pits in the skin and serving as organs of locomotion, or which are believed to have once possessed such organs and to have lost them.

Order 1. **Oligochaeta.**

Chaetopoda which have the chaetae arranged singly or in pairs and which have neither parapodia nor tentacles : the generative organs are definitely localized and the sexes are united in the same individual : development is practically entirely embryonic : the group inhabits fresh water or damp earth.

Ex. *Lumbricus, Allolobophora.*

Order 2. **Polychaeta.**

Chaetopoda which have the chaetae arranged in bundles of some size, almost always borne on conspicuous lateral out-growths of the body termed parapodia : the prostomium has, as a rule, tactile organs, known as tentacles and palps : there are no localized generative organs, ova and spermatozoa being developed from wide stretches of the coelomic wall; the sexes are separate : in the development a well-marked larval stage occurs : with few exceptions the group is marine.

Ex. *Nereis.*

Order 3. **Hirudinea.**

Chaetopoda in which chaetae and parapodia are absent and which move by means of a muscular sucker developed on the under surface of the posterior segments: there are no tentacles and the mouth acts as an anterior sucker: the coelom is reduced to capsules surrounding the genital cells and to a few narrow channels: the animals are hermaphrodite, and the genital pores single and median: the members of this order live on the juices of other animals, and there are both fresh water and marine species: development is entirely embryonic.

Ex. *Hirudo, Nephelis, Clepsine.*

CHAPTER VII.

PHYLUM ARTHROPODA.

ONE of the most striking features of the Annelida is the fact that they are segmented, that is to say their body is divided into a number of similar parts placed one behind the other like coaches in a train, each of which to a greater or less extent resembles the part in front of it. The likeness of the parts to one another varies. In some worms we might easily detect from which region of the body any given segment was taken. In the Earthworm, except in the region of the clitellum, there is little external difference; nevertheless if we consider the internal organs we can distinguish any of the first twenty segments from any other behind these and can easily arrange them in their proper order; but no matter how long the worm is, all the segments behind the twentieth resemble one another so closely that it is impossible to assign any to their right place, except the last of all (v. p. 89).

The animals included in the group of the Arthropoda are segmented like the Annelida, but with few exceptions *Arthropoda.* the number of segments is small and does not exceed twenty. The segments have also become more highly differentiated from one another in consequence of being modified to perform various functions, and they are more frequently fused together than is the case in the Annelids.

The Arthropoda have jointed outgrowths called limbs or appendages. These are always arranged in pairs, and at least one pair is modified so as to assist in holding and crushing the food. This character of possessing jointed limbs is what is indicated by the name Arthropoda (Gr. ἄρθρον joint; πούς foot).

The Arthropoda may be divided into three classes :—

I. The CRUSTACEA, which includes all the Crabs, Lobsters, *Classification.* Cray-fish, Barnacles, Wood-lice, etc., besides countless small forms such as the Water-flea, Cyclops, and many others which inhabit both salt and fresh water.

II. The ANTENNATA, which include all Arthropoda possessing one pair of feelers—antennae—and breathing by means of air tubes or tracheae. This group is divided into three sub-classes, viz. :

A. The Prototracheata, a group containing the genus *Peripatus*, an animal not found in Europe or in North America, but which must be mentioned because it seems to be a survival from an earlier age and because its structure has given us a clue to much that was obscure in the anatomy of Arthropods; it is in fact in many respects intermediate between the Annelids and the air-breathing Arthropoda.

B. The Myriapoda or Centipedes, the commonest British examples of which are the chestnut-coloured centipede *Lithobius forficatus* and the black "wire-worm"[1] *Iulus terrestris*.

C. The Insecta, the largest group in the Animal Kingdom. It contains about 250,000 named species, and includes all those creatures such as Beetles, Flies, Dragon-flies, May-flies, Moths, Bees, Ants, Wasps, etc., which we are accustomed to call insects.

III. The ARACHNIDA, including the Spiders, Harvestmen, Mites and certain larger forms such as the Scorpion, and *Limulus*, the King-crab.

If we go into an old garden and turn over a stone or look between the bark and the trunk of a decaying tree or examine the leaves, we may find representatives of each of the four larger classes mentioned above. The Crustacea may be represented by a Wood-louse (Fig. 79), the Myriapoda by a Centipede (Figs. 51 and 84), the Insecta by a Beetle (Fig. 92), and the Arachnida by a Spider (Fig. 53). If we compare these creatures one with another we shall see that they resemble each other in certain fundamental particulars.

To begin with they are all clothed in a hard coating consisting largely of the horny substance called chitin which Exoskeleton. does not form a simple chamber or house in which the body of the animal lies as a snail lies in its shell, but which is moulded accurately over all the body and even tucked into all the openings so that it forms an exact cast of the soft parts underneath. This covering is to be regarded as an exaggeration of the cuticle found in Annelids, and it is called an Exoskeleton in order to distinguish it from the internal framework of hard parts found in the Vertebrata.

---

[1] This is not to be confused with the larva of a beetle, *Elater lineatus*, which is also called a " wire-worm " by the British agriculturist.

120

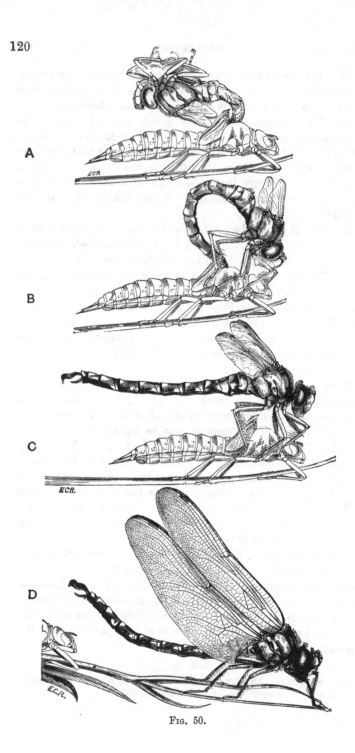

Fig. 50.

Fig. 50.

A. The anterior portion of the body of a Dragon-fly, *Aeschna cyanea*, freed
from the larval shell.   B. The tail being extricated.   C. The whole
body extricated.   D. The perfect insect, the wings having acquired
their full dimensions, resting to dry itself preparatory to the wings being
horizontally extended.

Were this hard exoskeleton of the same consistency all over the
body it would be impossible for the animal to bend its body or
to move at all, but at certain spots, as may be well seen between
the segments of a Centipede or between the members of a Beetle's
legs, the exoskeleton has remained soft like the leather joints in a
suit of mail-armour, and thus a certain amount of flexibility is
given to the whole body; for instance the Armadillo wood-louse
and the Pill-millepede can roll themselves up into spherical balls.

Not only is a hard exoskeleton a hindrance to unlimited
movement but it also interferes with growth. It is
impossible to increase in size when shut up in a hard
unyielding case. Now growth is one of the common characters
of all animals, and the way the obstacle presented by the exoskeleton
of Arthropods to growth is overcome is as follows. At certain
stated times the outer skin or ectoderm of the animal loosens itself
from the inside of the cuticle or investment, which splits or cracks,
usually along the middle of the back; through the opening thus
formed the body of the animal begins to appear, and gradually
withdrawing each limb from its case it works its way out. The
exoskeleton thus cast off forms a most accurate mould of the
animal which has left it, and even includes those portions which are
folded in at the mouth and anus and other openings of the body.
A dragon-fly emerging and freeing itself from its cast skin is shown
in Figure 50.

The skin of the animal when it steps out of its old casing
is quite soft, and it remains so for a varying time, a few hours in
the case of some insects, one to three days in a Cray-fish. During
this period the animal grows. After a longer or shorter time
during which the body remains soft and capable of extension the
secretions of the skin commence to harden, and very soon the
animal is again enveloped in a hard case which rapidly assumes the
colour and appearance appropriate to the species in question.

This moulting of the skin in the Arthropoda is termed the
ecdysis. It takes place at more or less regular times in each
species, in the Cray-fish three or four times or even oftener during

the first year, the period of most active growth, later but once
annually, usually about Midsummer; the Cockroach moults three
times during the first year, after which the moults are annual, but
it does not become adult till after the seventh ecdysis, when it is
four years old.

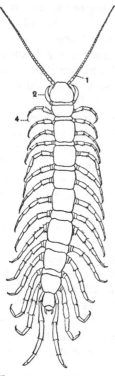

If we further examine our wood-louse,
Divisions of Body. centipede, beetle, and spider,
we shall notice at once that
they are all, like the Annelids, bilaterally
symmetrical; and it may as well be stated
at once that with few exceptions this is
true of the internal organs as well as of
the exoskeleton. Another feature common to them all is that they possess
jointed limbs or appendages. These may
occur in all the segments, as in the wood-
louse and centipede, or the limbs may be
reduced in number and confined to definite regions of the body, as in the beetle
and spider, but they always exist and are
always jointed.

In the body of the centipede we can
recognise but two regions (Fig. 51), a
head and a trunk; the trunk consisting
of a number of segments, the head apparently of a single rounded one whose
really composite nature is shown by the
fact that it carries not one but several
pairs of appendages.

The same may be said of the wood-
louse (Fig. 79), though here the trunk is
divisible into two parts by the character
of the appendages. The anterior part or
thorax bears walking legs, the posterior
part or abdomen plate-like appendages
which act as respiratory organs. In

Fig. 51. A Centipede, *Lithobius forficatus.* Dorsal
aspect × 12.

1. Antennae.   2. Poison
claws, 5th pair of appendages.   4. First pair of
walking legs.

neither of these creatures nor in the spider is there any constriction
between the head and the trunk, that is to say there is no neck.
In the beetle however there is a well-marked neck separating the
head from the rest of the body. The three following segments in
the beetle are again separated from those which come after and

form what is called the thorax. This part bears the three pairs of walking limbs and the two pairs of wings. The hindermost segments, often ten in number in insects, constitute the abdomen; this part of the body is devoid of jointed limbs, though doubtless the ancestors of insects once possessed them on all the segments.

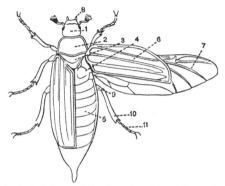

Fɪɢ. 52.   A Male Cockchafer, *Melolontha vulgaris*, seen from above and slightly enlarged.   After Vogt and Yung.

1. Head, stretched forward.   2. Prothorax.   3. Mesothorax.
4. Metathorax.   5. Abdomen.   6. Anterior wing (elytron) of right side, turned forward.   7. Posterior wing of right side, expanded.
8. Maxillary palps.   9. Femur of third right leg.   10. Tibia of third right leg.   11. Tarsus of third right leg.

In the Insecta the abdomen may be constricted off from the thorax as it is in wasps (Fig. 95), or there may be no constriction. If we now turn to the spider we shall see that the division of the body into regions has gone along different lines, and we can recognise only two principal parts, a so-called c e p h a l o - t h o r a x or p r o s o m a to which all the appendages are attached, and a stalked abdomen or fused m e s o - and m e t a - s o m a behind which is devoid of obvious limbs, though certain little knobs at its hinder end, from the summit of which are spun out the silken threads used in making the web, have been shown to be rudimentary limbs. The abdomen has lost all trace of external segmentation. In the harvestmen—Phalangids—long-legged creatures resembling spiders and found only during the summer months, the constriction between the cephalo-thorax and abdomen is absent, but the latter is distinctly divided into segments (Fig. 103).

Thus in the Arthropods the body is divided into segments, and these segments are not all equal and alike, but they have become variously modified and some of them have fused together, as in the

head of insects, and the abdomen of spiders, so that certain regions
of the body may be distinguished, and this is one of the most
characteristic features of the group.

In all Arthropoda certain of the appendages have lost the
Appendages. function of locomotion and are bent round and
brought into connexion with the mouth. These
mouth or oral appendages assist in catching and holding the food,
and to some extent in biting and tearing it into small pieces.
With the exception of the Arachnida, in which the arrangement

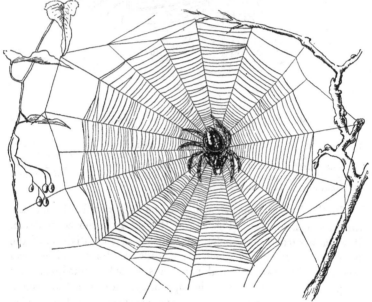

Fig. 53.   The Garden Spider, *Epeira diadema*, sitting in the centre of its web.
After Blanchard.

is somewhat different, and of *Peripatus*, which has only one pair of
jaw-limbs, the first pair of oral appendages is termed the mandible,
the second is the first maxilla and the third the second maxilla;
the last-named however in the Myriapoda retains the appearance of
a walking leg, although too short to be used for walking.   On the
other hand in Crustacea a varying number of the appendages immediately succeeding the second maxilla are often turned forward
and assist in the feeding.   When this is the case they are termed
maxillipedes (Figs. 54 and 56).

The modification which all these appendages undergo is similar in kind. The first stages of it are seen in the Arachnida. Here the first pair of appendages is always a pair of little claws placed in front of the mouth, the last joint shutting down on the next division of the limb like a knife-blade on the handle. After this pair come others, more leg-like, of which sometimes all and in any case one pair have inwardly directed projections on their lowest joints termed gnathobases, so that when the limbs are brought together they act like a pair of nut-crackers (Fig. 55). Or to take another example, as we pass from the segment of the Cray-fish or of a *Gammarus* which bears the great claws forwards through the maxillae to the mandible we find the outer parts of the limb dwindling in size, and the basal projection growing bigger (Figs. 54 and 56). In the mandible only a minute rudiment of the other joints remains, and is called the palp. In the mandible of the insect even this has disappeared. Limbs which have undergone these characteristic changes are called gnathites (Gr. γνάθος, a jaw).

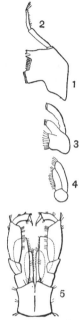

Fig. 54.   The mouth appendages of *Gammarus neglectus*. From Leuckart and Nitsche, after G. O. Sars.

1. The left mandible.
2. Its palp.
3. 1st maxilla of left side.
4. 2nd maxilla of left side.
5. Maxillipede of each side together forming an under lip.

If we cut open the body of an Cavities of Body. Earthworm, a Starfish or a Vertebrate, we lay open a chamber in which the alimentary canal and many other organs apparently lie. This chamber is the coelom or primary body cavity, which has no connexion with the blood system, though amoebocytes float in the fluid it contains. If we cut open the body of an Arthropod or of a Mollusc we also open up a chamber which may be spacious, as in an Insect or a Snail, or which may be much reduced and filled up by the various organs of the body and by muscles, as in a Cray-fish or a Mussel. This cavity however, as development shows, is not similar in its nature to the coelom of an Earthworm or a Vertebrate, and it further differs

in that it contains blood and is continuous with the cavity of the
heart and large blood-vessels. A special name has been given to
this cavity and it is termed the Haemocoel (blood-cavity). The
presence of this secondary body cavity or Haemocoel instead of a
Coelom introduces at once a peculiarity in the physiology of the
circulation of the Arthropoda. Instead of the oxygen-bearing and
food-carrying blood being conveyed all over the body in minute
capillaries which ramify in every part of every tissue, in the animals
in question the tissues are floating in and bathed by the blood,
which surrounds the organs on all sides and is kept in action by
the contraction of a muscular tube—the heart—which opens freely
into the body-cavity or Haemocoel.

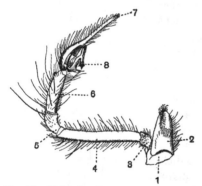

FIG. 55.   Pedipalp of *Tegenaria guyonii*, the
large house-spider.

1. Coxa.   2. Maxilla, the gnathobase.
3. Trochanter. 4. Femur.   5. Patella.
6. Tibia.   7. Tarsus.   8. Palpal organ.

A true coelomic cavity
(as is proved both by its
origin and its relation to
the excretory and repro-
ductive organs) is however
found in Arthropods in the
cavities of the reproductive
organs (Fig. 67) and in
certain vesicles connected
with the inner ends of
some of the excretory
organs, such as are found
in the coxal glands of
Arachnids. It is however
obviously much reduced
and takes a smaller part
in the economy of these

animals than it does for instance in the Annelida or Echinodermata.

The skin of an Arthropod like that of an Earthworm includes
Skin and Con- not only the ectoderm, but a firm support for the
nective Tissue. same called the dermis. The dermis in this case is
formed of well-developed connective tissue and this tissue also
forms an investment for every organ in the body, so that, as Huxley
remarks, if all the organs were dissolved away there would remain
a complete cast of them in connective tissue. The same statement
is true of Mollusca and Vertebrata: it is therefore important to
obtain a clear idea what connective tissue is.

Its groundwork is a jelly-like secretion intervening between
ectoderm and mesoderm, or mesoderm and endoderm, and therefore

to some extent comparable with the structureless lamella of *Hydra* or the jelly of a Medusa. Into this substance cells are budded from the adjacent layers, chiefly from the mesoderm. These cells add to the secretion, in which fibres soon make their appearance, crossing each other at various angles. These fibres are to be looked on as more or less solid precipitates. The cells are found often flattened against them and connected with neighbouring cells

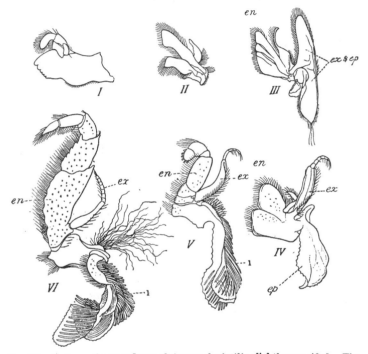

FIG. 56. Left mouth appendages of *Astacus fluviatilis*, slightly magnified. The other appendages are shown in Fig. 60.

I. Mandible. II. First maxilla. III. Second maxilla (Scaphognathite). IV. First maxilliped. V. Second maxilliped. VI. Third maxilliped. *en.* Endopodite. *ex.* Exopodite. *ep.* Epipodite. *ex & ep* in III form a scoop for circulating water over the gills.

by delicate protoplasmic threads. The connective tissue round the ends of a muscle is modified to form tendon. Here the fibres all pursue a parallel course and great tensile strength is the result.

Observation of development shows that connective tissue, haemocoel, and blood-vessels all generally arise from the same

rudiment in the embryo, which may be compared to the jelly of
Coelenterata. Blood is a portion of it where the jelly is fluid,
the fibres are not developed and the amoebocytes remain mobile.
In connective tissue the jelly becomes more solid, fibres are developed and the amoebocytes become stationary, being converted
into the so-called connective-tissue corpuscles. It is interesting to
note that under the abnormal circumstances of a wound the blood
of many animals can develop fibres; this property causes what is
known as clotting.

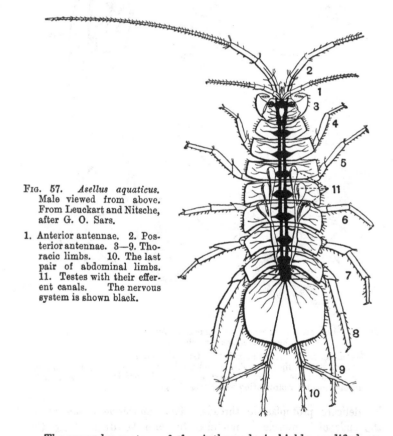

FIG. 57. *Asellus aquaticus.*
Male viewed from above.
From Leuckart and Nitsche,
after G. O. Sars.

1. Anterior antennae. 2. Posterior antennae. 3—9. Thoracic limbs. 10. The last
pair of abdominal limbs.
11. Testes with their efferent canals. The nervous
system is shown black.

The muscular system of the Arthropoda is highly modified as
The Muscular   compared with the primitive arrangement found in
System.   the Annelida. Instead of a continuous sheath of
muscle there are special bundles of muscular fibres for the purpose

of moving the various hard parts on one another. Each joint of each limb, for instance, is provided with a pair of muscles which move it on the next joint. One of these is called the flexor, or bender, and the other the extensor, or straightener. Here as in Annelida the muscles are derived from epithelial cells, but all trace of this origin is lost in the adult. In the muscle-cell in many cases the nucleus has divided, giving rise to several nuclei which are surrounded by unmodified protoplasm. All the rest of the proto-plasm is converted into fibrillae which consist of alternately dark and light stretches. Such muscles are said to be striped, and they have the power of contracting with much greater rapidity than muscles of the type found in Annelida. Muscles of the latter kind are called smooth muscles, and in Arthropoda exist in the wall of the gut and some other places.

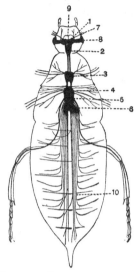

FIG. 58. View of nervous sys-tem of the Cockchafer, *Melo-lontha vulgaris.* After Vogt and Yung.

1. Cerebral ganglion.
2. Sub-oesophageal ganglion.
3. 1st thoracic ganglion.
4. 2nd thoracic ganglion.
5. 3rd thoracic ganglion.
6. Fused abdominal ganglia.
7. Nerves to antennae.
8. Optic nerves. 9. Origin of sympathetic nerves.
10. Abdominal nerves, a pair to each segment, which split into an anterior and pos-terior branch.

The nervous system of the Arthro-

Nervous System.   poda is built upon the same plan as that of the Annelids. It consists in its least modified form of a pair of closely ap-proximated supra-oesophageal ganglia forming a brain situated in front of the mouth, in the head in Insects, and in the anterior part of the body in those Arthropods which have no distinct head. This brain supplies nerves to certain sense organs, and gives off two stout cords, one of which passes to the left, and the other to the right of the oesophagus. These para-oesophageal (Gr. παρὰ, alongside) cords unite together behind the oesophagus, and where they unite they form a pair of sub-oesophageal ganglia which send nerves to some or all of the mouth appendages (Figs. 57 and 58). Behind this comes a chain of ganglia, normally one pair for each segment, which supply nerves to the organs and appendages of the segment in which they lie; each pair being connected with their successor by a double nerve-cord. These ganglia do not always remain distinct but show a tendency to fuse together. Thus in the Cray-fish the supra-oesophageal ganglion is shown by its development to be formed by the fusion of several pairs of ganglia; and the mandibles, both maxillae and the 1st and 2nd maxillipedes are supplied from a single ganglionic mass, the sub-oesophageal ganglion, which is the result of the fusion of five primitive ganglia; again in some Flies and in Spiders all the ganglia behind the mouth have fused into one large nervous mass situated in the thorax. This fusion takes place to a less extent in some beetles; for instance in the Cockchafer, *Melolontha vulgaris*, the thoracic ganglia remain distinct, but the abdominal have fused into one mass which has been drawn up into the thorax (6, Fig. 58).

The Crustacea are normally provided with two pairs of long feelers or antennae (though occasionally, as in the case of the Wood-louse, one pair may be entirely lost) and the Insects and Centipedes have a single pair, very conspicuous in some Butterflies and Beetles. These are organs of touch and frequently of smell; and in some cases, such as the Lobster, they also act as hearing organs. They are supplied with nerves from the brain. No such antennae exist in the Spiders or Mites, and this serves at once to distinguish the Arachnids from the other two classes.

The eyes of Arthropoda are peculiarly modified areas of the ectoderm. Over a certain area some of the ecto-derm cells become modified into visual sense-cells. In these there is a gelatinous rod developed from the outer end of the cell, and situated on the one side of it (Fig. 59 A); whilst from the base of the cell a nerve-fibre is developed. Usually several visual cells are pressed together in such a way that their rods cohere, and in this way a fluted spindle-shaped rod termed the rhabdome is built up (6, Fig. 59 A and E). This rod, like all such structures formed in visual cells, is cross-striped or in other words consists of layers of different densities. Other cells of the visual area remain as supporting cells, being longer than the surrounding ecto-derm cells but otherwise unmodified. Pigment—universally present in visual organs—is secreted either by these cells (4, Fig. 59 A) or by amoebocytes which have wandered out from the underlying dermis (8, Fig. 59 B). The cuticle covering the visual area becomes transparent and greatly thickened, and so acts as a condensing apparatus or lens. In the larger eyes each group of visual cells secreting a rhabdome is sharply marked off from the rest and covered by a separate lens or thickening of the cuticle. Such a group is

Eyes.

termed a retinula; and as it is surrounded by a sheath of pigment it can only be affected by light coming from an object directly in front of it and propagated parallel to its axis. Hence to an Arthropod possessing an eye with many retinulae the outer world will be presented as a mosaic of light and shade, each retinula giving an impression depending on the intensity of light in the field of view directly in front of it. Such a mosaic is an image, and this image

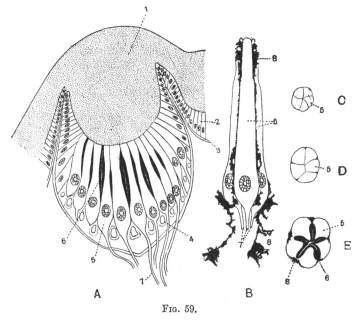

Fig. 59.

A. Vertical section through a lateral eye of a Scorpion, *Euscorpius italicus.* B. Diagram of retinula of a Scorpion's central eye.　C. D. E.　Transverse section of B taken at different levels.　From Lankester and Bourne.

1.　Cuticular lens.　2.　Epidermis of the general body-surface.　3.　Basement membrane.　4.　Epidermal cells which contain pigment.　5.　Nerve end-cells with nuclei.　6.　Rhabdome.　7.　Fibres of optic nerve. 8.　Pigment contained in connective tissue cells.

will be obviously the more detailed and definite the greater the number of retinulae in a given area. In some Arthropods, such as the common Fly, the eyes cover the greater part of the head. Eyes with numerous well-defined retinulae are known as compound eyes: and they usually present the appearance of numerous facets of hexagonal outline, owing to the fact that there is a lens corresponding to each retinula. In many cases the retinula

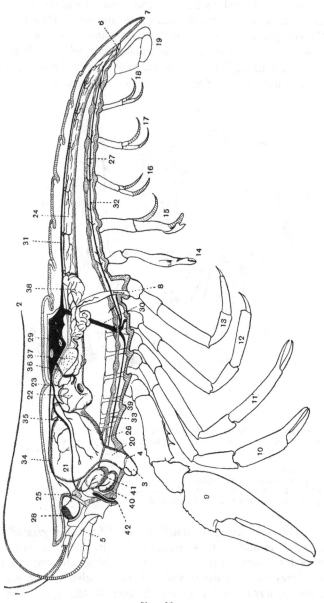

Fig. 60.

FIG. 60. Semi-diagrammatic view of internal organs, and some limbs of right
side of a Male Crayfish, *Astacus fluviatilis* × 1. Partly from Howes.

1. Antennule.   2. Antenna.   3. Mandible.   4. Mouth.   5. Scale
or squama of antenna (exopodite).   6. Anus.   7. Telson.   8. Opening
of vas deferens.   9. Chela.   10. 1st walking leg.   11. 2nd walking
leg.   12. 3rd walking leg.   13. 4th walking leg.   14. 1st abdominal
leg, modified.   15. 2nd abdominal leg, slightly modified.   16. 3rd
abdominal leg.   17. 4th abdominal leg.   18. 5th abdominal leg.
19. 6th abdominal leg, forming with telson the swimming paddle.
20. Oesophagus.   21. Stomach.   22. Mesenteron or mid-gut.
23. Cervical groove.   24. Intestine.   25. Cerebral ganglion.
26. Para-oesophageal nerve-cords.   27. Ventral nerve-cord.   28. Eye.
29. Heart.   30. Sternal artery.   31. Supra-intestinal artery.
32. Sub-intestinal artery in abdomen.   33. Sub-intestinal artery in
thorax.   34. Ophthalmic artery.   35. Antennary artery.   36. Hepatic
artery.   37. Testis.   38. Vas deferens.   39. Internal skeleton.
40. Green gland.   41. Bladder.   42. External opening of green gland.

is depressed beneath the general surface, and the adjacent ectoderm
cells meet above it. These cells secrete clear glassy rods which
cohere to form a crystalline cone. This also happens in some
simple eyes, such as the central eyes of the Spider; here the clear
rods remain unconnected, and the whole upper layer of ectoderm
cells is known as the vitreous layer. In the Cray-fish the compound
eyes are carried on the ends of moveable eye-stalks; and in Spiders
the eyes, which in these animals are always simple, are sometimes
elevated on a little prominence like a lighthouse, borne on the head
and thorax. The number and position of the eyes in the Spiders
are points of great use in identifying the various species.

As a rule the alimentary canal of the Arthropods is about as
long as the body, so that it is straight; the Insects
however form an exception to this rule, since in their
case the canal is longer than the body, and consequently has to
be coiled or twisted in order to tuck it away in the limited space.
It was mentioned above that the chitinous exoskeleton of Arthropods is tucked in at both the mouth (stomodaeum or fore-gut)
and anus (proctodaeum or hind-gut), and in many this lining
extends so far in as to leave but a small part of the alimentary
canal free from it. In some species the hinder end of the stomodaeum secretes deposits of chitin, as in the cockroach, or is
hardened by calcareous deposits, as in the lobster or cray-fish, and
thus teeth are formed which lie inside what has been termed the
stomach or gizzard. When the moult or ecdysis of the shell takes
place the linings of the fore- and hind-gut are also cast off.

The presence of chitin lining parts of the alimentary canal
serves to discriminate those parts of the canal which are to be

looked on as merely parts of the skin—the stomodaeum and proctodaeum—from the true endoderm. The stomodaeum is usually divided into a narrow portion leading from the mouth, called the oesophagus (20, Fig. 60), and an expanded portion containing teeth, called the stomach (21, Fig. 60). The proctodaeum is usually a straight cylindrical tube, called the intestine (24, Fig. 60). The small piece intervening between them is called the Mid-gut or Mesenteron; it alone corresponds to the human gullet, stomach, and intestine, and to the whole alimentary canal of a worm behind the pharynx. In it digestion is carried on and into it opens the so-called liver, i.e., one pair—rarely more—of glands consisting of great tufts of branching tubes lined by yellowish-brown cells. These secrete a fluid that assists in digestion. The food passes in part into these glands and some of it is there digested.

In most animals the heart is a muscular sac which opens into a Circulatory apparatus. system of tubes with muscular walls, called arteries, through which blood is driven to all parts of the body, finally passing into narrow tortuous passages, the so-called capillaries, whence it reaches the thin-walled veins through which it returns to the heart. Thus, excepting such fluid as soaks through the thin walls of the capillaries, the blood is entirely confined within definite channels which do not open into the body cavity. But in the Arthropods the state of things is different; the heart (Figs. 60 and 101), which lies in the middle line just below the skin of the back, opens by a series of slits called ostia into the body-cavity (haemocoel), and when the heart expands the blood which is in the body cavity enters these slits, but cannot pass out again through them when the heart contracts, as each slit has a valvular arrangement which prevents this. When the heart contracts the blood is therefore forced forwards and leaves the heart by a vessel—the aorta—or by vessels with various names, which sooner or later open again into the haemocoel, and so the circuit is complete. The part of the haemocoel in which the heart lies and into which the ostia open is called the pericardium, and it is separated from the remainder of the haemocoel by a horizontal septum called the pericardial septum, in which however there are perforations. The pericardium of Arthropoda thus contains blood, and is consequently widely different from the pericardium of Mollusca or Vertebrata, which is in both cases part of the coelom.

In those Arthropods which have a localised respiratory system, and in which the blood takes part in respiration, there is a more definite course for the blood than that sketched above. The extent to which blood-vessels with definite walls are developed varies in the different members of the group; thus the Scorpion and King-crab have a much more specialized circulatory system than

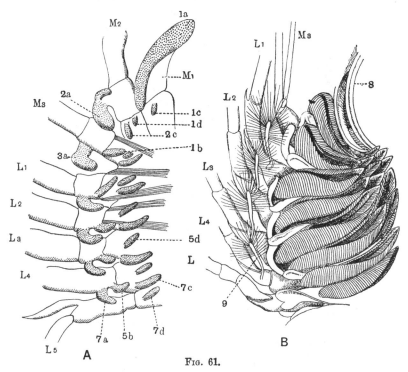

FIG. 61.

A. Left side of a Larva of the Prawn, *Penaeus*, to show the origin of the gills. Slightly magnified. From Claus. L_1 to L_5. The first to fifth ambulatory limbs. M_1 to M_3. The first to third maxillipeds. 1a, 2a, 3a, 7a. Podobranchs. 1b, 5b. Anterior arthrobranchs. 1c, 2c, 7c. Posterior arthrobranchs. 1d, 5d, 7d. Pleurobranchs. Of these rudiments of gills only nineteen develope. B. Left side of a fully-grown Prawn, *Penaeus semiculcatus*, to show fully-grown gills. Slightly magnified. 8. Exopodite of second maxilla, which flaps to and fro and so causes a current over the gills. 9. Exopodite of fourth ambulatory limb.

have the Spider and Mites. The blood is very rarely red, but is usually slightly tinged with a bluish colour by a substance acting in the same way as haemoglobin, but differing from it in composition.

In the lower and more simply organized Crustacea, such as the water-fleas, there are no special breathing organs, but the blood is able to absorb its oxygen and give off its

Respiratory apparatus.

carbonic acid as it courses under the thin skin. But in the larger
and more complicated Crustacea, such as the shrimp and lobster, a
special apparatus is present in the form of gills. These gills are
thin-walled extensions of the skin which project from the surface
of the body near the base of the limb or on the side of the thorax
into the surrounding water; inside them the blood flows to and fro
and a current of water washes them on
the outside. The gills are classified
according to their point of origin, being
termed podobranchs (1a, Fig. 61) when
they arise from the proximal joint of
the limb, arthrobranchs (1b, Fig. 61)
when they are outgrowths of the thin
membrane covering the articulation
between the appendages and the body,
and pleurobranchs (1d, Fig. 61) when
they arise from the side of the body
above the insertion of the appendage.
In order to increase the surface of the
gill, it is usually much folded or produced into a number of small processes.
In a lobster the gills are borne on the
sides of the cephalo-thorax, as the
fused head and thorax is called. There
are twenty on each side, and they are
protected from injury by a broad flap
called the branchiostegite, which
has grown down from the back and
formed a chamber between itself and
the side of the body; in this cavity the
gills lie concealed. At the front end of
this chamber lies a small paddle, which
is part of the second maxilla or scapho-

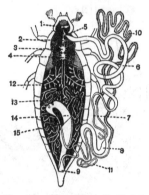

Fig. 62. View of male Cockchafer, *Melolontha vulgaris*,
from which the dorsal integument and heart have been removed to show the internal
organs. After Vogt and Yung.

1. Cerebral ganglion. 2. 1st
thoracic ganglion. 3. 2nd
and 3rd thoracic ganglia fused.
4. Fused abdominal ganglia.
5. Oesophagus. 6. Mid-gut.
7. Small intestine. 8. Colon.
9. Rectum. 10. Malpighian
tubules, brown portion with
caeca. 11. Malpighian tu-
bules, distal end. 12. Trachea
with vesicles. 13. Testes,
opening into coiled vasa deferentia. 14. Penis. 15. Single
vas deferens.

gnathite (Fig. 56 III), which throws out the water from the front
end of the gill chamber two or three times every second and thus
keeps a current of fresh water passing into this space behind. This
may be easily demonstrated by adding some coloured granules, such
as carmine or indian-ink, to the water in which lobsters are living.

Many of the Arthropoda breathe air and their respiratory
mechanism is very peculiar and unlike anything else in the animal
kingdom. Instead of the blood being taken to a gill or lung and

there purified and then driven with its oxygen all over the body to every organ and tissue, the air itself is introduced into the body and is carried by minute tubules to every tissue and cell (12, Fig. 62). Thus in these animals the blood has lost one of its main functions— the respiratory—and remains simply a nutritive fluid. The fine tubules through which the air travels are termed tracheae (Gr. τραχύς, τραχεῖα, rough, corrugated), and the three groups of Myriapods, Arachnids, and Insects are sometimes collectively termed the Tracheata, in contradistinction to the Crustacea, though there is reason to believe that the tracheae in Arachnida have arisen in a different way to those in the other two groups. The tracheae open to the exterior at certain definitely arranged pores termed stigmata,

usually found at the sides of the body. From these pores the tracheae pass inwards, dividing into smaller and smaller branches which ultimately end in the various tissues. Each trachea is really a pouch of skin tucked into the body, and hence is lined by chitin continuous with the exoskeleton covering the rest of the body. The tracheae are kept from collapsing by a thickened ridge of the chitinous lining which coils round inside the tube like a spiral of wire

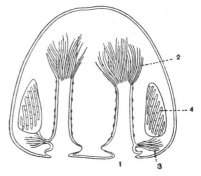

FIG. 63. Horizontal section through the abdomen of a Spider, *Argyroneta*. After MacLeod. Magnified.

1. Opening to exterior, tracheal stigma.
2. Terminal tracheae.   3. Lateral tracheae.   4. Lung books.

inside a water hose. Oxygen is thus absorbed by all the parts of the body directly from the air and not from the blood. This peculiar mode of respiration has had a profound influence on Insect structure.

A very primitive Arachnid, *Limulus*, the King-crab, which lives in the sea, breathes by means of what are called gill-books; these are piles of delicate leaf-like plates placed one over the other like the leaves of a book and attached to the posterior surface of the appendages of the hinder part of the body, which are flattened (Figs. 64 and 107). When these plates are moved up and down the leaves of the gill-books fly apart and the water gets in between them, and oxygen passes from it into the blood which circulates in the substance of the leaves. In the Scorpions, which are geologically the oldest and most primitive group of the Arachnids that live on

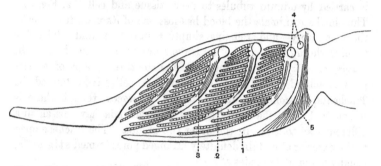

FIG. 64. Section through the operculum and gills of a King-crab, *Limulus*.
× about 16. The normal number of gills in a Limulus is five, the section from which this drawing is made showed only four.

1. Operculum.   2. 2nd gill-book.   3. Muscle which moves gills up and down.   4. Blood-vessels.   5. Muscle which raises the operculum.

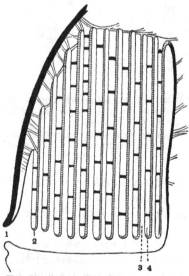

FIG. 65. Longitudinal section through the lung book of a Spider. Magnified. From MacLeod.

1. Opening to the exterior or stigma.
2. Free edge of the pulmonary leaves.
3. Space in which the air circulates.
4. Space in which the blood circulates.

land, the arrangement is the same, only the gill-books and the plate-like appendages which carry them are smaller, and the books are packed away into pits on the under side of the body, whilst the highly modified appendages extend horizontally below so as to floor in the pits, leaving only a slit through which air enters. This arrangement prevents the gill-books—now called lung-books—from drying up. In other Arachnids the gill or lung-book has been lost, and only the pit remains, and this is enlarged and burrows into the body, forming tracheae which may assume all the appearance of Insect tracheae. The Spiders form an interesting link, for some of their lung-books have been thus replaced by tracheae and others remain as in the Scorpion (Fig. 63). The tracheae of the Antennata have however developed in a different manner. In Peripatus, the oldest and most primitive member of the group, the stigmata or openings of the tracheae are scattered irregularly all over the surface of the body. Each leads into a short straight tube which ends in a bunch of diverging tracheae. In the Myriapoda there is usually only one pair to each segment, and the same is the case with the Insecta, but in them tracheae belonging to successive segments usually join so as to form longitudinal trunks which may even (as in the Flies) become swollen so as to form reservoirs of air. In some of the smaller Crustacea where the cuticle is thin the exchange of gases between the blood and the surrounding medium seems to take place all over the surface of the body. Another mode of exchange is the so-called anal respiration which is met with in many Phyllopods and Copepods, in *Gammarus* and *Asellus*, in the larvae of Decapods and in certain Insect larvae. In these animals the rhythmic contraction of the muscular walls of the rectum alternately pumps in and expels water carrying oxygen in and out of the anus. In the Insect larvae the walls of the rectum are richly supplied with tracheae.

The excretion of the nitrogenous products of katabolism from an animal's body is a function of fundamental importance. We have seen in the Earthworm that this is performed in a series of little tubes called nephridia, one pair in each segment, and the same is roughly true of Peripatus. But in the other Arthropods, with few exceptions, where the primitive segmentation is much changed and modified, such structures do not exist in each segment, but a single pair of excretory organs suffices for the whole body. In the simpler Crustacea each of these is a tube with glandular walls (i.e. walls composed of cells which are filled with excreta), and each tube opens at the base of the second maxilla, on each side. They are termed the maxillary or shell-glands (Figs. 69 and 70). In the larger and more complex forms corresponding organs open on the second antenna (40, Fig. 60). Here the organ consists of a network of parallel glandular tubes joining each other at intervals and opening into a thin bladder which communicates with the outside. The whole is called the antennary or green-gland. In a few cases both antennary and maxillary glands co-exist, at least for some time of the animal's life.

Nitrogenous excretion.

A typical nephridium has been already defined as a tube which opens at its inner end into the coelom. Since the excretory glands of Crustacea have in some cases at their inner ends thin-walled dilatations which are by most zoologists regarded as portions of the true coelom the Arthropoda may be said to possess modified nephridia.

In Insects and Myriapods the excretory system, like the breathing apparatus, is peculiar; the waste nitrogenous matter is taken up by certain tubules called Malpighian tubules, after a celebrated Italian anatomist named Malpighi (Figs. 62 and 83). These lie in the body cavity surrounded by the blood; they do not open directly into the exterior but into the front end of the proctodaeum, and through this their excreta leave the body.

In the Arachnids the excretory apparatus is of two kinds which may coexist. In Scorpions and Spiders there are Malpighian tubules superficially resembling those of Insects, only shorter and less numerous, but they open into the endodermic tube or mesenteron and are therefore endodermic, not ectodermic structures. There are also organs which, like the green- and shell-glands of the Crustacea, are to be regarded as modified nephridia, which open at the one end into a space which is a remnant of the coelom and at the other to the exterior; they are termed coxal glands, since they lie mainly in the coxal or proximal joints of the legs. The contrast between the ordinary earthworm with its numerous pairs of nephridia and the larger and more active crayfish with its single pair is a very striking one for which some explanation is sought.

Eisig experimenting on some of the marine Polychaeta discovered that the ectoderm performed part of the function of excretion. He fed the animals on a substance (indigo carmine) which was soluble and was consequently digested but which was got rid of by the excretory organs. This substance was found in the nephridia, and also in the ectoderm from which it was secreted into the cuticle and into the chaetae which, as has been shown (p. 91), are special developments of the cuticle. Hence we may conclude that in the Arthropoda with their enormous production of cuticle this function of the ectoderm has been so strengthened that nephridia have become superfluous. *Peripatus,* which alone retains nephridia in every segment, has indeed the thinnest cuticle of any Arthropod.

With few exceptions Arthropods are bisexual. The reproductive organs are comparatively simple. Both the ovary (Fig. 67) and the testis (Fig. 66) are continuous with their ducts, which, in the Crustacea, Arachnids and

Reproductive
Organs.

some Myriapods, usually open to the exterior on the under surface of the middle region of the body and at the posterior end of the body in Insects and one division of the Myriapods. The space

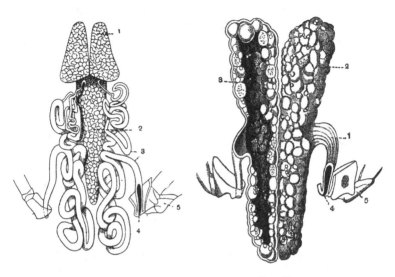

FIG. 66.   FIG. 67.

FIG. 66. Male reproductive organs of *Astacus fluviatilis* × about 2½. From Howes.   1. Right anterior lobe of testis.   2. Median posterior lobe of testis.   3. Vas deferens.   4. External opening of vas deferens.   5. Right fourth ambulatory leg in which the vas deferens opens.

FIG. 67. Female reproductive organs of *Astacus fluviatilis* × about 2. From Howes.   1. Right oviduct. The left oviduct is shown partly opened.   2. Right lobe of ovary.   3. Left lobe of ovary with the upper half removed to show the cavity of ovary or coelom into which the ripe ova drop.   4. External opening of oviduct.   5. Right second ambulatory leg on which the oviduct opens.

inside these glands, lined by the reproductive cells, is regarded as part of the coelomic cavity.

The foregoing account of the Arthropoda enables us to give the following definition of the group :—The Arthropods are bilateral animals with segmented bodies. The segments are not all alike and frequently fuse one with another ; some at least bear a pair of jointed limbs, of which those in the region of the mouth are modified to catch and bite the food. The nervous system consists of a supra-oesophageal mass or brain, a nervous ring round the oesophagus and a ventral chain of

Definition of the Group.

ganglia sometimes fused into a single mass. A heart is usually present above the alimentary canal and blood enters it through a series of paired valvular slits from the haemocoel or blood-cavity. The sexes are usually distinct. The coelom is much reduced.

This definition is in the main true of all Arthropods, whether insect, spider, centipede, or crab. We must now consider however how the various subdivisions of this great group may be distinguished one from another.

Class I. Crustacea.

The Crustacea are with few exceptions, such as the wood-louse, inhabitants of the water, and they breathe either through the general surface of the body or by means of gills. They have as a rule two pairs of antennae and these as well as their other jointed limbs are typically biramous, that is, they consist of a basal portion bearing two prolongations. They have at least three pairs of appendages converted into jaws.

The Crustacea are usually divided into two groups, the Entomostraca (Gr. ἔντομος, cut in pieces; ὄστρἄκον, a shell) and the Malacostraca (Gr. μἄλἄκός, supple); and each of these is again divided into four and three Orders respectively.

Sub-class A. Entomostraca.

This group may be regarded as a lumber-room for all Crustacea which are not included in the well-defined division Malacostraca, and the only character which can be attributed to all the members is that of not possessing the marks of Malacostraca.

For the most part they are small Crustacea of simple structure. The number of their segments varies within wide limits; some Ostracoda having only seven pairs of limbs, whilst in *Apus* there are sixty-eight pairs. The dorsal part of their head has, in many cases, grown backwards and downwards like a mantle to form a large hood or shell, termed the carapace, which may cover a large part of the body, and in some cases this becomes divided into two lateral halves hinged together like a mussel's shell. In many descriptions of Entomostraca the words "thorax" and "abdomen" are used to describe

VII.]   PHYLLOPODA.   143

regions of the body. Such terms are in strictness applicable only to the higher Crustacea, where the trunk is sharply differentiated into two regions distinguished by the character of their appendages. Amongst the Entomostraca however the appendages of the trunk form a uniform series : often it is true the last segments are devoid of appendages, and to these the term abdomen (16, Fig. 68) is usually applied, but to us this seems an unjustifiable and misleading use of a term which has an exact significance only amongst Malacostraca.

Entomostraca have no internal teeth in their stomach. As a rule the young are not like their parents but are larvae of a special kind called Nauplii; these after a number of ecdyses, during which the number of segments increases, grow up into adults.

The Nauplius possesses an oval, unsegmented body, a median simple eye, three pairs of appendages and a large upper lip. The first pair of limbs representing the first antennae of the adult are simple and unjointed, the other two pairs have a basal piece and two branches. The inner branch of one or both pairs has a hook for masticatory purposes. These two pairs of appendages become the second antennae and mandibles of the adult ; both are at first placed behind the mouth.

The Entomostraca consists of the following Orders :

Order I. Phyllopoda.

As the name implies the Phyllopoda (Gr. φύλλον, a leaf ; πούς, a foot) are characterized by possessing flattened leaf-like swimming limbs. Of these there are at least four pairs but there may be many more. The larger Phyllopods are not uncommon in Britain ; one genus, *Artemia*, taken at Lymington, flourishes in salt-pans in which the salt is so concentrated as to be fatal to other animals. *Branchipus* (Fig. 68) is devoid of the carapace and has an elongated heart extending throughout the body. It occurs in stagnant water, and has been recorded in several localities in the south of England. It is often found in the vicinity of Montreal, in Canada, in the pools of rain water which have accumulated in disused quarries. *Apus* is another of the larger forms which was formerly found in Britain but has not been met with for some years and is possibly now extinct in this country. It has a large carapace, and its flattened leaf-like appendages are regarded as primitive types of the Crustacean limb from which all the numerous modifications of the higher forms may be derived. Of these eleven pairs are situated in front of the genital opening and are often termed "thoracic," one pair being attached to each of the pre-genital segments of the trunk. Behind the genital opening there are fifty-two so-called "abdominal" pairs of legs, of which several pairs are attached to each post-genital segment except the last two or three.

The genera *Simocephalus* and *Daphnia*, common in ponds and ditches, both in England and America, differ from the foregoing in having fewer segments and in possessing a bivalved carapace which completely encloses the body. The first antennae, or, as they are generally called, the antennules, are small and simple, but the second antennae are very large and forked and project from the shell, and by their lashing movement carry the animal through the water (Figs. 69 and 70). The carapace is to a certain extent transparent, and through it the beating of the heart, the circulation of the blood and the movements of the thoracic leaf-like appendages may be made out. Within the substance of each valve of the carapace a coiled glandular tube may be detected ; this is the shell-gland or typical excretory organ of the Entomostraca which opens on to the exterior in the region of the second maxilla. The male (Fig. 69) is usually smaller than the female (Fig. 70), and is certainly very much rarer. The females lay two kinds of eggs, (i) unfertilized eggs, which develop in the space intervening between the dorsal side of the body and the shell which acts as a brood-pouch, and (ii) fertilized eggs, which are larger and become

Fig. 68. Dorsal view of female, *Branchipus* sp. found in a pond in Sussex × about 10.

1. Antennae. 2. Head. 3. Eyes. 4—14. The eleven "thoracic" limbs. 15. The caudal forks. 16. The fifth "abdominal" segment.

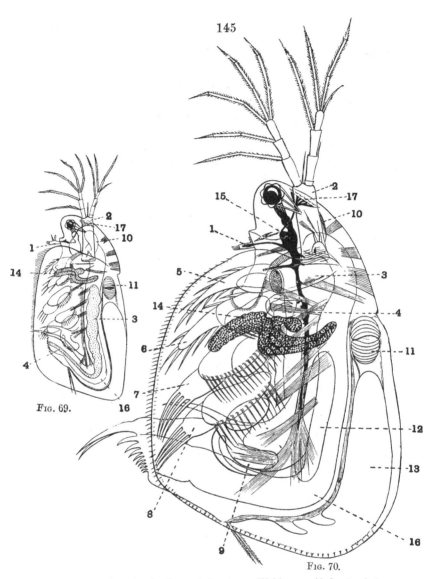

FIG. 69. Side view of male *Simocephalus sima*. Highly magnified. 1. Antennules. 2. Antennae. 3. Testis. 4. Vas deferens. 10. Hepatic diverticulum. 11. Heart. 14. Shell-gland. 16. Mid-gut. 17. Neck organ.

FIG. 70. Side view of female *Simocephalus sima*, magnified to the same extent as Fig. 69. From Cunnington. 1. Antennules. 2. Antennae. 3. Mandibles. 4. Maxillae. 5. 1st pair of legs. 6. 2nd pair of legs. 7. 3rd pair of legs. 8. 4th pair of legs. 9. 5th pair of legs. 10. Hepatic diverticulum. 11. Heart. 12. Ovary. 13. Brood-pouch. 14. Shell-gland. 15. Brain. 16. Mid-gut. 17. Neck organ.

surrounded by a special modification of the brood-pouch called the
ephippium. The nature of the eggs produced is regulated by
favourable or unfavourable conditions of life. At a suitable temperature and with a sufficiency of food and water, the unfertilized
eggs are produced in large numbers at short intervals. Periods of
drought or the cold of winter bring about the formation of eggs
which are fertilized and enclosed in the ephippium. Sheltered by
this case the eggs are enabled to withstand freezing or desiccation
and with a return of suitable conditions a young *Daphnia* hatches
out from each egg to continue the cycle of life.

The Phyllopoda are divided into two Sub-orders, viz. :—

Sub-order 1. **Branchiopoda.**

Long-bodied forms devoid of a brood-pouch and not using
the second antennae as swimming organs.

Ex. Apus, Artemia, Branchipus.

Sub-order 2. **Cladocera.**

Short-bodied Phyllopoda with a dorsal brood-pouch and
long second antennae.

Ex. Simocephalus, Daphnia.

The Branchiopoda live in fresh-water and as a rule in the stand-
ing water of pools and ponds; they are more rarely found in
brackish or salt water. They swim actively about by means of
the vibrations of their flattened limbs. As a rule aquatic animals
swim with the upper surface towards the surface of the water, but
the Branchiopoda seem very indifferent to this rule, and are quite
frequently seen swimming upside down. The Cladocera also frequently swim upside down, the genus *Daphnia* however usually in
a vertical position with the head uppermost.

Order II. Ostracoda.

This Order (Gr. ὀστρακώδης, shell-like) contains a great number
of species which do not differ greatly from one another. In form
they resemble *Daphnia*, but the head does not protrude from
between the valves of the carapace, and some of the internal organs
of the body, viz., the ovary or testis and branches of the liver, are
prolonged into the valves of the carapace. This latter is a very
characteristic structure, consisting, like the shell of the Mussel, of
two valves. It opens by an elastic ligament which tends to pull

the valves apart, and it closes by the contraction of a muscle which runs across the body from one valve to another. The whole body is included in the carapace, antennae and all.

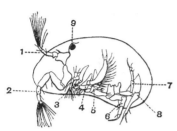

FIG. 71. Lateral view of *Cypris candida*. After Zenker.

1. Antennules.   2. Antennae.
3. Mandibles.   4. 1st maxillae.
5. 2nd maxillae.   6. 1st pair of legs.   7. 2nd pair of legs.
8. Tail.   9. Eye.

Ostracoda have fewer appendages than any other group of Crustacea ; besides the antennules, antennae, mandibles and two pairs of maxillae, they possess only two pairs of limbs, and these are stout and cylindrical in marked contrast to the appendages of the Phyllopoda. The hinder part of the body is rudimentary.

Two pairs of excretory organs have been described in some species of Ostracoda, the shell-glands common to all Entomostraca opening at the base of the second maxillae and a pair of antennary glands opening at the base of the second antennae. The last named are seldom found except in the Malacostraca.

Both pairs of antennae are jointed, unbranched appendages, which are used in swimming, another most important distinction from *Daphnia* and its allies.

The males differ from the females, which either (*Cypris*) lay their eggs on water-plants or (*Cypridina*) carry them about within their shells. The majority of species are found in the sea but others occur in fresh-water. They are flesh eaters, and as they exist in great numbers they fulfil the important duty of scavenging on a small scale, and thus they prevent the accumulation of dead organic matter in the water.

Order III. Copepoda.

This Order (Gr. κώπη, oar ; πούς, foot) is also a large one and its free swimming members exhibit a very characteristic structure and appearance. The body is of an elongated pear shape, and consists of a large round head and a tapering trunk of comparatively few segments. The carapace so characteristic of the preceding order is entirely absent. The head bears a single median eye in front, the lateral compound eyes so conspicuous in most Crustacea being

absent. Attached to the head are
five pairs of appendages, two pairs
of unbranched antennae, a pair of
mandibles and two pairs of ·max-
illae. The head is not separated
from the trunk by any constriction.
The latter bears four pairs of swimming feet of a typical forked pattern.
Each of these appendages is something like a ⋏. The base of the
limb consists of one or two joints
and is called the protopodite. It
splits at its free end into two
prolongations, the inner of which
is known as the endopodite
and the outer as the exopodite.
Both are flattened, consisting of
stout joints each of which bears
spines, and the whole forms a
convenient paddle. Each limb
is also joined to its fellow of the
opposite side by a transverse
moveable ridge so that the right
cannot move without the left.
By the simultaneous action of all
the limbs of the trunk the animal
is enabled to execute a series of
swift darts through the water;
by the action of the second antennae a slow, gliding movement
is carried out, whilst the maxillipedes by sweeping movements
search the water for food. A
forked limb is characteristic of
the Crustacea, and is not met
with in other groups of the
Arthropods. It appears over and
over again in all the Orders,
retaining its primitive form in
some instances, as in the abdominal appendages of a Cray-

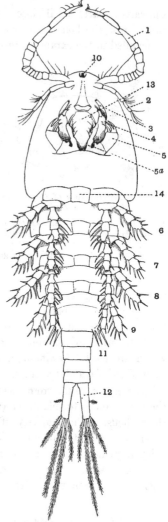

Fig. 72. Ventral view of male *Cyclops*
sp. Magnified.
1. Antennule.   2. Antenna.
3. Mandible.   4. 1st maxilla.
5. The two halves of the 2nd
maxillae sometimes called inner
and outer maxillipedes. 6–9. 1st–
4th thoracic limbs.   10.   Eye.
11. Bristles near male generative
opening.   12. Caudal fork.
13. Mouth.   14. Copula or
plate connecting the right and
left limb of each pair.

fish, but more often by the suppression of one part (usually the exopodite) and by the development and modification of others, the original form becomes masked and difficult to recognise. When

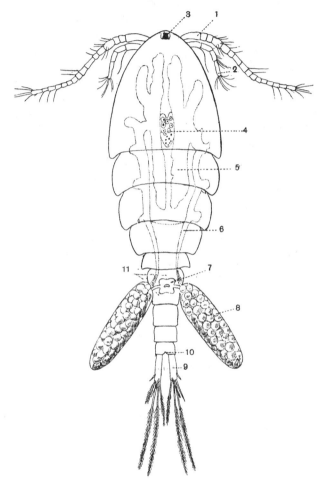

FIG. 73.   Dorsal view of female *Cyclops* sp.   Magnified.   Partly after Hartog.

1. 1st Antenna.   2. 2nd Antenna.   3. Eye.   4. Ovary.   5. Uterus.
6. Oviduct.   7. Spermatheca or pouch for receiving the spermatozoa
of the male.   8. Egg-sacs.   9. Caudal fork.   10. Position of anus.
11. Compound segment, consisting of the last thoracic (bearing the genital
opening) and the first abdominal.

both forks are conspicuously developed the limb is said to be biramous. The four or five last segments of the Copepod's body bear no appendages. The last is produced into two processes, forming a caudal- or tail-fork.

The sexes in the free-living species are not markedly different, but if we examine specimens of such a genus as *Cyclops*, which is common in our fresh-water pools, we shall find that in the breeding season the female carries about with her two egg-sacs (Fig. 73). These are attached to her body just behind the last pair of appendages and project freely at the side. Each egg-sac may contain four or five dozen eggs which are glued together by a cement-substance. Such egg-sacs are very characteristic of the Copepoda and are found even in the parasitic members of the order. Most of the latter live on fish, and some have acquired the name of "fish-lice." Their mouth appendages have lost their biting function and have become adapted for piercing the tissues of the host on which they live. Their segmentation is suppressed and their appendages are reduced and the body has grown out into all sorts of curious processes. The male is often much smaller than the female and as a rule retains the crustacean characters more than she does. Occasionally they are found on the skin of a fish, but more often they occur in the mouth and on the gills, sometimes half and sometimes almost wholly embedded in the flesh of their host.

Order IV. Cirripedia.

Some of the Copepods have become so modified by their parasitic habits that unless we were able to trace their development, including the larval forms through which they pass before becoming adult, we should have difficulty in assigning them to their proper place amongst the Crustacea. A somewhat similar modification occurs in the Cirripedia (Lat. *cirrus*, a tuft of hair ; *pes*, a foot) and is associated with a fixed or sessile habit of life. After passing through a variety of free-swimming larval forms the animal comes to rest and attaches itself by the anterior end of the body to a stone or rock, the bottom of a ship, or some other object submerged in the sea, and then becomes adult.

Like that of the Ostracods the body of a Cirripede is enclosed in a carapace consisting of two valve-like folds which have grown out from the region of the head, but these are usually strengthened by five calcareous plates, a right and left scutum, a right and left tergum and a median carina, and in *Balanus*, the common acorn-barnacle of our sea-shores, a further armour of triangular plates developes in an additional outer fold of skin which encircles the body. In *Lepas*, the barnacle usually found in clusters on the bottom of ships, which often seriously impedes their progress, this ring is absent, but the anterior end of the head bearing the first antennae at its end has grown out into a long stalk which lodges

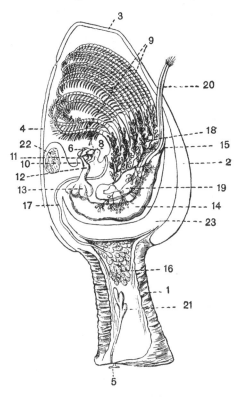

Fig. 74. A view of *Lepas anatifera*, cut open longitudinally to show the disposition of the organs. From Leuckart and Nitsche, partly after Claus.

1. Stalk.   2. Carina.   3. Tergum.   4. Scutum.   5. 1st antennae.
6. Mandible with "palp" in front.   7. 1st maxilla.   8. 2nd maxilla.
9. The six pairs of biramous thoracic limbs.   10. Labrum.   11. Mouth.
12. Oesophagus.   13. Liver.   14. Intestine.   15. Anus.   16. Ovary.
17. Oviduct.   18. Testes.   19. Vas deferens.   20. Penis.
21. Cement gland and duct.   22. Adductor scutorum muscle, which closes the carapace.   23. Mantle cavity, i.e., the space intervening between the carapace and the body.

some of the internal organs of the body. The second antennae though present in the larvae are lost in the adult. The rest of the body is enclosed within the carapace. Around the mouth are a pair of mandibles and two pairs of maxillae, and the thorax carries six pairs of biramous many-jointed limbs beset with numerous hair-like spines, the lashing of which kicks food particles towards the mouth (Fig. 74). These limbs are slender and flexible and thus differ from the corresponding limbs of Copepods.

Like some Copepods the Cirripedes are without a heart, and the existence of special respiratory organs is doubtful. Unlike other Crustacea they are, as a rule, hermaphrodite, the male and female reproductive organs being united in one individual. A few species are parasitic, chiefly on other Crustacea, and these have reached a very extreme stage of degeneration.

Sub-class B.   MALACOSTRACA.

The second of the two large groups into which the Crustacea are divided contains most of the more familiar forms, such as Crabs, Lobsters, Shrimps, Wood-lice, etc. For the most part the Malac-ostraca are larger than the Entomostraca and the number of their segments is a fixed one. With the exception of the first Order, Leptostraca, which is really a connecting-link between true Malac-ostraca and the lower forms, this number is nineteen and there are nineteen pairs of appendages. One of the most marked characters in the Malacostraca is the differentiation of the trunk into two distinct regions, the thorax and the abdomen. It is true, as is mentioned above, that many authors speak of an abdomen in the Entomostraca, but by this they mean with a few exceptions the hindermost segments which are devoid of limbs. In any En-tomostracan if we examine the series of limbs behind the jaws we shall find that they constitute a continuous series without any sudden change in their character. In a Malacostracan, on the other hand, we find an abrupt change at one point in the character of the limbs. The hinder limbs or swimmerets (pleopods) are markedly different from the front limbs, for whereas in the swim-meret both forks of the limb, endopodite and exopodite, are equally developed, in the last five limbs or the thorax (peraeopods) the endopodite is large and the exopodite is small or absent. It is this difference in character which defines the abdomen.

Although the division between the head and thorax is not always apparent, as a rule we may assign five segments to the head, eight to the thorax and six to the abdomen, which ends in an unseg-mented flap called the telson. The reason for this want of a definite boundary between head and thorax in the Malacostraca is that the carapace, which as we have seen is an outgrowth of the head, has become fused with the dorsal surface of the thoracic segments, whilst at the sides it forms freely projecting flaps, which since they cover the gills are known as branchiostegites (Gr. στέγω, to cover).

The excretory organ of the Malacostraca opens at the base of the second antennae and not as in the Entomostraca on the second maxilla. As a rule the typical larva—the Nauplius—of the last-named group is not present in the life-history of the Malacostraca, which may hatch out from the egg in a practically adult condition or may pass through several larval stages, the first of which is the Zoaea, a larva with many appendages, possessing eyes and in all ways more differentiated than the Nauplius.

Order I. Leptostraca.

The order Leptostraca (Gr. λεπτός, slight, small) contains but three genera, which are interesting because they form an intermediate stage between the Malacostraca and the Entomostraca. Like many of the latter they are provided with a bi-valve carapace which, unlike that of all other Malacostraca, is not fused with the thoracic segments. Behind the six appendage-bearing segments of the abdomen there come two more segments without limbs and the hindermost bears two diverging filaments constituting a "caudal fork," such as is commonly found amongst the Entomostraca. The thoracic limbs are flattened and leaf-like, as in *Apus*, but the mandible bears a three-jointed feeler or palp and the eyes are stalked,—both, on the whole, Malacostracan characters. The excretory organ opens on the second antenna, but in the larva the shell-gland or maxillary excretory organ is found and traces of it exist in the adult.

The order is marine, and very widely distributed throughout the ocean. Its members are capable of living and thriving in very foul water, so foul as to be fatal to most other animals. *Nebalia* is the best known genus.

Order II. Thoracostraca.

The Thoracostraca (Gr. θώραξ, a breast-plate) form a large group and contain many very different forms. They are placed together because the carapace has become fused with several of the thoracic segments, so as to form a region known as the cephalo-thorax, the exoskeleton covering which is not jointed and is not bivalved as in *Nebalia*.   The eyes of the Thoracostraca are compound and almost always are borne on moveable stalks.   The order is divided into four sub-orders.

Sub-order 1. Schizopoda.

This sub-order includes the lowest of the Thoracostraca.   The name is suggested by the circumstance that all the eight pairs of thoracic limbs are biramous; the first and sometimes the second pair are reduced in size and provided with gnathobases; they assist the manibles and maxillae and hence are termed maxillipedes.   It will occur to most observers that the thoracic feet of the Schizopod resemble the ordinary form of swimmeret or abdominal appendages

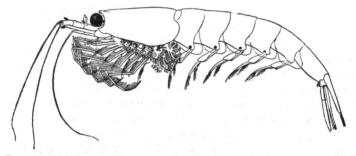

FIG. 75. *Nyctiphanes norwegica*, a Schizopod. Slightly magnified. From Watasé. The black dots indicate the phosphorescent organs. The gills are seen between the cephalothoracic and the abdominal appendages.

in the more familiar Lobster or Crayfish.   This is so; the swim-merets of a Schizopod are however sharply distinguished from the thoracic limbs by their smaller size.   It appears probable that the first step in the evolution of an abdomen was the reduction in size of the appendages so as to transform the hinder part of the body into a powerful swimming fin, and many Schizopods only use the abdomen in this way, since most of the swimmerets are very small and appear to be practically functionless.   The last one however is broad and assists the tail in its vigorous strokes. Some Schizopods have a series of phosphorescent organs which under certain conditions emit a pale but very perceptible light like that of a glow-worm. This light seems to be controlled by the animal but its use is not very clear.

There are very interesting differences amongst the genera composing the Schizopoda. The genus *Euphausia* for instance has long feathery gills attached to the basal joints of the thoracic legs and the eggs are not borne about by the mother but hatch out into Nauplii, which pass through a series of metamorphoses before becoming adult. In *Mysis* on the other hand the gills are few and simple and the eggs are borne under the thorax on flat plates termed oostegites, which project inwards from the hinder thoracic appendages. In these two genera we see the beginning of two tendencies which have led the descendants of primitive Schizopoda to differentiate themselves in two different directions. One group have taken to carrying the embryos about until they are fully developed; at the same time the gills are reduced and the carapace, which is essentially a gill-cover, tends to disappear. This group includes the Stomatopoda, Cumacea and Arthrostraca.

In the other group the gills and carapace are retained, and though the eggs are for a time carried about attached to the swimmerets the young one passes through a larval stage before becoming adult. This group includes the Decapoda.

Sub-order 2. **Decapoda.**

The Decapoda (Gr. δέκα, ten) derive their name from the circumstance that the first three pairs of thoracic appendages have become maxillipedes, that is to say have been modified so as to assist in mastication, leaving five pairs of large conspicuous limbs which have lost all trace of an exopodite, for prehension and locomotion.

This group includes the lobsters, cray-fish, shrimps, prawns, hermit-crabs and crabs, etc. In the division of the crabs, the Brachyura (Gr. βρᾰχύς, short; οὐρά, tail), the abdomen is reduced in size and turned up and closely applied to the under surface of the thorax, except when the animal is "in berry" and then the masses of eggs force the abdomen away from the thorax. As a rule crabs are broader than they are long and the breadth is partly due to the large gill chambers on each side of the body. The gills are really outside the body, but are in a special chamber bounded by the branchiostegite or free edge of the carapace.

The Anomura are in some respects intermediate between the foregoing and the following divisions. As in the crabs the abdomen is folded somewhat forwards but the tail-fin is not so much reduced. The last pair or last two pairs of the thoracic limbs are reduced and turned dorsalwards. Some species—the hermit-crabs—

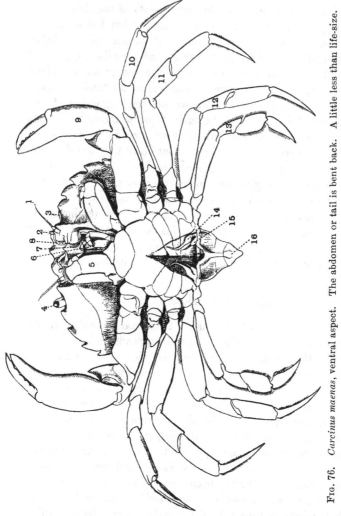

Fig. 76. *Carcinus maenas*, ventral aspect. The abdomen or tail is bent back. A little less than life-size. 1. Antenna. 2. Antennule. 3. Left eye reclining in its socket. 4. Right eye erected. 5. Right 3rd maxilliped in natural closed condition, its fellow is bent back to expose 6. 6. The mandible. 7. 1st maxilliped. 8. 2nd maxilliped. 9. Chela or great claw. 10, 11, 12 and 13. 1st, 2nd, 3rd and 4th ambulatory limbs. 14 and 15. 1st and 2nd abdominal legs which are modified in connexion with reproduction. 16. Position of anus.

shelter themselves in the empty shells of molluscs. In these cases the abdomen does not develope a hard covering as the animal is sufficiently protected by its lodging. It remains soft and acquires a spiral twist as it moulds itself to the interior of its borrowed shell.

In the lobsters, shrimps, etc., which form the division Macrura (Gr. μακρός, long ; ουρά, tail), the tail is relatively large and is not folded up against the thorax.

Some Decapods have left the sea and taken to live on land and this has in some cases involved a change of structure, the gills which breathe water being supplemented by the soft vascular lining to the gill-cavity covered by the branchiostegite which, like a lung, breathes air.

The group is for the most part marine, though the well-known fresh-water cray-fishes, *Astacus* in Europe and *Cambarus* in North America, form striking exceptions.

Sub-order 3. Stomatopoda.

The Stomatopoda (Gr. στόμα, a mouth) are a sharply defined group with few genera, and may be regarded as an offshoot from the primitive Schizopoda peculiarly specialized. The members attain a considerable size, some eight inches or more in length. The carapace is small and only covers the anterior five thoracic segments ; the appendages of these segments are turned forward towards the mouth and take part in feeding, and so are termed maxillipeds. They end in a claw, the last joint shutting down on the penultimate one like a knife blade into its handle. They are thus very different from the maxillipeds of Schizopoda or Decapoda, which are really limbs on the way to become jaws and have developed gnathobases. The maxillipeds of Stomatopoda are grasping, not chewing organs, and have undergone the same modification as the great claw of the lobster, which might just as reasonably be called a maxilliped. A fertile source of confusion in the study of the Arthropoda is the use of names like maxillipede, thorax, abdomen, etc., to denote different things in different groups. The last three thoracic limbs are for walking ; they are very feeble and retain a rudiment of the outer fork or exopodite. The abdomen is large and bears six pairs of flattened swimming limbs, each of which carries a gill in its outer branch.

Unlike most other Crustacea, *Squilla* and the other members of the group do not carry their eggs about with them, but lay them in the burrows in which they live, and by sitting over them and moving their abdominal limbs they keep up a current of water which aërates the eggs. They are exclusively marine and live buried in the sand or hidden in crevices of the rock. They move actively and are difficult to catch.

Sub-order 4. Cumacea.

The members of the sub-order Cumacea (Gr. κῦμα, a wave, or billow) are mostly small; they live in the sea on sandy bottoms at considerable depths, but come to the surface at night. They are especially interesting because to a certain extent they are intermediate in character between the Thoracostraca and the Arthrostraca. Thus their paired eyes are not stalked and are sometimes fused together to form a single eye; the carapace is reduced so as to leave several segments of the thorax uncovered (Fig. 77) and on some of the thoracic legs there is a small exopodite. They

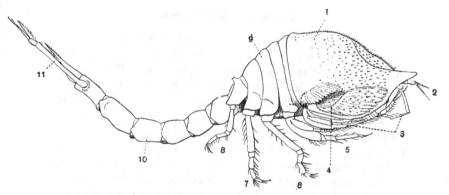

Fig. 77. Female *Diastylis stygia* × 6. After Sars. The carapace is represented as transparent to show the gill.

1. Carapace.   2. First antenna.   3. First leg.   4. Gill borne on first maxilliped.   5, 6, 7 and 8, second to fifth leg.   9. Free part of thorax.   10. Abdomen.   11. Appendage of the last segment of the abdomen.

have a single pair of gills borne by the first thoracic limb. In the female the abdomen, which is long, has lost all the limbs except the last pair.

Order III.   Arthrostraca.

The members of the Arthrostraca (Gr. ἄρθρον, a joint; ὄστρακον, a shell) have sessile eyes, i.e., without stalks. The carapace, which in most of the above-mentioned groups covers the segments of the thorax, is absent, and consequently seven of the latter are usually freely moveable on one another, the first and in rare cases the second thoracic segment remaining immoveably fused with the head.   They thus represent a further stage in the same process which we found going on in Cumacea and Stomatopoda.   Only one of the thoracic appendages is modified so as to form a maxilliped; there are consequently seven pairs of walking legs attached to the thorax.   In the female these legs bear inwardly directed processes which together form a brood pouch in which the eggs develope.

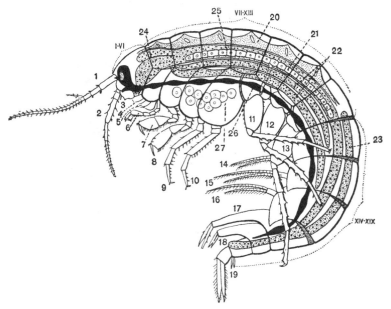

Fig. 78. *Gammarus neglectus.* Female bearing eggs seen in profile. From Leuckart and Nitsche, after G. O. Sars.

i–vi. Cephalothorax.   vii–xiii. Free thoracic segments.   xiv–xix. The six abdominal segments.   1. Anterior antenna.   2. Posterior antenna. 3. Mandibles.   4. 1st maxilla.   5. 2nd maxilla.   6. Maxilliped. 7–13. Thoracic limbs.   14—16. Three anterior abdominal limbs for swimming.   17—19. Three posterior abdominal limbs for jumping. 20. Heart with six pairs of ostia.   21. Ovary.   22. Hepatic diverticula.   23. Posterior diverticula of the alimentary canal.   24. Median dorsal diverticulum.   25. Alimentary canal.   26. Nervous system. 27. Ova in egg pouch, formed from lamellae on the coxae of the second, third and fourth thoracic limbs.

The Arthrostraca are mostly small animals, living in either salt or fresh water; they assume very different forms, some of them having a rudimentary abdomen.   They are divided into the sub-orders Amphipoda (Gr. ἀμφί, on both ends ; πόδα, feet), which are for the most part compressed or flattened from side to side

(Fig. 78) and carry their gills on their thoracic appendages, and the Isopoda (Gr. ἴσος, equal; πόδα, feet), which are depressed or flattened from above downwards (Figs. 57 and 79), and whose gills are the modified endopodites of the appendages of the abdomen. A typical example of the last named sub-order is the Hog-water Louse, *Asellus aquaticus* (Fig. 57), common in our ponds and streams, but many of the groups are parasitic and lose most of their characteristic Crustacean features. As in the case of the Decapods some genera of Isopoda have forsaken the sea for a life on land, amongst which the wood-lice, *Oniscus* and *Porcellio*, exhibit certain peculiarities usually associated with Insects; thus the mandible has no palp, one pair of antennae is usually lost, and there are certain tubular air passages believed to be respiratory burrowed in the abdominal endopodites which recall by their structure the tracheae of air-breathing Arthropods. This is another proof that any attempt to group together all animals possessing tracheae leads to absurdities.

FIG. 79. A Wood-louse, *Porcellio scaber* × about 2. From Cuvier.

Class II.   ANTENNATA.

Sub-class A.   PROTOTRACHEATA.

A short account of the genus *Peripatus* must be given, as this animal is of a very primitive nature and both its adult structure and the mode of its development throws much light upon the origin and anatomy of Myriapods and Insects and indeed on the Arthropods generally.

The different species of *Peripatus* are differently coloured, but they mostly possess a beautiful velvety coat. In shape they resemble caterpillars but carry two large antennae on their heads, and at the base of each antenna is an eye. On the under surface of the head is the mouth and tucked into it on each side is a toothed jaw. This is an appendage which has been modified so as to form a true gnathite. At each side of the mouth is a third pair of appendages, the oral papillae, from the tips of which a sticky slime can be ejected which entangles the insects and spiders on which the animal lives. The other appendages, which vary in number in the different species from seventeen pairs to over forty, have the form of soft cylindrical papillae ending in two claws and function as walking legs (Fig. 80). The anus is posterior, and at the base of each leg is a slit-like pore, the opening of a nephridium. The genital pore is in front of the anus.

FIG. 80.   *Peripatus capensis* × very slightly.   From Sedgwick.

The body cavity is a spacious haemocoel divided into three longitudinal compartments by two bands of muscles which run from its outer upper angle towards the middle ventral line. The lateral compartments are continuous with the cavities of the limbs and lodge the nephridia, the salivary-glands and the nervous system. The alimentary canal, slime-glands and generative organs lie in the middle compartment.

The mouth leads into a large muscular pharynx, such as is found in many Chaetopods. The salivary glands open near this. They are interesting structures, as development has shown that in origin they are derived from nephridia, a state of things which recalls the fact that in certain Oligochaets some of the nephridia open into the oesophagus. The pharynx leads by a short oesophagus into a roomy endodermic stomach which reaches back nearly to the anus, a short proctodaeum only being interposed (Fig. 81).

The structure of the heart and pericardium closely resembles that of the same organs in Myriapods and Insects.

The animal breathes by bunches of tracheae or short tubes which pass from the exterior into the tissues and convey air. Their external openings or stigmata are partly in two rows above and between the legs and partly scattered irregularly.

At the base of each leg is a nephridium which ends internally in a vesicle. Embryological research has shown that this vesicle is a remnant of the true coelom which is spacious in the embryo, but becomes displaced as development proceeds by the haemocoel. Enclosed in the proximal part of the leg there is a gland called the crural gland.

Peripatus is bisexual, and again embryology has demonstrated that the cavity of the sexual organs is coelomic. The male deposits its spermatozoa in packets in the body of the female. It is not known how they reach the ova but they are usually found in the ovary and possibly bore their way through the tissues, as they do in some leeches. The ducts of the reproductive organs are believed to be modified nephridia.

The nervous system consists of a brain and two ventral cords, which however do not approach one another but lie wide apart. They are connected by nine or ten transverse commissures in each segment (Fig. 81). Posteriorly the two ventral nerve-cords fuse above the proctodaeum, an arrangement which recalls what occurs in certain primitive Mollusca.

There are many species of *Peripatus*, which are by some authorities grouped into three or four genera. They are found in widely separated parts of the world and afford, as is often the case with archaic animals, an excellent example of "discontinuous distribution." They have been found in South America and the West Indies, in South Africa, in Australia, New Zealand and in some of the Islands of the Malay Archipelago and in Lower Siam.

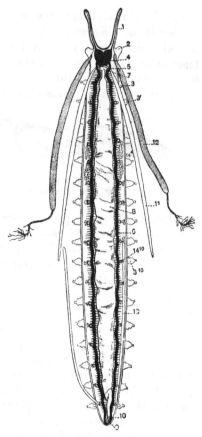

Fig. 81. *Peripatus capensis*, male, dissected to show the internal organs × 2. After Balfour.

1. Antennae, showing antennary nerve. 2. Oral papilla. 3, 3', 3¹⁰. 1st, 2nd and 10th leg of right side. 4. Brain and eyes. 5. Circum-oesophageal cord. 6. Ventral nerve-cord of right side, showing the transverse commissures. 7. Pharynx. 8. Stomach. 9. Anus. 10. Male generative opening. 11. Salivary glands. 12. Slime glands and reservoir. 13. Enlarged crural gland of the 17th leg. 14⁴, 14¹⁰. 4th and 10th nephridia of right side.

The development of *Peripatus* first definitely solved the problem of the nature of the various spaces in the Arthropod body and has also thrown much light on some of the peculiarities of insect embryology. Some species lay eggs and some produce living young, and thus the genus affords favourable opportunities for testing theories as to the effect of development within the body of the parent on the embryology of the offspring.

That the animal is a most interesting " missing link " becomes evident if we attempt to sum up the Annelidan and the Arthropodan features of its anatomy. Thus *Peripatus* resembles Annelida in the nervous system, the muscular pharynx, the structure of the eyes, the serially repeated nephridia, the shortness of the stomodaeum and of the proctodaeum, the thinness of the cuticle and the hollow nature of the paired appendages ; but in the indications of joints in the appendages, the reduction in size of the coelomic spaces, the presence of a wide haemocoel and of tracheae, the nature of the antennae and of the heart and pericardium, the position of the genital pore and the presence of true gnathites, *Peripatus* approaches the Myriapods and Insects.

In habits these animals are shy and inconspicuous, hiding under bark or stones and preferring a moist surrounding. They avoid the light and move with deliberation, testing the ground as they advance with their antennae.

Sub-class B. MYRIAPODA.

The Myriapoda (Gr. μυρίος, countless) are characterised by the possession of a head distinct from the rest of the body, bearing antennae, mandibles and one pair of maxillae, followed by a large number of segments bearing simple leg-like appendages.

Compared with the Crustacea or Insecta the group is a small one, yet it contains some thousands of species which, if we except a few small families, fall readily into two subdivisions (I) *Chilopoda*, and (II) *Diplopoda*. The subdivisions differ markedly from one another especially in the position of their reproductive openings which in the *Diplopoda* are on the third segment behind the head and in the *Chilopoda* are terminal. For this reason some naturalists break up the sub-class and associate the *Chilopoda* with the Insecta.

11—2

Order I. Chilopoda.

The very active, lithe, chestnut-brown, rather fierce-looking little centipede, *Lithobius forficatus*, which is very common during the summer months under the bark of old trees, under leaves and other rubbish, is a good example of the *Chilopoda* (Gr. χίλιοι, a thousand) or Centipedes. In the winter it buries itself in the soil. The female lays her eggs from June till August and hastens to cover each with a thin layer of earth; otherwise the egg is seized and devoured by her mate.

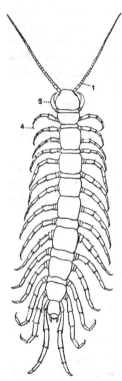

FIG. 82. A Centipede, *Lithobius forficatus*. Dorsal aspect × 12.

1. Antennae. 2. Poison claws (5th pair of appendages). 4. First pair of walking legs.

If we examine a little more closely one of these Centipedes, we shall see that the body is divided into a head followed by a very narrow segment and then by fifteen other segments of varying size. The head bears a pair of long antennae, the first appendages, which are constantly waving about. Close behind the point of origin of these antennae lie the eyes. If we turn the animal over and observe the under surface of the head we shall at once see a pair of large vicious-looking claws—the poison claws or fifth pair of appendages. The tip of each of these is pierced and, as it strikes the prey, a drop of poison is squeezed out which soon kills any insect or larva the Centipede wishes to eat. Although the tips of the poison claws are turned forwards beneath the head, yet these appendages really spring from the first segment of the trunk, which is enlarged and is known as the basilar segment. If we separate with a pair of mounted needles these poison claws we shall see attached to the head the fourth pair of appendages, sometimes called the second maxillae, though they resemble legs and have undergone little modification. They are reduced in size and each has a blunt functionless gnathobase. In front of these the third pair of appendages or first maxillae take the form of a

lobed plate, being united with one another in the middle line.
These cover in their turn the second pair of appendages or mandibles which, like those of Insects and unlike those of Crustacea,
consist simply of the blade, there being no palp or feeler.

The fifteen segments which succeed the one carrying the poison
claws each bear a pair of seven-jointed running legs ending in
a pair of claws.

Near the base of some of the legs—not of all—in the soft skin
uniting the hard dorsal and ventral plates of chitin,
Internal
structure.
is an oval opening. This leads into a chamber from
which the tracheae pass off. These tubes divide and
subdivide into smaller tubes, which run all over the body and traverse
all the tissues, entering even the smallest cell. They are lined with
a cuticle of chitin and are kept from collapsing by the presence of a
fine spiral thickening of this cuticle which gives them a very characteristic appearance when seen through a microscope. They are full
of air and constitute the respiratory apparatus of the Centipedes.

The existence of such a breathing apparatus, which is confined
to Peripatus, the Myriapods, the Insects and certain of the
Arachnids, is associated with certain striking features in the internal anatomy of the body. In most animals the oxygen of the
air is taken up by the blood and carbon dioxide is given out at
certain fixed points called gills or lungs, and the vascular system is
arranged so as to drive the blood through these specialized respiratory organs. The blood takes the oxygen to the various tissues
and takes from them the carbon dioxide which is removed from the
body at the same centres. In the Tracheata however the air is
itself conveyed by means of the tracheae to all the cells of the body
and the gaseous exchange takes place on the spot. The blood has
in the Tracheata lost one of its chief functions, the respiratory one,
and exists chiefly as a nutritive fluid bathing the alimentary canal
and taking up from it the soluble food which it conveys to the other
tissues. It is kept in circulation by a contractile heart which lies
along the middle dorsal line of the animal. In *Lithobius* this heart
has a pair of ostia or openings in each segment, into which the
blood from the pericardium pours, only to be sent out of the heart
again at its anterior end into the general cavity of the body, for
here the heart has an opening and there is no system of smaller
vessels or capillaries.

One of the peculiarities associated with the above-mentioned
method of breathing is the nature of the excretory organs which rid the

body of its nitrogenous waste. In *Peripatus* we find more or less typical nephridia and we meet with modifications of these in the coxal-glands of some Arachnids, but in the Myriapoda and in the Insecta these organs are wanting and their place is taken by certain outgrowths from the proctodaeum, called after an Italian anatomist Malpighian tubules. In *Lithobius* there are two such tubules, blind at their free end, and at the other opening near the hind end of the alimentary canal (6, Fig. 83). Their walls contain traces of uric acid and urates which they have taken up from the blood and which they presumably excrete through the alimentary canal.

The last-named organ is a straight tube which runs from one end of the body to the other. A pair of salivary glands pour their secretion into it near the mouth, but no other digestive glands exist. *Lithobius* is carnivorous, living chiefly upon insects, their larvae and on earthworms.

A large part of the space in the head is occupied by the bilobed brain which supplies the antennae and the mouth appendages. This brain is con-

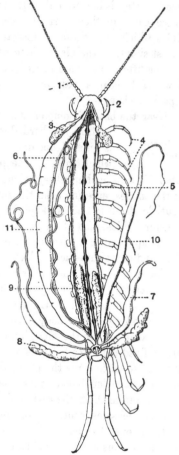

FIG. 83. *Lithobius forficatus*, dissected to show internal organs × about 2. After Vogt and Yung.

1. Antenna.   2. Poison claw.
3. Salivary gland.   4. Walking legs.
5. Ventral nerve-cord.   6. Malpighian tubule.   7. Vesicula seminalis.   8. Small accessory gland.
9. Large accessory gland.   10. Unpaired testis.   11. Alimentary canal.

nected by means of para-oesophageal commissures with a long ganglionated ventral cord which supplies nerves to the legs and the rest of the body (Fig. 83).

Myriapods are bisexual, the ovary and testis are continuous with

their ducts which open to the exterior on the ventral surface of the
last segment.

Order II. Diplopoda.

The second large subdivision of the Myriapoda is well illustrated
by the black " wire-worm," *Iulus terrestris*, very commonly found in
Great Britain curled up under stones or burrowing in the soil, where
it is said to do much damage by gnawing the tender roots of
plants, for all Diplopods (Gr. διπλόος, double) are vegetarians.

The wire-worm is a black, shiny cylindrical animal with an
enormous number of legs, in spite of which its movements are much
slower than are those of *Lithobius*.   The terga or dorsal shields are
in this sub-group very much enlarged, whilst the sterna or ventral
shields are very much reduced, and thus it comes about that the
bases of each pair of legs, instead of being separated by the width
of the body, are close together.   Another peculiarity in this subdivision is that each tergum corresponds with two segments and
that each apparent segment bears two pairs of legs and has the
internal organs also duplicated.   This double arrangement however
only begins at the fifth segment behind the head.

Fig. 84. *Iulus terrestris*, sometimes called the " Wire-worm."   From Koch
× about 3½.

1. Antennae.   2. Eyes.   3. Legs.   4. Pores for the escape of the
excretion of the stink-glands.

The appendages on the head are :—(i) Short, usually clubbed-
shaped antennae; (ii) mandibles; (iii) a single pair of maxillae
fused into a lobed plate.

Both the first two segments behind the head bear but one pair
of legs, the third has no legs but carries the opening of the
generative ducts.   The fourth free segment has one pair of legs, the
remainder two pairs.

The female *Iulus* lays its eggs, some 60 to 100, in an earthen
receptacle she has prepared beneath the surface of the ground.

Sub-class C. Insecta.

The immense group of Insects far outnumbers in species any other group of animals and in all probability exceeds in number all the species of the rest of the animal world. New insects are constantly being discovered and although some quarter of a million have already been named and to some extent described, it is believed that at least as many more remain unrecorded.

In spite of these numbers Insects are as a whole a uniform group and show less diversity in size and structure than many of the smaller groups, as for instance the Crustacea or the Mollusca. Probably their great number and small range of structural variation is not unconnected with the fact that they have found a new medium in which to pass some part, at any rate, of their life. The other group of animals which have taken to flying and lead an aërial life—the birds—show a somewhat similar range of species accompanied by a uniformity of structure which in their case is even more marked.

Insects may be characterized by their body being divided into three distinct regions, the head, the thorax and the abdomen (Fig. 85). The head bears one pair of antennae and three pairs of gnathites. The thorax consists of three enlarged segments, each of which bears on the ventral surface a pair of legs and the two hindermost of which bear on the dorsal surface a pair of wings. The abdomen consists of a varying number of segments, ten being perhaps the usual number but fewer often occur. The abdomen bears no appendages except at the posterior end, where a pair of rod-like outgrowths—the anal cerci—are often found.

External features.

Owing to the similarity of Insects to one another and their great number the study of them has become a very special branch of Zoology, which is termed Entomology. The necessity of extremely detailed study is due to the same cause and a great number of technical terms are in use for describing the numerous structures which build up the body of the Insect.

In this short book it will only be possible to indicate a few points about the anatomy of Insects and we will take as a type the common Cockroach because it is both a generalized form and not too small for dissection.

The common Cockroach of the British kitchen is *Stylopyga orientalis*, but a larger form, *Periplaneta americana*, is often met

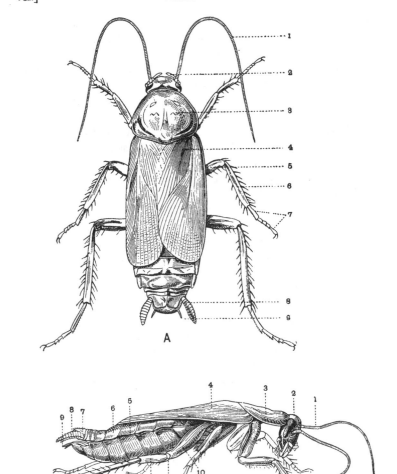

Fig. 85. *Stylopyga orientalis*, male × 2. A. Dorsal view. B. Side view. From Kükenthal.

A. 1. Antenna.   2. Palp of first maxilla.   3. Prothorax.   4. Anterior wings.   5. Femur of second leg.   6. Tibia.   7. Tarsus.   8. Cerci anales.   9. Styles.

B. 1. Antenna.   2. Head.   3. Prothorax.   4. Anterior wing. 5. Soft skin between terga and sterna.   6. Sixth abdominal tergum. 7. Split portion of tenth abdominal tergum.   8. Cerci anales.   9. Styles. 10. Coxa of third leg.   11. Trochanter.   12. Femur.   13. Tibia. 14. Tarsus.   15. Claws.

with on ships and from them makes its way to the docks; it is also often found in zoological gardens, etc. *Phyllodromia germanica*, a small species, is becoming increasingly common in England.

The whole body of the Cockroach is covered by a chitinous covering which varies in thickness, from the black hard head to the thin whitish areas which exist at the joints and which permit movement of the harder parts on one another. Except in the head the segments of the body can be detected externally, and, as in other members of the Arthropoda, the segmentation affects some only of the internal structures, such as the heart, the tracheae, the muscles and the nervous system, the other organs of the body not being influenced by it.

The head of a Cockroach is a flattened structure placed at right angles to the axis of the body. It is oval in outline, its upper edge being considerably broader than the lower. It is loosely jointed to the thorax by a neck which permits considerable movement (Fig. 85). This neck enters the head near its upper edge and below it the head hangs free. On the upper and outer edge of the head are a pair of kidney-shaped, facetted eyes of a shining black colour, on the inner curve of which the antennae or feelers have their origin. These are long whip structures, often as long or longer than the body; they are made up of many joints and during life are in active movement, now stretched downward as if trying the ground on which the creature moves and now waving aloft as if testing the air.

The mouth is on the lower edge of the head and is covered in front by a small moveable flap called the labrum or upper lip. At the sides it is protected by the first and the second pairs of appendages, and behind the fusion of the right and left third pair it forms a plate called the labium, which completes the boundaries of the mouth behind (Fig. 86).

If the first pair of mouth-appendages or mandibles be removed from the head and examined through a lens, each is seen to be a single-jointed stout jaw with a toothed inner edge which bites against the corresponding part of its fellow. It is characteristic of the mandibles of all Antennata to have no palp or remnant of the distal joints of the limb, such as is almost universally present in Crustacea.

Behind the mandibles and like them situated on each side of the mouth, are the first maxillae. Each consists of a number of joints and each joint has a special name. Like the typical

gnathite of other Arthropods we may regard them as consisting
of a limb-like appendage with out-growths from the basal joints
biting against corresponding processes—gnathobases—of the
fellow appendage. There are two of these gnathobases, the hard
pointed lacinia and an outer portion, the softer galea (B, Fig. 86).
The lowest joints form an L-shaped hinge which, when opened out,
protrudes the jaw. The outer portion of the first maxilla is many-
jointed and is sensory in function, constantly touching and testing
the ground as the animal moves about. It is called the maxillary palp.

The second maxillae are united across the median line and thus
constitute a fold or plate called the labium, which bounds the

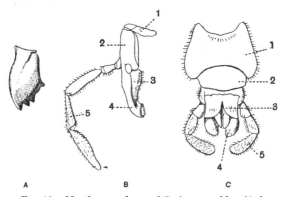

Fig. 86. Mouth-appendages of *Stylopyga*. Magnified.
A. Mandible.　B. 1st maxilla.　1. Cardo.　2. Stipes.　3. Lacinia.
4. Galea.　5. Palp.　C. Right and left 2nd maxillae fused to form
the labium.　1. Submentum.　2. Mentum.　3. Ligula, corresponding
to the lacinia.　4. Paraglossa, corresponding to the galea.　5. Palp.

mouth behind as the labrum bounds it in front (C, Fig. 86). Each
half may be resolved into elements similar to those of the first
maxillae, the fused basal joints of the pair of appendages form the
mentum and sub-mentum, the galea being represented by the
paraglossa, whilst the inner gnathobase corresponds with the
lacinia and is termed the ligula. As in the case of the first
maxilla the outer joints of the appendage which have a tactile
function are termed the labial palp.

The thorax is built up of three segments, the pro-, meso- and
meta-thorax. The skeleton of each segment con-
Thorax.
sists of a dorsal hard piece, the tergum, united
with a corresponding ventral piece, the sternum, by a soft in-

tervening pleural membrane. The tergum of the first thoracic segment is clearly visible, but the meso- and meta-terga are concealed by the wings. The two pairs of wings are formed by folds of the skin arising from the terga of the meso- and meta-thorax respectively, and in a state of rest conceal the dorsal surface of the animal behind the prothorax. The front pair are termed elytra; they are hard and horny, one overlaps the other, and they probably serve more as protectors to the delicate hind wings than as organs of flight. The posterior wings are thin and membranous and are of greater area than the elytra, and they constitute the effective organs in the rare flight of the Cockroach. At rest they are folded like a fan and concealed by the elytra. In *Stylopyga orientalis* the wings—as is not uncommon amongst Insects—are rudimentary in the female.

The ventral plate or sternum of each thoracic segment bears a pair of legs by means of which the Cockroach scuttles rapidly about.

Each leg consists of a number of joints, viz., a thick flat coxa applied to and articulating with the sternum; a minute triangular joint, the trochanter; a stout joint called the femur; a more slender one termed the tibia, armed with spines; then a piece consisting of five short joints called the tarsus, with a whitish hairy patch under each joint which acts as a sole; and finally a pair of terminal claws (Fig. 85). These names, as is too often the case in Zoology, were suggested by fanciful and misleading comparisons with the parts of the limb of a vertebrate.

The abdomen consists of ten segments and here the terga and sterna can be easily seen as they are not obscured by the insertion of the wings and legs. The eighth and ninth terga are however both tucked under the seventh and are not readily seen until the animal is artificially stretched. The tergum of the tenth or last segment stands out from the hind end of the animal and is cleft into two lobes. The sterna are equally distinct but the first is small. The abdomen is broader in the female than in the male, and the seventh sternum is shaped like the bow of a boat and projects backwards, hiding the posterior sterna and supporting the lower surface of a roomy pouch or cavity in which the egg-case is formed. In the male the seventh sternum conceals the eighth and ninth.

Abdomen.

The stigmata leading to the tracheae are placed in the soft pleural membrane connecting terga and sterna.

The abdomen is usually regarded as being without appendages, but a pair of jointed cerci anales emerge below the edge of the tenth tergum in each sex, and in the male the ninth sternum bears a pair of anal styles. The claim of these structures to be reckoned as appendages of the same rank as the antennae, the gnathites and the legs, was at one time not generally conceded, but appears to be now fairly established for the cerci anales.

In the soft tissue between the tenth tergum and the last visible sternum, at the hind end of the body, is placed the anus, below it the single genital pore is situated. The anus is supported by certain thickened plates in the skin, known as podical plates, and around the genital orifices are arranged certain rods and bars, symmetrical in the female but asymmetrical in the male, whose functions and meaning are obscure, but which are connected with the processes of copulation and of egg-laying. As we have seen, this region is in the female enlarged into a genital sac by the growth and modification of the seventh sternum and the tucking into the space so formed of the skin carrying the eighth and ninth sterna. The opening of the oviduct is on the eighth sternum and on the ninth is the single opening of the receptaculum seminis or spermatheca, which consists of two pouches of unequal size composed of inturned ectoderm : these are always found full of spermatozoa in the fertilized female. The spermatozoa apparently leave the spermathecae when the eggs are being laid and fertilize the ova whilst they are in the genital sac. Two glands, consisting of branching tubes,—called colleterial glands—open separately behind the spermatheca. They secrete a fluid which hardens to form the egg-capsule in which the eggs are laid. This is moulded in the genital sac and may often be seen half-protruding between the distended seventh sternum of the mature female.

When the skin is removed from the dorsal surface of the Cockroach the cavity laid open is not a coelom but a

Internal structure.

haemocoel, and it is largely filled by a loose white tissue, known as the fat-body, which surrounds the various internal organs. If the alimentary canal be disentangled from this it is at once evident that in Insects, unlike other Arthropods, the intestine is longer than the body and the larger portion of it which lies in the abdomen is coiled in order to stow it away. Like the digestive tube of other Arthropods a large part of its length consists of the stomodaeum and proctodaeum. The former consists of an oesophagus which quickly passes into a large crop in which the food is stored for a time. The lining of both these regions

bears hairs and the muscles in their walls are striped. The crop
is followed by a gizzard, which bears internally six hard chitinous
teeth, and behind them are fine hairs which act as strainers, so
that only finely divided food can pass on into the mesenteron or
chylific ventricle, as the part of the alimentary canal is called
which alone is lined by endoderm and is capable of absorbing
nourishment. This tube is produced in front into seven or eight

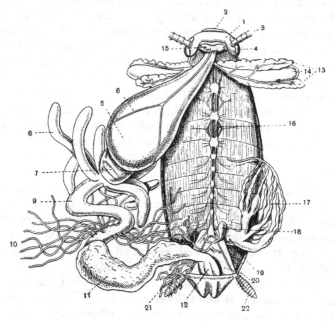

FIG. 87. A Female Cockroach, *Stylopyga*, with the dorsal exoskeleton removed
and dissected to show the viscera. Magnified about 2.

1. Head.   2. Labrum.   3. Antenna cut short.   4. Eye.   5. Crop.
6. Nervous system of crop.   7. Gizzard.   8. Hepatic caeca.   9. Mid-gut
or mesenteron.   10. Malpighian tubules.   11. Colon.   12. Rectum.
13. Salivary glands.   14. Salivary receptacle.   15. Brain.
16. Ventral nerve-cord with ganglia.   17. Ovary.   18. Spermatheca.
19. Oviduct.   20. Genital pouch in which the egg-cocoon is found.
21. Colleterial glands.   22. Anal cercus.

pouches, the so-called hepatic diverticula. The mesenteron, the
limit of which is marked by the insertion of the Malpighian tubules,
is succeeded by the intestine or proctodaeum, a long coiled tube
which enlarges posteriorly and opens by the anus. The enlarged
portion is called the rectum (Fig. 87), the anterior coiled portion
the colon.

Lying along the crop on each side is a pair of branched glands
—the salivary glands—and a bladder termed the salivary
reservoir. All three are provided with long ducts. Those of the
two glands on each side unite to form a single tube which then
receives the duct of the reservoir, and the common ducts of the two
sides open behind the mouth but in front of the second maxillae.
The saliva converts starch into sugar. The secretion of the hepatic
diverticula emulsifies fats and turns insoluble proteids into the
soluble forms (peptones). This secretion seems to pass forward into
the crop and there true digestion is effected. The digested and
dissolved food then passes through the filter of the gizzard into the
mesenteron, where absorption of the nutritious parts is effected, the
undigested portions passing on to the intestine and so out of the
body. The cells of the mesenteron undoubtedly exert some action
on the products they absorb, though we are ignorant of its precise
nature, but in the end they pass on the products of digestion—altered
no doubt—to the blood which everywhere bathes the alimentary
canal, and by it the new material is conveyed all over the body.

The blood is kept in circulation by the heart which lies in the
middle dorsal line close under the skin; in fact it can be seen
through the skin in the region of the abdomen. It is a long vessel
made up of thirteen chambers corresponding with the three thoracic
and ten abdominal segments, each chamber opening into the one in
front and the whole somewhat resembling a row of funnels fitted one
into another. At the broader hinder end of each chamber is a pair
of ostia or holes through which the blood enters from the pericardium, and there are valves which prevent the blood being forced
out of the ostia or forced backward when the heart contracts, so
that its only course is to flow forward. The pericardium is separated
from the rest of the haemocoel by the pericardial septum, in which
however there are certain holes which permit an interchange of contents between the two cavities. When at rest the pericardial septum
is arched upwards, but it is pulled outwards and flattened by the
periodic contraction of certain muscles attached to its sides called
the alary muscles, thus enlarging the pericardium and causing an
inflow of blood into it from the rest of the haemocoel. This blood
enters the heart when it is relaxed and the ostia are open, and hence
by the alternate contractions of the alary muscles and the muscular
wall of the heart the circulation is maintained.

The anterior end of the heart, called the aorta, opens by a
trumpet-shaped orifice into the haemocoel and the blood pours out

of this and bathes all the organs of the body. It thus takes up the soluble food which has left the alimentary canal and conveys it to those parts of the body where it is needed, and in a similar way it yields up its superfluous fat to be stored up in the fat bodies and gives up its waste nitrogenous materials to the Malpighian tubules, whence they are passed out of the body. It is noticeable that just as water-plants are as a rule not of a compact shape, but finely subdivided so as to expose as large a surface as possible to the surrounding medium, so in the body of the Cockroach the various organs are long, tubular or diffuse structures offering a large surface to the nourishing and purifying blood.

The heart contracts almost as frequently as the normal human heart, i.e., about seventy-two times a minute when the Cockroach is at rest, but at other times its rate of contracting varies a good deal. The blood is colourless and contains amoeboid corpuscles. It is slightly alkaline.

It is obvious from the above account that in the Cockroach the blood is mainly a means of conveying nutriment to the organs and taking certain waste matter from them, and that unlike what is usual in other animals, its respiratory function is at a minimum. Owing to the nature of the tracheae, the air with its oxygen is taken directly to each organ, almost to each cell, without the intervention of the blood.

The tracheal system opens to the outer air by ten pairs of oval pores or stigmata. These lie in the soft integument between the terga and sterna, one pair just in front of the mesothorax, one pair just before the metathorax and eight pairs just in front of each of the first eight abdominal segments, so that they seem to be intersegmental in position. These openings lead into tubes or tracheae which soon bifurcate and divide. The larger branches have a definite and symmetrical arrangement. There are dorsal arches running up towards the heart and ventral arches descending towards the nerve-cord. These arches are connected with one another longitudinally by trunks which run on each side of the pericardium and the nerve-cord. Large trunks also are given off to the alimentary canal. It follows that should one stigma become blocked the organs which its tracheae supply are still provided with air. The finer branches become smaller and smaller until they become veritable capillaries which penetrate every tissue.

The tracheae being full of air present a glistening silvery appearance which is unmistakeable. They are prevented from

collapsing by the presence of a spiral thickening of the chitinous lining
which runs round the interior of the tube just as the wire spiral
strengthens the india-rubber tube of some kinds of garden hose.
Respiration is effected by the alternate arching up and flattening of
the abdomen, resulting in an alternate increase and diminution of
its volume, which, since the blood is incompressible, secures an
alternate inrush and expulsion of air from the tracheae. In all
probability only the contents of the larger tracheae are affected by
this process ; by diffusion the oxygen from this " tidal air " is handed
on to the finer tracheae.

In Insects the tracheal mode of respiration reaches its highest
development and it is accompanied by a correlative diminution and
simplification of the circulatory apparatus, the oxygen of the air
being conveyed directly to and the carbon dioxide removed directly
from the cells to the outer air, whilst the blood loses its respiratory
function. It is doubtful how far this state of things is connected
with the peculiar disappearance of the coelom—which presumably
exists only in the cavity of the reproductive organs—because a
similar replacement of the coelom by a haemocoel is found in
Crustacea and Mollusca, where the blood is respiratory and the gills
are compact and active organs ; but probably it is correlated with
the modification which the excretory system undergoes and this
again is undoubtedly influenced by the state of the body-cavity.

Peripatus, the most primitive Arthropod, has typical nephridia,
each of which opens internally, not into a general coelom, but into
a small sac which is really a special part of the coelom not communicating with any other coelomic space but belonging only to
the nephridium which opens into it. The main body-cavity is a
haemocoel. Mollusca also retain a pair of typical nephridia or in the
case of *Nautilus* two pairs. These organs open into special coelomic
spaces which are as a rule small, the more spacious cavities of the
body being haemocoelic. Crustacea probably retain nephridia, those
of the fourth segment—shell glands—being persistent in the Entomostraca, whilst those of the second segment—green glands—
persist in the Malacostraca. It is not absolutely proved but it
seems probable that the inner ends of these glands represent
coelomic spaces. But even amongst Crustacea, in certain Amphipods
for instance, we find the transference of the function of the nephridia
to outgrowths of the alimentary canal. In Arachnids the nephridia
or coxal-glands show very varying degrees of development, but on the
whole they are tending to die out and to be replaced by Malpighian

tubules or diverticula of the intestine. Finally in Myriapods and Insects, where the true coelom is reduced to its minimum and where the tracheal mode of respiration attains a very high degree of development, all traces of nephridia have disappeared and the waste nitrogenous matter is excreted solely by the Malpighian tubules.

These are very fine long caecal diverticula,—so fine as to be but just visible to the unaided eye,—60—80 in number, arranged in six bundles which open into the beginning of the narrow part of the proctodaeum from which they are outgrowths. They float in the blood, winding about amongst the abdominal viscera (Fig. 87). They contain crystals, probably of urate of soda, which are taken up from the blood and which leave the body through the intestine.

The nervous system of the Cockroach is constructed on the same plan as that of the other segmented Invertebrates with which we have had to do. There is a large supra-oesophageal ganglion or brain giving off commissures, which encircle the oesophagus and unite below in a sub-oesophageal ganglion. Together these occupy a considerable portion of the cavity of the head. The supra-oesophageal ganglion supplies paired nerves to the eyes and to the antennae and is thus the sensory centre. The sub-oesophageal ganglion supplies the mandibles and both pairs of maxillae. From it two cords pass backward and bear three pairs of ganglia in the thorax and six in the abdomen. This difference between the number of nerve ganglia and the number of segments is carried to a much greater extent in some Insects where, as in Spiders, all the post-oesophageal ganglia tend to fuse into a common nervous mass in the thorax (Fig. 58).

The only specialized sense organs are the eyes and the antennae. The eyes have fundamentally the same structure as those of the Crustacea : the antennae are the seat of the senses of smell and taste, and are in addition very delicate tactile organs. The maxillary palps are also tactile and are constantly touching and testing the ground on which the Cockroach is moving.

Cockroaches are bisexual. The ovaries in the female consist of
Reproduction. two sets of eight tubes, each of which has developed from a coelomic sac in the embryo. They unite at their anterior end into two cords which pass to the dorsal wall of the thorax and become attached to the pericardial septum, and at their posterior end they fuse into two short oviducts which join to form a small uterus (Fig. 87). Each of the sixteen tubes contains cells, some of which become ova, and as they approach the oviduct the ova

become arranged in a single row. At the same time they increase greatly in size by the deposition of yolk in the egg, so that an ovum just before it leaves the body is of considerable size.

The eggs are apparently fertilized after leaving the uterus by spermatozoa which emerge from the spermatheca (in which they have been deposited by the male) situated behind the opening of the uterus. Whilst still in the genital pouch the fertilized eggs are surrounded by the secretion of the colleterial glands which open behind the spermatheca, and this secretion hardens into the egg-capsule or cocoon.

The paired testes of the male are functional only during youth and as they diminish in size after they cease to function, they are only to be found with difficulty. They lie concealed by the fat body below the terga in the region of the fourth, fifth and sixth abdominal segments. They look somewhat like elongated bunches of cherries, their translucent colour strongly contrasting with the opaque white of the fat body. Two vasa deferentia lead from the testes to a pair of large reservoirs called vesiculae seminales, which together form the "mushroom-shaped gland." Into these at an early age the cells destined to form spermatozoa pass from the testes and there they undergo their further development. The mushroom-shaped gland opens to the exterior by means of a short muscular tube, the ductus ejaculatorius, which has its orifice just below the anus. The name of the gland is derived from its form; it has a thick stalk surrounded by a crown of branches. Fertilization and oviposition take place during the summer.

Sixteen eggs are laid in each egg-capsule, and for some seven or eight days, until the mother finds some warm and secluded hiding-place to deposit her load, she carries about the capsule half-protruding from the genital pouch. When the embryos in the eggs are fully formed, which takes about twelve months, it is said that they secrete some fluid, probably saliva, which dissolves the upper part of the capsule and so permits of their escape. In *Phyllodromia germanica* the mother is said to take a part in freeing her offspring from their temporary imprisonment. When they first appear they are white with dark eyes, but the integument soon thickens and darkens. They have no wings, but in other respects they resemble their parents and thus there is no metamorphosis such as occurs in the Butterflies and many other Insects. They run actively about, devouring any starchy food they can find, and when in time they grow too large for their coat of mail it splits and a soft Cockroach extricates itself therefrom. The integument soon hardens again. This casting of the skin or ecdysis takes place seven times, and after the seventh moult, when the insect is four years old, it is adult.

The Insects are usually subdivided into eight Orders, which are mainly based (i) on the structure of the gnathites ; (ii) on the nature of the wings; (iii) on the amount of Metamorphosis which the life-history of the Insect presents. A short account of each of the criteria is therefore subjoined.

The Mouth-Parts (Gnathites) of Insects.

The mouth-parts of Insects can in almost all cases be resolved into a pair of mandibles which never bear palps and two pairs of maxillae which are usually provided with palps. In the different orders of Insects and in different members of the orders these mouth-appendages show many modifications and are put to a very great variety of uses. One or other part may be suppressed and disappear, others may coalesce, as is the case with the right and left second maxillae of the Cockroach, but as a rule traces of all the gnathites may be found though often much altered. The mouth-parts of insects have been grouped as follows : (i) *BITING*. This kind of mouth is found in the Aptera, the Orthoptera, the Neuroptera and the Coleoptera (v. below), in which orders there is as a rule no very great difficulty in recognizing the various parts which have been described in the Cockroach. As an example of the great variety presented by the mandible, those of the male Stag-beetle, *Lucanus cervus*, may be mentioned. In this animal the mandibles may equal in length the whole of the rest of the body. (ii) *SUCKING*. This kind of mouth is found in the Lepidoptera or Butterflies. Here the mandibles are rudimentary, but the gnathobases of the first maxillae are much elongated and frequently coiled into what is sometimes termed the proboscis. Each half is grooved and so applied to the other as to form a tube, and in some cases the two halves of the tube are locked together by minute hooks. The palp of this maxilla is absent or rudimentary. The labium composed of the second maxilla is an important structure in the larva or caterpillar, as it forms the spinnerets through which the silk of the cocoons is excreted, but in the adult it is practically absent although its palps persist as large hairy structures. The hollow tube formed by the maxillae is well adapted to suck up the fluids on which the Lepidoptera live. The suction is performed by a

powerful muscular sac called the suctorial-stomach, which is a
lateral outgrowth of the oesophagus and communicates with it.
(iii) *PIERCING AND SUCKING.* The Diptera or Flies possess both
sucking and piercing organs, which are as a rule somewhat unequally
developed. The basal portion of the labium or second maxilla is
much elongated and takes the form of a fleshy protuberance which
to some extent ensheaths the other parts. In the house-fly,
Musca domestica, the piercing organs are fused to the labium and
act only as supporting rods for it, but in the Gnat or Mosquito
they are free and reach a high degree of development. These parts
are overlaid by a somewhat enlarged labrum and between this and
within the grooved labium the pointed stylets lie. Amongst these
a sharp style called the hypopharynx situated behind the mouth
may be distinguished. Two pairs of lateral stylets, identified by
some as the modified mandibles and first maxillae, also exist and
the maxillary palp acts as a sensory organ. The Hemiptera or Bugs
have a very similar set of gnathites in correspondence with their
habit of boring into animals or plants to feed on their juices.
In this Order the mouth-parts when not in use are bent under the
body and lie along the under surface of the thorax. The labium
is jointed and its edges are curved so as to form an incomplete tube,
only the base of which is partly covered by the labrum. Within
this groove four sharply pointed styles—the mandibles and first
maxillae—work. They are as a rule finely toothed like little saws
and are well adapted for piercing the skin. There are no palps.
(iv) *BITING AND SUCKING.* The Hymenoptera or Bees and Wasps
have mandibles not unlike those of a Cockroach and use them for
biting and moulding their food of pollen and the wax they secrete
from their bodies. The laciniae of the maxillae form blade-like
structures and their palps have much diminished in size. The labial
palps however are large, and the conjoined median outgrowths from
the labium (corresponding to the ligula of the Cockroach, 3, Fig.
86 c) form a kind of grooved tongue along which the nectar in the
flowers is sucked up.

It thus appears that although the mouth-parts of Insects are
highly modified in connexion with the kind of food they live on and
the modes in which they obtain it, nevertheless the various mouth-
parts have a common ground plan, and although the authorities
differ as to details a fundamental similarity runs through these
appendages in the different Orders.

Wings of Insects.

The wings of Insects are folds of the integument, flattened so that the two sides are in contact on their inner surfaces. At certain places however slits are left and through these tracheae pass taking air to the wings. The ectoderm of both upper and

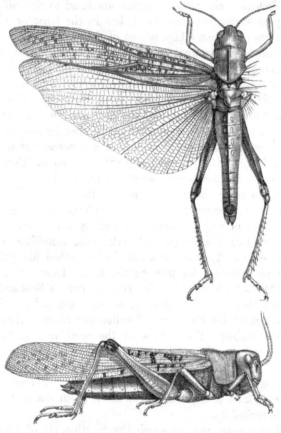

FIG. 88.   *Pachytylus migratorius.*   A Grasshopper.   Natural size.

under layers is also thickened along certain lines and the presence of these thickenings divides the wing up into a number of areas. These lines are usually called wing-veins or wing-nerves, but as they are neither veins nor nerves it is better to call them nervures. The presence and disposition of the nervures is of the highest importance in classification. Wings may be thin, membranous, and transparent, as in the Grasshopper (Fig. 88) or Dragon-fly (Fig. 50), where there is an enormous number of nervures, or in the Flies and Bees, where there are few nervures, or they may be thickened and strongly chitinised as in the front wings of Beetles. In this last group the anterior wings are called elytra (Gr. ἔλυτρον, a cover) and they always meet together in a median longitudinal line, so that when they are closed the insect appears to be wingless (Fig. 92): in some few cases they have fused together so that the posterior or flying wings are rendered useless. In one great division of the Hemiptera one half of the anterior wing is horny and strongly chitinous, the other and posterior half membranous. In the Cockroach, as we have seen, the anterior wings tend to become horny and are of little use in flight (Fig. 85). The posterior wings of the same insect when at rest are folded together something like the leaves of a shut fan and many species in several of the orders fold up their hind wings when not in use. The tucking away of these wings under the small elytra is a complicated affair in some Insects such as the Earwig, where the nippers at the end of the body are said to aid in the process. Many Insects however, such as the Dragon-flies, Plant-lice, Butterflies and Moths, Flies and Bees and numerous others, do not fold up their wings but either bear them erect or lying depressed on the body. In some cases the wings are quite transparent, but in the Moths and Butterflies they are covered with a dense fur of flattened scales which can readily be brushed off as a fine powder or dust (Fig. 91). It is these scales which give rise to the beautiful and in some cases gorgeous colouring of the Lepidoptera.

Two pairs of wings are present as a rule, but the order Diptera has only the anterior pair, the hind wings being replaced by certain stalked structures called balancers or halteres. In the Hymenoptera, the two wings of each side are clamped together by means of hooks on the hind wing which fit into a ridge on the hinder edge of the front wings (Figs. 93, 94, and 95). The two wings of each side thus move as one. It is not uncommon to find isolated species in which the wings are not developed, for instance, the females of many Plant-lice and some Moths; while Fleas, which are sometimes placed amongst or near the Order Diptera, never possess wings, though their absence is compensated for by a special development of the powers of jumping.

Metamorphosis.

A very large number of animals that live in the water, whether in the sea or fresh water, hatch out from the egg in a larval condition. That is to say, the being which leaves the egg is very unlike the adult in structure and habits, but by growth and a series of accompanying changes in time passes over into the adult which is capable of reproducing. These changes constitute the metamorphosis. In the life-history of land animals such larval stages are rare and indeed hardly exist outside the Insecta and the Amphibia. As we have seen, the egg of a Cockroach gives rise to a young Cockroach which differs but little from the adult and gradually grows into it by a series of small changes, but which never at any time undergoes a long period of profound rest. But if we consider the case of a Moth or Butterfly we shall find that the egg does not give rise to an animal resembling a minute Butterfly but to a worm-like larva or caterpillar, which has no wings and in other respects is very unlike the Insect which produced the egg. This caterpillar as a rule eats voraciously and grows rapidly with little change in form until its fourth ecdysis, when a sudden change occurs, and the so-called pupal stage supervenes. In this stage with a few exceptions the animal now called a pupa (Lat. *pupa*, a puppet) is motionless and ceases to feed. It may be uncovered and protected only by the hardened integument or it may be enclosed in a casing or cocoon. In the case of the Silkworm Moth and some others, this is constructed of silk. During the pupal stage, the animal undergoes a profound change, many of its organs and tissues being broken up and new ones constructed. When this process is completed the pupa casts its skin, makes its way out of the cocoon and emerges as an imago (Lat. *imago*, an image) or perfect insect.

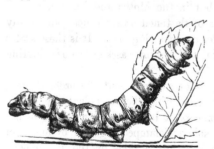

Fig. 89.   Larva of *Bombyx mori*, the Silkworm.   Life size.

The various orders of Insecta differ in the degree to which metamorphosis occurs. In the Aptera there is no metamorphosis and the development is said to be direct. In Orthoptera and Hemi-

ptera there is no quiescent pupa stage and the chief difference
between the larva and adult is the absence of wings. Here the
metamorphosis is said to be incomplete. The
same is true of most of the Neuroptera, a
very varied assembly of Insects, in some forms
of which however, e.g. the Caddis-flies, a pupal
stage exists. Amongst the Lepidoptera, Coleoptera, Hymenoptera and Diptera there is a well-
marked pupal stage, and these orders are said
to have complete metamorphosis. Various names
have been given to the larvae of Insects without
very precise definitions. Those of the Lepidoptera are usually called Caterpillars. They are
often gaudily coloured and bear tufts and
bunches of hair. Besides the three pairs of legs
which are found on the three segments following
the head and which correspond with the legs of

Fig. 90. Cocoon of
Bombyx mori, from
which silk is spun.
About life size.

the imago, certain of the abdominal segments bear fleshy stumps
called abdominal legs. The larvae of some of the Saw-flies (Hymenoptera) have a similar bright colouring and resemble Caterpillars, and
like them feed exposed on leaves, etc. The larvae of Beetles and
most Hymenoptera are as a rule hidden underground or in galls
or wax-comb. They are whitish in colour and unattractive, and

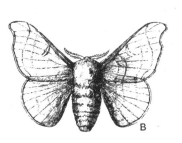
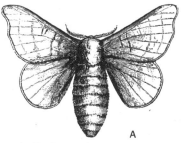

Fig. 91. Silk-worm moth, *Bombyx mori*.
A. Female.   B. Male.

are often termed grubs, whilst the footless white larvae of the
Diptera, which are for the most part deposited in some organic
substance—whether alive or not—are usually called maggots.

In the following account of Insect Classification we can only
indicate the chief characters of each Order and mention the names
of one or two common members of each.

Order I. Aptera.

Wingless Insects, with scales and hairs covering the body.
The mouth-parts are adapted for biting. They move by running or
by springing by aid of a caudal style which is kept bent forwards
under the abdomen and retained in this position by a ventral hook.
When released from this hook the recoil of this style hurls the
insect into the air. The segments of the thorax are not fused
together and there is no metamorphosis.

Not all wingless Insects belong to this Order. The name
Aptera (Gr. ἄπτερος, wingless) refers to the belief that the ancestors
of these Insects never had wings and that thus they represent a
lower stage of evolution than the rest of the sub-class.

For the most part the Aptera are minute Insects living in
retired spots under leaves or rubbish, in root-gutters, etc., but
they are widely distributed over the world. One of the best known
is the Silver-fish, *Lepisma*, which hides in disused cupboards, old
chests of drawers, sugar barrels, etc. It runs with great rapidity.

Order II. Orthoptera.

The Orthoptera (Gr. ὀρθός, straight; πτερόν, a wing) have mouth-
parts adapted for biting. The anterior wings are as a rule stiff,
and when the Insect is at rest one overlaps the other, and both
usually cover and conceal the large membranous hinder wings with
which the creature flies. There is an incomplete metamorphosis,
the young being at first without wings.

This Order is a very varied one and doubts exist as to whether
it is a natural one. It includes the Cockroach, whose anatomy has
already been described ; the Earwig, *Forficula* ; the *Mantis* ; the
leaf and stick insects, *Phyllium* and *Phasma* ; the Grasshopper
(Fig. 88), *Acridium*, *Pachytylus* ; the Locust, *Locusta* ; the Cricket,
Gryllus ; and many others.

Order III. Neuroptera.

The Neuroptera (Gr. νεῦρον, a tendon and hence a nervure) have
biting mouth-parts. Both pairs of wings are membranous and used
in flight, and the " veins " of the wings form a more or less close
network. Metamorphosis complete or incomplete.

This Order, like the preceding, contains many families which,
except as regards the structure of the wings, have little resemblance
to one another. The following are a few of the more widely known

species: the White-ant, *Termes*; the May-fly, *Ephemera*; the Dragon-fly, *Libellula* or *Aeschna* (Fig. 50); the Ant-lion, *Myrmeleo*; the Aphis-lion, *Hemerobius*; the Caddis-fly, *Phryganea*, which in some respects approaches the Lepidoptera; and the *Thrips*, an insect which injures corn crops and certain flowers and is sometimes elevated to the position of a distinct Order.

Order IV. Coleoptera.

The Coleoptera (Gk. κολεός, a sheath) have mouth-parts adapted for biting. The anterior wings are hard and horny and fit together in the middle line with a straight suture. The hind wings are membranous and folded. The metamorphosis is complete, that is, there is an active larval stage (Fig. 92), followed by a quiescent stage during which extensive changes in the internal anatomy take place.

Unlike the preceding, this order is clearly defined and its members are on the whole very like one another. It has always been a favourite order with Ento-mologists and the number of species named and described is far greater than in any other Order. It in-cludes all the Beetles. For the most part these Insects are dull in colour but their firm exoskeleton gives them a very definite outline and renders their preservation and identification comparatively easy,

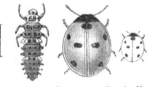

Fig. 92. In the centre *Coccinella septempunctata*, the Lady-bird × about 2½, with the larva to the left × about 2½, and the adult beetle, natural size, to the right.

which may to some extent account for their popularity with collectors.

Order V. Hymenoptera.

The Hymenoptera (Gr. ὑμενό-πτερος, membrane-winged) have mouth-parts adapted for biting and sucking. The ligula of the labium is long and grooved, whilst the paraglossae are small.

Fig. 93. *Formica rufa*, the Wood-ant.

1. Female.   2. Male.   3. Neuter.

The mandibles are well-developed and the laciniae of the first maxillae large. The four wings are alike, membranous in texture, and the hind wings are hooked on to the anterior in such a

Worker-bee.   Queen-bee.   Drone.

FIG. 94. *Apis mellifica*, the Honey-bee.

way that the two wings of each side move together. They differ from the wings of the Neuroptera in possessing fewer veins. The metamorphosis is complete.

This group comprises the Ants, Bees and Wasps. Many of them live in highly complex communities and in their social habits and general intelligence they reach a level which is only surpassed by man himself. The group includes the Wood-wasp, *Sirex*; the

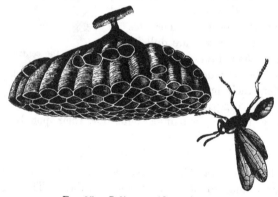

FIG. 95. *Polistes tepidus* and nest.

Saw-fly, *Tenthredo*; the Gall-fly, *Cynips*; the *Ichneumon*; the Ant, *Formica*; the Wasp and Hornet, *Vespa*; the Humble-bee, *Bombus*; and the Honey-bee, *Apis*.

Order VI. Hemiptera.

The Hemiptera (Gr. ἡμι, half) have mouth-parts arranged for piercing and sucking. The basal part of the labium is elongated and tubular and the mandible and first maxilla form sharp pointed styles. The two pairs of wings may be alike or may differ and the anterior pair are in some cases half horny and half membranous. The metamorphosis is incomplete, there being no quiescent stage.

The members of this Order present very great divergence both of form and of size ; they are colloquially known as Bugs and Lice. Amongst the commoner forms are the Water-boatman, *Notonecta* ; the Water-scorpion, *Nepa* ; the Bed-bug, *Acanthia* ; the *Cicada*, remarkable for its chirping noise ; the Frog-hoppers, including the Cuckoo-spit, *Aphrophora* ; the Plant-louse, *Aphis* ; the *Phylloxera*, which destroys vines ; Scale Insects and lice.

Order VII. Diptera.

The Diptera (Gr. δι-πτερά, two-winged) have mouth-parts arranged for piercing and sucking. The only difference in this respect from the Hemiptera consists in the fact that the sucking tube is partly formed by the labrum and that the first maxillae retain palps. Only one pair of wings, the anterior, are present ; the posterior are represented by a pair of short knobs called balancers or halteres (Fig. 97). The metamorphosis is complete.

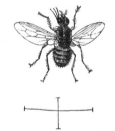

FIG. 96. *Glossina morsitans*, the Tsetse-fly.

The Diptera or Flies form one of the largest of the Insect Orders, probably as large as the Coleoptera, although at present the number of species of Beetles named and described is far greater than that of Flies. Amongst the commoner genera are the Gnats and Mosquitoes, *Culex* ; the Daddy-long-legs, *Tipula* ; the Gall-fly, *Cecidomyia* (Fig. 97); the Horse-fly, *Tabanus* ; the Bot-fly, *Oestrus* ; the common House-fly and Blue-bottle, *Musca*, and many others. The Flea, *Pulex irritans*, which is wingless but endowed with considerable powers of jumping, is sometimes placed in a Sub-order of the Diptera and sometimes in a separate Order.

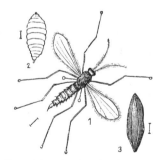

FIG. 97. *Cecidomyia destructor*, the Hessian-fly.

1. Insect. 2. Larva. 3. Pupa, or "flax seed." All magnified.

Order VIII. Lepidoptera.

The Lepidoptera (Gk. λεπίς, a scale, πτερόν, a wing) have mouth-parts adapted for sucking only. The two pairs of wings are similar in appearance and covered with scales (flattened spines) which give rise to the beautiful pattern on the wings but are easily rubbed off. None of the wings fold up and when not in use are either held erect or are depressed on each side of the body. The metamorphosis is complete.

This Order is very clearly defined and the members show a marked resemblance one to another. It includes the Butterflies and Moths, and all of them exhibit a very definite and complete metamorphosis. The eggs give rise to worm-like larvae known as caterpillars, which consume much food, generally of a vegetable nature (Fig. 89). After a considerable time, varying from a few weeks to three years, the caterpillar comes to rest, and in such cases as the Silk-worm Moth, *Bombyx mori*, surrounds itself by a case or cocoon spun by itself, which furnishes the material silk (Fig. 90). Within this cocoon, or in some species without forming a cocoon, the caterpillar forms a pupa, and whilst in this state it undergoes a very thorough reorganization and gradually the mature Insect is built up ; after a certain time this emerges and occupies its comparatively short life in the propagation of its species (Fig. 91). The female usually deposits its eggs on or near the plants which serve as food for its offspring.

Class III. ARACHNIDA.

The third large group of the Arthropoda is a very varied one and contains many animals which differ markedly in their structure one from another. Perhaps the most distinctive features of the

External features. Arachnida (Gk. ἀράχνη, a spider ; εἶδος, shape) are (i) There are no true gnathites. No appendage loses all other functions and becomes exclusively a jaw, although the proximal joints of several are prolonged inwards towards the mouth and help to take up food ; in a word some of the limbs have developed gnathobases ; (ii) The most anterior appendages are never antennae but always a pair of nippers, termed chelicerae ; (iii) The active catching and walking legs of the fore part of the body or prosoma are strongly contrasted with the plate-like modified limbs of the middle part of the body or mesosoma when the latter exist, but in many cases these have disappeared and in others have become so modified that they are no longer recognisable as limbs. Nearly all Arachnids moreover agree in having the anterior end of the body, the prosoma[1] as it is called, marked off from the rest and covered by a single piece, the carapace. The rest of the body or abdomen is in some forms differentiated into two regions, the mesosoma and metasoma, but in other cases this distinction does not exist ; it may be segmented or it may not. The prosoma bears six pairs of appendages and of these the last four are usually walking legs. The appendages of the abdomen are connected with the respiratory function and are much modified, often—in the terrestrial forms—forming floors for the respiratory chambers. The breathing apparatus may be tracheal, or in a few marine forms branchial, or may take the form of respiratory chambers, the last named and the gills having a peculiar form found only amongst Arachnida. They consist of "books" of thin superposed lamellae attached to the posterior aspect of an appendage. When modified for breathing air these "books" are called lung-books. When, as is the case in *Limulus*, they breathe oxygen dissolved in water they are called gill-books. The genital orifice is usually on the anterior end of the abdomen and ventral : the group is bisexual. Many different Orders are included in the Arachnida, the best known being perhaps those which include the Spiders, the Harvestmen, the Mites and the Scorpions. The last named are found only in warm climates and Mites are too small for investigation with the naked eye, so that we will take the Spider as an example of Arachnid structure.

Order I. Araneida.

Spiders belong to the Order Araneida (Lat: *aranea*, a spider), in which the abdomen is unsegmented and soft. The second pair of appendages, the pedipalpi, are leg-like and modified in the male in connexion with the fertilization of the female. The abdomen bears certain modified appendages called spinnerets, on which

---

[1] The name cephalothorax is often applied to this region, but the term is too misleading to be used. The cephalothorax of Decapod Crustacea includes the first thirteen segments of the body: the prosoma of Arachnida only includes six, and therefore corresponds roughly to the "head" of the higher Crustacea. Similar criticism might be launched against the use of the word "abdomen," but here the error is too deep-rooted for correction since the term is used in describing both Crustacea and Insecta, and in each case in a different sense.

open the glands, the secretion of which produces the Spider's web. If we examine such a Spider as *Epeira diademata*, which is common enough in English gardens, sitting on or near its wheel-shaped web

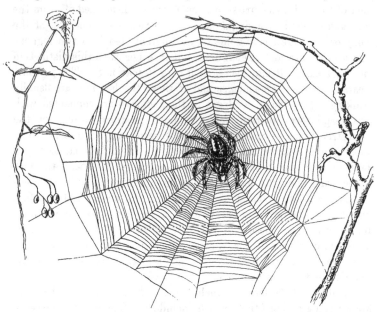

Fig. 98.   The Garden Spider, *Epeira diademata*, sitting in the centre of its web.
After Blanchard.

Fig. 99.   Front view of the head of a Spider, *Textrix denticulata.* Magnified.   From Warburton.

1. Head.   2. Eyes.
3. Basal joint of chelicerae.
4. Claw of chelicerae.

(Fig. 98), we notice that behind the prosoma there is a slender waist and that this is followed by a large swollen abdomen with no outward trace of division into segments, or into meso- and meta-soma.

There are six pairs of appendages, and it is at once noticeable that there are no antennae or feelers to act as sensory organs. Their function is to some extent taken over by the long walking legs. The first pair of limbs are called chelicerae; in *Epeira* these are two-jointed, the terminal joint being pointed and folded down against the basal joint except when being used (Fig. 99). This pair of appendages contains poison glands and the poison

External structure.

escapes through an opening at the point of the second joint. By means of it the Spider can kill insects and seriously hurt larger animals.

The second pair of appendages in the Arachnida are called pedipalpi (Fig. 100). In *Epeira* they resemble the walking legs, but in the male at the final moult the last joint becomes altered and forms a hollow sac—the palpal organ—which plays an important part in fertilizing the female.

Then follow four pairs of walking legs each with seven joints and terminated with two or three claws; in some species they are provided with a pad of short hairs called a scopula, which helps the animal to run on walls and ceilings.

The mouth is very minute, for the Spider does not swallow solid food but sucks the juices of its prey. It lies between the bases of the pedipalps, and the basal joint of each of these appendages has a cutting blade termed the "maxilla" (2, Fig. 100). It is a common feature of the Arachnids that the basal joints of one or more of the pairs of appendages are produced inwards towards the mouth and act as jaws, but the modification never goes so far as to obscure the limb-like form of the appendage and so produce a true gnathite.

On the ventral surface of the abdomen just behind the waist is situated the genital opening, protected by a plate which is the result of the fusion of a pair of appendages, and on each side of this is the slit-like orifice of a lung-book. The lung-books are very remarkable structures. Each opens to the exterior by a pore through which the air enters, and consists of a sac the cavity of which is largely occupied by a number of thin plates in the substance of which the blood circulates, and is thus

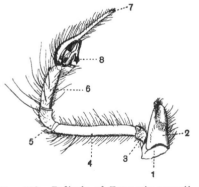

FIG. 100. Pedipalp of *Tegenaria guyonii*,
the large house-spider.

1. Coxa. 2. Gnathobase, the so-called "maxilla." 3. Trochanter. 4. Femur.
5. Patella. 6. Tibia. 7. Tarsus.
8. Palpal organ.

brought into close relationship with the air which passes in and out between the neighbouring plates; the sac is floored in by a special plate which is a modified appendage (Figs. 63 and 65). Such a breathing apparatus is peculiar to the Arachnida. In some Spiders we find a second pair of lung-books placed behind the others, and in other species this second pair is replaced by a pair of tracheae recalling the respiratory mechanism of the Myriapods or the Insects (Fig. 63). They have however been independently developed, and probably owe their origin to the sac of a lung-book from which the lamellae have disappeared.

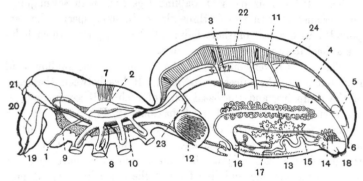

Fig. 101.   Diagram of a Spider, *Epeira diademata*, showing the arrangement of the internal organs × about 8. From Warburton.

1. Mouth.   2. Sucking stomach.   3. Ducts of liver.   4. Malpighian tubules.   5. Stercoral pocket.   6. Anus.   7. Dorsal muscle of sucking stomach.   8. Caecal prolongation of stomach.   9. Cerebral ganglion giving off nerves to eyes.   10. Sub-oesophageal ganglionic mass.   11. Heart with three lateral openings or ostia.   12. Lung-book.   13. Ovary.   14. Acinate and pyriform silk glands.   15. Tubuliform silk gland.   16. Ampulliform silk gland.   17. Aggregate or dendriform silk glands.   18. Spinnerets or mammillae.   19. Distal joint of chelicera.   20. Poison gland.   21. Eye.   22. Pericardium.   23. Vessel bringing blood from lung sac to pericardium.   24. Artery.

Near the hinder end of the abdomen are four tubercles or spinnerets, and if these be pushed aside, two more,

Spinnerets.

shorter in length, come into view. These are the organs which form the web and they have been shown to be vestiges of abdominal appendages. They are very mobile and are pierced at their ends by hundreds of minute pores through which the silk exudes as a fluid, hardening on exposure to the air (18, Fig. 101).

The silk is secreted by a large number of glands which have their exit at the above-mentioned pores. Of these in *E. diademata* there are five different sorts and each secretes a special kind of thread; for the various lines in a Spider's web differ considerably one from another, in accordance with the use they are put to. The circular lines are sticky and help to catch insects for the Spider's food, the radial lines are stout and form a framework for the support of circular lines; the threads with which the Spider binds up its captured prey differ from these, and there is still another kind of thread with which it constructs its cocoons, and each kind of line is supplied from different sets of glands.

The dissection of a Spider requires much care, since the organs almost fill the body and are completely embedded in the large masses of the digestive and reproductive glands. The oesophagus, which leads from the mouth, opens into a strong sucking "stomach," which is really like the stomach of the Cray-fish. a stomodaeum. This is attached by muscles to the chitinous exoskeleton, and when the muscles contract its cavity is enlarged and thus a sucking action is induced at the mouth (Fig. 101). Behind this is an endodermic portion of the alimentary canal which gives off certain caeca or blind tubes, followed by an intestine which traverses the abdomen and is further provided with a number of ducts which collect the products of a very capacious digestive gland or "liver." The intestine, which is also lined by endoderm, is followed by a short proctodaeum, the proximal portion of which is swollen up into a pouch called the stercoral pocket. It ends in an anus situated close behind the spinnerets.

Spiders possess two kinds of organs which excrete waste nitrogenous material : (i) the Coxal glands, which are true nephridia, i.e., glandular tubes running between a reduced coelom and the exterior, and (ii) Malpighian tubules, a pair of simple pouches opening into the endodermal intestine and thus in their origin differing from those of Insects. The coxal glands are better developed in some species, such as the common House-spider, *Tegenaria derhamii*, than is the case in *E. diademata*, where they are very degenerate and where their functions seem to have largely passed to the Malpighian tubules. In fact these structures are an interesting example of a set of organs degenerating and of their functions being assumed by another set.

The heart of the Spider is of the same general type as that of Myriapods ; it is a tube with paired slit-like openings—ostia—at the sides, through which the blood enters to be driven out again through certain rather ill-defined vessels to circulate in the spaces between the various organs.

The nervous system is concentrated ; there is a bi-lobed ganglion

13—2

above the oesophagus which gives off nerves to the eyes and the chelicerae ; this is connected by two lateral cords, which pass one on each side of the oesophagus, with a large nervous mass situated in the thorax. From this, nerves pass off to supply the remaining five pairs of limbs and two nerves arise which pass backward and

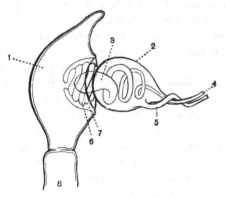

Fig. 102.   Diagrammatic view of Palpal Organ.

1. Tarsus.   2. Bulb.   3. Vesicula seminalis.   4. Opening of vesicula seminalis.   5. Conductor.   6. Haematodocha.   7. Alveolus.

supply the abdomen. The only conspicuous sense organs of Spiders are the eyes, which are "simple"; of these in *E. diademata* there are four large eyes arranged in a square on the top of the head and two small ones on each side of the square. This number, eight, is not uncommon in Spiders, where both the number of eyes and their disposition are much used in systematic classification.

The male, as is not uncommon amongst the Araneida, is smaller than the female. The ovaries and testes lie in the abdomen and have the form of a network of tubes, a form characteristic of Arachnida ; the spermatozoa are conveyed to the palpal organs of the pedipalpi of the male and by them introduced into pouches, the spermathecae of the female. The eggs are fertilized before they are laid, which latter event usually takes place in October, when they are enclosed in cocoons of yellowish silk. The young are hatched out in the following spring and at once begin spinning. By means of the minute threads they secrete they weave a kind of nest about the size of a cherry-stone which hangs suspended from some twig or leaf. At the least disturbance the hundreds of young Spiders in the nest begin to disperse ; the spherical nest breaks up as into dust, but when the disturbance is at an end the minute Spiders, so small as to be almost invisible, re-assemble and again form their little spherical nursery.

The number of species of Spiders is very great and their habits are very diverse and well worthy of study.

Order II.   Phalangida.

The Phalangids (Gr. φᾰλάγγιον, a venomous kind of spider) or Harvestmen are in common talk usually classed with Spiders, but

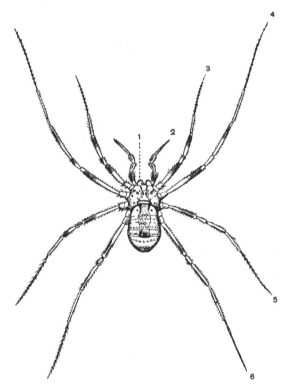

Fɪɢ. 103.   A Phalangid or Harvestman, *Oligolophus spinosus*, adult male × 2.

1.   Chelicerae.   2.   Pedipalps.   3, 4, 5 and 6.   First, second, third, and fourth legs.

they differ from the latter in having no waist, that is, the abdomen is not separated from the prosoma by a constriction, and they breathe entirely by tracheae. They have four long and very slender pairs of legs, which easily break, and their eyes are some-

times elevated above the surface of the head on a tubercle like a
look-out tower.   The abdomen is distinctly divided into segments.

As a rule these creatures are nocturnal and are usually met with
in dark corners or amongst the stalks of hay or grass.   Their long
legs enable them to steal with a gliding spring upon their prey,
for the most part insects or spiders, for they are carnivorous.   They
are dull in colour, grey, brown or blackish, as becomes an animal
that loves the dusk.   About twenty-four species have been recorded
in Great Britain.   Phalangids die down as winter sets in, but the
eggs last through the cold weather and give rise to a new generation
in the spring.

Order III. Acarina.

The Acarids (Gr. ἀκᾰρές, a morsel) or Mites form an enormous
order whose function in life is to a large extent to play the

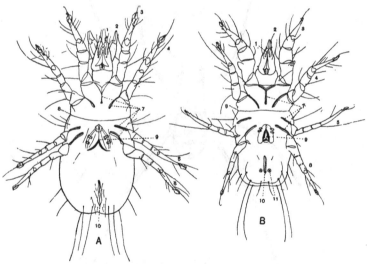

Fig. 104.   *Tyroglyphus siro*, seen from the ventral side. A. Female. B. Male.
Magnified.   From Leuckart and Nitsche.

1. Pedipalpi.   2.   Chelicerae.   3, 4, 5, 6.   First, second, third and fourth
walking legs.   7.   Chitinous thickenings supporting legs.   8.   Furrow
round body.   9.   Reproductive opening, flanked by two suckers on each
side.   10.   Anus.   11.   Suckers at side of anus.

scavenger, and the terrestrial forms confer the same benefits on the
dwellers on the Earth that the Ostracoda and many of the smaller
Crustacea do on the aquatic fauna.   Many of them however have
adopted parasitic habits and cause disease amongst larger animals,

while some induce the formation of galls and other deformities amongst plants. Most of the Mites, as their name indicates, are of minute size; but the female Ticks, belonging to the family Ixodidae, which live amongst the undergrowth of forests on the look-out for some vertebrate prey, can when they become attached to their hosts—man, cattle, or even snakes—by distending their bodies with the blood they suck, swell out to the size of hazel-nuts.

Anatomically they are difficult to characterize. Like the Phalangids, they have no waist, and when special breathing organs are present they take the form of tracheae ; they differ however from the Phalangids in never showing signs of segmentation. The chelicerae may be clawed or chelate, like a lobster's claws (Fig. 104), but they often take the form of piercing stylets and the pedipalpi may form a sheath to protect them.

The number of species is very great ; amongst the commoner forms may be mentioned, *Tetranychus telarius*, often known as the Red Spider, which spins webs under leaves in which whole colonies shelter. This species is believed to do great damage in hot-houses. *Tyroglyphus siro*, the Cheese-mite, which burrows in decaying cheese, and the genus *Phytoptus*, which causes the conical galls on lime-trees, maples, etc., are also familiar.

Order IV. Scorpionida.

Scorpions are not found in Great Britain, though they are common on the Continent of Europe around the Mediterranean basin and generally in warm climates. They retain a more marked segmentation than is the case with the other Arachnids we have considered. The abdomen is very long, distinctly segmented and differentiated into two portions ; (*a*) the mesosoma, consisting of seven segments of the same diameter as the prosoma, bearing the respiratory appendages ; (*b*) the metasoma, a much narrower part, consisting of five segments and a curved spine like a tail at the apex of which is the opening of a poison gland. The mesosoma has six pairs of appendages. The first of these forms the genital operculum, a plate bearing on its posterior aspect the genital pore in both sexes ; the second are " pectines," curious comb-shaped structures, whose exact function is not yet determined, but which are morphologically reduced and thickened gill-books. The third, fourth, fifth and sixth segments bear each a pair of lung-books, and it has already been explained that the floors of these are formed of highly modified

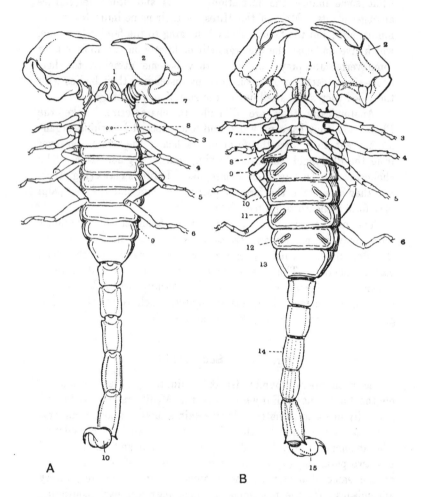

FIG. 105. A. Dorsal view of an Indian scorpion, *Scorpio swammerdami* × ⅔.
B. Ventral view of the same × ⅔.

A. 1. Chelicera.   2. Pedipalp.   3, 4, 5, 6. 3rd to 6th appendages, or
walking legs.   7. Lateral eyes.   8. Median eyes.   9. Soft tissue at
side of body, pleura.   10. The poison sting or telson.

B. 1—6 as in A.   7. The genital operculum.   8. The pectines.   9, 10,
11, 12. The four right stigmata leading to the four lung-books.   13. The
last segment of the mesosoma.   14. The third segment of metasoma.
15. The telson. In each case the metasoma, which is usually carried bent
forward over the meso- and pro-soma, has been straightened out.

plate-like appendages which in the adult have lost all trace of
their origin from limbs. The seventh segment of the mesosoma
shows no traces of limbs and tapers to join the first segment of
the metasoma. At the posterior end of the fifth metasomatic
segment, on the ventral surface, is situated the anus, and behind
this is a conical pointed joint which contains the poison glands
and which forms a very efficient and powerful sting. The whole of
this tail is very mobile and the sting can readily be directed to any
point. In life the tail is usually borne turned forward over the
body so that the sting threatens the head.

Both the chelicerae, which are small and short, and the pedipalpi,
which are long and six-jointed, end in nippers, the latter recalling
the appearance of the claws of a lobster. The four pairs of walking
legs end in claws.

The mouth is very minute, for like the Spiders Scorpions only
suck the juices of their prey. They feed for the most part on
Insects and Spiders. The basal joints of the first two pairs of
appendages, like those of the pedipalps in Spiders, are all produced
towards the mouth, forming gnathobases which probably help to
hold their food.

Scorpions usually hide under rocks and stones during the day,
being often very intolerant of heat, but they creep out as dusk
comes on and run actively about. The Scorpion is viviparous, the
young being born in a condition resembling their parents.

Order V. Xiphosura.

A very peculiar aquatic Arachnid called *Limulus*, or popularly
the "King-crab," inhabits the warm seas on the Western side of the
Pacific Ocean and along the shores of the Western Atlantic. It is
a littoral form, that is to say, it lives not far from the shore; it
burrows in sand or mud at a depth of from two to six fathoms, often
lying with only its eyes, which are on the top of the body, exposed.

The shape of the body is something like a half-sphere with a
piece cut out and a long spine is attached to the truncated side.
This spine has given the name Xiphosura (Gr. ξίφος, a sword; οὐρά,
a tail) to the Order. The half-sphere is hinged, and the part in
front of the hinge is the prosoma; the rest is the abdomen or
meso- and meta-soma. On the upper surface of the half-sphere are
a pair of simple eyes near the middle line, and there is a pair of
compound eyes situated further back nearer the edge. The under
surface of the half-sphere is partially hollowed out and concealed in

this hollow on each side of the middle line of the prosoma are six pairs of appendages. The most anterior of these are typical nipper-like chelicerae, the next is not specially modified to form a pedipalp, but it and the remaining four pairs are walking legs. All of them

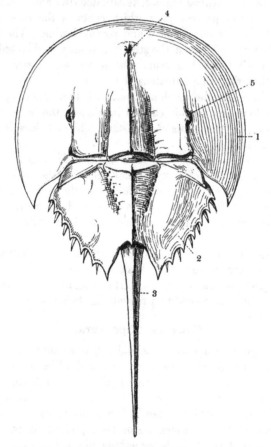

Fig. 106.   Dorsal view of the King-crab, *Limulus polyphemus* × ½.

1. Carapace covering prosoma.   2. Meso- and meta-soma.   3. Telson.
4. Median eye.   5. Lateral eye.

send inwards a spiny gnathobase, which helps to form the border of the mouth. The sixth pair of limbs end in some flattened blade-like structures which assist in digging and burrowing in the sand and in extracting the worms which form the principal item of the diet of the King-crab. The seventh appendages, or the first on the mesosoma, take the form of a flattened plate or operculum which bears the reproductive pores on its posterior surface. It is bent back and underlies the eighth, ninth, tenth, eleventh, and twelfth pairs of appendages, which are also plate-like and each of which bears on its posterior surface a gill-book. There is a striking similarity

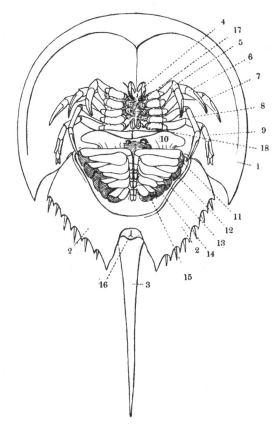

Fig. 107.   Ventral view of the King-crab, *Limulus polyphemus* × ½.

1. Carapace covering prosoma.   2. Meso- and meta-soma.   3. Telson.
4. Chelicera.   5. Pedipalp.   6, 7, 8, 9.   3rd to 6th appendages, ambulatory limbs.   10.   Genital operculum turned forward to show the genital aperture.   11, 12, 13, 14, 15.   Appendages bearing gill-books.
16. Anus.   17. Mouth.   18.   Chilaria.

between these organs and the "lung-books" of the Scorpion; the latter, however, do not project, but are sunk in pits. The metasoma terminates at the anus, but behind it a long sword-like tail projects. This post-anal tail corresponds with the swollen stinging tail or telson of the Scorpion. It is used by the animal to right itself when it is upset by the motion of the waves.

A curious plate of fibro-cartilage to which muscles are attached lies inside the body near the ventral surface. It is formed of modified connective tissue in which a cheesy material termed chondrin has been deposited in the ground substance, and is largely built up of interlacing tendons of muscles so that it acts as an internal supporting structure or endoskeleton. It is called the endosternite. Possibly it was a feature of primitive Arthropoda, as similar endosternites occur in many other Arachnida and in some of the more primitive Crustacea.

The internal anatomy differs in many points of detail from that of the Spider, but in essentials there is a fairly close resemblance. Unlike the Scorpion, *Limulus* lays eggs and these are fertilized in the water and pass through what may be termed a larval stage.

In many respects Limulus seems to be related to the extinct *Eurypterina*, whose fossil forms are so abundant in the Upper Silurian and Old Red Sandstone formations; and like some species of *Limulus* they attained a great size, two feet or more in length being not uncommon. The Eurypterines were aquatic and indeed seem to form an intermediate stage between the Scorpion and *Limulus*, and confirm us in the conclusion drawn from the anatomy of *Limulus* that this animal retains in many points the habits and structure of the marine ancestors of Arachnida.

Phylum ARTHROPODA.

Bilaterally symmetrical Coelomata whose coelom has undergone great change. Segmented animals with the segments usually arranged in groups. Paired hollow and jointed limbs on some of the segments.

Class I. CRUSTACEA.

Aquatic Arthropods usually breathing by gills, with two pairs of antennae. A limb-bearing thorax usually fused with the head and followed by a segmented abdomen which may be limbless.

Sub-class A. ENTOMOSTRACA.

Small, simple Crustacea with varying number of segments. The stomach has no teeth. The larva is a Nauplius.

Order 1. Phyllopoda.

Long-bodied and usually well segmented with a shield-like shell protecting head and thorax and sometimes abdomen; with leaf-like swimming appendages.

Sub-order i. Branchiopoda.

Large-bodied forms with no dorsal brood-pouch, second antennae not enlarged for swimming, numerous swimming appendages.

Ex. *Apus, Branchipus, Artemia.*

Sub-order ii. Cladocera.

Small, short forms with bivalved shell, a dorsal brood-pouch and enlarged swimming second antennae.

Ex. *Daphnia, Simocephalus.*

Order 2. Ostracoda.

Usually small forms with body unsegmented. At most seven pairs of appendages and a rudimentary abdomen, all shut up in a bivalve shell.

Ex. *Cypris, Cypridina.*

Order 3. Copepoda.

Usually elongated and clearly segmented but often much modified by parasitism. Four or five pairs of biramous thoracic appendages.

Ex. *Cyclops, Argulus.*

Order 4. Cirripedia.

Sessile animals whose not clearly segmented body is enclosed in a fold of skin strengthened by calcareous plates. Usually five biramous thoracic appendages. Hermaphrodite as a rule.

Ex. *Lepas, Balanus.*

Sub-class B. MALACOSTRACA.

Large Crustacea as a rule with five segments in the head, eight in the thorax and six in the abdomen. Nauplius larva very rare.

Order 1. Leptostraca.

Bivalve shell covering the eight free thoracic segments but not fused with them, abdomen of eight apparent segments with anal forks. Thoracic limbs leaf-like.

Ex. *Nebalia.*

Order 2. **Thoracostraca.**

All or most thoracic segments fused with head and covered by a cephalothorax. Eyes as a rule stalked.

Sub-order i. **Schizopoda.**

Eight pairs of biramous thoracic appendages. Eyes stalked.

Ex. *Mysis.*

Sub-order ii. **Decapoda.**

Thoracic segments fused with head. Last five thoracic appendages uniramous and used for walking. Eyes stalked.

Division *a*. **Macrura.**

Abdomen long.

Ex. *Astacus.*

Division *b*. **Brachyura.**

Abdomen short.

Ex. *Cancer, Carcinus.*

Sub-order iii. **Stomatopoda.**

Cephalothoracic shield short. Five pairs of maxillipeds. Abdomen large and bearing gills on its appendages.

Ex. *Squilla.*

Sub-order iv. **Cumacea.**

Four or five free thoracic segments. Two pairs of maxillipeds. Eyes sessile.

Ex. *Cuma, Diastylis.*

Order 3. **Arthrostraca.**

Seven, rarely six, free thoracic segments. No cephalothoracic shield. Eyes sessile.

Sub-order i. **Amphipoda.**

Body laterally compressed. Gills on thoracic appendages.

Ex. *Gammarus.*

Sub-order ii. **Isopoda.**

Body dorso-ventrally compressed. Gills on abdominal appendages.

Ex. *Asellus, Porcellio, Oniscus.*

Class II. ANTENNATA.

A single pair of antennae and with tracheal respiration.

Sub-class A. PROTOTRACHEATA.

Soft, caterpillar-like bodies with numerous pairs of appendages. Nephridia present.

Ex. *Peripatus.*

Sub-class B. MYRIAPODA.

Terrestrial, with head well marked off from body, which consists of many similar segments bearing six- or seven-jointed appendages.

Order 1. **Chilopoda.**

Animal flattened dorso-ventrally, bases of legs wide apart: to each tergum corresponds one pair of legs: the segment following the head has a large pair of poison claws: genital opening between the last pair of legs.

Ex. *Lithobius.*

Order 2. **Diplopoda.**

Animal cylindrical, bases of legs close together: to each tergum behind the fourth correspond two pairs of legs: no poison claws: genital opening on the third segment behind the head.

Ex. *Iulus.*

Sub-class C. INSECTA.

Body divided into three regions, head, thorax and abdomen. Head bears the antennae and three pairs of persistent mouth parts; thorax three pairs of walking appendages and usually two pairs of wings; abdomen as a rule without appendages.

Order 1. **Aptera.**

Wingless insects with hairy and scaly bodies ending in anal filaments. No metamorphosis.

Ex. *Lepisma.*

Order 2. **Orthoptera.**

Jaws biting, wings usually unalike. Metamorphosis incomplete.

Ex. *Forficula, Stylopyga, Phasma, Acridium, Gryllus.*

Order 3. Neuroptera.

Jaws biting, sometimes sucking. Wings alike, membranous, with many nervures. Metamorphosis varies.

Ex. *Termes, Ephemera, Libellula, Phryganea.*

Order 4. Coleoptera.

Jaws biting. Anterior wings hard and curving together with a median, straight suture. Metamorphosis complete.

Ex. *Coccinella, Melolontha.*

Order 5. Hymenoptera.

Jaws biting and licking. Four membranous wings with few nervures. Metamorphosis complete.

Ex. *Formica, Apis, Vespa.*

Order 6. Hemiptera.

Jaws piercing and sucking. Wings alike or different. Metamorphosis incomplete.

Ex. *Acanthia, Cicada, Aphis.*

Order 7. Diptera.

Jaws piercing and sucking. Hind-wings reduced, front-wings membranous. Complete metamorphosis.

Ex. *Culex, Musca.*

Order 8. Lepidoptera.

Jaws sucking. Four similar wings covered with scales. Metamorphosis complete.

Ex. *Bombyx.*

Class III. ARACHNIDA.

No antennae and no true gnathites. Prosoma of six appendage-bearing segments followed by a meso- and meta-soma whose appendages are when present usually much modified.

Order 1. Araneida.

Meso- and meta-soma soft, unsegmented. Four to six spinnerets, two to four lung-books.

Ex. *Epeira.*

Order 2.　**Phalangida.**

No waist between pro- and meso-soma which latter with meta-soma is segmented.　Tracheate.

Ex.　*Oligolophus.*

Order 3.　**Acarina.**

No waist.　Minute and often reduced forms mostly tracheate.

Ex.　*Tyroglyphus, Tetranychus.*

Order 4.　**Scorpionida.**

Meso-soma seven segmented in adult, meta-soma five segmented and ending in a post-anal poisonous telson.　Four lung-books.

Ex.　*Scorpio.*

Order 5.　**Xiphosura.**

Shield-shaped carapace covers prosoma.　Meso- and meta-soma fused.　Gill hooks.　The telson forms a spine.

Ex.　*Limulus.*

CHAPTER VIII.

PHYLUM MOLLUSCA.

MOLLUSCA (Lat. *mollis*, soft) is the name which is given to one of the largest and most important phyla of the animal kingdom. In it are included not only our terrestrial snails and slugs and many fresh-water species but also the oysters, mussels, periwinkles, whelks and countless other species of "shell-fish," bivalve and

General description. univalve, which crowd the rocks laid bare at low-water around our coasts: and in addition to these, the extraordinary Octopuses, Squids and other forms of Cuttle-fish belong to the same great phylum. The name Mollusca seems to have been suggested by the fact that the members of the phylum do not possess any internal hard parts such as are found in Man and other vertebrates. This softness of internal constitution is shared by other classes with no relation to the Mollusca, as for instance the great group of the Arthropoda. The Arthropods however possess a horny covering which closely invests them and following every irregularity of their outlines, so that it seems a real part of themselves. This is the exoskeleton or cuticle, which constitutes one of the great differences between them and the Mollusca. The latter, it is true, possess also an exoskeleton composed principally of calcareous matter, but this adheres only to a part of the surface. It is usually very thick and easily detached, and so it is frequently looked on as a separate thing from the animal and is known as the shell. The shell is to be looked on as a secretion produced by a part of the skin only: this part of the skin, which almost always projects from the rest of the body as a flap, is called the mantle. The space between the mantle-flap and the rest of the body is known as the mantle-cavity. The mantle-cavity shelters the gills or organs of respiration, and into it open the kidney or kidneys and the anus, and usually also the genital ducts.

Class I. GASTEROPODA.

In order to fix our ideas we may take the common English garden snail, *Helix aspersa*, which has also established Description of Snail. itself throughout considerable areas in North America, or, if procurable, the larger *Helix pomatia*, which on account of its size is easier to dissect, as a type of the Mollusca. In Lower Canada the genus *Helix* is not very abundant, and the

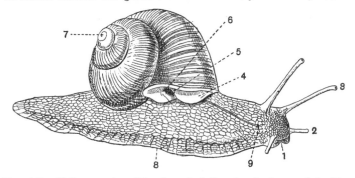

FIG. 108. *Helix pomatia.* Side view of shell and animal expanded. From Hatschek and Cori.

1. Mouth. 2. Anterior tentacles. 3. Eye tentacles. 4. Edge of mantle.
   5. Respiratory pore. 6. Anus. 7. Apex of shell. 8. Foot.
   9. Reproductive aperture.

largest species, *Helix albolabrus*, is rather small for convenient dissection. *Limnaea stagnalis*, the large river-snail, is however common and easy to obtain, and its structure is similar in its main outlines to that of *Helix*.

The shell is coiled into a spiral form; the body contained in it consists of a visceral hump, coiled like the shell and closely adhering to it, and of a portion which we call the head, neck, and foot, which can be drawn within the opening of the shell if the animal is alarmed, but which under ordinary circumstances is quite outside it. The snail is devoid of anything in the nature of legs,—an important character of the Mollusca as contrasted with the Arthropoda,—but the part of the body next the ground is a flat muscular surface called the foot. By means of wave-like contractions of the longitudinal muscular fibres of this organ the snail moves along, always preparing the ground for itself by depositing a layer of slime on it. This slime is poured forth from a gland which opens in front of the foot, just beneath the mouth

(14, Fig. 111). The foot is one of the most important organs of the Mollusca; it takes different shapes in the different groups but always assists locomotion. In the pond-mussel, for instance, it is shaped like a wedge, in order to force a path through the soft mud at the bottom of the ponds in which the animal lives. The different shapes which the foot assumes afford the chief basis for the classification of Mollusca.

The head of the snail bears two pairs of feelers, or tentacles, which are hollow outgrowths of the body-wall (2, 3, Fig. 108): these when irritated are protected by being pulled outside in, and so are brought into the interior of the body. The first or shorter pair are supposed to be the chief seat of the sense of smell: the second and longer pair have at their tip a small pair of black eyes. These eyes are merely minute sacs, the walls of which are made of light-perceiving cells, connected at their bases with a nerve which leads to the brain; in the cavity of the vesicle is a horny lens which nearly fills it up. The eyes of nearly all the Mollusca are con-structed on the same plan, but in the Cuttle-fish not only is the vesicle large and spacious and the lens proportionately smaller, but there is in addition a series of folds of skin surrounding the place where the eye comes to the surface, which constitute an outer chamber, and outside this, eyelids, so that the whole organ acquires a superficial similarity to the human eye.

If we carefully pick away the shell of the animal and lay bare the visceral hump, brushing away any mucus which may adhere to the body, we shall see on the right side of the animal a round hole (5, Fig. 108). A bristle passed through this reaches into a large cavity separated from the outside by an exceedingly thin wall. This space is nothing but the mantle-cavity, which, as explained above, is the space comprised between the projecting mantle flap and the rest of the body. The peculiarity about the snail is that the mantle edge has become fused to the back of the neck so as to shut the mantle-cavity off from the exterior, leaving only this little hole of communication. The mantle-cavities of the marine allies of the snail, such as the whelk and periwinkle, are not so completely shut off, inasmuch as in them the mantle flap merely lies against the neck but is not fused to it, and inside the mantle-cavity there is a gill. This gill consists of a hollow axis bearing on one or both sides a close set row of thin plates inside which the blood circulates and receives oxygen from the water by diffusion. Fresh supplies of water are drawn into the mantle-cavity by the action of myriads of cilia which cover the gill. A gill of this nature is called a ctenidium, owing to its comb-like appearance (Gr. κτενίδιον, a small comb). Now, since the snail breathes air, not water, it has lost the gill, but to compensate for the loss it has changed the whole mantle-cavity into a lung. The floor of the mantle-cavity, really the back of the neck, is arched and composed of muscles: when these contract the floor flattens and thus the mantle-cavity is enlarged and air is drawn in.

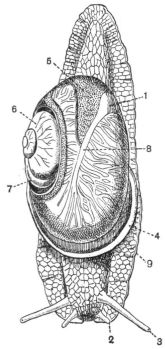

The blood is contained in large vessels running in the thin roof of the mantle-cavity: these are clearly seen when the mantle flap is clipped away from the neck and turned over to the right (8, Fig. 109, and Fig. 110). These vessels are seen to all converge to the heart, which consists of two small oval sacs placed end to end. That into which the vein enters is thin-walled and is called the auricle: the other thicker one is called the ventricle (Fig. 110); it is the more muscular of the two and drives the blood through two arteries to the body. One of these passes up to the visceral hump, and the other forward to the head and neck. In Molluscs which have gills the auricle always receives the blood from the gill: when there is one gill, as is the case with nearly all the univalves, there is only one auricle: but where, as in the bivalves and cuttle-fish, there are two or even four gills (as in *Nautilus*) there are likewise two or four auricles. The heart is surrounded by a space called the pericardium, which really corresponds to the body-cavity or coelom of Vertebrates, Annelids and Echino-

Fig. 109. *Helix pomatia.* The animal seen from the dorsal side after removal of the shell. From Hatschek and Cori.

1. Auricle of the heart receiving pulmonary vein.   2. Anterior tentacles.   3. Eye tentacles.   4. Edge of mantle.   5. Nephridium.   6. Liver.   7. Albumen gland.   8. Pulmonary vein.   9. Foot.

derms, for into it the excretory organ opens and in the embryo the genital cells are budded from its wall. Other large spaces existing in the head and neck have no connexion with the coelom but are really parts of the blood system. Since there are no regular veins, except those which run in the mantle-roof, the arteries open into irregular spaces. It will be remembered that the space called pericardium amongst the Arthropods is really a blood space and that the heart opens into it by openings called ostia: the coelomic character of the pericardium of Mollusca is then another distinguishing

Internal structure.

feature of the group. It opens by a narrow ciliated passage, the reno-pericardial canal, into the kidney, which is seen in the mantle-roof beside the pericardium (5, Fig. 109). The kidney looks like a solid yellow organ; but in reality it is a vesicle into the cavity of which numerous folds project, covered by the peculiar cells which have the power of extracting waste products from the blood, which flows in spaces in the kidney wall. The kidney communicates with the exterior by a narrow thin-walled tube, the ureter, which runs along the right side of the body and opens on the lip of the respiratory opening, just above the opening of the anus (10, Fig. 112).

The kidney in Mollusca varies a good deal in structure, but is always built on the same funda-

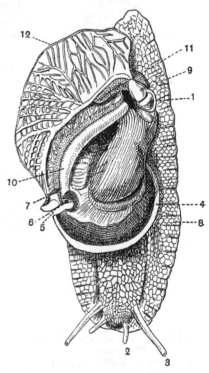

Fig. 110. *Helix pomatia*, with the upper wall of the pulmonary chamber cut open and folded back. From Hatschek and Cori.

1. Ventricle. 2. Anterior tentacles. 3. Eye tentacles. 4. Cut edge of mantle. 5. Respiratory pore. 6. Anus. 7. Opening of ureter. 8. Foot. 9. Auricle receiving pulmonary vein. 10. Rectum. 11. Nephridium. 12. Upper wall of pulmonary chamber.

mental plan as that of the Snail. Where there are two gills there are likewise two kidneys. Often there is no ureter, but the kidney opens directly to the exterior, as in the cuttle-fish, the whelk (*Buccinum*), the limpet (*Patella*). In the cuttle-fish instead of irregular spaces there are regular veins in its walls and the folds covered with special cells are only developed over the course of these veins.

Turning now to the digestive system of the snail we notice several very interesting peculiarities. The mouth is situated in front, beneath the small pair of tentacles, and there is a curved horny bar, the jaw, in the roof of the mouth. Against the jaw works a rasp-like tongue, called the r a d u l a, the surface of which is a horny membrane covered with myriads of minute, recurved teeth. Underneath this membrane there are certain small pieces or cartilage to which muscles are attached which pull the cartilages and the membrane covering them alternately downwards and forwards and upwards and backwards, so that the tongue is worked against the jaw. Thus the snail is enabled to tear pieces out of

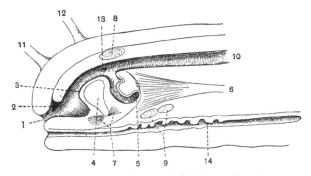

Fɪɢ. 111. Inner view of right half of head of *Helix*, to show the arrangement of the radula × 2.

1. Mouth.   2. Horny jaw.   3. Radula.   4. Cartilaginous piece supporting radula.   5. Radula sac from which radula grows.   6. Muscle which retracts the buccal mass.   7. Intrinsic muscles which rotate the radula.   8. Cerebral ganglion.   9. Pedal and visceral ganglia. 10. Oesophagus.   11. Anterior tentacle.   12. Eye tentacle.   13. Orifice of duct of salivary gland.   14. Mucous gland which runs along foot and opens just under the mouth.

the leaves on which it feeds (Fig. 111). A similar organ is found in all Mollusca, except the Bivalves or Lamellibranchiata, and the number, shape and arrangement of the teeth are an important help in classification. The horny membrane is continued backward into a little blind pouch, called the radula sac: here is its growing-point, where new teeth are continually being formed as the old ones wear away. In the limpet (*Patella*), this radula sac is extra-

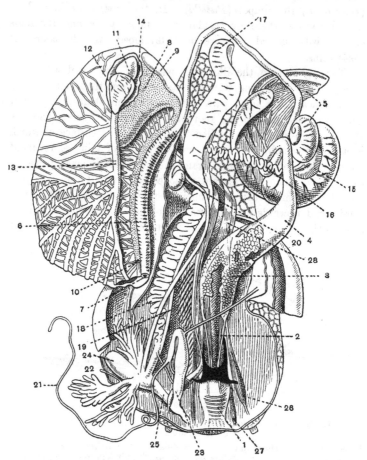

FIG. 112. *Helix pomatia*, opened and the viscera exposed.
From Hatschek and Cori.

1. Pharynx.   2. Oesophagus.   3. Salivary glands, with duct.
4. Stomach.   5. Liver.   6. Rectum.   7. Anus.   8. Kidney.
9. Inflated commencement of ureter.   10. Opening of ureter to exterior.
11. Ventricle.   12. Auricle.   13. Pulmonary vein.   14. Opening of nephridium into pericardium.   15. Ovo-testis.   16. Common duct of ovo-testis.   17. Albumen gland.   18. Female duct.   19. Male duct.   20. Spermatheca.   21. Flagellum.   22. Accessory glands.
23. Penis.   24. Dart sac.   25. Vagina.   26. Eye tentacle retracted.   27. Anterior tentacle retracted.   28. Muscles which retract the head, pharynx, tentacle, etc.

ordinarily long, attaining a length two or three times greater than that of the body. In the cuttle-fish the radula is present and the jaw is developed into upper and lower beaks, like those of a parrot, with which the animal tears its prey to pieces. The Bivalves have lost all trace of both jaws and radula: they live on the microscopic organisms brought to them in the currents of water which they produce, and so they do not need to masticate their food.

The radula sac and the muscles and cartilages belonging to the radula, form a swelling which is called the buccal mass. Behind this comes the oesophagus or gullet, which appears narrow by comparison, but its cavity is really as large as the space inside the buccal mass. The gullet soon widens out into the first stomach or crop, which is used for storing the food. On the outside or surface of this two branching whitish structures are seen, the salivary glands. They secrete a juice which runs forwards through two tubes, the salivary ducts, opening into the buccal mass. The saliva mingles with the food as it is being masticated. The crop is situated in the hinder part of the neck, and behind it the alimentary canal passes under the mantle-cavity and up into the visceral hump. The great mass of this hump is occupied by a brownish looking organ, called the liver. This, like the similarly named organ in the Arthropoda, is a great mass of tubes lined by cells of a deep brown colour: the tubes join together and eventually open by two main tubes, one above and one below, into a dilatation of the alimentary canal. This swelling is the true digestive stomach. It is probable that the "liver" assists digestion by preparing a fluid which is poured into the stomach: its function is thus not the same as that of the human liver. In fact it must be confessed that the name liver has been recklessly given by the older naturalists to any brown-coloured organ found near the stomach of an Invertebrate. The part of the alimentary canal behind the true stomach is called the intestine. It takes a turn in the liver substance and then runs out of the visceral hump along the right side of the body to open by the anus, which, as we have seen, is placed just behind the respiratory opening.

The central nervous system resembles that of the Annelida in being made up of ganglia, each of which might be compared to a miniature brain, connected together by means of commissures, that is, bands of nerve fibres. The two largest ganglia, which are placed above the oesophagus one at each side and connected by a commissure, are called the supra-oesophageal or cerebral ganglia,

or sometimes the brain (1, Fig. 113); but there is no reason to think
that they are any more important to the animal than the others.
Underneath the oesophagus there is what at first sight seems to
be a compact nervous mass, connected with the supra-oesophageal
ganglia by a commissure on each side forming a nerve collar
(Fig. 113). Closer inspection shows that this mass is perforated
by a hole through which passes the
great anterior artery from the ventricle,
and that from both the lower and
upper halves a separate nerve comes
off to go to the cerebral ganglia. Thus
the apparently simple nerve collar consists of two commissures on each side
united in a common sheath. Between
them a minute nerve passes down, to
end finally in a minute membranous
sac lined by ciliated cells and cells
with sense hairs and containing fluid
in which a little ball of carbonate of
lime floats. This sac, the otocyst, is
the only other important sense-organ,
besides the eyes, which the snail possesses. It is difficult to dissect, but if
the small bivalve *Cyclas* be taken,
the shell opened and the foot cut off
and slightly compressed, or if one of
the transparent Molluscs, such as
Pterotrachea, which float at the surface of the sea, be examined, it is
perfectly easy to see both otocysts
with the microscope. It used to be
supposed that the function of this organ was to perceive sound, but

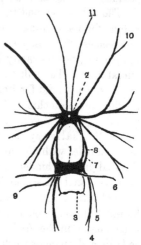

FIG. 113. View of nervous system of *Helix pomatia*.

1. Cerebral ganglion. 2. Pedal,
pleural, and visceral ganglia.
3. Buccal ganglia. 4. Nerve
to lips. 5. Olfactory nerve.
6. Optic nerve. 7. Pleuro-
cerebral commissure. 8. Pe-
do-cerebral commissure. 9.
Genital nerve. 10. Nerve to
mantle. 11. Nerve to vis-
cera.

whilst it is probable that some vibrations of the air affect it, it is
nearly certain that, like the otocysts of Medusae and Arthropoda,
its main function is to enable the Mollusc to keep its balance by
allowing it to perceive whether it is leaning on one side or not.
As the snail changes its position the little ball inside rolls about
and affects different parts of the wall of the vesicle, and hence
probably different fibres in the nerve which supplies it.

Not all Mollusca possess eyes, but all, except perhaps the
Oyster, which never moves, possess otocysts. The experiment has

been made in the Cuttle-fish of cutting them out, and it is then found that the animal loses its power of keeping its balance in the water and tumbles about.

To return to the central nervous system. In the pond-snail, *Limnaea*, the hinder part of the sub-oesophageal nervous masses consists of no less than five ganglia, strung together on a short loop of nervous fibres, which is called the visceral loop. Of these a pair nearest the head are called the pleural ganglia, the next are called the visceral ganglia, and the one at the end the abdominal ganglion (Fig. 115). The front and lower part of the sub-oesophageal nervous mass consists of the pedal ganglia, which send nerves exclusively to the foot. Pleural and visceral ganglia can be recognised in the young snail, but they become indistinguishably joined in the adult. In other Molluscs, such as the Sea-hare (*Aplysia*), or the Ear-shell (*Haliotis*), the visceral loop is long and the ganglia widely separated. In these animals it can be seen that the pleural ganglia send nerves to the sides of the body, and that from the visceral ganglia nerves come off which go to the base of the gill or gills. At the base of each gill

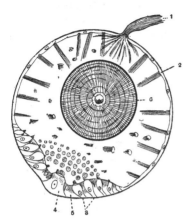

FIG. 114. Optical section through the auditory vesicle or ear of *Pterotrachea friederici*, a transparent pelagic Mollusc × about 150. After Claus.

1. Auditory nerve.   2. Ciliated cells.
3. Auditory cells.   4. Large central auditory cell.   5. Supporting cells.
6. Otolith.

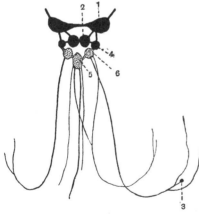

FIG. 115. Nervous system of *Limnaea*. After Lacaze-Duthiers.

1. Cerebral ganglion.   2. Pedal ganglion.
3. Osphradial ganglion.   4. Pleural ganglion.   5. Abdominal ganglion.
6. Visceral ganglion.

there is a patch of thickened skin, called an osphradium (Gr. ὀσφραίνομαι, to smell), provided with numerous sense-cells, which enables the animal to test the water which enters its mantle-cavity. Of course no such organ exists in the Snail. The muscles of the radula are supplied by nerves from a special pair of small ganglia placed on the buccal mass—the buccal ganglia—connected with the supra-oesophageal ganglia.

We thus see that the nervous system of the snail consists of a pair of supra-oesophageal ganglia connected by commissures with (*a*) a pair of pedal ganglia supplying the muscles of the foot with nerves, (*b*) an extremely short visceral loop, the ganglia on which are so closely placed as to become practically confluent with each other, whence nerves go to all parts of the body, and (*c*) a small pair of buccal ganglia supplying the buccal mass. The nervous systems of all Mollusca are built on this plan : in the bivalves, however, where there is no radula, not only are the buccal ganglia absent, but the pleural and cerebral are fused with one another, and, as the visceral loop is long, we find

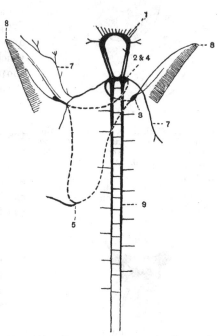

Fig. 116. Nervous system, osphradium (olfactory organ) and gills of *Haliotis*. After Lacaze-Duthiers.

1. Cerebral ganglion.   2. Pedal ganglion.
3. Osphradial ganglion.   4. Pleural ganglion.   5. Abdominal ganglion.
7. Nerves to mantle.   8. Gills.   9. Pedal nerves.

three widely separated pairs of ganglia,—cerebro-pleural, pedal and visceral—the last named often termed "parieto-splanchnic," in different parts of the body. The Cuttle-fish have a closely massed nervous system, like the snail, which is protected in a kind of rudimentary skull, made of cartilage.

The only organs of the snail which remain to be mentioned are the reproductive organs. These are exceedingly complicated in this Mollusc, both sexes being united in the same individual, a condition of affairs which is known as hermaphroditism. The essential genital organ is the ovotestis, a small yellowish patch of delicate tubes spread out on the surface of the liver, on the inner side of the uppermost coil of the spire (Fig. 112). This organ produces both eggs and spermatozoa and both travel down a single tube. Before the duct reaches the neck it receives the secretion of a large organ, called the albumen gland. This secretion consists of a fluid which has proteids in solution and is of high nourishing value. Beyond the albumen gland although externally simple the duct is divided by a septum into two passages, one for the eggs and one for the spermatozoa, and still further on it becomes completely divided into two separate tubes. The female portion opens to the exterior by a thick-walled muscular part, the vagina, into which a tuft of tubes—the mucous glands—opens. The vagina also receives the opening of an organ called the spermatheca, which is a round sac at the end of a long duct in which the spermatozoa received from another individual are stored up. In addition to this, a thick-walled sac called the dart-sac also communicates with the vagina. In this sac is found a calcareous rod which is thrown out into the body of another individual about the time of fertilization. The male duct opens also into a muscular organ called the penis, which can be partly everted, that is, turned inside out, and so protruded. The function of this organ is to transfer the spermatozoa to another individual; it has a blind pouch projecting inward beyond the place where the male duct enters it called the flagellum; in this the spermatozoa are massed together into bundles called spermatophores. Both penis and vagina have a common genital opening far forward on the right side of the neck (Fig. 108).

Few Mollusca have such complicated generative organs as the snail. One large group of marine snails, the Opisthobranchiata, resemble *Helix* in being hermaphrodite, but none possess the dart-sac, and in many the generative opening is placed further back and connected with the opening of the penis by a groove called the seminal groove. Hence the penis is obviously derived from a muscular pit on the side of the head into which the spermatozoa trickled and was at first unconnected with the generative opening.

In another group of marine snails, the Prosobranchiata, there is a separation of the sexes and the albumen gland is absent. The penis is not a sac which can be turned inside out, but a projecting lobe of the body, often of great size. In the most primitive Mollusca—the Solenogastres—the genital organ remains throughout like a thickening of the wall of the pericardium or coelom; the eggs and spermatozoa drop into the pericardium and find their way out by the nephridia, just as is the case with Annelida.

This is the case also in Cephalopods, where, however, there were originally four kidneys, and the one or two which serve as generative ducts are specialized for this purpose; thus the duct is in the male prolonged into a papilla which serves as the penis. A commoner case is for the generative organ to be closely connected with one kidney and to burst directly into it. This is found in the simpler Prosobranchiata, such as the Limpet (*Patella*), the Ear-shell (*Haliotis*) and their allies. In *Nucula* and the simplest bivalves there are two generative organs and they open into both kidneys; in the Pond-mussels (*Anodonta* and *Unio*), and the more modified forms, they open independently close to the kidney opening. There is little doubt that in all Mollusca the tube conveying away the generative products was originally a kidney or a part of one.

Having got some idea of the arrangement of the organs of the snail we must proceed to consider certain points about the form of the body considered as a whole. If we except the genital opening, the head and neck of the snail are exactly bilaterally symmetrical in their outer form; on each side there is a taste-tentacle and an eye-tentacle and the mouth and the opening of the mucous gland are exactly in the middle line. Most of the ordinary animals we see—birds, quadrupeds, fishes, insects, worms, etc.—are bilaterally symmetrical with regard to the exterior and many with regard to the whole body. The peculiarity of the snail is that, while it follows the ordinary rule as far as the head, neck and foot are concerned, it departs from it with respect to the visceral hump and the included organs. The shell is, as we all know, spiral, but this shape is due to the shape of the visceral hump contained within it, by the activity of the skin of which the shell is produced. This spiral shape again is simply due to one side being longer than the other, and it is connected with the shortness of the right side that we find the opening of the vent on the right side. In all bilaterally symmetrical animals this opening is situated in the middle line, but in some of the marine allies of the snail—the whelk, limpet and others forming the group

Asymmetry of body.

Prosobranchiata—the inequality of the two sides of the visceral
hump is carried to such an extent that the anus is brought right
round so as to open nearly over the middle of the neck ; and where,
as in the Ear-shell (*Haliotis*), there are two gills, the left becomes
pushed over to the right side and the gill belonging to the right
side becomes displaced to the left.   Since the visceral ganglia are
connected with the bases of the gills, one side of the visceral loop
becomes pulled over the other in consequence of the displacement
of the gills (Fig. 116).   This condition of the nervous system is
called the streptoneurous condition; it exists in all the groups
which are ordinarily termed "sea-snails," i.e., Prosobranchiata, and
though most of these have only one gill, the twisting of the visceral
loop may be regarded as a proof that they originally had two.   In
another large division of the sea-snails, the Opisthobranchiata, the
shell is generally small or has quite disappeared, and since where
this has taken place there seems to have been a tendency to undo
the twisting, the anus becomes pushed back to nearly the middle
line and the visceral loop becomes straightened out and shortened.
There is reason to believe that this last process has gone on in the
snail, though it has kept its shell.   It appears then that the curious
spiral form of part of the body and the inequality of the sides
have something to do with the possession of a large shell by a
crawling animal.   We do not understand very clearly how the one
thing has brought about the other, but we can understand that
there would be a tendency in a tall visceral hump to topple over
to the one side or the other and thus exercise a greater strain on
one side than on the other.   Certain it is, at any rate, that the
only existing Mollusca which possess large coiled shells and yet are
bilaterally symmetrical, are the pearly *Nautilus* and another rare
Cuttle-fish (*Spirula*), which do not crawl but swim.

The class or primary division of the Mollusca to which the snail
belongs is called the Gasteropoda, on account of the flat smooth
foot or crawling surface which they all possess (Gr. γαστήρ, the belly ;
πούς, ποδὸς, foot).   The shell is typically composed of a single piece,
never of paired pieces ; and from this circumstance is derived the
general term "univalve" often applied to the Gasteropoda by
collectors ; in one small division of the class (the Isopleura) the
shell is represented by eight pieces placed one behind the other in
the middle line.

Class II.   Lamellibranchiata = Pelecypoda.

The characters mentioned at the end of the last section sharply
separate the Gasteropoda from another class of
Mollusca, the Lamellibranchiata or Pelecypoda, to
which the common mussel and innumerable marine
forms, such as the oyster, clam, cockle, etc., belong.   The Molluscs
belonging to this class have a shell composed of two similar pieces,
the right and left valves, united by a horny flexible piece, the hinge
(Fig. 117).   The foot is typically formed like a wedge or axe-head,

Fresh-water Mussel.

Fig. 117.   Shell containing *Anodonta mutabilis*, and behind it the inner face of
an empty left shell.

1. Points of insertion of the anterior protractor (above) and retractor muscles
(below) of the shell.   2. Point of insertion of the anterior adductor muscle.
3. Point of insertion of the posterior protractor of the shell.   4. Point of
insertion of the posterior adductor muscle.   5. Lines formed by successive
attachments of the mantle.   6. Umbo.   7. Dorsal siphon.   8. Ventral
siphon.   9. Foot protruded.   10. Lines of growth.

whence the name Pelecypoda (Gr. πέλεκυς, a hatchet), and is used
as a plough to force a way through the mud in which the creatures
live.   There are many species of pond- or river-mussels in North
America: *Anodonta cygnaea* is perhaps the commonest in England,
but in places *Unio pictorum* is abundant; *A. cygnaea* occurs in
Canada and the United States and in these countries *Unio com-
planatus* is also common.   Any one of these forms will serve our

purpose.   The shell is about four inches long and two inches high,
and is covered with a black horny layer, the so-called
The Shell.
periostracum.   The shell is apt in places to be
eroded by the action of the carbonic acid in the water.   Underneath it is a thick slightly translucent layer of crystals of carbonate
of lime, called the prismatic layer.   The inner part next the
mantle is composed of thin layers placed one above the other.
This is the mother-of-pearl or nacreous layer, which in many
Molluscs has an iridescent sheen, owing to its action on light.
These three layers are also present in the shell of the snail and in
all other Molluscan shells, but they are very easily made out in the
shell of the pond-mussel.   To the periostracum the colour of the
Molluscan shell is mainly due.   The periostracum and prismatic layers
are formed by the edge of the mantle and if destroyed they cannot be
replaced.   The nacreous layer is deposited by the whole surface of
the mantle.   If by chance a grain of sand gets wedged in between
the mantle and the shell it is apt to become covered with layers of
mother-of-pearl, and in this way a pearl is formed.   The more costly
pearls however arise within the soft parts of the body, usually
encysting around some parasitic larva.   The shell is marked by a
series of curved lines running parallel to one another.   These lines
mark the limits of growth attained in each year, the amount intervening between two lines being the amount of growth accomplished
in a year.   It will be seen that the common focus around which the
curves run is not in the centre of the hinge line, but decidedly
nearer one—the anterior—end.   This common focus is called the
umbo, and it represents the shell with which the *Unio* started life
(6, Fig. 117).

As might be expected from the shape of the shell, the mantle
has the form of two great flaps hanging down at the sides of the
body.   The flaps have a free edge in front, below and behind, but
pass into the general wall of the body, with which they fuse, above.
The edges of the mantle flaps are very much thickened and closely
adherent to the shell; as stated above, it is by these edges alone
that the periostracum and the prismatic layer are formed.

The hinge is strictly speaking part of the shell; it is secreted
by the ectoderm of the back of the animal between the two mantle
lobes.   When the valves of the shell are pressed closely together
the hinge is bent out of shape and by its elasticity it tends to
throw the valves apart; hence when a mussel is dead the valves
always gape.

The two valves in *Unio* articulate with one another by means

of teeth. There are a pair of stout teeth a little in front of the
umbo, on the left valve, working on either side of one tooth on the
right valve ; these are called the cardinal teeth. A long ridge
on the right valve, working between two ridges on the left valve,
is called the lateral tooth. *Anodonta* derives its name (Gr. ἀν-,
not ; ὀδούς, ὀδόντος, a tooth) from the circumstance that the shell is
devoid of teeth.

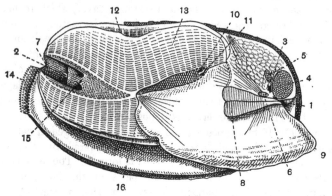

FIG. 118. Right side of *Anodonta mutabilis* with the mantle cut away and the
    right gills folded back × about 1. From Hatschek and Cori.

1. Mouth.   2. Anus.   3. Cerebro-pleural ganglion.   4. Anterior adductor
muscle.   5. Anterior protractor muscle of the shell.   6. Retractor
muscle.   7. Dorsal siphon.   8. Inner labial palp.   9. Foot.   10. External opening of nephridium or organ of Bojanus.   11. Opening of
genital duct.   12. Outer right gill-plate.   13. Inner right gill-plate.
14. Ventral siphon.   15. Epibranchial chamber, the inner lamellae of
the right and left inner gills having been slit apart.   16. Posterior protractor muscle.

When the shell is removed from the animal the cut ends of the
fibres of two large muscles are seen. These muscles, which run

The Body.   transversely from the one valve to the other, are called
           the anterior and posterior adductors respectively,
and it is by means of them that, when danger threatens, the animal
closes the valves and shelters foot, gills and body, within. Just
behind the anterior adductor is a pair of small muscles running into
the foot, and these are the anterior protractors of the shell. A
similar pair, the posterior protractors, are found just in front of the
posterior adductor, and by the combined action of the four the shell is
drawn forward, the foot being (relatively) fixed in the mud (Figs. 118,
121). The foot is thrust forth by the forcing of blood into it, through
the contraction of the muscles which underlie the skin in various parts
of the body. The retractors (of the shell) enable the animal to

move backwards when necessary. A small group of muscles running from the mantle to be attached to the shell near the umbo pull the shell downward and help to plough a furrow in the mud. The animal moves by forcing out the foot and wedging it in the mud in front and then drawing the body after it.

At the sides of the body on each side we find the branchia or gill, or c t e n i d i u m, which as in the Gasteropoda consists of a hollow axis bearing two rows of plates. The ctenidium is, however, highly

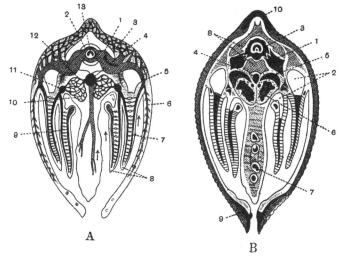

FIG. 119.   A. Diagrammatic section through *Anodonta* to show the circulation of the blood.   B. Section through *Anodonta* near the posterior edge of the foot. From Howes.

A. 1. Right auricle.   2. Ventricle.   3. Keber's organ.   4. Vena cava. 5. Efferent branchial trunk.   6. Efferent pallial vessel.   7. Efferent branchial vessel.   8. Branchiae.   9. Afferent branchial vessel. 10. Efferent renal vessel.   11. Afferent branchial trunk.   12. Afferent renal vessel.   13. Rectum.

B. 1. Right auricle.   2. Epibranchial chamber.   3. Ventricle.   4. Vena cava.   5. Non-glandular part of the kidney.   6. Glandular part of the kidney.   7. Intestine in foot.   8. Pericardium.   9. Shell. 10. Ligament of shell.

modified in *Unio*. The axis is attached high up to the side of the body in front but projects freely into the mantle cavity behind. The plates have become narrowed so as to form long filaments, and the ends of each row are bent up and are, in the case of the outer row, fused to the mantle lobe. The bent-up ends of the inner row are joined to the foot in front and to the corresponding parts of the ctenidium of the other side behind, but in the middle

they are free, at least in some species (Fig. 119). Successive filaments of one row are welded together into a plate, called a lamella,
by the fusion of their adjacent edges, leaving only occasional holes
for the percolation of the water, so that individual filaments appear
like ridges on a ploughed field (Figs. 118, 120). The descending
and bent-up ends of the same filament are tied together by cords or
narrow plates of tissue traversing the space between them. These
cords and plates are called interlamellar junctions, since they
unite two lamellae. The pieces of tissue uniting the filaments are
called interfilamentar junctions, or collectively, subfilamentar
tissue. Gill-plate is the name given to the whole mass composed
of one row of V-shaped filaments: there is thus an outer and an
inner gill-plate on each side, and each gill-plate has two lamellae
formed from the descending and ascending limbs of the filaments,
respectively (Fig. 119). It is this peculiar modification of the
ctenidia which has suggested the name Lamellibranchiata for the
class.

Each V-shaped filament is clothed on its outside, that is, the side
looking away from the concavity of the V, with high ectoderm cells
carrying powerful cilia. By the action of these a strong indraught
of water is produced, the current entering between the posterior
borders of the mantle lobes, which normally gape slightly. On this
current the animal depends both for respiration and for nutrition,
since the food consists entirely of the minute animals and plants
swept in with the water. The normal position of the Mussel is to
have the anterior end deeply embedded in the sand or mud and the
posterior end protruding; the animal moves only when for some
cause the water becomes unsuitable for its purposes.

Since the upturned ends of the inner rows of filaments of both
ctenidia are united behind the foot, a bridge is formed dividing the
mantle cavity into an upper or epibranchial division and a lower
or hypobranchial. The gaping opening between the mantle lobes
at the posterior end is similarly divided into an upper portion, the
dorsal siphon, and a lower, the ventral siphon. Since it is the
outer lower surfaces of the filaments which are clothed with cilia,
it is into the ventral siphon and hypobranchial chamber that the
current passes. The lips of both siphons—especially the ventral
siphon—are plentifully beset with small papillae, which are sensitive
to light and shade. If the shadow of the hand be allowed to pass
over them the mantle edges are instantly drawn together and the
siphons thus closed. In the scallop (*Pecten*) similar papillae are

developed into well-formed eyes. Part of the water passes through the small holes left between the gill-filaments and so into the epibranchial chamber and escapes by the dorsal siphon, carrying with it the matter cast out from the kidneys and the anus. As it percolates through the gills the blood which circulates in these organs receives oxygen and gets rid of its carbonic acid.

A large part of the water, however, pursues a different course. In front of the gills there are situated two organs called labial palps, on each side of the anterior part of the animal (8, Fig. 118).

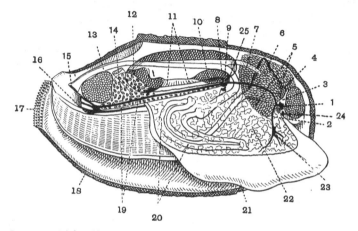

Fig. 120. Right side of *Anodonta mutabilis*, dissected to show the viscera × about 1. From Hatschek and Cori.

1. Cerebro-pleural ganglion.   2. Cerebro-pedal commissure.   3. Oesophagus.   4. Anterior protractor muscle.   5. Liver.   6. Stomach.   7. Aorta.   8. External opening of organ of Bojanus or nephridium.   9. Nephrostome or internal opening of the same.   10. Pericardium.   11. Right auricle.   12. Posterior end of ventricle passing into posterior aorta.   13. Rectum.   14. Glandular part of nephridium.   15. Anus.   16. Opening of epibranchial chamber.   17. Ventral siphon.   18. Edge of shell.   19. Cerebro-visceral commissure.   20. Intestine.   21. Foot.   22. Reproductive organs.   23. Pedal ganglion of right side.   24. Mouth.   25. Opening of the reproductive organ.

These are triangular flaps, an upper and lower on each side, the surfaces of which are covered with grooves clothed with abundant cilia on the sides turned towards one another. The two superior labial palps are connected by a narrow ridge crossing above the mouth; the two inferior labial palps by a similar ridge beneath it. The mouth thus lies at the bottom of a trough,—the lips of which are formed by the superior and inferior labial palps respectively.

The mouth is situated beneath the great anterior adductor muscle which projects beyond it like a forehead. The action of the labial palps is to direct a large portion of the incoming current into the mouth, and thus the animal obtains its food.

The alimentary canal shows a considerable resemblance to that of the snail. No trace, however, of radula, buccal mass, crop or salivary glands, is to be seen. A short oesophagus leads at once into the stomach, which is a wide sac receiving right and left the ducts of the two lobes of the liver. The intestine runs vertically down into the foot, makes several loops there and then turns back and reaches nearly to the point from which it started, i.e., the hinder end of the stomach. Thence it pursues a straight course through the pericardium and over the posterior adductor muscle, to end in an anal papilla which projects into the epibranchial mantle-cavity. For part of its course the ventral wall is infolded towards the cavity so as to produce a ventral typhlosole comparable to the dorsal typhlosole in the worm. The straight concluding portion of the intestine is called the rectum. The pericardium is situated in the mid-dorsal line posterior to the stomach. The fact that it surrounds the rectum is the consequence of its origin as a pair of sacs in the embryo lying to the right and left of the intestine, which later meet above and below this organ.

There are, as mentioned above, two kidneys or nephridia in the mussel. These, frequently termed the organs of Bojanus, are dark coloured bodies situated beneath the floor of the pericardium on either side of the vena cava. Each consists of a U-shaped tube lying horizontally, with one limb placed vertically above the other and the bend directed backwards. The deeper limb is the active part; it has numerous folds projecting into it which are covered with dark cells. It opens into the pericardium in front by a curved slit lined with powerful cilia which produce an outward current. This of course is the reno-pericardial duct such as has been already described in the snail. The outer and upper limb is wide and smooth-walled and opens into the deeper limb beneath the posterior adductor. In front it opens to the exterior through a pore with thick lips placed just above the place where the upturned ends of the inner row of filaments are attached to the foot (Fig. 120).

The kidney, since it is a tube lined with excretory cells and communicating internally with the body-cavity, is a nephridium comparable to that of the worm, *Lumbricus*. In the worm the function of the internal opening is to convey to the exterior the fluid in the body-cavity, which contains excretory matter thrown out by the cells lining the coelom. The anterior end of the pericardium of the Mussel has a brownish red colour and is produced into numerous little pockets lined by peculiar cells, which are excretory in function (Fig. 119). This portion of the pericardial wall is called Keber's organ, and the excreta thrown out by it pass down the reno-pericardial canal.

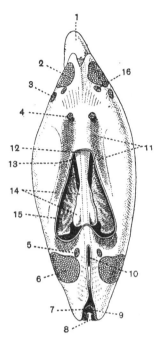

FIG. 121. Dorsal view of *Anodonta mutabilis*, with the upper wall of the pericardium removed to show the heart × about 1. After Hatschek and Cori.

1. Foot. 2. Anterior adductor muscle. 3. Retractor muscle. 4. Depressor muscles. 5. Posterior protractor muscle. 6. Posterior adductor muscle. 7. Dorsal siphon. 8. Ventral siphon. 9. Anus. 10. Split between left and right mantle lobes through which larvae at times leave the epibranchial chamber. 11. Keber's organ. 12. Rectum traversing ventricle. 13. Nephrostome or internal opening of organ of Bojanus. 14. Ventricle. 15. Left auricle. 16. Anterior protractor muscle.

The heart consists of a ventricle which surrounds the rectum, and two flat triangular auricles, the broad bases of which are inserted into the wall of the pericardium just over the place where the bent-up ends of the outer filaments of the ctenidium are attached to the mantle. From the ventricle blood is driven forwards by an anterior aorta dorsal to the rectum, and backwards by a posterior aorta ventral to the rectum. From these arteries it finds its way into a multitude of irregular spaces in the foot and the other portions of the body, and eventually reaches a vessel, called the vena cava, lying under the floor of the pericardium in the middle line, between the right and left upper limbs of the two kidneys. From the vena cava the blood streams out through many channels in the wall of the kidney and reaches the axis of the ctenidium, whence it makes its way into the filaments, especially those of the outer row. From the upturned edges of these last it reaches the mantle and from this the auricle.

Some blood is sent to the lobes of the mantle, as in the snail, and through the thin skin absorbs oxygen ; this blood is returned direct to the auricle without passing through the gill ; from this fact it appears that the mantle lobe as well as the gill is a respiratory organ.

In the nervous system the cerebral and pleural ganglia on each side are generally regarded as coalesced, but a distinct pleural ganglion has been observed in some cases on the cerebro-visceral commissure anterior to the pericardium. There is a long visceral loop, ending in two closely conjoined visceral ganglia, placed beneath the posterior adductor (Fig. 120). On either side of these, just where the axis of the ctenidium becomes free from the body, is a thickened patch of yellow ectoderm—the osphradium. This is a peculiar sense-organ, the function of which, it is believed, is to test the water passing over the gill as to suitability for respiration. There is a pair of large otocysts in the foot.

Mussels are male and female : their productive organs are paired, and consist on each side of a bunch of tubes spreading through the foot. The ducts are continuous with the walls of the ovary or of the testis. They open by slit-like orifices just in front of the opening of the nephridia on each side of the foot. The spermatozoa are swept out by the water passing through the dorsal siphon and are sucked in by the inhalant currents of female indi-viduals. The eggs when cast out are detained between the two lamellae of the outer gill-plate and there fertilized. They develope into peculiar larvae called Glochidia, provided with a sticky thread or byssus. A bivalve shell is developed but not the foot. When a fish passes by the mother expels the Glochidia from the gills, and they seize hold of the tail or fins of the fish and embed themselves therein. They develope there for some weeks and change gradually into the adult form. They show a remarkable sensitiveness to the presence of fish, but if they fail to attach themselves to one they fall to the bottom of the water and perish.

Lamellibranchiata as a group have very uniform habits : the principal points in which they differ from one another are (1) the degree of complexity which that all-important organ the ctenidium has attained, and (2) the extent to which the animal is able to burrow.

The simplest forms, such as *Nucula*, have ctenidia like those of a Gastropod, a fact which suggests the view now generally held, that the Lamellibranchs are descended from some primitive type of Gastropod.

In others, such as the Sea-mussel (*Mytilus*), the ctenidia have the same external appearance as those of *Unio*, but the filaments are very loosely united with one another and their upturned ends are not fused to the mantle. The foot is small and tongue-shaped, the animal never burrows and rarely moves, being fixed by a cord of mucus called the byssus, secreted by a gland in the hinder part of the foot.

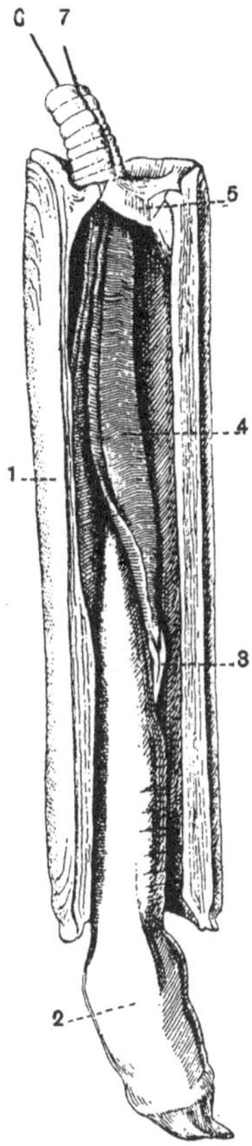

In the Oyster (*Ostrea*) the foot has disappeared and the animal passes its life resting on one side. In the Scallop (*Pecten*) the foot has also atrophied, but the animal is able to swim through the water by flapping the valves of the shell. The Cockle (*Cardium*) has a large and powerful foot by which it is enabled to execute leaps.

Mya (sometimes known as the Clam, though this term is applied to many Bivalves) and its allies burrow deeply in the sand and have the edges of the mantle behind drawn out into two long tubes closely apposed to one another, termed the dorsal and ventral siphons. By means of these tubes they keep up a connexion with the surface, so that the currents of water are not interrupted. Similar tubular funnels, though not so much drawn out, are seen in the Razor-shell (*Solen*) (Fig. 122). *Pholas* and some others are able to burrow in rock; this is said in some cases to be effected by an acid secretion poured out by the flat disc-like end of the cylindrical foot. *Teredo*, the ship-worm, burrows in timber; the siphons are very long and covered with a shelly deposit; the original valves of the shell are very small compared to this secondary shelly tube. This animal is very destructive to submerged wooden structures; a wooden pile supporting a pier in

FIG. 122. *Solen vagina*, the Razor-shell, the shell is opened and the posterior part of the mantle is torn to admit of this.
1. Shell. 2. Foot. 3. Labial palps. 4. Gills. 5. Torn portion of mantle. 6. Bristle in ventral siphon. 7. Bristle in dorsal siphon.

Vancouver was in eighteen months reduced to a mere spongework
of wood by its ravages.

Class III. CEPHALOPODA.

A third class of the Mollusca, very differently constituted from
the Lamellibranchiata, is that of the Cuttle-fish, or Cephalopoda.

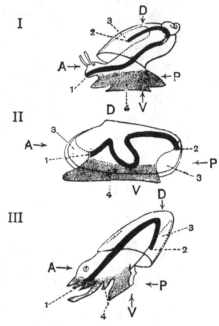

This paradoxical name,
literally "head-footed"
(Gr. κεφαλή, head ; πούς,
ποδός, a foot), is suggested
by the circumstance that
the foot has grown for-
ward and upwards at each
side of the head, and that
these two extensions have
met and coalesced, so to
speak, on the back of the
neck. The edges of this
part of the foot, which
may be called the fore-
foot, are drawn out in-
to strap-like processes,
which are the arms by
which the animal seizes
its prey. The edges of
the hinder part of the
foot, on the other hand,
have become bent round
and joined beneath the
animal, so as to form a
tube, the funnel, through
which water is ejected
from the mantle-cavity.

The best known Brit-
ish Cepha-

Sepia.

lopoda are

FIG. 123. Diagrams of a series of Molluscs
to show the form of the foot and its
regions and the relations of the visceral
hump to the antero-posterior and dorso-
ventral axes. After Lankester.

I. A Prosobranch Gastropod.   II. La-
mellibranch.   III. A Cephalopod.
A. Anterior surface.   P. Posterior surface.   D. Dorsal surface.   V. Ventral
surface.   1. Mouth.   2. Anus.
3. Mantle-cavity.   4. Foot.

the Squid, *Loligo forbesi*, often caught by trawlers, to whom it is
known as the 'ink-fish'; *Sepia officinalis*, the cuttle-fish, taken in
the southern waters, is abundant in the Mediterranean, where it is a
favourite article of food ; *Moschites cirrosa* with eight arms and a
single row of suckers ; *Polypus* (or *Octopus*) *vulgaris* almost con-

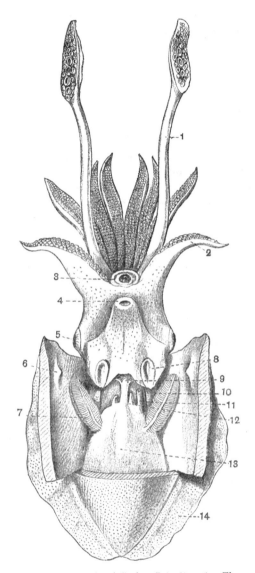

FIG. 124. Posterior view of male of *Sepia officinalis* × 1. The mantle-cavity has been opened to expose its contents.

1. Long arm half protruded. 2. A short arm; this one is hecto-cotylized. 3. Lips surrounding horny jaws; mouth. 4. External opening of funnel. 5. Eye. 6. Cartilaginous knob on mantle which fits into the socket 8. 7. Gill. 8. Socket for 6. 9. Anus. 10. Depressor muscle of the funnel. 11. Reproductive pore. 12. Right kidney papilla. 13. Visceral mass. 14. Fin.

fined to the south coast and commoner on the French shores and in the Mediterranean. Another squid, *Illex illecebrosa*, is common in the Gulf of St Lawrence and on the shores of the eastern United States. The body of *Sepia* appears to be composed of a swollen head separated by a neck from a tapering trunk. When closely examined, however, the body is seen to be nothing but a long pointed visceral hump, like that of the snail, but it is not twisted and is unprotected by an external shell. The mantle, as in the snail, is a skirt-like fringe of skin, the space between its inner surface and the visceral hump forming the large mantle-cavity. In order to compare the animal with the snail it must be placed with the point of the hump projecting upwards and backwards (Fig. 123). The so-called head includes the true head, with two enormous eyes of almost human aspect, surrounded by the fore-foot. The fore-foot is drawn out into eight short pointed arms, thickly covered on their inner sides with stalked suckers (Fig. 124), and two very long arms bearing suckers only at their expanded ends. These latter can be pulled back into two large pits situated at their bases, and when so retracted they are completely hidden from view.

The sucker is a cup with a horny rim which keeps the opening from collapsing. In its base there is a swelling, which is the end of a muscle running into the stalk, and by the contraction of this, when the cup is applied to any object, a partial vacuum is produced. By means of the suckers the Squid can take a firm hold of its prey.

The hind-foot is the tube, known as the funnel (Figs. 124 and 127). The posterior end of this is overlapped by the hind end of the mantle; in other words, it projects into the mantle-cavity. The mantle is very muscular; by the contraction of longitudinal muscles running towards the apex of the hump the mantle-cavity is widened; by the contraction of circular muscles it is narrowed, and by the alternate action of these two sets of muscles, water is sucked in and forced out of the mantle-cavity.

When, however, the mantle-cavity is contracted, two projecting pegs on the inner side of the mantle fit into sockets on the outer side of the funnel (6 and 8, Fig. 124). No water can then escape over the free edge of the mantle, and all is ejected in a narrow and forcible stream through the funnel. The funnel itself is muscular and by contraction aids the process; there is a valve-like projection inside it which prevents water from being driven back into the mantle-cavity (Fig. 127).

Since water is sucked in gently and ejected forcibly the animal
is propelled in the opposite direction, that is backwards, by the
reaction of the stream against the surrounding water. *Sepia* can
however also swim gently forward by wave-like undulations of the
two lateral fins. These fins are flaps of skin projecting from the
sides of the visceral hump (14, Fig. 124).

It has been said that the visceral hump is unprotected by an

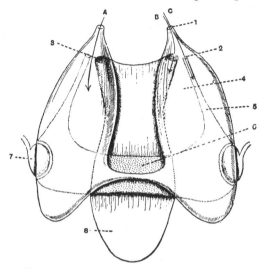

FIG. 125. A diagram showing the relation of the kidneys to the pericardium
in *Sepia*.

1. External opening of the kidney into mantle-cavity.   2. Internal opening
of the kidney into the pericardial coelom.   3. Opening of the right kidney
into the dorsal sac and hence into the left kidney.   4. Left kidney,
ventral portion.   5. Reno-pericardial canal.   6. Pericardium (part of
the coelom).   7. Branchial heart.   8. Dorsal sac common to both kidneys.
A. Arrow passing into left kidney by external opening from mantle cavity.
B. Arrow passing into right kidney through external opening into median
lobe.   C. Arrow passing into external opening and then into internal
opening, and so into pericardial coelom. The extension of the coelom in
which the generative cells arise is not shown in this diagram.

external shell. This is not strictly true. On the anterior surface
of the hump there is an oval plate-like shell completely hidden in
a sac formed by the meeting over it of upturned flaps of skin (14,
Fig. 127). From its upper surface project an innumerable number
of delicate calcareous plates parallel to one another, the spaces
between them being filled with gas so as to give lightness.

In one or two living Cephalopoda, as for instance the Pearly
Nautilus (*Nautilus pompilius*), and in very numerous extinct forms,

there was a large tubular external shell which might be straight or coiled, but which always had the peculiarity of having a large number of septa or transverse plates dividing up its cavity into chambers, only the last of which contained the visceral hump, the rest being filled with gas (Fig. 130). In *Sepia* it is supposed that the chambers have become so small and shallow that the last one simply appears as a plate situated on part of the surface of the hump. The other chambers contain gas as in *Nautilus*.

Sepia possess two well-formed ctenidia, each consisting of an axis bearing two rows of thin plates. The axis is suspended from the body by a membrane, and the ctenidia project forwards and downwards instead of backwards as in the Lamellibranchiata (7, Fig. 124).

As in Lamellibranchiata, there are two kidneys which open by little papillae placed just in front of the bases of the ctenidia (12, Fig. 124). Just inside the papilla is a narrow opening, the lips of which are folded so as to make it appear like a rosette. This is the internal opening of the kidney: it leads into a lateral prolongation of the pericardium, which is the reno-pericardial canal (5, Fig. 125). The kidney has the form of a wide sac and may perhaps be compared to a U-shaped kidney, like that of *Unio*, in which the two limbs have become merged in one another.

The wall of the kidney is smooth, except over the course of the large veins which run beneath its upper and inner wall. Here the epithelium is folded and consists of tall cells which are actively engaged in extracting excreta from the blood, as is shown by the rows of granules with which they are filled. From the anterior ends of the kidney two outgrowths project which immediately fuse into one and constitute a great pouch called the dorsal sac (8, Fig. 125) stretching upwards and backwards just underneath the shell. This peculiar extension gives to the two kidneys the form of M, between the median V and outer limbs of which lies the pericardium. Posteriorly the cavities of the right and left kidneys also communicate with one another.

The pericardium is a wide sac lying between the dorsal sac of the kidneys and their ventral parts. At the sides it gives off the reno-pericardial canals, whilst the stomach and intestine project into its roof.

The ventricle of the heart is a spindle-shaped sac lying transversely (1, Fig. 126). Into its two ends open the tubular thin-walled auricles which receive the blood from the ctenidia. From its

anterior wall a powerful artery, the anterior aorta, is given off
running forward above the oesophagus to the head, and a smaller
artery or posterior aorta goes backwards and upwards to supply the
stomach and genital organ.   These arteries have regularly formed
branches from which the blood enters definite veins.   This formation
of well-defined channels for the blood is characteristic of the Cephalopoda.   Of these veins the most important are : (1) the anterior
vena cava ; this is a channel in the mid-ventral line in front of the

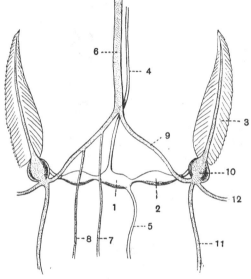

FIG. 126.   View of heart and chief blood-vessels of *Sepia cultrata*.
Partly after Parker and Haswell.

1. Ventricle.   2. Auricle.   3. Ctenidium.   4. Anterior aorta.   5. Posterior aorta.   6. Anterior vena cava.   7. Vein from ink-sac.
8. Genital vein.   9. Branchial vein.   10. Branchial heart.   11. Right
abdominal vein.   12. Vein from the mantle.

kidneys, which forks and sends a branchial vein over each kidney to
the base of each gill ; (2) the abdominal veins ; these are a pair of
large channels which come from the mantle, especially the upper
part of it, and join the forks of the vena cava just before they enter
the gills ; (3) the genital vein ; this is a trunk draining the genital
organ, it runs along the ventral wall of the dorsal pouch of the
kidney and joins the right fork of the vena cava ; (4) a vein from the
ink-sac joins the same fork, and (5) on each side a smaller vein from
the mantle joins the main venous system at the branchial hearts

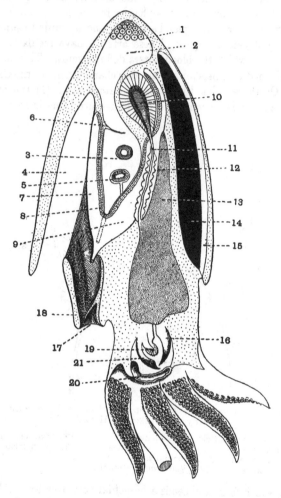

FIG. 127. Left half of a *Sepia*, showing the relation to one another of some
        of the principal viscera, × about 1. Diagrammatic.

1. Ovary.   2. Genital part of the coelom.   3. Pericardial part of the
coelom; the reference line touches the heart.   4. Mantle-cavity.   5. Section of intestine.   6. Incomplete septum between genital and pericardial
coelom.   7. Ventral limb of kidney.   8. Glandular tissue of kidney.
9. Dorsal limb of kidney.   10. Stomach.   11. Liver duct.
12. "Pancreatic" caeca.   13. Liver.   14. Shell.   15. Shell sac.
16. Dorsal horny jaw.   17. Anterior opening of funnel.   18. Valve
in funnel.   19. Radula.   20. Lips.   21. Ventral horny jaw.

(Fig. 126). Where these veins come in contact with the kidney wall the special excretory tissue mentioned above is developed.

A peculiar feature in the circulatory system is the presence of a pair of muscular swellings of the forks of the vena cava just before they enter the gills. These are the branchial hearts, the function of which is to drive the blood into the gills, whereas the auricles drain it out of them. Each branchial heart projects on the one side into the kidney and on the other side into the reno-pericardial canal, and the epithelium of the latter where it covers the heart is greatly thickened so as to form a cushion, the function of which is excretory. This, like Keber's organ in the river-mussel, is a remnant of the primitive excretory function which probably all the cells of the coelom once possessed.

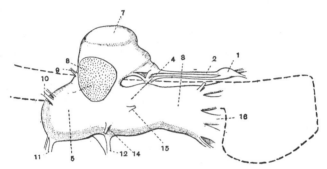

Fig. 128. Lateral view of the central nervous system of *Sepia officinalis*. Magnified. From Cheron.

1. Upper buccal ganglion.   2. Nerves connecting buccal ganglion with cerebral ganglion.   3. Brachial ganglion.   4. Infundibular ganglion. 5. Pleural ganglion.   7. Supra-oesophageal ganglion.   8. Cut end of optic nerve.   9. Superior ophthalmic nerve.   10. Pallial nerves. 11. Visceral nerve.   12. Anterior nerve to the funnel.   14. Auditory nerve.   15. Inferior ophthalmic nerve.   16. Nerves to the arm. The dotted outline represents the buccal mass and the oesophagus.

The alimentary canal of *Sepia* is constructed on very much the same plan as that of the snail. The mouth is situated in the centre of the arms and surrounded with a frilled lip (Fig. 127). There is a large buccal mass containing the radula and there are a pair of powerful jaws shaped like a parrot's beak, moveable on one another, of which the ventral is the larger. There are two salivary glands and a long narrow oesophagus but no crop. The oesophagus widens behind into the stomach, which receives, as is usual in Mollusca, the ducts of the liver. With the stomach is connected a side pouch spirally coiled. The liver is enormous, occupying all the anterior portion of the visceral hump; the ducts traverse the dorsal extension of the kidneys, and are in this position covered externally with excretory tissue which by the older naturalists was termed "pancreatic caeca" (12, Fig. 127) from a mistaken comparison with the human pancreas. The intestine is slightly bent on itself and ends in an oval papilla. A peculiar sac, the ink-bag, the cells lining which secrete the pigment known as Indian ink or Sepia, opens by a long duct on this papilla (Fig. 129). When the Cuttlefish is alarmed it ejects this ink and darkens the water so much as completely to escape from view.

The nervous system consists of ganglia even more closely massed than in the case of the snail. The supra-oesophageal ganglia form one rounded mass; they are produced at the sides into the very much larger optic ganglia which are in close relation to the eyes (Fig. 128). The pedal ganglion is divided into an anterior ganglion called the brachial, and a posterior ganglion sometimes called the infundibular. The brachial ganglion supplies a stout nerve to each arm, and each nerve swells out into a small ganglion just where it enters the arm (Fig. 129). The infundibular ganglion supplies the funnel.

Posterior to the infundibular ganglion are the two pleural ganglia fused together. These give rise to a visceral loop which supplies the various internal organs and the gills. From the same ganglion two short nerves run to the mantle and terminate in the two large stellate ganglia, which underlie the skin and supply nerves to all the muscles of the mantle (Fig. 129). The buccal mass is supplied by two ganglia, superior and inferior, each representing a pair joined by a minor nerve collar running round the oesophagus. The inferior ganglion corresponds to the buccal pair in the snail, the superior is a separated part of the cerebral.

It has been already stated that *Sepia* possesses complicated eyes. In the embryo these are like the eyes of the snail, merely sacs lined by visual cells and containing a transparent horny secretion which serves as a lens. In fact, in the embryo, the sac is at first a pit which gradually closes up. Immediately over the spot where the first pit closed a second pit is formed in which a second horny lens is formed just over the first one, so that the lens consists of two pieces, and as in the eye of the Vertebrate, there is an anterior and a posterior chamber in the eye separated from one another by the lens.

Into the anterior chamber a circular fold projects called the
iris, fulfilling exactly the same function as the iris of the human eye
or the diaphragm in a photographic camera. Outside and around
the eye a circular fold acts as an eyelid.

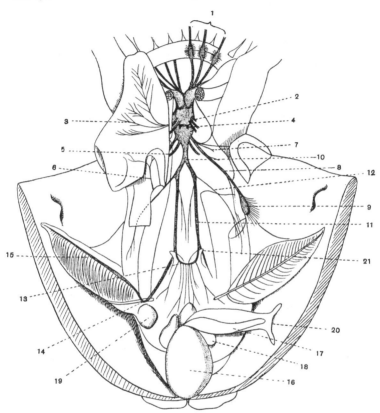

Fig. 129. *Sepia officinalis* dissected to show the nervous system, ventral view.
From Cheron.

1. Three nerves to the arms dissected out. 2. Auditory nerve. 3. Anterior
nerve to the funnel. 4. Nerve to vena cava. 5. Posterior nerve
to funnel. 6. Continuation of this nerve. 7. Accessory nerve to
mantle. 8. Left nerve to mantle. 9. Stellate ganglion. 10. Common trunk of the visceral loop. 11. Left branch of the visceral loop.
12. Nerve to muscles. 13. Nerve to viscera. 14. Ganglion on
branchial heart. 15. Nerve of ctenidium. 16. Ink-bag. 17. Duct
of ink-bag. 18. Left nidamental gland. 19. Branchial heart.
20. Position of anus. 21. External opening of kidneys.

The cerebral, pedal and pleural ganglia are surrounded by very
tough connective tissue, in which the fibres, although still visible,

have a cheesy consistence, the tissue being called fibro-cartilage. In this way a kind of skull is formed, which sends a scoop-like extension on either side over the back of the eye, covering the optic ganglia. The edges of the scoop pass into ordinary connective tissue round the rest of the eye. This tissue forms the wall of the eyeball.

The otocysts are branched and embedded in the ventral wall of the skull. They receive nerves from the cerebral ganglia, but as the nerve fibrils traverse the pedal ganglia they appear to arise from the latter.

The genital organ in *Sepia* occupies the apex of the visceral hump. It is, as examination of young specimens shows, a thickening of the wall of the coelom, behind the pericardial coelom. The space into which the eggs and spermatozoa are shed is only shut off from the pericardium by an incomplete partition, so that in *Sepia* a state of things persists throughout life which is found only in the embryos of some other forms (Fig. 127).

The genital duct is present only on one side, the left, and is produced into a prominent papilla which in the male is used as a penis (Fig. 124). The outermost part of the duct in the male is a wide pouch in which the spermatozoa are welded into masses and enveloped in cylindrical cases called spermatophores. Just beyond this the duct receives the excretion of two glands termed prostate glands, and then narrows into a very fine tube which opens internally into the coelom. Since in some cuttle-fish the genital duct is paired and in *Nautilus* there are two pairs of kidneys, while the genital ducts of that animal appear to be portions of the kidney ducts split off, it has been suggested that the genital duct of *Sepia* is all that is left of a missing pair of kidneys.

The oviduct is simple, but in the female there are four glands, the nidamental glands, situated on the wall of the mantle cavity just outside the kidneys (18, Fig. 129). From the secretion of these glands tough egg-shells resembling india-rubber cases are made. The egg is about the size of a pea and when the young cuttle-fish emerges it is already like the adult.

Cuttle-fish feed chiefly on crabs, shrimps and other Arthropoda, using their beaks to break the hard shell. Some are large enough to attack men and this circumstance has given rise to many legends. Gigantic species are sometimes cast dead on the shores of Nova Scotia, the length of body being ten feet, and of the arms over fifty feet. The phylum Mollusca finds its climax in the cuttle-fishes.

Modern cuttle-fish have a fairly uniform structure. The two long tentacular arms are absent in the Octopoda. In *Loligo* the shell is horny, in *Polypus* it is entirely absent. In *Ommatostrephes*, the common cuttle-fish of the Gulf of St Lawrence, the anterior chamber of the eye is open and the lens bathed with sea-water.

Nautilus is a remarkably interesting cuttle-fish, widely different from Sepia and the others, but closely allied to most of the extinct forms. The arms are short, broad, ill-defined lobes, the suckers being represented by tentacles with raised ridges round their bases. There is a large external shell coiled

Nautilus.

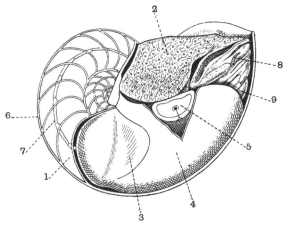

Fig. 130.   Side view of *Nautilus pompilius* × ½.   After Graham Kerr.   Half the shell has been removed to expose the animal and the chambers of the shell.

1. Last completed chamber of the shell.   2. Hood part of foot.   3. Shell muscle.   4. Mantle cut away to expose (5) the pin-hole eye.   6. Outer wall of shell, some of which is cut away to show the chambers.   7. Siphon.   8. Tentaculiferous lobes of the foot.   9. Funnel.

forwards in the median plane over the animal's head, so to speak (Fig. 130). The visceral hump is enclosed in the last chamber and from the apex of the hump a membraneous tube called the siphuncle is given off, which runs through all the other chambers, piercing the septa. There is a fold of the mantle turned back over the anterior edge of the shell, the first foreshadowing of the shell sac of *Sepia*.

There are four gills and four kidneys and four auricles in the heart. Thus *Nautilus* shows traces of segmentation. The papillae of the posterior kidneys are split, one half leading directly to the reno-pericardial canal, the other into the sac-like kidney. The reno-pericardial canals thus open directly to the exterior, and the genital ducts are in such a relation to the anterior kidneys as to make it probable that they are the reno-pericardial canals belonging to these, which have acquired independent communication with the exterior.

Phylum MOLLUSCA.

Mollusca are classified as follows:

Class I. Gasteropoda.

Mollusca with a flat foot adapted for crawling. There is a buccal mass and radula; distinct pleural ganglia are present and the shell is never composed of paired pieces.

Sub-class I. Isopleura.

Bilaterally symmetrical forms with a shell composed of eight median plates situated in a longitudinal series. Numerous pairs of ctenidia.

Ex. *Chiton.*

Sub-class II. Anisopleura.

Asymmetrical forms, with the left side of the visceral hump long in comparison to the right, the anus, kidneys and ctenidia being shifted forwards.

Division I. Streptoneura.

The arms and ctenidia shifted so far forward that the visceral loop is pulled into the shape of an eight and the gill is anterior to the heart.

Order 1. **Aspidobranchiata.**

Usually two kidneys, two auricles, two ctenidia. The axis of the ctenidia free and both rows of plates present.

Ex. *Haliotis, Patella.*

Order 2. **Pectinibranchiata.**

One kidney, one auricle, one ctenidium in which the axis is adherent to the mantle and only provided with one row of plates.

Ex. *Buccinum,* the Whelk.

Division II. EUTHYNEURA.

The visceral loop is untwisted and the gill is posterior to the heart.

Order 1. **Opisthobranchiata.**

Marine forms with a ctenidium and mantle.

Ex. *Aplysia.*

Order 2. **Pulmonata.**

Land and fresh-water forms breathing air, having the mantle cavity converted into a lung and the ctenidium aborted.

Ex. *Helix.*

Class II. SOLENOGASTRES.

Degenerate worm-like Mollusca devoid of shell and foot and with a ventral ciliated groove. There is a rudimentary radula and the genital organs burst into the pericardium, the nephridia serving as genital ducts.

Ex. *Proneomenia, Neomenia.*

Class III. SCAPHOPODA.

Mollusca with a tubular shell and mantle and a long cylindrical foot ending in three processes. A buccal mass and pleural ganglia are present. The genital organ opens into the left nephridium.

Ex. *Dentalium.*

Class IV. LAMELLIBRANCHIATA (PELECYPODA).

Mollusca with a shell composed of two valves united by a hinge, and a mantle of two lobes. The foot is usually wedge-shaped and the plates of the ctenidia are fused to form gill-plates. No buccal mass. Pleural ganglia fused with the cerebral.

Sub-class I. PROTOBRANCHIATA.

Small Lamellibranchiata with a simple ctenidium like that of Gasteropoda, and large labial palps.

Ex. *Nucula.*

Sub-class II. FILIBRANCHIATA.

Lamellibranchiata in which the filaments of the ctenidium are loosely united with one another and their bent-up ends are not united to the mantle.

Ex. *Mytilus.*

Sub-class III. Eulamellibranchiata.

Lamellibranchiata in which the filaments are welded into a "lamella" or plate and their bent-up ends are joined to the mantle.

Ex. *Unio, Anodonta.*

Class V. Cephalopoda.

Mollusca in which the front part of the foot surrounds the head and is drawn out into sucker-bearing arms whilst the hind portion of the foot forms a muscular tube. The ganglia are massed together and protected by a skull: there is a buccal mass with a radula and two jaws.

Sub-class I. Tetrabranchiata.

Cephalopoda with four ctenidia, four kidneys, four auricles, a large external shell, no suckers and very short arms.

Ex. *Nautilus.*

Sub-class II. Dibranchiata.

Cephalopoda with two ctenidia, two kidneys and two auricles. The shell enveloped in the mantle and the arms are long and provided with suckers.

Order I. **Decapoda.**

Dibranchiata with two long and eight short arms.

Ex. *Ommatostrephes, Sepia.*

Order II. **Octopoda.**

Dibranchiata with eight arms of equal length.

Ex. *Polypus (Octopus).*

CHAPTER IX.

Phylum Echinodermata.

The class of animals known as the Echinodermata comprises the well-known Star-fish or Five-fingers, the equally well-known Sea-urchins, the less familiar Sea-cucumbers and Brittle-stars, lastly the graceful Feather-stars. The name is derived from two Greek words, ἐχῖνος, which means hedgehog (and was also used for the sea-urchin), and δέρμα, the skin. The prickles and spines with which many members of this Phylum are covered constitute a very prominent feature in their appearance. Spines, it is true, are sometimes absent, but in every case, whether this is so or not, the skin contains a skeleton consisting either of plates or of rods, and the spines are merely rods belonging to this skeleton projecting outwards and still covered by the skin which they push before them.

Class I. Asteroidea.

The most familiar of all the British Echinoderms is probably the common star-fish, *Asterias rubens*, which may be found at low water on almost any part of the coast where shell-fish, its favourite food, abound. Very similar species, *Asterias vulgaris* and *Asterias polaris*, abound on the American coast, the first-named on the New England coast, the second further north in the Gulf of St Lawrence. The species represented in Fig. 131 belongs to a different genus, *Echinaster*, but in all essential features of its anatomy it agrees with *Asterias*. *Echinaster sentus* is common on the N. American coast. The name "star-fish" denotes the shape. The body is produced into five arms or lobes which are arranged like the spokes of a wheel round the centre of the body or disc, on the under side of which the mouth is situated. These arms are termed radii, and the re-entrant angles between them interradii.

The star-fish creeps about with its mouth downwards : its
Skeleton.   motion is effected by means of numerous delicate
              semi-transparent tentacles.   These are situated in
five grooves which run along the under side of the arms and converge towards the mouth, where they merge into a depression

Fig. 131.   Oral view of *Echinaster sentus* with tube-feet extended × about 1.
From Agassiz.

surrounding that opening.   These grooves are termed the ambulacral grooves : the tentacles situated in them are called the
tube-feet, and the depressed space round the mouth in which
all the grooves unite is called the buccal membrane (Lat. *bucca*,
the cheek) or the peristome (Gr. περί, around, and στόμα, mouth).

So far as we have yet seen the Echinoderms seem to differ from Coelenterates, which are also radiate animals, in the details of the arrangement of the organs rather than in any fundamental features. The skeleton no doubt is peculiar in being embedded in the skin: but the spicules of the Alcyonaria occupy a similar position, although they rarely cohere to form the definite rods and plates like those characteristic of Echinoderms. When the soft parts are dissolved away from one of these rods or plates by caustic potash it is seen to consist of a delicate network of calcium carbonate; and it is found by observation of the developing young that such a plate is formed by a little heap of cells coming together and secreting a limestone rod between them: this rod then branches at both ends and the branches bifurcate again so that the twigs of the second or third degree approach each other and joining form a mesh, and this process of bifurcating and joining is repeated until the plate or spine is built up. The growth of the primitive rod into the meshwork is rendered possible only by the growth of the cells which shed out the calcium carbonate. These cells remain throughout life, more or less modified, as a kind of living network interpenetrated by the skeletal one.

When however we cut a star-fish open we see that the animal apparently consists of two sacs placed one within the other. The innermost sac or alimentary canal opens

Coelom.

in the centre of the upper surface by a minute opening, the anus, through which undigested matter is thrown out, and on the under surface by the mouth. The space or sac which apparently surrounds the digestive cavity of the star-fish is a true coelom: like the coelom in a segment of an annelid it has been formed by the union of two sacs which in the embryo lay right and left of the digestive tube. From its walls the muscles are developed, the generative cells, and also the cells which give rise to the skeleton. Between the outer wall of the body-cavity and the true external skin which corresponds to the ectoderm, there is a mass of more or less gelatinous substance exactly corresponding to the jelly of a Medusa or the connective tissue of an Arthropod, which constitutes the substance of the body-wall. Into this material wander cells budded from the wall of the coelom: these cells from their power of movement and appearance can be recognised as amoebocytes. It is from these cells that the skeleton is formed in the way we have described above: some of them, however, retain their primitive character and wander about, probably carrying food to the various parts of the body-wall or perhaps carrying away waste matter. A third section again secrete the fibres which bind the various rods constituting the skeleton together and thus form a simple and primitive variety of connective tissue.

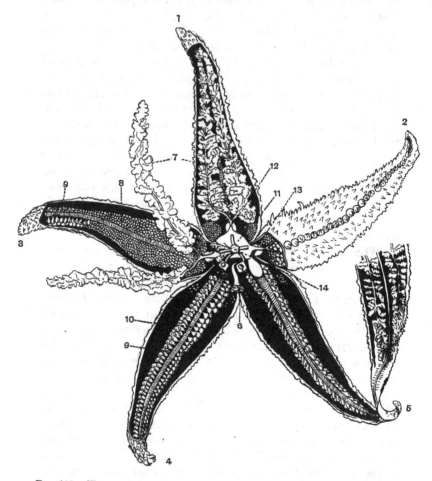

Fɪɢ. 132. The common Star-fish, *Asterias rubens*, dissected to show motor, digestive and reproductive systems. After Rolleston and Jackson.

1—5. The five arms.　　6. Madreporic plate and canal.　　7. Arborescent "hepatic" caeca, two in each arm.　　8. Generative organs.　　9. Ampullae of tube-feet.　　10. Ambulacral plates, inner surface.　　11. Pyloric sac.　　12. Duct leading from pyloric sac to pyloric caeca.　　13. Stomach bulging into arm.　　14. Anus.

The alimentary canal can most easily be examined by carefully
cutting away all the upper parts of the five arms in
one piece, cutting along both sides of each arm, then
raising the upper part of the animal and clipping through the
intestine near the anus. By this means the animal is separated
into an upper and lower half and all the internal organs are dis-
played in one piece or the other. The alimentary canal is then
seen to consist of several regions clearly distinguished from one
another. It begins with an exceedingly short gullet which passes
at the lips into the buccal membrane already mentioned : the gullet
widens out above into an exceedingly loose baggy stomach produced
into ten short pouches, two situated in the beginning of each arm.
Above the stomach and communicating with it by a wide aperture
lies a flattened pentagonal bag, called the pyloric sac, and from
each of the five angles of this sac there is a tube given off which
runs into each arm, where it is soon divided into two parallel sacs,
each produced into a multitude of little, short pouches. These sacs
are called the pyloric caeca: caecum, Latin "blind," being a convenient zoological term for a blind pouch. The pyloric caeca are
tied to the upper side of the arm, each by two bands of transparent
membrane called mesenteries. From the centre of the pyloric
sac a short straight tube runs to the upper surface of the animal
where it opens by a minute anus : this tube is called the rectum,
a name, as we have seen, commonly given to the last portion of the
digestive tube. The rectum has attached to it two branched
tubes of a brown colour which open into it, called the rectal
glands.

The reason of the division of the digestive sac into various
parts is of course the different uses to which they are put in the
life of the animal ; and we may stop for a moment to enquire what
these uses are.

Star-fish feed chiefly on bivalve shell-fish, such as mussels,
cockles and clams, though they will attack almost any animal.
Their mode of seizing their prey is very curious. If they are
attacking a bivalve, they bend all their five arms down round it,
thus arching up the central portion of the body. Then the stomach
is pushed out,—this being rendered possible by the turning inside
out of its edges, which as we saw above, are loose and baggy—and
wrapped around the fated mollusc. The pushing out is effected by
the contraction of some muscle fibres in the body-wall : these tend
to diminish the space which the coelom occupies, and as this is

filled with incompressible fluid, the stomach must be pressed out. After some time has elapsed the star-fish relaxes its hold and it is then seen that the shell of the mollusc is completely empty and as clean as if it had been scraped with a knife. It was long a puzzle how the star-fish succeeded in forcing its victim to relax its muscles and allow the valves to open. It was supposed that the stomach secreted a paralysing poison, but it has been conclusively proved that this is not the case, but that the star-fish drags the valves of its victim apart by main force, often actually breaking the adductor muscles. The pull exercised by the suckers is not nearly strong

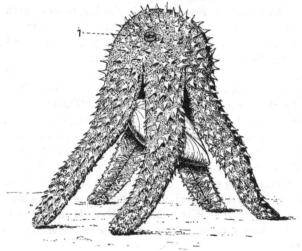

FIG. 133. Star-fish, *Echinaster sentus*, in the act of devouring a Mussel.
1. Madreporic plate.

enough to open the valves at once, but the star-fish has staying power and eventually the mussel is slowly forced open. The cells lining the stomach include a large number of goblet-cells (*v.* p. 102) swollen by drops of clear fluid; those of the pyloric sac, on the other hand, present a different appearance. These are full of minute granules and recall the appearance of the cells in other animals which contain the active digestive principle. Hence it seems reasonable to suppose that the mussel is digested by the secretion of the pyloric sac and its appendages, which flows downwards into the stomach. Any portion remaining undigested is expelled through the rectum; no food ever penetrates into the pyloric caeca.

The locomotion of the star-fish is effected in the following manner. The tube-feet which crowd the ambulacral grooves are

during life continually extended and retracted. At their ends are
flat circular discs, and these discs are pushed against
the stone or rock or whatever else the star-fish is
clinging to. Then by the contraction of their muscles the centre
of the disc is pulled upwards, and so it is made to adhere in exactly
the same way in which a boy makes a leather "sucker" adhere to a
stone. When once the disc is firmly fixed the contraction of the
tube-foot draws the animal after it.

Water-vas-cular system.

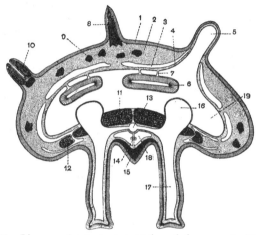

FIG. 134. Diagram of a transverse section of the arm of a Star-fish.

1. Ectoderm.　　　2. Jelly.　　　3. Peribranchial space in the skin.
4. Peritoneal lining of body-cavity.　　　5. A branchia.　　　6. Pyloric
caecum.　　7. Mesentery supporting a caecum.　　8. Spine.　　9. Ossicle
in skin.　　10. Pedicellaria.　　11. Ambulacral ossicle.　　12. Adambulacral ossicle.　　　13. Radial trunk of water-vascular system.
14. Radial septum separating the two perihaemal spaces.　　　15. Radial
nerve-cord, a thickened band of ectoderm with a plexus of nerve-fibrils
underlying it.　　　16. Ampulla of tube-foot.　　　17. Tube-foot.
18. Perihaemal space.　　　19. Coelom.

We found that in order to examine the alimentary canal it was
advisable to divide the star-fish into an upper and a lower half. If
we now cut away the tube-feet and look at the roof of the ambulacral groove from which they project, it will be seen that the
groove is roofed in by a double series of calcareous rods, meeting
each other at an angle like the beams of a church-roof (11, Fig. 134).
These are called the ambulacral ossicles. They can be drawn
together by muscle fibres running from one of a pair to its
fellow just under the spot where they meet. By this action the
ambulacral groove is narrowed; and at the same time, inwardly

projecting spines lining its edges are made to meet, so that the tube-feet are entirely protected by a trelliswork of spines. These spines are attached to rods, called the adambulacral ossicles, firmly bound to the outer edges of the ambulacral ossicles (12, Fig. 134). Inside the animal, between the ambulacral plates, a series of pear-shaped transparent bladders tensely filled with fluid project into the coelom (Figs. 132 and 134). These are really the swollen upper ends of the tube-feet and are termed ampullae. They act as reservoirs into which the fluid contents of the lower part of the foot are driven when the longitudinal muscles of the tube-foot contract. The bladder-like upper end of the foot has only circular muscles, and when these contract the fluid is driven back into the lower part of the tube-foot and it is expanded. The tube-feet, though from the above description it would seem as if each was capable of acting without the others, are really all parts of one system : they are connected by short transverse tubes, with a canal running along the whole length of the arm immediately under the ambulacral ossicles, called the radial water-vessel (13, Fig. 134). This canal and its branches can easily be seen in microscopic sections of the arms of young star-fish, or they can readily be demonstrated by cutting off the tip of the arm of a fully-grown specimen, finding the end of the radial tube on the cut surface and inject-ing it with coloured fluid by means of a fine pipette. The five radial tubes are connected with each other by a ring-shaped canal lying just within the peristome, which is called the water-vascular ring. There are nine small pouches called Tiede-mann's bodies projecting inwards from the ring canal. In these are formed the amoebocytes which are found floating in the fluid of the canal, and which arise by budding from the wall of each pouch. From the ring canal also in one interradius, where the tenth Tiedemann's body if it existed would be found, a tube is given off which leads to the upper surface of the disc, where it opens by a sieve-like plate, pierced by numerous minute pores, called the madreporite (Figs. 132 and 133). This vertical tube receives the awkward name of the stone-canal because its walls are stiffened by calcareous deposit ; its cavity is reduced to a mere slit by the projection into it of an outgrowth of its wall shaped in section like a Υ with coiled ends, which is also strengthened by lime. Although, as we have said, the cavity of the stone-canal is a mere slit, yet it is lined by long narrow cells carrying most powerful cilia. In many species of star-fish, although not in *Asterias*, stalked sacs resembling greatly enlarged ampullae are attached to the water-vascular ring. These appear to act as reservoirs of fluid for it: they are known as Polian vesicles after Poli, the naturalist who first described them.

Now since all the movements of a tube-foot can be accounted for by the action of the longitudinal muscles of its lower part and the circular muscles of the ampulla, the question arises as to what is the purpose of this apparatus of radial and circular tubes, stone-canal and madreporite? There is one interesting little mechanism which supplies a valuable clue to the answer to this question. This is a pair of valves placed in the tube-foot at the entrance of the transverse canal, which unites it with the radial tube. These valves swing open into the tube-foot when the pressure in the radial tube is greater than the pressure in the tube-foot, but when the pressure in the latter is the higher they close, so that under no circumstances can water escape from the tube-foot into the radial canal. So it appears that there is an arrangement which allows fluid to pass into the tube-foot but which prevents its return, and this implies that under ordinary circumstances there must be a loss of fluid from the tube-foot. We must in fact suppose that when the tube-foot is driven out by the contraction of the ampulla, the contained fluid slowly transudes through its thin walls and the loss is supplied from the radial canal. The pressure in the radial and circular canals is kept up by the action of the cilia in the stone-canal, by means of which a slow but steady current is produced, setting in from the outside through the madreporite.

The function of the whole system of tubes therefore is to keep the tube-feet full of fluid and thus tense and rigid, so that they can perform their functions properly.

The nervous system of the star-fish is one of the most interesting features in its anatomy. The ectoderm consists of long delicate cells bearing flagella and interspersed with goblet-cells similar in appearance to those lining the stomach. The slime which these cells manufacture covers the surface of the animal and no doubt protects it from the attacks of bacteria and microscopic algae. But the chief point of interest is that at the bases of the long delicate cells there is an indescribably fine tangle of delicate nerve-fibres which are doubtless outgrowths of some of the cells. Here and there a nucleus is seen amongst them which belongs to a neuron—that is, an ectoderm cell which has lost its connection with the rest and has become pushed down into the

Nervous system.

fibrillar layer. The ectoderm all over the body is therefore underlain by a nervous sheath and is very sensitive, but there are certain
places where the nervous sheath becomes very much thickened and
it is these areas which constitute the true sense-organs and the
central nervous system.

Isolated sense-cells, that is, cells having a stiff protruding hair,
are scattered all over the surface; but the only spot where they are
collected in groups so as to form true sense-organs is on the tips of
the tube-feet. The tube-feet are then practically the only sense-
organs, and since the radial water-tube ends at the tip of an arm in
a freely projecting tentacle, we might regard the whole radial tube
as a huge, branched, sensitive tentacle. There is the more justification for doing this when it is found that the radial tube with its
freely projecting tip is in the young star-fish quite independent of
the outgrowth of the body called the arm, and only secondarily
becomes applied to it. At the base of the end-tentacle there is a
thick cushion of nervous matter in which are excavated a number
of ectodermal pits lined by cells containing orange pigment.
These pits are organs of vision: and it has been experimentally
shown that a star-fish deprived of these organs is insensible to
light.

The central nervous system consists of five thick bands of
nervous tissue situated one above each ambulacral groove underneath the radial water-tube (Fig. 134). They are termed the
radial nerve-cords and are joined by a circular band of a similar
nature, called the nerve-ring, lying under the water-vascular ring.
Intervening between the radial nerve-cord and the radial water-tube
there are two canals lined by flattened cells and separated from one
another by an imperfect septum (14, Fig. 134). They are called
the radial perihaemal canals and are outgrowths from the
coelom. From their upper walls are derived the muscles which
move the ambulacral ossicles on one another: from their lower walls
a layer of ganglion cells and nerve-fibres, which may be termed the
coelomic nervous system in order to distinguish them from the
main mass of ganglion cells and fibres which are derived from the
ectoderm. This coelomic nervous system, which is very thin in
the star-fish, seems to serve as the channel by which impulses from
the radial nerve-cord reach the ambulacral muscles. The five pairs
of radial perihaemal canals are connected with one another by a
circular canal lying above the nerve-ring called the outer perihaemal ring.˙ Inside this is another circular canal called the inner

perihaemal ring, which is an expansion of the foot of the axial sinus (see p. 260).

The upper or aboral surface of the star-fish is provided with two most interesting groups of organs, pedicellariae and dermal branchiae. The former are minute pincers, composed of two or rarely three blades moving on a basal piece.

Pedicellariae.

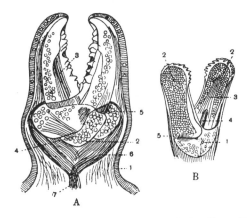

Fig. 135. Pedicellariae from *Asterias glacialis*. From Cuénot.

A. Crossed form × 100.   1. Ectoderm.   2. Base of left " jaw."   3. Muscle closing the " jaws."   4. Basal ossicle.   5. Muscle opening the " jaws."   6. Fibrous band connecting the basal ossicle with one of the rods of the skeleton.   7. Fibres of peduncle.

B. Straight form.   1. Basal ossicle.   2. " Jaws."   3. Adductor muscle.   4. Muscle closing the " jaws."   5. Muscle opening the " jaws."

These close when the skin of the back is irritated; their main purpose appears to be to keep the surface of the animal clear from zoophytes and other small encrusting organisms. They cover the thickened bases of the blunt spines with which the back is beset. Larger pedicellariae are scattered in the inter-spaces between the spines and are distinguished from the smaller by the fact that the blades do not cross as is the case with these. The larger kind are also found on the adambulacral spines.

The pedicellariae are probably little spines of the second order. In the small blunt-armed star-fish, *Asterina gibbosa*, there are no true pedicellariae, but the plates on the back bear small spines arranged in twos or threes, which act somewhat like pedicellariae when the skin is irritated.

The dermal branchiae (5, Fig. 134) are conspicuous in a star-fish when alive; they are very difficult, on the other hand, to detect in preserved specimens. They are in fact thin spots on the body-wall, where it consists only of the ectoderm and the wall of the coelom—closely apposed, the jelly, fibres and skeletal rods being absent. These spots project like little finger-shaped processes and their purpose is to facilitate respiration. The fluid in the coelom or body cavity being separated from the external water by a very thin membrane, the dissolved oxygen is able to pass from the one fluid to the other with great ease.

There is no localized excretory organ in the star-fish or indeed in any Echinoderm. Throughout the phylum so far as is known this function is performed by the amoebocytes which float in the coelomic fluid and have been produced by the budding of the cells forming the wall of the coelom. When charged with excreta the amoebocytes endeavour to make their way out. This in the star-fish they effect by accumulating at the base of the dermal branchiae and working their way through the thin body-wall and so escaping into the ocean.

The organs of sex in the star-fish are very simple. Both kinds of germ cell are aggregated in great feather-shaped glands situated in pairs in the bases of the arms and opening in the angles between the arms or in the inter-radii. The ten ovaries in the female and ten testes in the male are connected by a circular cord of immature germ cells called the genital rachis running round the disc just dorsal to the coelom. This is embedded in the wall of a tube called the aboral sinus which like the other spaces in a star-fish, apart from those of the digestive canal, is an outgrowth of the coelom. The rachis is in turn connected with a pillar of similar cells running alongside the stone-canal which used to be called the heart, under a mistaken idea of its function, but which we shall term the genital stolon. The genital rachis is formed as an outgrowth from the genital stolon and the latter is an outgrowth from the coelomic wall, so that the genital cells are derived from the coelomic cells as in other Coelomata. The genital stolon is interposed between the general coelomic cavity of the animal and a special division of the same which is called the axial sinus, and which runs parallel to the stone-canal. The axial sinus is derived from the anterior portion of the coelom in the larva. Underneath the madreporite there is still

Branchiae.

Reproductive organs.

another division of the coelom completely shut off from the rest, which may be termed the madreporic vesicle. It apparently represents a rudimentary second water-vascular system, since in exceptional cases it may develope the rudiments of radial canals. The genital stolon projects into the axial sinus; it has a brown colour which no doubt suggested the connection with the blood-system to the earlier anatomists, but true blood-vessels do not exist in Echinodermata. The ova and spermatozoa are thrown out into the water by pores situated on the under or oral surface at the base of the arms and unite with each other there. The young lead a free-swimming existence, and are so unlike the star-fish that no one would ever dream of suspecting that the two had anything to do with each other. As however these peculiarities are fundamentally the same in each of the groups of the Echinoderms they will be dealt with later when the characters of these other groups have been studied.

The other species of star-fish, which are all grouped together in the Class Asteroidea (Gr. ἀστήρ, a star; εἶδος, shape), differ but little in really fundamental points from *Asterias rubens*. Pedicellariae may, as we have seen, be absent; the arms may be short so that the shape almost becomes that of a pentagon and the arrangement of the plates and spines constituting the skeleton may vary very much. In one family, the Astropectinidae, there is no anus, the rectum ending blindly, and the tube-feet have pointed ends. These star-fish do not climb but run over the surface of the sand.

The number of arms is most often five, but not only do individual variations from this rule occur in species where five is the normal number, but species and even genera and families are characterised by having a larger number: the common Sun-star, *Solaster papposus*, for instance, has from eleven to thirteen arms.

Class II. Ophiuroidea.

The next order of Echinoderms is termed the Ophiuroidea (Gr. ὀφίουρος, serpent-tailed; εἶδος, form) or the Brittle-stars. These like the star-fish have a body with five arms diverging from a central disc on all sides like the conventional representation of a star. The arms are, however, sharply marked off from the central disc, and they do not, as in the true star-fish, insensibly merge into it, but are continued along grooves on the under surface to the immediate neighbourhood of the mouth: further they are exceedingly long and flexible and totally unlike the stiff arms of *Asterias* or its allies.

The habits of the animal too are very different from those of the star-fish. Instead of creeping slowly along by the action of the tube-feet it springs along by muscular jerks of the arms, sometimes

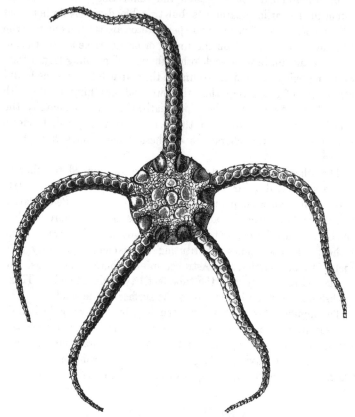

FIG. 136.   Dorsal, upper or aboral view of *Ophioglypha bullata* × about 2½.
From Wyville Thompson.

pushing with four arms and seizing hold in front with one, sometimes pushing with three and hauling itself along with two. The name Brittle-star is derived from the readiness with which, if irritated, the animal will snap off an arm.

As might naturally be expected, the most striking differences from the star-fish are seen in the arms. No ambulacral groove is

apparent: the arm being encased in a cuirass consisting of four series
of plates, an upper row, two lateral rows each bearing a row of spines
on its edge, and an under row (Fig. 137). On close inspection the
short pointed tube-feet may be seen protruding from minute pores
at the sides of the under row of plates. A thin section of the arm
reveals the fact that there really is a space corresponding to the
ambulacral groove of the star-fish, but that by the approximation of
its edges it has become closed off from the outer world so that it
forms a canal, the so-called epineural canal (4, Fig. 137). Above
this canal, at the spot one would term the apex of the ambulacral

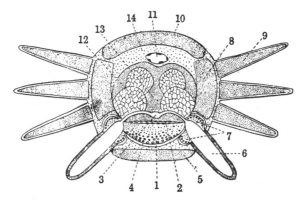

Fig. 137. Section through an arm of an *Ophiuroid*. Diagrammatic, magnified.

1. Radial nerve-cord.   2. Radial perihaemal canal.   3. Radial water-
vascular canal.   4. Epineural canal.   5. Ventral plate.   6. Tube-
foot.   7. Pedal ganglion.   8. Lateral plate.   9. Spine.
10. Dorsal plate.   11. Coelom.   12. Longitudinal muscle.
13. "Vertebra."   14. Soft tissue supporting plates.

groove in a star-fish, there is a ridge of nervous matter covered on
the lower side by cells exactly resembling the skin cells covering
the nerve ridges in *Asterias*. This is the radial nerve, and above
this again is a large rounded disc of calcareous matter, the so-called
vertebra. This really corresponds to a pair of ambulacral plates
which have become fused together. So much might be inferred
from the fact that the radial water-tube runs in a groove on its
under surface, and it is clearly proved by examining young specimens. Each vertebra is very short, and it not only has rounded
knobs and cups in order to enable it to slide on its successor and
predecessor, but is connected to each of them by four great muscles,

by the contraction of which the arm is moved in any direction
(Fig. 137). If the two side muscles contract the arm is moved
toward that side, if the two upper, upwards, and so on. These
muscles are the seat of the chief activities of the animal, and it is
not surprising to find that a pair of large nerves comes off between
each two vertebrae to supply them, and that where these nerves
are given off the nerve-cord is thickened and the nerve-cells
increased, so that a string of ganglia is produced strongly recalling
the ventral nerve-cord of the Earthworm. Between the vertebrae
and the radial nerve-cord there is a single canal (2, Fig. 137), representing the pair of radial perihaemal canals in a similar position in

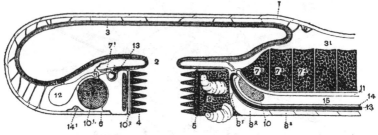

FIG. 138. A diagrammatic vertical section of an *Ophiuroid*. After Ludwig.
The circumoral systems of organs are seen to the left, cut across, their
radial prolongations cut longitudinally, to the right.

1. Body-wall.   2. Mouth.   3. Coelom.   3¹. Coelom of the arm.
4. Mouth papillae.   5. Torus angularis.   6. Oral plate.   7¹. 1st ambulacral ossicle.   7², 7³, 7⁴. 2nd to 4th ambulacral ossicle or "vertebrae."
8¹, 8², 8³. 1st to 3rd ventral plate.   9. 1st oral foot.   10. Transverse muscle of the 2nd joint.   10¹. External interradial muscle.
10². Internal interradial muscle. (The line should point to the dotted
tissue.)   11. Water-vascular system: to the left the circumoral ring,
to the right the radial vessel.   12. Polian vesicle.   13. Nerve-ring and
radial nerve: the ganglia on the latter are not shown.   14. Genital
rachis.   15. Radial perihaemal canal.

Asteroidea. From the ventral wall of the canal the coelomic
nervous system is formed: and it is by the greater development of
this system where the nerves to the ambulacral muscles are given
off that the ganglionic swellings of the nerve-cord are produced.
The vertebrae and these muscles nearly completely fill the arm,
leaving only a small canal above the vertebrae (11, Fig. 137): this
is an outgrowth of the body cavity or coelom, but there is no
branch of the alimentary canal continued into it, as was the case
with the star-fish.

The digestive sac is here a simple flattened bag lined by cells
somewhat like those lining the pyloric sac of the star-fish. There

is no anus, and the edges of the stomach cannot be pushed out. How then, it may be asked, does the Brittle-star eat and of what does its food consist?

It must be confessed that, in spite of their quick movements and highly developed nervous system, Brittle-stars belong in general to the great army of mud-eaters and scavengers. Where they live— usually at the bottom of sea pools and at such depths of the ocean as to be in still water—the mud or sand is impregnated with decaying animal and vegetable matter, and the Brittle-stars shovel this material into their mouths by means of the two pairs of tube-feet of each arm which lie nearest the mouth and are called the

Fig. 139. Oral view of part of the disc and arms of *Ophioglypha bullata* × 4½. From Wyville Thompson.

oral tube-feet. The interradii between the arms project inwards over the mouth, as the mouth-angles; these are lined along their edges and at their tips with broad blunt spines called teeth and mouth papillae, so that they form an efficient strainer and prevent coarse particles entering the stomach (Figs. 138 and 139). The calcareous plate at the apex of each mouth-angle which bears these spines is called the torus angularis (5, Fig. 138).

We saw that in the star-fish the whole surface is covered with a sensitive skin, but that the tube-feet act as sense-organs as well as being locomotor in function. In the Brittle-stars the sole purpose of the tube-feet is to serve as sense-organs; they are often covered with little warts consisting mainly of sense-cells with their delicate hairs sticking out all round, just like the batteries of cnidoblasts in Hydra, and in all cases there is a special nervous swelling surrounding the base of each tube-foot called the pedal ganglion (7, Fig. 137). As, however, these tube-feet have lost their power of attaching themselves by a sucking action to objects and hence are of no use for locomotion, the ampullae have disappeared; and as the action of the ampullae is probably the chief cause of the loss of fluid in the tube-feet of the star-fish, in the Brittle-star, where the loss must be very small, the stone-canal is excessively narrow and the madreporite instead of being a regular sieve has two pores only, rarely more. It is very curious to find that the madreporite is on the underside of the animal; in the young Brittle-star it is on the edge of the disc, but in each interradius the upper surface grows more rapidly than the ventral and so it is forced round on to the underside. To the water-vascular ring are attached four large Polian vesicles, the interradius occupied by the stone-canal alone being without one. The tube-feet are the only sense-organs, in a sense still more real than is the case with star-fish, for in the Ophiuroidea the rest of the ectoderm, after having given rise to a cuticle, has disappeared, the solid mail of plates which the animal possesses appear to render it impervious to sensations of contact.

The organs of sex are very simple; they are situated in the disc in the interradii and consist, in each interradius, of several short pouches. These open into ten sacs, called the genital bursae; one pair being placed in each interradius. These sacs are merely invaginations of the ectoderm which does not here disappear as over the rest of the body; they are lined by ciliated cells which keep up a constant current of fresh water pouring into them and thus they fulfill the same function as the "dermal branchiae" of Star-fishes.

Class III. Echinoidea.

The general appearance of the dried skeleton of a Sea-urchin or Echinoid, is familiar to most people, but many would fail to recognize any resemblance to a star-fish in the slightly flattened sphere covered with spines. If, however, we are fortunate enough to see one living, we at once perceive that along five meridians the sphere is beset with beautiful semi-transparent tube-feet, ending in suckers,

exactly like those of the star-fish. In fact, the Sea-urchin might be
described as a star-fish in which the upper surface had shrunk to
insignificant proportions, being represented by a small patch of
leathery skin at the upper pole: or, if we regard the whole radial
tube with its tube-feet as one immense branched tentacle and the
arm as its support, we should say that the arm had been again

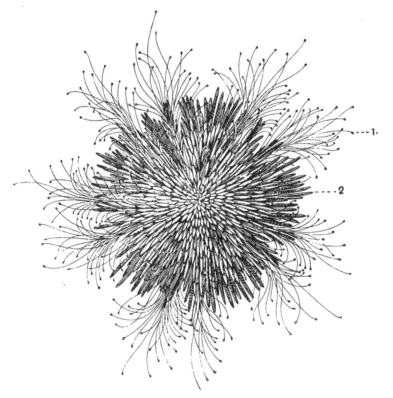

Fig. 140. *Strongylocentrus dröbachiensis* × 1. Aboral surface. From Agassiz.
1. Expanded tube-feet.   2. Spines.

merged in the body so that the radial tube was bent back in a
curved course. As a matter of fact the end of the radial tube
projects very slightly beyond the general surface and bears at its tip
a mass of pigment which corresponds with the eye of the star-fish,
though no eye-structure has been detected in it. This is situated
near the upper end of the body, just outside the small area of
leathery skin mentioned above.

The skeleton of the Sea-urchin is a cuirass of plates fitting edge to edge, with two openings. Of these the upper (already referred to) is covered with leathery skin and has the small anus in the centre of it and is called the periproct (Fig. 148): it is this area which corresponds to the whole upper surface of the star-fish. The other opening is in the centre of the lower surface and is likewise covered by flexible skin; it surrounds the mouth and is called the peristome (Fig. 147).

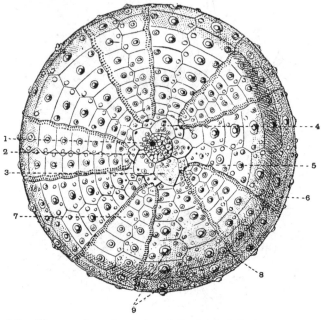

Fig. 141. Diagram of an aboral view of the dried shell of an Echinoid × 1. The spines, pedicellariae and tube-feet have been removed.

1. Anus. 2. Leathery skin round anus, periproct. 3. Madreporic plate. 4. Genital plate with genital pore. 5. Ocular plate with eye. 6. Line of junction of ambulacral and interambulacral plates. 7. Ambulacrum. 8. Pores through which tube-feet protrude. 9. Bosses which bear the spines.

The cuirass itself is called the corona and consists of twenty strips, each made up of a row of plates. Corresponding to each tube-foot area or radius there are two rows of so-called ambulacral plates, and each intervening area or interradius is similarly covered by two rows of large plates. As in Ophiuroids, there is no ambulacral groove visible from the outside: it is represented by the epineural canal, immediately inside which there is the radial nerve-cord.

It is necessary to bear this in mind when the term ambulacral plate is used; the so-called ambulacral plates of an Urchin do not correspond to the similarly named plates in the star-fish, as they do not roof in the ambulacral groove, but form a floor for it. Inside the nerve-cord there is a single radial perihaemal canal as in the Brittle-star (2, Fig. 145 B). As the plates of the skeleton are not movable on one another nothing corresponding to the ambulacral muscles of the star-fish exist at least over most of the radius, and the radial perihaemal canal is separated from the general coelom only by a thin septum in which the radial water-tube is embedded. For the same reason there is no recognizable coelomic nervous system.

If the continuous cuirass of the Sea-urchin and the closed ambulacral groove remind one of an Ophiuroid, the resemblance ends there; for in the Urchin the ectoderm, consisting of long slender cells with a tangle of nerve-fibres at the base, is spread over the whole surface outside the skeleton, just as in a star-fish. This sensitive layer controls, it is found, the movements of the spines, which are among the most important organs of the Urchin. These spines, unlike the spines of the star-fish or Brittle-star, have hollow bases, which articulate with smooth rounded bosses on the plates (B, Fig. 145). They are tied to these bosses by a sheath of muscle-fibres, so that by the special contraction of any side of the sheath they can be moved in any direction. The skin covering the sheath has developed a specially thick nervous layer.

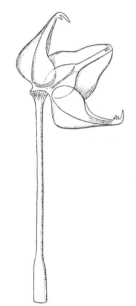

Fig. 142. A glandular or gemmiform Pedicellaria from *E. esculentus* × 18. From Chadwick.

Sea-urchins such as we have been describing live on stony or rocky bottoms, over which they slowly creep by means of their tube-feet. The spines are pressed against the substratum and keep the animal from rolling over under the pull of the tube-feet and also help to push it on. The spines are usually of two distinct sizes, longer primary spines, and shorter secondary spines. The forest of spines has a kind of undergrowth of pedicellariae. These are of several

kinds and are much more highly finished organs than those of
star-fishes; they have a long stalk, which is partly stiffened by
a delicate calcareous rod, and the jaws are three in number. One

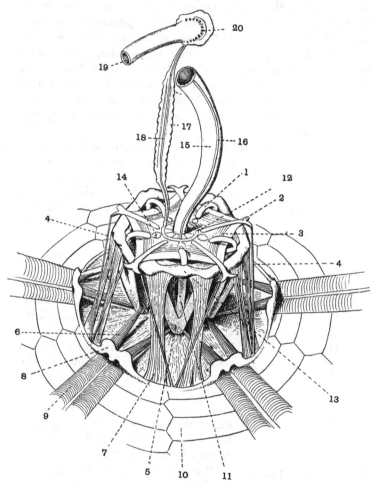

Fig. 143. Aristotle's Lantern of *E. esculentus* × 2. Partly from Chadwick.

1. Upper end of tooth enveloped in lantern membrane.   2. Radius.
    3. Transverse muscle of radii, elevator.   4. Depressor muscles of radius.
    5. Jaw.   6. Retractor muscles of the jaws.   7. Protractor of jaws.
    8. Auricula.   9. Ampullae of tube-feet.   10. Inter-ambulacral plate.
    11. Tooth.   12. Circular water-vascular vessel.   13. Epiphysis.
    14. Polian vesicles.   15. Oesophagus.   16. Ventral "blood"-vessel.
    17. Genital stolon.   18. Stone-canal.   19. Rectum.   20. Madre-
    poric plate.

kind has short stumpy jaws, each with a poison bag at its base and
a stiff stalk; these are doubtless weapons of defence and enable the
Urchin to give any unwelcome visitor which may come too close a
warm reception. Such pedicellariae are called gemmiform (Fig. 142).
Another kind, termed tridactyle, has long jaws and a flexible
stalk. It was supposed that these helped the animal to climb by
seizing hold of waving fronds of sea-weed till the tube-feet could
get a hold, but this is proved not to be the case. It has been

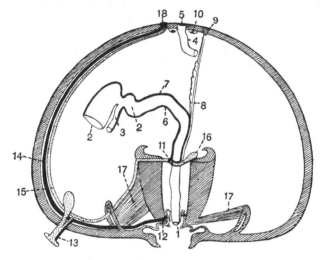

FIG. 144.   Diagrammatic vertical section of a Sea-Urchin.   From Leuckart,
after Hamann.

1. Mouth.   2. Intestine cut short.   3. Siphon.   4. Rectum.   5. Anus.
6.   Ventral "blood"-vessel on intestine.   7.   Dorsal "blood"-vessel
on intestine.   8. Stone-canal.   9. Madreporic plate.   10. Genital
rachis.   11. Water-vascular ring.   12. Nerve-ring.   13. Tube-foot
with ampulla.   14. Radial nerve.   15. Radial water-vessel.   16. Polian
vesicle.   17. Muscles; those on the left pull Aristotle's Lantern outwards, those on the right retract it.   18. Ocular plate.

shown that a gentle movement in the water excites the tridactyle
pedicellariae while a stronger movement calls the gemmiform into
activity.   Besides these there are two other kinds of pedicellariae.
It seems most probable that these elaborate organs are for the
purpose of protecting the sea-urchin against the attacks of certain
animals which in their absence would either fix themselves on
the skin of the Echinoid or even burrow into it.   The number
and variety of these organs are an indication of the danger that exists
from this source.

The Urchin is provided with five white chisel-like teeth, each of which slides on a pair of grooved pieces called alveoli, meeting in a point below. Each pair of alveoli meet in a point where they clasp the tooth. Above they are united by two pieces called epiphyses (13, Fig. 143) which meet in an arch. A pair of alveoli with their epiphyses are often spoken of as a jaw, and adjacent jaws are joined by stout, inwardly projecting rods called rotulae. The whole apparatus of five jaws has received the name of 'Aristotle's Lantern.' This can be pushed out or pulled in by muscles attached to arches called auriculae, rising from the inner side of the skeleton (8, Fig. 143). Through the auriculae the radial water-tube and nerve pass, and thus they correspond in position to the ambulacral plates of star-fish.

The food of the Urchin consists ordinarily of seaweed which it gnaws with its teeth. No doubt the little worms and molluscs always found in abundance on the surface of the weed add a flavour to the repast. The alimentary canal is exceedingly unlike those of the Echinoderms so far studied. The gullet ascends vertically between the teeth and passes into the intestinal tube which runs in a spiral right round the body and then turns sharply back and describes one turn of a spiral in the opposite direction, after which it bends inwards and runs straight up to the anus (Fig. 146). For the first part of its course a small tube, the so-called siphon, runs parallel to it, opening into it at both ends.

The water-vascular ring is situated above the masticatory apparatus and is thus widely separated from the nerve-ring, which is situated below it : the radial tubes, in consequence, run downwards along the "lantern" before bending outwards under the auriculae (see Fig. 144). The water-vascular ring bears small pouches which have been termed Polian vesicles. They seem, however, to correspond to Tiedemann's bodies in an Asteroid. The first pair of tube-feet in each radius are different to the rest, in that they are short and not capable of extension, and that their discs are oval. These tube-feet protrude through the peristome and are called the buccal tube-feet; they function as tasting organs, and are thrown into violent excitement if a piece of eatable matter is put near them (Fig. 147).

In describing the Asteroidea it was mentioned that the genital stolon or "dorsal organ" had been mistaken by former authors for a heart, and that true blood-vessels were unknown amongst Echinodermata. If by blood-vessel is meant a tube with well-defined

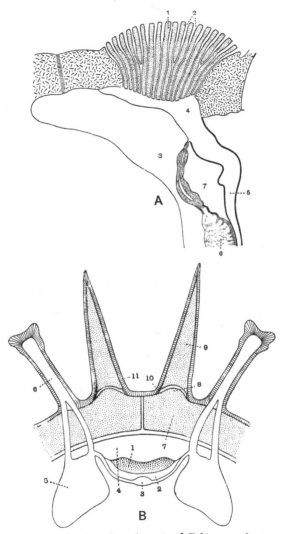

Fig. 145. Sections through parts of *Echinus esculentus*.

A. A section at right angles to the plane of the Madreporic plate × 16. From Chadwick. 1. Madreporic plate. 2. Pores in the same. 3. Madreporic vesicle. 4. Ampulla of madreporic plate or dilatation of stone-canal into which the pores open. 5. Madreporic tube. 6. Genital stolon. 7. Axial sinus.

B. A section at right angles to an ambulacral area. 1. Radial nerve-cord. 2. Radial perihaemal canal. 3. Radial water-canal. 4. Epineural canal. 5. Ampulla. 6. Cavity of tube-foot. 7. Ambulacral plate. 8. Boss for articulation of spine. 9. Spine. 10. Muscles which move the spine. 11. Ectoderm.

walls in which there is a definite circulation of fluid this is strictly
true.

It must be remembered, as was pointed out in the chapter on
Arthropoda, that blood-vessels and connective tissue have been
derived from the same primitive tissue, which may be compared to
the jelly of Coelenterata. Now Echinodermata probably represent a
stage before the evolution of either blood-vessels or proper connective
tissue. Apart from the plates of the skeleton the substance of the
body-wall has little more consistence than the jelly of an *Aurelia,*

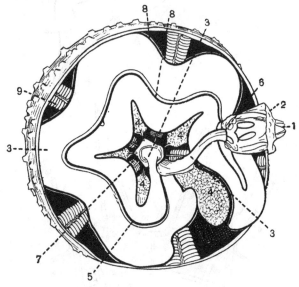

Fig. 146. View of Sea-Urchin, with part of the shell removed to show the
course of the alimentary canal. From Leuckart, after Cuvier.

1. Mouth surrounded by five teeth (displaced).   2. Lantern of Aristotle.
3. Oesophagus, coiled intestine and rectum.   4. Ovaries with oviducts.
5. The siphon.   6. "Blood-ring."   7. Fold of peritoneum supporting
genital rachis.   8. "Blood-vessel" accompanying intestine.   9. Ampullae
at base of tube-feet.

and readily degenerates into slime. The ground substance has
remained so fluid that it is still traversed by amoebocytes which
carry excreta to the exterior. In Echinoidea along two tracts, one
situated on the same side of the oesophagus as the stone-canal and
the other on the opposite side, the jelly intervening between the
inner wall of the coelom and the oesophagus has undergone the first
stage in the change to a blood-vessel. The fibres are scantily

developed and the amoebocytes are present in immense numbers, whilst the ground substance has become more fluid and probably contains proteids, since it stains with carmine like protoplasm. These tracts are termed dorsal and ventral "blood"-vessels. The dorsal vessel is on the side next the stone-canal. These tracts have not the form of tubes, but are networks of irregular passages devoid of proper walls. They accompany the alimentary canal throughout most of its course and it seems as if the products of digestion were accumulated in them. A so-called "blood-ring" of similar character surrounds the oesophagus just above the water-

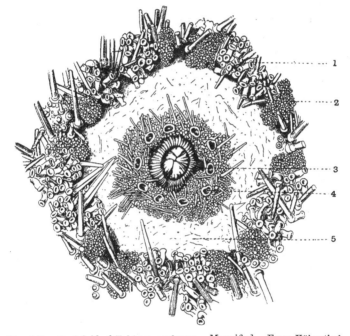

Fɪɢ. 147.   Oral field of *Echinus esculentus*.   Magnified.   From Kükenthal.

1.   Ambulacrum with tube-feet and spines.   2.   Gill.   3.   Teeth of Aristotle's lantern.   4.   Buccal tube-feet.   5.   Soft membrane around mouth, the Peristome.

vascular ring and into this the two "vessels" open. A similar ring has been described in Asteroidea and Ophiuroidea : in some species of the former class a tract of similar substance appears to run down the arm just above the nerve-cord in the septum separating the two perihaemal canals, and the name of these canals (Gr. περί, around ; αἷμα, blood) has been suggested by this circumstance.

Breathing, as one might expect, is carried out wherever the body-wall is thin enough to allow the oxygen to diffuse through, that is to say by the tube-feet and by the peristomal membrane. The tube-feet, as in the star-fish, are provided with large ampullae which project freely into the great spacious body cavity. Oxygen thus taken into the fluid filling the tube-feet can be passed into the body cavity through the ampullae, and there is a curious arrangement to facilitate this. Where the tube-foot passes through the skeleton it is split into two parallel tubes which reunite below (B, Fig. 145): so that on the dried shell we see on the ambulacral plate several pairs of pores, each pair corresponding to a single tube-foot. As the cells lining the inside of the latter are ciliated, the splitting of the tube is apparently to facilitate the separation of the upward and downward currents of water.

The peristome has ten branched pouches, situated one pair in each interradius and projecting outwards. These are the gills, but it is unreasonable to suppose that all the breathing is done by them. They communicate not with the general body cavity, but with a part of it, called the lantern coelom, shut off from the rest by a septum stretched between the teeth and jaws. Embedded in the upper wall of this are certain rods called radii, which are connected with each other and the auriculae by muscles, and by means of these the upper wall of the lantern coelom can be raised or depressed and so the pressure inside altered. When these rods are depressed water is driven out into the gills and there absorbs oxygen. When they are raised the water is sucked back into the lantern coelom and the oxygen passes through the thin wall of the latter into the general coelom.

The organs of sex are alike in external appearance in both sexes (Fig. 146). They have the form of five great bunches of tubes hanging down into the body cavity and opening by five small holes placed in plates called the genital plates, forming the summit of the interambulacral series on the corona and situated just outside the periproct (Figs. 141 and 148). In the young Sea-urchin there is a genital rachis connecting them together, and throughout life there is a genital stolon alongside the stone-canal. The genital stolon is relatively much larger in Echinoidea than in Asteroidea, and surrounds the axial sinus in the lower part of its course, so that this space appears like a cavity excavated in the stolon. Hence by one author the axial sinus and stolon were mistakenly described as a nephridium with glandular walls, and the madreporic vesicle, which is here much enlarged and extends parallel with the axial sinus, was called by the same author the "accessory kidney."

The Echinoidea are divided into three Orders:—

I. The ordinary Sea-urchins, such as we have described, constitute the Order ENDOCYCLICA, which live chiefly on rocky and stony ground. The other two Orders live in sand or mud and have undergone singular modifications in order to fit them for this kind of life.

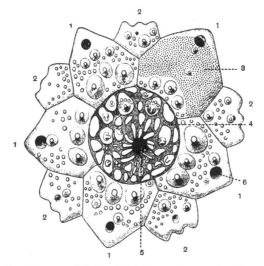

FIG. 148. Aboral system of plates of *Echinus esculentus* × 4. From Chadwick.
1. Basal plate. 2. Radial plates. 3. Madreporite. 4. Periproct, or region around anus. 5. Anus. 6. Genital pore.

They are termed the Irregular or Exocyclic Sea-urchins, because whereas the anus has become shifted from the upper pole of the body down one side to the edge, or even to the under-surface of the more or less flattened body, the madreporite and genital plates still retain their position. In both Orders the tube-feet of the upper part of the ambulacra are the main breathing organs, and are greatly flattened and expanded at the base, while the pores through which they pass are arranged in two converging curves on each ambulacrum, the figure produced being compared to a petal of a flower, hence the name applied to them, viz., petaloid ambulacra. The special characters of these two Orders are as follows:

II. CLYPEASTROIDEA or Cake-urchins. They live at or near the surface of the sand. They still retain their teeth, which are placed almost horizontally, and they use them as spades to shovel the sand into the mouth. All the spines covering the upper surface are ciliated and so a constant current of water sweeps over the expanded tube-feet which act as gills. In addition to these tube-feet the whole aboral surface, radii and interradii included, is covered with a multitude of minute tube-feet provided with suckers. Similar tube-feet are found on the oral surface, but they are confined to the radii. This immense multiplication of tube-feet is of course due to the small purchase that any one of them is able to get on such a yielding material as sand. In a word, the animal moves itself by a multitude of minute pulls instead of by a lesser number of stronger pulls as do the Endocyclica. There are calcareous pillars stretching from the upper to the lower surface of the shell or test, apparently to enable them to withstand rough usage, since in many cases they live within reach of the breakers. The best known British species is *Echinocyamus pusillus*, a little oval Sea-urchin about the size of a pea, whence the common name applied to it, viz. Pea-urchin. On the east coast of North America one species, *Echinarachnius parma*, the Sand-dollar, is very common; this is an extremely flattened Urchin of circular outline, the shape and size of which have suggested a comparison with the famous silver-dollar of the United States currency.

III. SPATANGOIDEA or Heart-urchins. These live buried at depths of a few inches to a foot beneath the surface of the mud, and the body is more or less oval or egg-shaped, slightly flattened underneath. The mouth is sometimes in the centre of the under-surface and sometimes nearer one end, and is usually crescentic and always without any trace of jaws. These Urchins have usually only four of the ambulacra "petaloid"; the fifth has a few long tube-feet with expanded fringed discs. In the case of the familiar British species *Echinocardium cordatum* it is known that the Urchin extends these tube-feet from its burrow right up to the surface of the sand and collects with them decaying organic matter lying on the surface. This is pushed within reach of the buccal tube-feet and so reaches the mouth.

The Spatangoidea do not use their tube-feet to walk with, but move by means of spines which are provided with flattened tips, and so the small tube-feet present in such multitudes in the Clypeastroidea are absent. Besides these spines they possess peculiar lines of very small spines covered with cilia, which cause a current to pass over the gill-like tube-feet. Such rows of ciliated spines are termed fascioles.

CLASS IV. HOLOTHUROIDEA.

The fourth group of the Echinoderms is termed the Holothuroidea or Sea-cucumbers, and consists of animals of a more or less sausage-shaped form, with the mouth at one end and the large anus at the other.

These animals have undergone the same essential modification as the Sea-urchins, the arms having been re-absorbed into the body so that the radial tubes run down the side of the body and end near the vent. The nervous system also is situated beneath the surface, the ambulacral groove being represented by the epineural canal. The skin, as in Echinoidea, retains its well-marked ectoderm with nervous sheath.

They are however distinguished by some most marked characteristics :—1. The skeleton has almost entirely disappeared, being represented only by grains and prickles of various shapes completely buried in the skin. 2. The muscular system of the body-wall is most powerfully developed: there is a pair of strong longitudinal muscles running inside each radial tube, and transverse muscles run across each interradius. 3. The buccal tube-feet are highly modified and are the means by which the animal feeds itself. 4. The anus is wide and the concluding portion of the intestine termed the cloaca is strongly muscular, and it is used as a breathing organ, water being sucked in at the anus and thrown out again. 5. The stone-canal does not reach the exterior, but terminates in a sieve plate hanging down into the interior of the body.

The breathing by means of the anus is carried out by certain organs called gill-trees. These are two great branched tree-like outgrowths of the hinder part of the intestine, reaching right through the body cavity to near the mouth (Fig. 149). Water is taken in by the anus and forced up into the finest branches of these and no doubt diffuses through into the body cavity under the pressure set up by the contraction of the muscles of the anus. Hence it is that the animal is able to dispense with an external madreporite, and also to obtain the fluid necessary to keep its tube-feet tense from its own body cavity. From the water-vascular ring one or more long-stalked Polian vesicles hang down into the body cavity.

The muscular body-wall has a very curious effect on the economy of the animal. When it is irritated it contracts the muscles, and

280

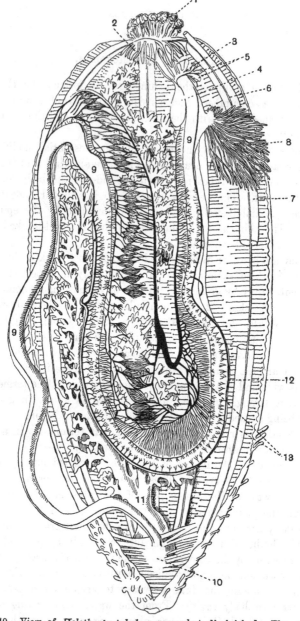

FIG. 149. View of *Holothuria tubulosa* somewhat diminished. The animal is opened along the left dorsal interradius and the viscera are exposed. After Ludwig.

1. Tentacles.　　2. Ampullae of tentacles.　　3. Water-vascular ring.
4. Polian vesicle.　　5. Stone-canals.　　6. Radial water-vessel.
7. Radial longitudinal muscle partly cut away.　　8. Reproductive organ.　　9. Alimentary canal.　　10. Cloaca.　　11. Respiratory trees.
12. Ventral "blood-vessel."　　13. Dorsal "blood-plexus."

since the fluid in the body cavity is practically incompressible, the effect is to set up a tremendous pressure. As a result of this the wall of the intestine near the anus tears and a portion or the whole of the intestine is pushed out. The gill-trees are the first to go, and in some species the lower branches of these are covered with a substance which swells up in sea-water into a mass of tough white threads in which the enemies of the animal are entangled. A lobster has been seen rendered perfectly helpless as a consequence of rashly interfering with a Sea-cucumber. These special branches are termed Cuvierian organs.

A Holothuroid is only temporarily inconvenienced by the loss of its internal organs. After a period of quiescence it is again furnished with the intestine and its appendages. Some species, which are able to pull in the mouth end of the body with their tentacles, when strongly irritated snap off even this, and yet are able to repair the loss.

The intestine is a simple looped tube which has three limbs. One limb runs down towards the anus, the next turns up again towards the mouth and then bends back into the final limb which goes towards the anus. These limbs are attached by mesenteries to different interradii of the body, the first to that which in the ordinary position of the animal is mid-dorsal, the next to the left ventral, and the third to the right ventral (Fig. 149).

Accompanying the alimentary canal are so-called dorsal and ventral "vessels" similar to those of the Echinoidea, and there is also a "blood-ring" like that described in the same class. In Holothuroidea the ventral vessel is close to the alimentary canal but the dorsal vessel is borne on a little ridge projecting from the intestine. The alimentary canal is enswathed by minor branches of the network of which the dorsal and ventral vessels form merely the large trunks. The whole system thus assumes a very complicated appearance, but even here it has been shown that there is no circulation nor even a proper wall to the spaces. The longitudinal vessels indeed often do not appear to communicate with the blood-ring.

The buccal tube-feet form a crown of from ten to twenty-five great branched tentacles, and their different shapes are used to classify the various families of the Sea-cucumbers. Most species feed on sand or mud, but one Order can be described only as anglers. In them the tentacles are long and delicately branched so that they resemble pieces of sea-weed. The animal stretches them out, and they become the resting-place of numbers of the minute animals which swarm in sea-water. When one tentacle has got a sufficient freight it is bent round and pushed into the mouth which is closed on it. It is then forcibly drawn out through the closed lips so that all the living cargo is swept off.

The organs of sex are similar in nature to those of the Urchins, but are represented by only one mass of tubes which all unite in a common opening near the tentacle region, and it is in this region that the stone-canal opens in the one or two rare cases where it still opens to the exterior. Hence it appears that whereas in the irregular Sea-urchins the genital openings and madreporite have remained fixed while the anus has been shifted, here the anus has remained in its original position while the genital opening has been shifted towards the mouth.

The Holothuroidea are divided into the following five Orders.

1. Elasipoda: Sea-cucumbers whose tentacles have shield-shaped ends drawn out into short processes, devoid of gill-trees, with the tube-feet of the upper surface of the body modified into stiff respiratory processes. Live only at great depths in the ocean.

2. Aspidochirotae: Sea-cucumbers with shield-shaped ends to the tentacles,—these have also large ampullae so that they can be individually retracted. With gill-trees and often Cuvierian organs.

3. Dendrochirotae: Sea-cucumbers with long delicately-branched tentacles without ampullae. The whole front end of the body can be pulled in by means of special muscles. Gill-trees present.

4. Molpadidae: Sea-cucumbers with tentacles unbranched or with two or four small lateral branches, and no other tube-feet except a circle of papillae round the anus. Gill-trees present.

5. Synaptidæ: Sea-cucumbers in which the tentacles have two rows of short branches. No tube-feet except these, the radial canals having also disappeared. No gill-trees. The body-wall is thin and transparent and oxygen can diffuse through it.

Class V. CRINOIDEA.

The last group of the Echinoderms is termed the Crinoidea (Gr. κρίνον, a lily), animals long familiar to collectors of fossils under the name of Lily-encrinites. They differ from other Echinoderms in that from the centre of what corresponds to the upper or aboral surface of other orders, there springs a jointed stalk by which the animal is moored to the substratum.

Animals of this type were much more common in past times than now. Large masses of limestone are actually made up of their skeletons. The modern order of Crinoidea includes a few species surviving at great depths in the ocean, and about the mode of life of these we know little. There are, however, besides these a number of species not sharply marked off from each other assigned to a family, the Comatulidae, containing two genera, *Antedon* and *Actinometra*, which live at moderate depths in the ocean and which have been thoroughly studied. These however are exceptional in that they break off from the stalk when they are mature and swim about by muscular movements of the long arms. The stump of the lost stalk forms a knob called the centro-dorsal ossi-cle, which is provided with grasping processes called cirri, by means of which the animal can temporarily attach itself.

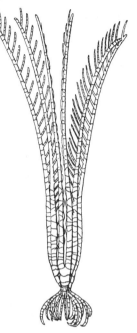

Fig. 150. *Antedon acoela*, Car. A young individual × 1½. After Carpenter.

We may select for our type the common *Antedon rosacea*, which can easily be captured by the dredge at moderate depths. This animal has a small disc and ten extremely long arms. It reminds one of a star-fish, in the fact that on the oral sides of these arms there are open grooves converging to the mouth, and that the skin lining these grooves is modified to form nervous bands uniting in a ring round the mouth. These ambulacral grooves are further lined by powerful cilia which cause currents of water carrying small animals to flow towards the mouth, and thus the animal is fed. The tube-feet are small and apparently of use only as gills, those springing from the grooves on the disc alone retaining their sensory function.

The skeleton is peculiar. The ventral side of the body is

covered by a leathery skin, but on the aboral side there are first the
centro-dorsal ossicle, a round knob representing the uppermost
joint of the stem, and then five rows of plates, called radials,
radiating from it. These five rows show us that here, as in most
other Echinoderms, we have to do with five primitive arms or
radii; these radii, however, bifurcate the moment they become

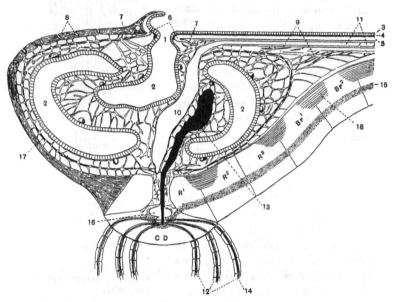

Fig. 151. Transverse section through the disc and base of an arm of
Antedon rosacea. After Ludwig.

1. Mouth.   2. Various sections of alimentary canal.   3. Epithelium of
ambulacral groove.   4. Nervous layer of ambulacral groove.   5. Radial
water-canal.   6. Circumoral water-vascular ring.   7. Stone-canals.
8. Pore-canals.   9. Trabeculae traversing the coelom.   10. Axial
coelom communicating with 11.   11. Coelom of arm.   12. Cirri.
13. Genital stolon giving off 14.   14. Branches of genital stolon in
cirri.   15. Radial nerve of dorsal nervous system.   16. "Chambered
organ" in centre of dorsal nervous system.   17. Calcareous rods
developing into trabeculae.   18. Muscles connecting radial and brachial
plates.   CD. Centro-dorsal piece. R^1, R^2, R^3. First, second and third
radial plate. $Br.^1, Br.^2$ First and second brachial plate.

free from the disc, and so there are ten arms: the uppermost
plate in each of the five rows having a double facet, on to which
fits the lowest of the rows of plates supporting the arms.

The arms really bifurcate again and again, but in each case one
of the forks does not develope further and forms a pinnule. If in
the case of any of the bifurcations both forks were to develope we

should have an increase in the number of arms, and indeed species of *Antedon* with twenty, forty and even one hundred arms are known.

There is no madreporite, but the whole of the upper soft integument is riddled with isolated pores which lead into the body cavity and are lined by ciliated cells. The water-vascular ring has hanging down from it a large number of stone-canals, which also open freely into the body cavity. Only one pore and one stone-canal exist in the stalked young, but their position, comparatively near the mouth, is utterly different from that in any other Echinoderm. In the adult the cavity of the coelom is traversed in every direction by cellular cords called trabeculae.

The anus is situated on the ventral side of the body in an interradius, the alimentary canal being coiled in a simple spire in the disc. We have spoken above of the ambulacral grooves being lined by nervous cells, like those forming the radial nerve in star-fish. This is indeed so, but the Crinoid possesses another and much more important nervous system. From the body cavity five canals are given off which penetrate the stalk. These canals swell up in the substance of the centro-dorsal ossicle into chambers, and in the permanently stalked forms like *Rhizo-crinus* or *Pentacrinus* they form similar chambers wherever the stalk bears cirri. The coelomic cells which form the walls of these canals develope great masses of nervous fibrillae. In *Antedon* of course only the five uppermost chambers remain when the stalk disappears—they are termed collectively the chambered organ—and the nervous lining of these

Fig. 152.   *Rhizocrinus*
× about 2¼.   From Sars.

constitutes a kind of brain (Fig. 151). This "brain" is separated
from the body cavity of the calyx by a shelf-like fold strengthened
by a calcareous plate called the rosette, which represents a circle
of five plates alternating with the columns of radials clearly seen
in more primitive crinoids. From this brain cords go off to the
cirri, and five great cords run upwards perforating the radial rows of
plates and eventually bifurcating pass into the arms perforating the
plates which form the skeleton of the latter. These cords are at
first tubular outgrowths from the
brain, the cells forming the walls
of which become converted into
nervous matter. It has been experimentally proved that it is this
nervous system which controls the
muscles moving the arms, and
that if the whole soft part of
the disc including the ambulacral
nervous system be removed the
animal swims just as well as
before.

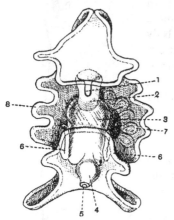

Fig. 153. Ventral view of a Larva of
a Holothurian taken at Marseilles
× about 100. From Joh. Müller.

1. Mouth.   2. Oesophagus.
3. Stomach.   4. Rectum.
5. Anus.   6. Coelomic sac.
7. Rudiment of water-vascular system.
8. Ciliated band.

The organs of sex are rounded
masses found in the pinnules and
are really, as in Asteroidea,
Ophiuroidea and Echinoidea,
swellings on branches of a genital
rachis. There is also a genital
stolon, which however has no
connection with any of the stone-
canals, but rises from the rachis
to the centre of the dorsal wall of the coelom. The young are
carried on the pinnules for some time and have a very short free
swimming life, very soon settling down and developing into little
pentacrinoids with a jointed stalk. The name "pentacrinoid"
is suggested by their resemblance to *Pentacrinus*. These stalked
young present interesting features in the skeleton found in many
living and fossil Crinoidea but absent in the adult *Antedon*. Thus
the mouth is guarded by five inter-radial valves each supported
by an oral plate, and the rosette is represented by five ossicles.

Leaving aside the Crinoidea, the development of which is known
only in one case and is there evidently much modified, the eggs of

the other four groups of Echinoderms develope into free swimming
animals which for periods varying from a fortnight to six weeks lead
a free life at the surface of the ocean. These young are called
Dipleurula larvae and they are, as mentioned, utterly different to
adult Echinoderms : unlike these, but like most other animals, they
are bilaterally symmetrical (Fig. 153). They swim by means of a
powerful longitudinal ciliated ring, drawn out into a number of
arms or processes. They possess a complete alimentary canal,
consisting of oesophagus, stomach and rectum, while the coelom is
represented by two sacs lying one at each side of the digestive
tube. These sacs are, as a study of the early development teaches,
portions of the alimentary canal budded off from the rest. One of
the most interesting features in the development is the fact that
these sacs undergo transverse division in the same way as do the
germinal bands of an Annelid. On each side three segments are
formed. The most anterior on each side often coalesce to form a
median sac into which the originally single madreporic pore opens
on the left side : a portion of this sac becomes the axial sinus of
the adult. The middle sections on each side in exceptional cases
develope each into the rudiments of a water-vascular system showing
that this structure was originally paired. That on the right side,
however, normally remains small, whilst the left one takes on the
form of a wreath with five projections, by the union of the two ends of
which the water-vascular ring with the rudiments of the five radial
canals is formed. This is called the hydrocoele. The most posterior
divisions form the body cavity of the adult ; the left one grows in a
ring-shaped manner, encircling, as with a wider ring, the ring of the
hydrocoele, while through the centre of both rings the oesophagus
of the adult grows. The oesophagus of the larva is usually cast
off, but sometimes, as in the Ophiuroidea, it is directly converted
into that of the adult by being shifted to the left before the rosette
of the water-vascular system becomes a ring.

The Dipleurulae of the Asteroidea fix themselves at the conclusion
of their larval life by the anterior end of the body, using the prae-
oral lobe as a stalk. The fixed stage is omitted in other Eleutherozoa,
but the larva of *Antedon rosacea*—the only Crinoid whose development is known—also converts the prae-oral lobe into a stalk. But
in the case of the Asteroidea the body becomes bent on the stalk in
such a way that the stalk springs from close to the mouth of the
adult. The stalk is eventually absorbed and the star-fish commences
its adult life, breaks loose from its attachment and moves away.

In the Crinoid, on the other hand, the mouth becomes rotated away from the stalk, and the latter seems to spring from the aboral surface.

The whole of this development seems to point to the conclusion that the radially symmetrical Echinodermata are derived from a bilaterally symmetrical ancestor with traces of metameric segmentation ; that the acquisition of radial symmetry was due in the first instance to the assumption of a fixed mode of life, followed by the dwindling of the organs of the right side and the compensating greater growth in those of the left. The Crinoidea seemed to have retained the original mode of feeding by means of the current produced by cilia, and thus their mouth became shifted upwards away from the stalk into a position favourable for the capture of floating prey. In the Asteroidea and the other Eleutherozoa food is obtained by seizing it with the tube-feet, and hence in these the mouth was bent downwards so that the stalk seems to spring from the oral surface.

This development is not only interesting on account of the extraordinary metamorphosis which the young undergo, but also on account of the fact that whilst the adult is utterly unlike any of the other Coelomata, the structure of the young is reconcilable with the fundamental structure of Annelida and Mollusca, etc. The only plausible explanation of this is to be found in the hypothesis that the young represent in a rough sort of way the ancestor from which the Echinoderms were derived.

When the Vertebrata are dealt with it will be pointed out that the larvae of the most primitive forms bear a striking resemblance to those of the Echinodermata, and that in the embryos of many Vertebrata the coelom undergoes at first a similar division to what occurs in the Dipleurula, suggesting the conclusion that the highest groups in the animal kingdom are also sprung from the same ancestor as gave rise to the Echinoderms.

The Phylum Echinodermata is classified as follows :

Sub-Phylum A. Pelmatozoa.

Echinodermata which are fixed to some foreign object during the whole or part of their existence by a jointed stalk springing from the centre of the aboral surface.

Class I. Crinoidea.

Pelmatozoa with five long arms which repeatedly fork. The genital organs are borne in the tips of the branches.

Sub-Phylum B. Eleutherozoa.

Echinodermata which are free during the whole of their adult existence and rarely fixed even during the larval condition. When the immature form is fixed the stalk springs from the oral surface near the mouth and is not jointed.

Class I. ASTEROIDEA.

Eleutherozoa with arms (free radii) containing outgrowths of the alimentary canal and open ambulacral grooves. The arms have feebly developed muscles and locomotion is effected entirely by the tube-feet.

Ex. *Asterias, Echinaster.*

Class II. OPHIUROIDEA.

Eleutherozoa with arms sharply marked off from the central disc. The arms do not contain outgrowths of the alimentary canal and have closed ambulacral grooves. They have highly developed muscles, and locomotion is entirely effected by the arms, the tube-feet being purely tactile.

Ex. *Ophioglypha.*

Class III. ECHINOIDEA.

Eleutherozoa in which the arms have coalesced with the body, the radii being arranged like meridians on a sphere. The ambulacral grooves are closed. The body has a complete armour of closely adjusted plates and the spines are movably articulated with these and assist in locomotion.

Order 1. Endocyclica.

Echinoidea in which the anus is in the centre of the aboral pole and teeth are present.

Ex. *Echinus.*

Order 2. Clypeastroidea.

Echinoidea in which the anus is excentric, the dorsal tube-feet are flattened and teeth are present.

Ex. *Clypeaster, Echinocyamus, Echinarachnius.*

Order 3. **Spatangoidea.**

Echinoidea with an excentric anus and flattened dorsal tube-feet but without teeth.

Ex. *Spatangus.*

Class IV. HOLOTHUROIDEA.

Eleutherozoa resembling Echinoidea in the anus and ambulacral grooves; but with rudimentary skeleton, highly developed muscular body-wall and greatly enlarged buccal tube-feet by means of which all the food is obtained.

Ex. *Holothuria.*

CHAPTER X.

Phylum Brachiopoda.

Brachiopods (Gr. βρᾰχίων, the arm; πούς, ποδός, the foot) are true Coelomata and retain the coelom in a primitive and typical condition. Like the Mollusca, they are not segmented, and the only trace of a repetition of parts in the group is in a genus called *Rhynchonella* in which the nephridia are repeated, so that we find two pairs. A similar repetition of the same organs occurs amongst the Mollusca in *Nautilus*.

Brachiopods are exclusively marine. They have a shell consisting of two valves, so that at first sight they appear to resemble the Pelecypoda, but in Brachiopods the shells are placed ventrally and dorsally, and not on the two sides of the animal as in Mussels. In a few cases such as that of the primitive genus *Lingula* the two valves of the shell are nearly alike in size and shape and consist largely of horny matter or chitin. In most cases however the shell is calcareous, and since in Brachiopoda, as in most bilaterally symmetrical animals, the two sides resemble one another whilst the back and front are unlike, each valve of the shell is symmetrically shaped, but the dorsal valve differs from the ventral, the latter being usually the larger. In a few cases, such as that of *Crania*, a British form common in certain localities, the ventral valve is flat and attached by its whole surface to the substratum ; all that is then seen is the arched dorsal valve. Since in the overwhelming majority of Pelecypoda the two valves of the shell are similar in appearance, while each is asymmetrical in shape, the umbo being situated near the anterior end, it is easy to distinguish at a glance the shells of the Pelecypoda from those of the Brachiopoda.

The posterior end of the body terminates in a stalk which in *Lingula* helps to keep the animal in the holes of the sand in which it lives. In other forms the stalk is shorter and it is firmly glued

19—2

to a rock so that when it has once fixed itself a Brachiopod cannot
change its place of residence
(Fig. 157). Each valve of
the shell is lined by the body-
wall of the animal, but the
body does not occupy the
whole of the space between
the two valves ; it is produced
into two folds or flaps called
the mantle-flaps, which are
each hollow and contain extensions of the coelom. These
secrete the larger part of
the valves of the shell. In
Lingula and some others the
free edges of these mantle-
folds, lying parallel with the
free edges of the shells, bear a

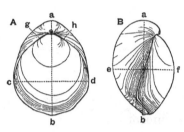

Fig. 154. *Terebratula semiglobosa*, a fossil
Brachiopod shell × ⅔.

A. Dorsal view.   B. Lateral view.
a. posterior, b. anterior end. The line
between a and b is called the length, it
traverses the aperture through which
the stalk projects. The line between
c and d is the breadth, and between e
and f the thickness, and between g and
h is the hinge-line.

number of chaetae which recall those of the Chaetopoda. It is
by no means certain that the shell of Brachiopods is an external
secretion like that of Mollusca : it seems possible that it is really
deposited in the connective tissue under the ectoderm.

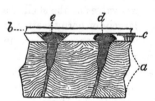

Fig. 155. A section through the
shell of *Waldheimia flavescens*.
Magnified.

a. Prismatic layer formed in connective tissue.   b. Epidermal
layer.   c. Outer calcareous
layer.   d & e. The expanded
outer ends of the tubes which
traverse the shell.

In most of the thick-shelled forms
the shell is traversed by processes of
the mantle, which nourish it, so that
in dried Brachiopods the shell seems
perforated with a number of pores.

If we slightly open the valves of
the shell of a living Brachiopod (so as
to avoid tearing the tissues) and look
in, we shall see between the ventral
and dorsal mantle-folds the anterior
body-wall of the creature. This sometimes runs almost horizontally across
between the space within the valves,
but often slopes obliquely from the
ventral to the dorsal valve of the shell. Part of this wall is modified
to form two long ridges, the ends of which project freely and are called
the arms ; they are coiled and are beset with tentacles (Fig. 156).
Running close to the origin of the tentacles is a little lip or flange
so placed that the two form a groove or gutter. The groove is

lined with cilia and so is the inner face of each tentacle. The whole of this apparatus is called the lophophore. It might be described as a ring of tentacles, the ends of which are drawn out so as to form the arms. It is never quite so simple as the above account would lead one to suppose, for the ring is often produced into two minor lobes forming the lesser arms situated between the main ones, and in many genera the two main arms are raised up from the level of the body-wall and each is twisted into a spiral. The dorsal shell may be prolonged into a series of plates and even into elaborate bands and loops which serve to support such lophophores.

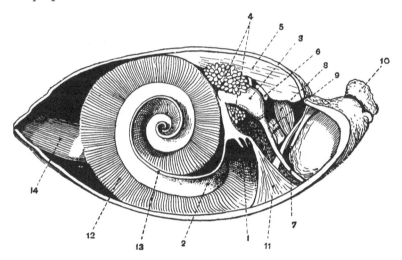

Fig. 156. View of the inner side of the right half of *Waldheimia australis*. From a dissection by J. J. Lister.

1. Mouth.   2. Lophophore.   3. Stomach.   4. Liver tubes.   5. Median ridge on dorsal shell.   6. Heart.   7. Intestine, ending blindly. 8. Muscle from dorsal valve of shell to stalk.   9. Internal funnel-shaped opening of nephridium.   10. Stalk.   11. Body-wall.   12. Tentacles. 13. Coil of lip.   14. Terminal tentacles.

The mouth lies on the middle line at the bottom of the gutter between the lip in front and the tentacles behind. Internal structure. The lophophore is thus an organ for catching food and passing it into the mouth. The cilia which cover the inner surface of the tentacles and line the gutter set up small whirlpools in the water so that minute animals and algae becoming involved in these are swept into the mouth. In many species the tentacles can be protruded between the valves of the shell, and thus the area they affect is enlarged.

The mouth leads into a simple stomach which ends in a short intestine. Both stomach and intestine are ciliated. A digestive gland called the liver consisting of a number of branching tubes opens on each side into the stomach, and as is the case in the Crustacea much of the digestion takes place inside these glands. In the genera of Brachiopoda which have a hinged shell the intestine ends blindly, but in those which have no hinge there is an anus which may open in the middle line as in *Crania,* or on the right side of the body as in *Lingula.*

On the dorsal surface of the stomach is a small, muscular, contractile vesicle, the heart (Fig. 156). This gives off a number of vessels, amongst others one which passes to each tentacle, which therefore possibly act as respiratory organs.

The chief part of the nervous system retains its primitive relation to the ectoderm. Just in front and just behind the mouth there are thickenings of the ectoderm forming a supra- and sub-oesophageal ganglion respectively. The latter contrary to the usual rule being much the larger (Fig. 157). They are connected by two lateral cords and give off a number of nerves, one of which runs to each tentacle. No sense organs, such as ears or eyes, are known; and indeed the fixed

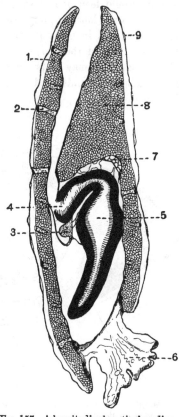

FIG. 157. A longitudinal vertical median section through *Argiope neapolitana.*

1. Ventral shell.   2. Canal containing blood-vessel.   3. Sub-oesophageal nerve ganglion.   4. Mouth. 5. Stomach.   6. Stalk.   7. Plexus of blood-vessels.   8. Median crest on dorsal shell.   9. Membrane which has separated from shell during the process of decalcification.

Brachiopod, whose "strength is to sit still" and sweep little particles of food towards its mouth, has but little need of specialized sense-organs.

The cavity of the coelom is reduced by the presence of the alimentary canal, the digestive gland and the heart, but it is still spacious. It is partially divided up by certain mesenteries which support the alimentary canal, and it is traversed by several pairs of muscles. Some of these muscles run from valve to valve, and when they contract close the shell, others being situated behind the hinge, so that when they contract the valves slightly open. Others again run from the valves to the inner surface of the stalk and their contraction bends the body one way or another and may even serve to slightly rotate it.

There is—except in one genus—but one pair of nephridia. These are short tubes which open by large trumpet-shaped openings (9, Fig. 156) into the coelom ; while their external openings are situated at the sides of the body behind the lophophore. The cells lining the nephridia are some of them ciliated, while others are crowded with coloured granules. As already mentioned, the genus *Rhynchonella* possess two pairs of nephridia.

As a rule, in Brachiopods the sexes are separated. The cells destined to form ova or spermatozoa are derived from those lining the body cavity. At certain places, usually four in number, the coelomic cells multiply and build themselves up into ovaries or testes according to the sex. When they are ripe they fall off into the coelom and make their way to the exterior through the nephridia. The spermatozoa are cast into the water by the male, and the female must bring them within the valves of her shell by the action of the current set up by the cilia, because the eggs are almost certainly fertilized as soon as they leave the nephridia. The eggs develope in certain brood-pouches situated at the sides of the animal which are formed by a bulging in of the body-wall. A larva is ultimately formed which leaves the body of the mother and swims about in the sea by means of a band of cilia. As it is extremely minute, although it swims quickly it does not get very far, and this probably accounts for the fact that Brachiopods are usually found in large numbers in one place.

Brachiopods are found in all seas. About eleven genera have
been dredged around the British Isles, most of them
Distribution in comparatively shallow water. *Lingula* is usually
and Classifica- found between tide-marks or in shallow water; it lives
tion.

in a tube in the sand, and the bristles round the mouth of the shell doubtless serve to keep out particles of sand which might otherwise injure the animal. It is found along the East coast of America, in the Pacific and other places. One species of Brachiopod, *Terebratula wyvillei*, has been dredged from a depth of close upon 3000 fathoms.

Perhaps the chief interest of the group is that it includes an enormous number of fossil forms which had a very wide distribution. The extinct forms far surpass both in variety and number the existing forms. Some species have lived on,—as far as we can judge from the shell,—unchanged from the time when the earliest fossil-bearing rocks were laid down. They may thus claim to be one of the oldest groups with which we are acquainted.

The Brachiopoda are classified as follows :—

Class I. ECARDINES.

Shell with no hinge and no internal skeleton. The alimentary canal has an anus.

Ex. *Lingula, Crania.*

Class II. TESTICARDINES.

Shell with hinge and internal prolongations, chiefly calcareous. No anus.

Ex. *Terebratula, Argiope, Waldheimia, Rhynchonella.*

CHAPTER XI.

PHYLUM POLYZOA.

THIS group includes a great number of species, the individuals of which are so small as to be barely visible to the naked eye, but they are with hardly an exception colonial in their habits and the colonies usually attain a fair size. These colonies take many shapes, some branching like a tree, others being flattened like a leaf, while others again are discoidal; often they are encrusting, that is to say they form a layer on some sea-weed or rock, for the majority of them are marine.

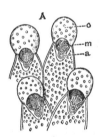

FIG. 158. Portions of two Polyzoan colonies. Magnified.

A. *Smittia landsborovii*. a. Avicularium.
   m. Orifice of cell. o. Ooecium or
   pouch in which the egg developes.
B. *Tubulipora plumosa*. After Hincks.

If one of these colonies be dried so that the organic matter shrivels up, a hard skeleton remains, and this is then seen to consist of a number of chambers or " cells," each of which opens to the exterior by an orifice, and, as a rule, communicates with its neighbours. The skeleton may be calcareous, or chitinous, or the colony may be gelatinous in consistency. The dried cell may be open or closed by a lid termed the operculum.

During life each of these cells lodges part of the body of a person of the colony, the "cell" being indeed the cuticle; part however of the person is not clothed with cuticle and is normally stretched out above the cell—the opening in the dried cell is in fact the place where the flexible part of the person begins. At the end of this flexible part is the mouth surrounded by a ring of ciliated

tentacles : on one side is the anus. The flexible part is termed
the polypide, and the cell the zooecium. If the polypide be
retracted, which occurs when it
is irritated, the anterior end is
inverted and forms the tentacle sheath in which the
tentacles lie. The operculum
when present is a movable fold
of the body-wall thrown back
when the polypide is pushed out,
and covering the opening of the
tentacle sheath when it is retracted.

The animal within the cell
has a U-shaped alimentary canal,
the anus being situated not far
from the mouth, but it is separated from it by a ring of tentacles in the centre of which the
mouth lies. This crown of
tentacles, called the lophophore, is not always circular,
but may be drawn out into a
horseshoe shaped structure (Fig.
159), and in the species in which
it undergoes this modification
there is a small projection, called
the epistome, which overhangs
the mouth, being situated inside
the tentacle ring. The tentacles
are ciliated, and the action of the
cilia brings food to the mouth,
which leads into an oesophagus
also ciliated, and this enlarges
into a rounded stomach usually
produced into a caecum (Fig.
159). From this a small intestine,
parallel with the oesophagus,
leads to the anus. From the

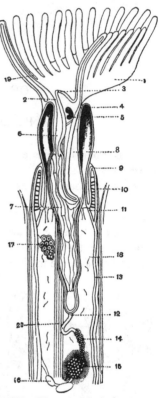

Fig. 159. View of right half of *Plu-
matella fungosa*, slightly diagrammatic. After Allman and Nitsche.

1. Lophophore. 2. Mouth. 3. Epistome. 4. Anus. 5. Nerve ganglion.
6. Oesophagus. 7. Stomach.
8. Intestine. 9. Edge of fold of
body-wall. 10. Wall of tube.
11. Muscles. 12. Funiculus.
13. Body-wall. 14. Testis.
15. Testis, more mature. 16. Stato-
blast. 17. Ovary. 18. Spermatozoa free in body-cavity. 19. Tentacles. 20. Retractor muscle.

aboral side of the stomach a cord of mesodermic tissue, called the
funicle, usually passes to the body-wall.

The body cavity is regarded as truly coelomic. It contains a fluid in which cells float; it is traversed by the funicle and by numerous strands of mesodermic tissue. The funicle may be the remains of a median mesentery which once separated the coelomic sacs of the two sides. Some of the cells of its walls give rise to reproductive cells, and the body cavity opens to the exterior in certain individuals which possess an ovary by a short tubular duct, the so-called inter-tentacular organ. This functions as an oviduct and has been regarded by some authors as a modified nephridium. A portion of the body cavity is separated from the rest by a horizontal septum, and forms a space at the base of the tentacles. This may open into the other part or may be completely shut off from it.

No heart or blood-vessels are present. It is possible that some of the nitrogenous waste matter may be got rid of by means of the inter-tentacular organ, but it has also been suggested that these waste products are stored up in certain cells on the funicle. From time to time the tentacles, alimentary canal and nervous system of an individual undergo degeneration and form a brown mass, called the brown-body, which forms a conspicuous feature in the colony. After a time the body-wall, which has not disintegrated, forms a new set of organs and the brown-body may come to lie in the stomach of the reconstituted individual. Thence it passes to the exterior through the anus. It is thought that much of the waste matter which has accumulated in the body of the animal is, in this way, eliminated. In certain Phylactolaemata there are a pair of small nephridia opening between the mouth and the anus.

A nerve-ganglion lies between the mouth and anus, situated in that part of the body cavity which runs round the base of the lophophore.

Polyzoan colonies are usually hermaphrodite. The testes are as a rule formed by the multiplication of the coelomic cells which lie at the side of the body-wall, while the ovary originates from the funicle or from the body-wall. They may be found in the same individual or in different individuals of the same colony. As a rule the eggs develope within some part of the parent colony, but they may be laid, escaping from the body cavity through the inter-tentacular canal, and then they pass at once into the sea-water. More usually the early stages of development are passed through in the tentacle-sheath or in a special pouch called an ooecium (Fig. 158), or in certain " cells " which contain rudimentary individuals. A free

swimming larval form is usually found, which after a time comes to
rest and by budding forms a new colony.

Just as in the colonies of Hydrozoa we found different
individuals set apart to perform different functions, so in Polyzoa
we find a similar specialization. Certain individuals may be
modified to accommodate and protect the developing egg, but
perhaps the most remarkable modifications are the Vibracula and
Avicularia of the Cheilostome Polyzoa. The vibracula are long
hair-like processes which sweep
through the water; the avicularia
are two snapping jaws provided
with powerful muscles, like the
claws of a lobster or the beak
of a parrot (Fig. 160). They are
modifications of a "cell" and its
operculum. The avicularia occasionally catch worms, crustacea
and other animals whose presence might interfere with the
colony, and by their action they
probably prevent the larvae of
encrusting animals settling on the
Polyzoan colony.

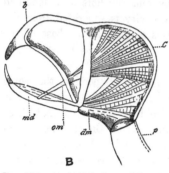

Fig. 160. An Avicularium of *Bugula*.
Magnified. From Hincks.

b. Beak.   *c.* Chamber representing
the body-cavity of the modified individual.   *dm.* Muscle which opens,
om. muscle which closes the mandible on the beak.   *md.* Mandible,
the operculum of the modified cell.
p. Stalk.

Besides the sexual method of
reproduction just mentioned certain internal buds termed stato-
blasts are formed in the group Phylactolaemata. Masses of
cells arise from the funicle and become enclosed between two
watchglass shaped chitinous shells whose edges are kept together
by a special ring of cells. As a rule, the Phylactolaemata die
down during the winter, but the statoblasts persist and when spring
recurs give rise to new colonies. A somewhat analogous process
ensures the perpetuation of the species in certain fresh-water
Ctenostomata.

Polyzoa are widely distributed throughout the sea, many occurring in shallow water, but others have been dredged at great
depths. The Phylactolaemata and a few genera from other subdivisions are fresh-water. Fossil forms are numerous and the
Coralline Crag, a tertiary deposit, takes its name from the large
number of coral-like calcareous forms, sometimes described as
"corallines," which are found in it.

There is a small class of Polyzoa with a solid body, i.e., no coelomic space, and with both ends of the alimentary canal included in the ring of tentacles, termed the **Entoprocta**. In this Class the body consists of a stalk and a "cup." The edges of the latter are fringed with a short row of ciliated tentacles surrounding a disc on which both the mouth and the anus open. When irritated these tentacles are bent inwards and the contraction of a sphincter muscle causes the edge of the disc to be drawn over them exactly as happens in a Sea-anemone. Sometimes in *Pedicellaria* the "cup" falls off and a new one is formed. Beneath the disc is situated a nerve-ganglion and the genital organs are continuous with a short duct which opens in the centre of the disc. The excretory system is either one or two blind, ciliated canals opening between the mouth and the anus.

All other Polyzoa are grouped together as **Ectoprocta**, and these are subdivided into

Order I. **Gymnolaemata.**

With a few exceptions marine and having a circular lophophore. Devoid of an epistome.

Suborder (1), **Cheilostomata.** An operculum covers the orifice of the cell. Avicularia and vibracula often present. Skeleton with more or less calcareous deposit in it.

Suborder (2), **Cyclostomata.** Cells tubular and ending in circular mouths. No operculum. Calcareous skeleton.

Suborder (3), **Ctenostomata.** Body-wall soft. The orifice of the cell is closed by the coming together of a fringed membrane.

Order II. **Phylactolaemata.**

Fresh-water forms with a horseshoe shaped lophophore (except *Fredericella*), an epistome, and statoblasts.

CHAPTER XII.

PHYLUM CHAETOGNATHA.

THE Chaetognatha (χαίτη, hair; γνάθος, jaw) are small cylindrical animals which swim at the surface of the sea. The name is suggested by the circumstance that at the sides of the mouth are two rows of curved movable bristles by means of which they seize their prey (e, Fig. 161).

The body has a small rounded head in front and tapers to a tail posteriorly; it is provided with one or two pairs of flat, lateral expansions termed fins; the general shape resembles that of a torpedo, if we leave the head out of account. The head is surrounded by a fold of the skin forming a hood which is most prominent at the sides and dorsal surface. Within the hood the head bears a row of sickle-like hooks whose points when at rest converge around the mouth, but are capable of being widely divaricated. The head also bears one or more rows of stout spines whose number and arrangement are of importance for the system of classification (f, Fig. 161).

The coelom is well developed and contains a fluid in which cells float. In strictness there are three pairs of coelomic sacs separated from one another by transverse and longitudinal partitions. In the head the coelomic space is practically obliterated by the great development from its walls of the muscles which move the hooks. The coelom of the trunk and tail is further divided into right and left halves by a vertical mesentery, which in the trunk region supports the alimentary canal (Fig. 162). This mesentery is pierced by numerous small holes.

The skin is covered by an epithelium more than one layer thick, some of the cells of which are modified to form sense-organs, while others project from the surface of the body and are known as adhesive cells (Fig. 162). Beneath the epithelium is a thin layer

FIG. 161. A ventral view of
Sagitta   hexaptera × 3½.
From O. Hertweg.

a. Mouth.   b. Intestine.
c. Anus.   d. Ventral
ganglion.   e. Movable
bristles   on   the head.
f. Spines on the head.
g. Ovary.   h. Oviduct.
i. Vas deferens. j. Testis.
k. Vesicula seminalis.

of jelly called the basement membrane,
and beneath this a layer of muscles.
Anteriorly the muscles are broken up into
numerous bundles which fill the cavity
of the head, but in the trunk and tail the
muscles form four distinct bundles, bilaterally arranged, two dorsal and two ventral.

The nervous system consists of a
dorsally placed ganglion in the head
which gives off two lateral nerves; these
pass round the alimentary canal and end
in a ventral ganglion situated (Figs. 161
and 162) near the centre of the body.
The cerebral ganglion gives off nerves to
the eyes, the olfactory organ, muscles, etc.,
and both it and the ventral ganglion
are connected with a tangle of nerve-
fibrils lying at the base of the ectoderm. A pair of eyes exist on the
upper part of the head, and behind
the eyes an organ to which an olfactory
function has been assigned. This consists of a ring of modified ciliated epithelium. Clumps of isolated tactile cells
with long hairs surrounded by supporting
cells are scattered over the body and fins
(Fig. 162).

The alimentary canal is simple and
straight. The mouth—with one exception—is ventral and it leads into a
pharynx which traverses the head; this
passes into an intestine lined by a single
layer of ciliated epithelium amongst
which are some glandular cells. The anus
is situated at the junction of the trunk
and the tail and—with one possible exception—is ventral.

In *Spadella marioni*, the exception
mentioned in the preceding paragraph,
there is a glandular structure in the head
which may be connected with the excre-

tion of waste nitrogenous material, but no other excretory organ is
known and no special respiratory or circulatory organs exist.

The Chaetognatha are hermaphrodite. The paired ovaries
(Figs. 161 and 162) lie in the trunk region of the body cavity
supported by a lateral mesentery. When mature they almost fill
the cavity. The oviduct traverses the ovary. It is not known how
the ova make their way into it, but spermatozoa are sometimes
found inside it, so that it acts as a receptaculum seminis. The
oviducts open to the exterior at the upper surface of the lateral fin,
just where the trunk passes into the tail.

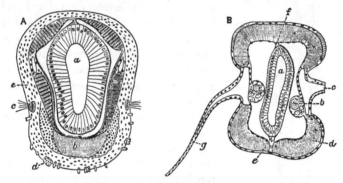

Fig. 162. A. Transverse section through a *Spadella cephaloptera* in the region
of the ventral ganglion × about 200.   B. Transverse section through a
Sagitta bipunctata in the region of the ovary × about 120.

a. Intestine.   *b.* in A. Ventral ganglion.   *b.* in B. Ovary.   *c.* in A. A
ciliated sense-organ.   *c.* in B. Base of the left fin which has been cut off.
d. in A. Adhesive cells.   *d.* in B. Left ventral muscle.   *e.* in A. Ectoderm.   *e.* in B. Ventral mesentery.   *f.* Dorsal mesentery.   *g.* Right fin.

The median mesentery of the trunk region is continued through
the tail, dividing its cavity into two ; and in each of these lateral
cavities the cells lining the body-wall are heaped up and form a
testis (*j*, Fig. 161). The cells divide up into spermatozoa, which
float in the coelomic fluid and are kept in motion by some of the
ciliated cells lining the body cavity. The spermatozoa escape
through a pair of short vasa deferentia which open on the one hand
into the coelom and on the other to the exterior on the tail.
Each has on its course a well-marked vesicula seminalis (Fig. 161).

The ova are transparent and pelagic. The cells destined to form
the reproductive organs of the adult are early set apart and
distinguishable. The development is entirely embryonic, no larval
form being recognizable.

The Chaetognatha consist of three genera, *Sagitta*, *Spadella* and *Krohnia*, amongst which some twenty-three species are divided. The genera differ from one another chiefly in the arrangement of the armature on the head and in the disposition of the fins, which are always horizontally placed and supported by fine skeletal rods. A caudal fin exists in addition to one (*Spadella* and *Krohnia*) or two (*Sagitta*) pairs of lateral fins.

The Chaetognatha are with hardly an exception pelagic, that is they live near the surface of the sea, and as is usually the case with animals which frequent the surface of the ocean they are transparent. At certain times of the year they are found in incredible numbers swimming on the surface by the muscular contraction of their bodies, their fins acting as balancers and having no movement of their own. At other seasons they descend and are taken at depths varying from 100 to over 1000 fathoms. The cause of their descent is unknown. Their food consists of Infusoria, small larvae and small Crustacea.

The zoological position of the Chaetognatha is obscure. They show no relationship with any of the larger groups ; possibly their nearest existing allies are to be found amongst certain aberrant Nematoda, such as *Chaetosoma*, but at present too little is known to make any close comparison possible.

CHAPTER XIII.

Introduction to the Phylum Vertebrata, Sub-Phyla Hemichordata, Cephalochordata and Urochordata.

THE Vertebrata comprise almost all the larger animals, including
Man. The name simply means jointed (Lat. vertebra,
a joint, and especially a bone of the spinal column),
and refers to the possession of a jointed internal
axis as the main part of the skeleton. In the lowest forms this
axis is not developed, but in place thereof there is a smooth elastic
rod, which has received the name of notochord, literally back-
string (Gr. *νῶτον*, back ; *χορδή*, string). In all the members of the
phylum this notochord is present at some stage of development,
although in the higher forms it subsequently becomes surrounded
and obliterated by the jointed rod or vertebral column. Hence
the name Chordata, which has been proposed for the group, is really
more appropriate ; but as the term Vertebrata has been sanctioned
by long usage it is inadvisable to depart from it.

Besides the possession of the notochord there are two other
features by which the Vertebrata are distinguished. They all
possess at some period of their lives slits in the wall of the front
part of the alimentary canal. These slits in the lower forms allow
the water which is taken in at the mouth for purposes of respiration
to escape, and hence they are called gill-slits. Further, the
nervous system takes on the form of a dorsal strip of sensitive
skin—the medullary plate, which becomes wholly or partly
enrolled to form a tube, the neural canal or spinal cord.

There are in all about 32,000 known species of Vertebrata,
including all the more familiar animals—fish, frogs, reptiles, birds
and mammals ; so that the word animal to the ordinary mind
generally calls up the idea of a vertebrate. Nevertheless the
number of species is not much more than half that of the Mollusca
and is not a tenth that of the described species of Arthropoda.

Sub-Phylum I. HEMICHORDATA.

The most primitive members of the phylum are certain worm-like forms, in which it has taken special research to discover traces

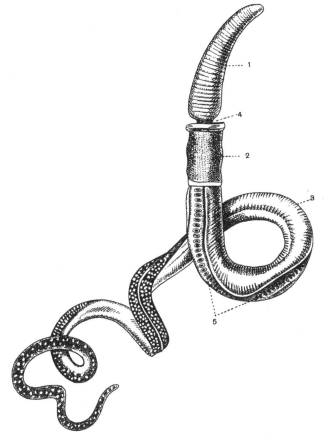

FIG. 163.   A *Dolichoglossus kowalevskii* × 1.   From Spengel.
1. Proboscis.   2. Collar.   3. Trunk.   4. Mouth.   5. Gill-slits.

of Vertebrate structure. They are marine and live in mud, passing it through their intestines and extracting nutriment from the organic matter it contains : thus they feed and move forwards by the same process. These animals are termed Hemichordata (Gr. ἥμι, half) on account of the short rudimentary notochord which they possess. Sometimes they are called Enteropneusta (Gr. πνεῦμα,

breath) because, like all Vertebrata, they use the anterior portion of
the gut for breathing. There are several genera, *Dolichoglossus*,
Chlamydothorax, *Glossobalanus* and others; Balanoglossus, although
used as a generic name, may also be used as a semipopular name
for any member of the Enteropneusta.

The body of the animal is divided into three portions: (1) a
conical anterior part in front of the mouth, the proboscis; (2) a
swollen cylindrical portion immediately behind the mouth, the
collar; and (3) a long trunk, at the end of which is the anus.

The proboscis contains one, and the collar and trunk each a pair
of special sections of the coelom or body cavity. The coelomic sacs
of the proboscis and collar communicate with the exterior by ciliated
tubes, the proboscis and collar pores (Figs. 164 and 165). The
cilia produce currents setting inwards: thus the collar and proboscis are kept swollen up and tense with water and form efficient
burrowing instruments. If a Balanoglossus be removed from the
water and laid upon damp sand it is incapable of burrowing and
wriggles helplessly about. As soon however as it is covered by
water the proboscis and collar are seen to dilate and become stiff,
and the proboscis is then inserted into the sand, soon followed by
the collar, whilst the trunk is dragged passively after them. As the
walls of both proboscis and collar are highly muscular the water can
be expelled through the pores and the volume of these regions
of the body diminished, but the action of the cilia soon swells them
up again. On the hinder wall of the proboscis cavity there is a
puckered membrane richly supplied with blood-vessels, which is
called the glomerulus and appears to act as a kidney. When the
water in the cavity has become impregnated with excretory products
it is expelled as explained above by a muscular contraction.

The alimentary canal runs straight from the mouth on the
anterior surface of the collar region to the posterior end of the
trunk; there is neither stomodaeum nor proctodaeum. In most
species in the anterior part of the trunk the canal has an 8-shape in
section, being partially constricted into two tubes, an upper or
branchial into which the gill-slits open, and a lower or oesophageal
along which the mud is passed which the animal has swallowed for
food. The notochord is a hollow tube of cells surrounded by a
tough membrane much thickened beneath (Fig. 164). This tube
opens into the alimentary canal in the collar region and projects
forward into the proboscis as a support for this organ, which is
attached by a very narrow neck to the collar. The whole skin is

sensitive, since there is everywhere a layer of nerve-fibrils underlying the ectoderm cells, but this fibrillar layer is especially thickened
along the mid-dorsal and mid-ventral lines of the trunk, these two
regions being connected by a ring of nervous tissue immediately
behind the collar.   The dorsal thickening alone is continued into
the collar region, and here it becomes rolled up so as to constitute
a short neural tube (Fig. 164) which becomes detached from the

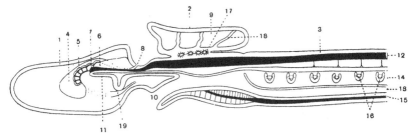

FIG. 164.   Longitudinal vertical section through the middle line of *Glossobalanus*.
Diagrammatic.

1. Proboscis.   2. Collar.   3. Trunk.   4. Proboscis cavity.   5. Glomerulus.
6. Pericardium.   7. Heart.   8. Proboscis pore.   9. Collar cavity.
10. Mouth.   11. Notochord.   12. Dorsal blood-vessel.   13. Oesophageal portion of alimentary canal.   14. Branchial region of alimentary
canal.   15. Ventral blood-vessel.   16. Gill-slits showing external and
internal openings; the outlines of the external openings are dotted.
17. Central nervous system.   18. Dorsal roots of nervous system.
19. Ventral pocket of proboscis cavity.

ectoderm and assumes a deeper position, it may retain however a
connection with the ectoderm through several strings of cells with a
fibrillar sheath, known as dorsal roots (18, Fig. 164).

There are very numerous gill-slits opening into the alimentary
canal, in the front part of the trunk region ; they ought rather to be
called pouches with a small outer and a large inner opening.   The
inner opening of each pouch is divided almost into two by a tongue
projecting down from its dorsal edge, the so-called tongue-bar.
This tongue-bar is specially richly supplied with blood-vessels and
so may be regarded as the principal respiratory organ.   The blood-
vessels are destitute in most cases of any proper wall: they are
as it were mere crevices between the epithelial walls of the gut,
coelom and skin.   There is however a well-defined contractile dorsal
channel running forward into the kidney, the contractility being
confined to the front end in the proboscis where there is a closed
sac with muscular walls, which pulsate rhythmically, situated above
the blood-vessel.   The sac is termed the pericardium, and the

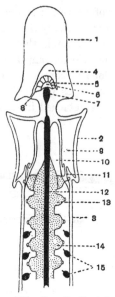

FIG. 165. Longitudinal horizontal section through *Glossobalanus*. Diagrammatic.

1. Proboscis. 2. Collar. 3. Trunk. 4. Proboscis cavity. 5. Glomerulus. 6. Pericardium. 7. Heart. 8. Proboscis pore. 9. Collar cavity. 10. Perihaemal cavity. 11. Collar pore. 12. Dorsal blood-vessel. 13. Alimentary canal. 14. Branchial sac with external opening. 15. Reproductive organs.

dilated part of the blood-vessel below it, the heart. This dorsal vessel communicates with a ventral vessel in the trunk region by two descending curved vessels at the sides of the collar. Each of the coelomic cavities of the trunk sends forwards into the collar region a narrow tongue lying at the side of the blood-vessel. These tubes from their relation to the vessel are called perihaemal tubes (Gr. περί, around ; αἷμα, blood).

The sexual organs or gonads are mere packets of cells in the gill region and behind it, developed from the wall of the trunk coelom (Fig. 165). Each when ripe forms its own opening through the body-wall.

One point of interest attaching to the Hemichordata is that they may commence life as free-swimming larvae, resembling the larvae of the Echinodermata, and suggesting the thought that perhaps two such different groups as the Vertebrata and Echinodermata may have descended by different paths from the same simple free-swimming ancestors.

Sub-Phylum II.   CEPHALOCHORDATA.

Leaving the Hemichordata we next come to some small fish-like animals, the Cephalochordata, which were formerly all included under the name *Amphioxus*, and indeed there is no very strong reason for breaking up this old genus. The name *Amphioxus* (ἀμφί, at both ends ; ὀξύς, sharp) refers to the shape of the body, which is long, flattened and pointed at both ends. It is remarkable that we here meet for the first time with a shape very common among Vertebrates, but almost absolutely unknown elsewhere in the animal kingdom, viz. a laterally compressed form with narrow ventral and

dorsal regions and deep sides.   It is common to find animals with
broad backs and bellies and narrow sides, but only Vertebrates show

FIG. 166. *Amphioxus lanceolatus* from the left side, about twice natural size.
After Lankester.   The gonadic pouches are seen by transparency through
the body-wall; the atrium is expanded so that its floor projects below the
metapleural fold; the fin-rays of the ventral fin are indicated between the
atrial pore and anus.   The dark spot at the base of the fifty-second
myotome represents the anus.

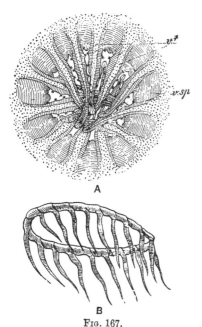

A

B

FIG. 167.

A. Velum of *Amphioxus* seen from the
inside of the pharynx.   After Lankester.

v.sp. sphincter muscle of velum.

v.t. velar tentacles lying across the oral
opening.

B. Oral cartilages of *Amphioxus.* After
J. Müller. The basal pieces lie end
to end in the margin of the oral
hood, and each basal piece sends up
an axial process into the corresponding oral cirrus.

the reverse condition.   In consequence of this peculiar
shape *Amphioxus* falls on its
side when it ceases moving.
It burrows in the sand, lying
with its mouth just protruding, and as its lips are fringed
with ciliated rods (Fig. 167)
a current is produced which
brings new water to the gills
and with it small swimming
organisms which serve as food.
At night *Amphioxus* often
leaves its burrow and swims
about, returning instantly to
the sand if alarmed.   It can
burrow with either the head
or tail.

The notochord is a
smooth cylindrical rod lying
above the gut and running
from end to end of the animal.
It consists of cells, much of
whose body is changed into
a gelatinous substance, and
which are surrounded by an
exceedingly firm membrane
termed the chordal sheath.
In the embryo the notochord

first appears as a groove in the dorsal wall of the gut, so that we may say that the notochord of the Hemichordata retains a form which is passed through in development by that of *Amphioxus*.

In the very young embryo also an indication is seen of the division of the body into the same three regions as we found in the Hemichordata. Just as in the embryo of Balanoglossus so here, the embryonic gut gives rise to five outgrowths from which the coelom of

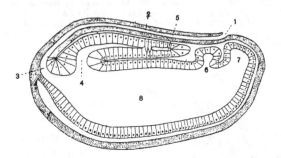

Fɪɢ. 168.   Diagrammatic longitudinal section of an embryo of *Amphioxus*.

1.   Neuropore—anterior opening of the neural canal.   2.   Neural canal.
3.   Neurenteric canal.   4.   Coelomic groove.   5.   Somite divided off
from coelomic groove.   6.   Collar-cavity.   7.   Head-cavity.   8.   Alimentary canal.

the adult is derived. These outgrowths are (1) a median anterior unpaired pouch, the so-called head-cavity corresponding to the proboscis-cavity of Balanoglossus; (2) an anterior pair of pouches, the collar-cavities, corresponding to the similarly named sections of the coelom of Balanoglossus; and finally (3) a pair of groove-like extensions of the dorso-lateral angles of the gut-cavity, called the coelomic grooves, developed only at the hinder end of the gut. From the last-named the coelom of most of the body arises, and they correspond to the trunk-coelom of Balanoglossus (Fig. 168). The proboscis or prae-oral region is however very small and bent down ventrally; its cavity becomes more or less obliterated in the adult. Dorsally the collar region is narrow from before backwards, but it extends obliquely downwards and backwards, causing a slight ridge to appear at the side of the body. This ridge grows out on each side into a flap, which meets its fellow beneath the ventral wall of the body and thus they enclose a space, the so-called atrial cavity (Figs. 170, 171). This still communicates with the exterior through the atrial pore. The gill-slits, which occur in the front part of the trunk region, open into this atrial cavity. The atrial flaps, enclosing the atrial cavity, are an obvious arrangement by which the slits are saved from being choked with the sand in which the creature lives. In the Hemichordata the hinder end of the collar region may extend over one or two gill-slits; the arrangement in *Amphioxus* may be regarded as a further development of this state of affairs. As in the case of Hemichordata, the slits of *Amphioxus* are divided into two, by a tongue-bar reaching from the upper margin almost to the bottom of the slit. Each slit thus becomes U-shaped (Figs. 169, 172).

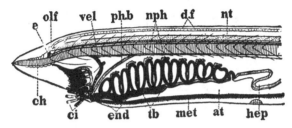

Fig. 169. Anterior region of young *Amphioxus* from left side. After Willey; the renal tubules inserted after Boveri.

at. Atrial cavity. *ci.* Oral cirri. *ch.* Notochord. *d.f.* Dorsal fin-chambers. *e.* Eye-spot. *end.* Endostyle. *hep.* Outgrowing liver; the index line passes through one of J. Müller's "renal papillae." *met.* Metapleural fold. *nph.* Nephridia. *nt.* Spinal cord. *olf.* Olfactory pit. *ph.b.* Peripharyngeal ciliated band. *tb.* Tongue-bars. *vel.* Velum.

The upper and anterior portion of the collar-cavity becomes separated from the rest: its inner walls thicken and develope into a powerful longitudinal muscle which forms the first myotome (Gr. μῦς, mouse, muscle; τόμος, a division).

The trunk coelomic cavity breaks up from the beginning into a series of pouches called somites, each of which subsequently divides into an upper and an under part. The inner walls of the upper parts undergo a similar change to that experienced by the corresponding part of the collar-cavity forming a series of myotomes. The name myotome is given to each of the metamerically arranged bundles of muscle-fibres. Each myotome is separated from the next by a connective-tissue partition. In *Amphioxus* the myotomes of the right side alternate with those of the left, so that the centre of a myotome on one side is opposite the connective-tissue partition on the other. Each is V-shaped, and they are arranged so ⋘. Hence in a transverse section several myotomes are seen on each side of the body. Thus we have two great series of longitudinal muscles broken up into myotomes, one on each side of the animal, by the alternate contraction of which powerful side-strokes of the

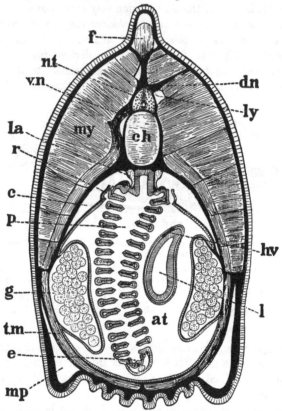

Fig. 170. Diagrammatic transverse section through pharyngeal region of a female *Amphioxus*. After Lankester and Boveri, from R. Hertwig.

at. Atrial cavity.   *c.* Dorsal coelom, separated from atrial cavity by the double-layered membrane known as the *ligamentum denticulatum*.   *ch.* Noto-chord.   *d.n.* Sensory nerve.   *e.* Endostyle, below which is the endostylar coelom containing the ventral aorta.   *f.* Fin-ray of dorsal fin.   *g.* Gonadic pouch containing ova.   *h.v.* Hepatic vein lying in the narrow coelomic space which surrounds *l,* the liver or hepatic coecum. *l.a.* Left aorta separated from the right aorta by the hyperpharyngeal (epibranchial) groove.   *ly.* Lymph-space.   *mp.* Metapleural fold. *my.* Longitudinal muscles of myotomes; over against the dorsal coelom these muscles are arranged vertically, and form the rectus abdominis of Schneider.   *n.t.* Spinal cord.   *p.* Pharynx.   *r.* Nephridium. *t.m.* Transverse or subatrial muscles.   *v.n.* Motor spinal nerve, the fibres of which have the appearance of passing directly into the muscle-fibres.   N.B. The connective tissue (cutis, notochordal sheath, etc.) and the coelomic epithelium are indicated by the black lines.

flat body propel the animal forwards.

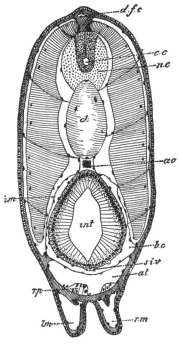

FIG. 171. Transverse section through post-pharyngeal region of young *Amphioxus*, to show groups of renal cells in floor of atrium. After Lankester and Willey.

ao. Aorta. *at.* Atrial cavity. *b.c.* Body-cavity (coelom). *c.c.* Central canal of nerve-cord (*n.c.*). *ch.* Notochord. *d.f.c.* Fin-cavity. *i.m.* Intercoelic membrane. *int.* Intestine. *l.m.* and *r.m.* Left and right metapleural folds. *r.p.* One of J. Müller's "renal papillae." *s.i.v.* Sub-intestinal vein.

The elasticity of the notochord acts like a fly-wheel in storing the force during the latter part of each stroke and reinforcing each stroke at its commencement. The cavity of the upper division of the somite persists throughout life, and is known as the myocoel (15, Fig. 179), the fold separating it from the cavity of the lower division being termed the intercoelic membrane (*i.m.* Figs. 171 and 177).

The lower portions of the somites fuse with one another and form a continuous body cavity round the hinder part of the alimentary canal (*b.c.* Fig. 171). In front, owing to the presence of gill-slits, there are formed a right and a left dorsal coelomic canal, and a ventral coelomic tube, or endostylar coelom, the dorsal and ventral portions communicating with one another by spaces in the gill-bars, that is, in the pieces of body-wall intervening between the gill-slits. These spaces are termed branchial coelomic canals. Although at first formed from backward prolongations of the collar region, the atrial flaps soon become invaded by the lower ends of the trunk myotomes; the ventral muscle however running across the under surface of the atrial cavity (Fig. 170), which by contracting diminishes its size and thus expels water, originates from the walls of the lower portions of the collar cavity.

The mouth is originally some distance behind the anterior end,

and on the left side, so that there is a prae-oral portion of the body
which in the embryo is occupied by an anterior division of the

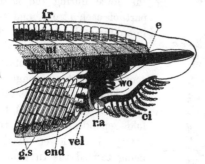

Fig. 172. Anterior portion of body of young
transparent *Amphioxus*. After J. Müller, slightly altered.

ch. Notochord.　*ci.* Oral cirri.　*e.* Eye-spot.
end. Endostyle.　*f.r.* Fin-rays.　*g.s.* Gill-
slits; the skeletal rods of the gill-bars are
indicated by black lines.　*nt.* Spinal cord, with
pigment granules near its base.　*r.a.* Downgrowth from right aorta lying to the right
of *vel.* the velum, with velar tentacles projecting back into pharynx.　*w.o.* Ciliated
epithelial tracts on inner surface of oral
hood.

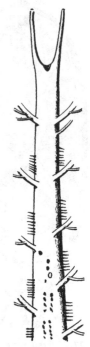

Fig. 173. Anterior portion of
spinal cord of *Amphioxus*;
seen from above. After
Schneider.

coelom corresponding to the proboscis
cavity of the Hemichordata. Subsequently however the atrial flaps extend
right to the anterior end, so that a new
terminal mouth is formed leading into a
chamber which is clothed by ectoderm
and which is therefore to be regarded as
the stomodaeum. The opening of the
stomodaeum now forms the apparent
mouth, and the lips of this secondary
mouth grow out into rods supported by
gelatinous material and covered with cilia,
the so-called oral cirri, the function

Between the first pair of
cranial nerves is seen the
eye-spot; one of the
branches of the second
pair of cranial nerves
sometimes arises directly
from the spinal cord as
shown on the right; far-
ther back are seen the pigment spots of the nerve-
cord.

of which has been already explained (Fig. 167). The walls of
the stomodaeum are known collectively as the oral hood. The
position of the primary mouth is still marked by a projecting lip,
the velum, which is produced into ten or twelve delicate tentacles.
These form a filter to prevent coarse material from reaching the
alimentary canal.

The nervous system is a simple tube with thick walls and very narrow cavity. It is almost as extended as the notochord, and lies above it. It does not however quite reach the front end of the body. Its extreme front tip is called the cerebral vesicle; it has a wide cavity with thin walls, so that the total diameter is not increased. There is a pit reaching down to it from the external

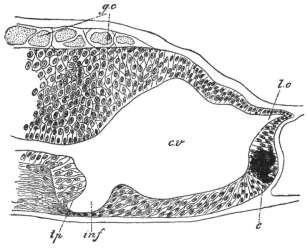

FIG. 174. Median vertical section through the cerebral vesicle of *Amphioxus*. After Kupffer.

c.v. Cavity of cerebral vesicle. *e.* Eye-spot. *g.c.* Dorsal group of ganglion-cells. *inf.* Infundibulum. *l.o.* Olfactory lobe. *tp.* Tuberculum posterius.

skin, possibly a rudimentary olfactory organ (Fig. 174), and in the wall of the vesicle itself is a mass of pigmented cells forming an eye-spot. In the young larval *Amphioxus* this part of the nerve-tube remains for a considerable time as an uncovered medullary plate, and one is inclined to imagine that it corresponds to the sensitive nervous surface of the proboscis in the Hemichordata, since in the larvae of these animals there is a sensitive plate with two eye-spots at the apex of the prae-oral lobe. In the wall of the nerve-tube are to be found two kinds of nerve-cells, viz.,(*a*) ordinary small nerve-cells, the processes of which soon pass outwards into the peripheral nerves, and (*b*) very large nerve-cells, the processes of which extend almost throughout the entire length of the nerve-tube. The processes of the latter kind of cell are called "giant fibres" ($g.f^1$ and $g.f^2$, Fig. 175): they appear to have to do with

coordinating the muscular movements of the animal. Besides the
nerve-cells, as in all nervous systems, there are a certain number of
supporting cells (*s.f*, Fig. 175). In the embryo of *Amphioxus* the
whole wall of the nerve-tube consists of a single layer of cells, all of
which abut on the cavity of the tube ; many of these cells become

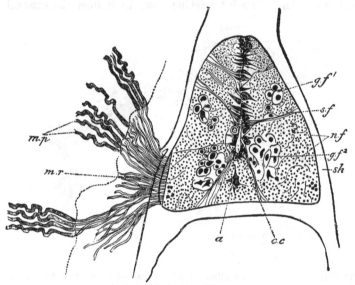

Fig. 175. Transverse section through the spinal cord of *Amphioxus* in the
middle region of the body. After Rohde.

a. Giant fibre.   *c.c.* Central canal.   *g.f*[1]. Giant nerve-fibres, which traverse
the spinal cord from before backwards.   *g.f*[2]. Giant nerve-fibres, which
traverse the spinal cord from behind forwards.   *m.p.* Muscle-plates, i.e.
terminations of the nerve-fibres on the muscles.   *m.r.* Motor nerve-fibres.
n.f. Longitudinal nerve-fibres cut across.   *s.f.* Supporting cells.
sh. Sheath of nerve-cord.

afterwards transformed into small round nerve-cells, and recede from
the cavity, assuming a more peripheral position : but others retain
their connection with the cavity and become drawn out into fibre-
like supporting cells. From the nerve-cord are given off two kinds
of nerves, but not at the same level, so that in a transverse section
one kind only is seen. These are:—(1) sensory nerves, going
directly to the skin and having a dorsal origin ; (2) motor nerves,
going to the myotomes. The nervous tube and the alimentary
canal at first both reach to the extreme posterior end of the body
and here are connected by a vertical tube, the neurenteric canal.
On the course of this tube the anus is formed. As development
proceeds the anus slowly shifts forwards and the neurenteric canal

becomes a solid string of cells and disappears. Thus is initiated the formation of a tail, by which term is denoted a portion of the body devoted entirely to locomotion and freed from all part of the gut, being filled only with muscles. The tail of *Amphioxus* acquires only a very limited development, but it soon becomes surrounded by a tail-fin, at first merely made up of the enlarged skin cells, but soon becoming a flap containing gelatinous material. A similar fold along the middle line of the back forms the dorsal fin, in which, in the larva, there are a series of metamerically arranged cavities lined by distinct epithelium, probably derivatives of the myocoelic cavities (*d.f.c.* Fig. 171). There are also low fin folds projecting from the sides of the atrial cavity and constituting the lateral or metapleural fins (Fig. 170), and a median ventral fold between the atrial pore and anus, called the ventral fin (Fig. 166). The dorsal and ventral fins are stiffened by a number of gelatinous rods more numerous than the myotomes of the corresponding regions of the body.

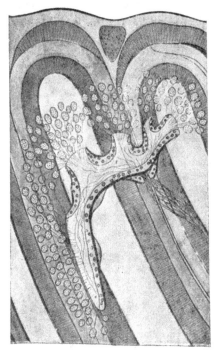

Fig. 176. *Amphioxus.* Nephridium of the left side, with the neighbouring portion of the pharyngeal wall, as seen in the living condition. The round bodies in the wall of the tubule represent carmine granules. Highly magnified. After Boveri.

The alimentary canal of *Amphioxus* is a perfectly straight tube consisting of stomodaeum or mouth gut, pharynx or branchial gut, and intestine or digestive gut. The pharynx has along both dorsal and ventral middle lines grooves lined with cilia connected with each other by a circular groove just inside the velum or true mouth. The ventral groove is called the endostyle or hypopharyngeal groove, the upper groove is termed the hyperpharyngeal, and the connecting groove the peripharyngeal band (Figs. 169, 170, 172). The function of these grooves is curious.

The endostyle produces a cord of mucus which is worked forwards by its cilia and pressed up the sides of the peripharyngeal bands. Here it is caught by the inrushing current of water produced by the cilia of the oral cirri and swept back along the hyperpharyngeal groove to the opening of the intestine, entangling in its passage the small plants and animals carried by the water ; the latter of course escapes into the atrial cavity through the hundred or so long narrow gill-slits. The intestine is prolonged forward on the right side of the pharynx into a blind pouch, the so-called liver (*l*, Fig. 178), which probably secretes a digestive juice.

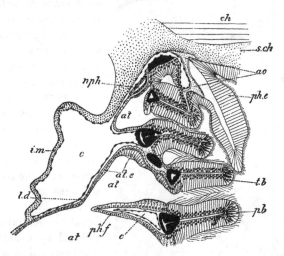

FIG. 177.   Portion of transverse section through the pharynx of *Amphioxus*, to show position of excretory tubule. After Weiss.

ao. Left aorta.   *at.* Atrial cavity.   *at.e.* Atrial epithelium.   *c.* Coelom.
ch. Notochord.   *i.m.* Intercoelic membrane.   *l.d.* Dorsal wall of atrial cavity.   *nph.* Nephridium.   *p.b.* Gill-bar.   *ph.e.* Epithelium of hyperpharyngeal groove.   *ph.f.* Fold attached to gill-bar containing branchial coelomic canal.   *s.ch.* Sheath of notochord.   *t.b.* Tongue-bar.

The execretory organs of *Amphioxus* are small and have only recently been discovered.   We have seen that in the region of the pharynx the coelom has become reduced to a narrow canal, beneath the pharynx, and to two dorsal canals at the sides of the notochord (Figs. 170, 177).   These latter canals have been described as having at the level of each tongue-bar a wide funnel leading into a short tube connecting them with the atrial cavity.   The edges of the funnel are supported by strings crossing the coelomic canal and inserted in its wall.   Goodrich has recently re-examined these organs and according to his account they do not open into the coelom but end internally in branches beset with solenocytes, or cells provided like the choanocytes of Sponges with a collar inside which a flagellum flickers. The same author has described similar structures in the Polychaeta, and holds them to be radically distinct from the wide-mouthed, simple nephridia of the Mollusca and of some Chaetopoda which he terms coelomoducts. These tubes are the nephridia, and taken collectively constitute the kidney. It has been proved that carmine injected into the animal is excreted by them. Besides these a number of thickened patches of the atrial epithelium, discovered by Johannes Müller and called by him renal papillae, are thought to assist in excretion (Figs. 169 and 171).

The blood system is exceedingly simple. The blood from the elementary canal is brought back by a sub-intestinal vein, which like a broad river is often subdivided into two or three parallel channels which then reunite with one another; in a word it is more a plexus than a tube. It runs to the tip of the liver on its outer side, returns on its inner side, and pursues its course then as a single channel under the pharynx, where it is called the ventral aorta. In this region it is contractile, deriving its muscles from the walls of the ventral coelomic tube. Vertical branchial vessels called arterial arches are given off; these ascend in the gill septa, that is, the portions of the wall of the pharynx intervening between the gill-slits. Arriving at the dorsal line of the pharynx these vessels empty into two longitudinal vessels, the dorsal aortae, which further back unite into one (Figs. 170 and 171).

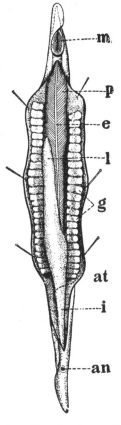

Fig. 178. *Amphioxus* dissected from the ventral side × 2. After Rathke, slightly altered.

an. Anus. *at.* Position of atrial pore; the extension of the atrium behind this point is indicated by the dotted line passing over to the right side of *i*, the intestine. *e.* Endostyle. *g.* Gonadic pouches. *l.* Liver. *m.* Entrance to mouth with the oral cirri lying over it. *p.* Pharynx.

The tongue-bars also

contain vessels emptying into the dorsal aortae; these communicate with the branchial vessels through what are called synapticulae, that is, cross pieces tying the tongue-bar to both sides of the gill-slit which it divides (Fig. 172).

Both gill-bar and tongue-bar are strengthened with rods of gelatinous tissue. These are the precursors of the visceral arches, which form such an important part of the skeleton in the

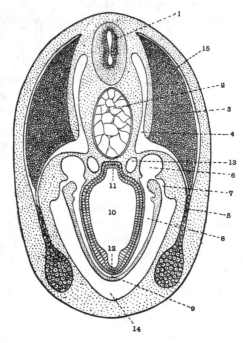

Fig. 179. Diagrammatic transverse section of *Amphioxus* to show the relation of the excretory and genital organs.

1. Nerve-cord. 2. Notochord. 3. Myotome. 4. Hollow sclerotome.
5. Gonadic pouch. 6. Dorsal coelomic canal. 7. Nephridium.
8. Branchial coelomic canal in gill-bar. 9. Endostylar coelom.
10. Pharynx. 11. Hyperpharyngeal groove. 12. Endostyle.
13. Dorsal aorta. 14. Atrial cavity. 15. Myocoel.

higher Vertebrata. Similar gelatinous tissue forms the rays of the dorsal and ventral fins, the sheaths of the notochord and nerve-cord and the dermis. It differs from ordinary connective-tissue in that although it consists of a ground substance with a deposit of fibres in it, it contains no amoebocytes or "connective-tissue corpuscles." The fibres in ordinary connective-tissue are largely if not entirely produced by the metabolic activity of the amoebocytes; but in *Amphioxus* it appears that they are produced by the cells of the coelom, ectoderm and endoderm which adjoin the connective-tissue. Thus the sheath of the notochord is deposited partly by the cells of the notochord, but chiefly by a hollow outgrowth from the myotome called the sclerotome (4, Fig. 179), the fin-ray by a coelomic sac (*d.f.c.* Fig. 171) which disappears in the adult, the rods of the gill- and tongue-bars largely by the epithelium of the pharynx. The connective-tissue of the Cephalochordata is therefore in a peculiarly interesting primitive condition; and that of the Hemichordata has the same structure.

The reproductive organs are very simple in construction. The sexes are separate, and ovaries and testes closely resemble each other in external appearance (Fig. 178). They take the form of squarish masses, called gonadic pouches, embedded in the outer walls of the atrial cavity. When ripe they burst into the atrial cavity, the eggs escape through the mouth or atrial pore, the spermatozoa through the atrial pore. The fertilized egg developes into a free-swimming larva of a remarkable form. There are no atrial folds covering the gills, but one set of slits is developed long before the other, and the mouth appears on the left side. It has been proved that the sexual organs are outgrowths from the lower ends of the myotomes, and remain throughout life connected with these by strings of cells (Fig. 179).

The Hemichordata and Cephalochordata are found all over the tropical and temperate regions of the world wherever a suitable substratum is found. The Hemichordata burrow in mud rich in decaying matter, but the Cephalochordata prefer clean sand, their food as we have seen consisting of swimming organisms.

Sub-Phylum III. TUNICATA or UROCHORDATA.

By many the group of Tunicata or Urochordata would be considered the lowest portion of the phylum Vertebrata, and if we had regard only to the adult structure this could not well be denied, for in the adult hardly a trace of the Vertebrate relationship is discernible. But the Tunicate commences life as a larva showing a more developed structure in several important points than *Amphioxus* possesses at any period of its life-history, and hence we must regard the simple organization of the adult as a degraded rather than a primitive condition of affairs.

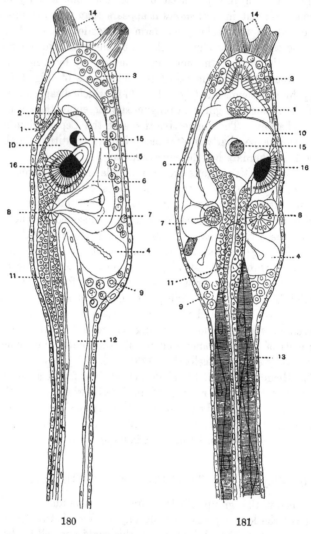

180   181

FIG. 180. Side view of the anterior end of a larva of *Ascidia* which has been
free-swimming for two days × 375.   FIG. 181. Dorsal view of the same.
After Kowalewsky.

1. Mouth.   2. The connection of the brain with the stomodaeum.
3. Endostyle.   4. Intestine.   5. Branchial cavity.   6. 1st gill-slit.
7. 2nd gill-slit. 8. Atrial opening.   9. Blood corpuscles.   10. Cavity
of brain.   11. Dorsal nerve-tube.   12. Notochord.   13. Muscles.
14. Fixing organs.   15. Otocyst.   16. Eye.

The typical Tunicate larva is often called the Ascidian tadpole
because its form recalls that of the well-known larva
of the frog. It attains a length of about a quarter of
an inch, and consists of a small round trunk and a thin vertical
tail four or five times as long as the trunk. The tail is the organ
of locomotion, and is provided with a sheet of muscles on either side
by the alternate contraction of which powerful side-strokes are
executed and the animal is propelled forward (13, Fig. 181). The
tail is stiffened by a well-developed notochord—which does not
extend into the trunk, hence the name "Urochordata" (Gr. οὐρά,
tail; χορδή, a string). A uniform flap of skin, a continuous
fin, forms a border to the tail. The trunk contains the enlarged
pharynx which opens by a narrow mouth in front : and laterally
communicates with the exterior by two ciliated openings—the gill-
slits. Its ventral wall is swollen out into a pocket which causes the
under lip to protrude as a bulky chin. In this pocket we find
developed a ciliated groove, the endostyle, having the same position
as the organ similarly named in *Amphioxus*. On the chin outside
are three peculiar warts which secrete a sticky slime and which are
used by the larva to fix itself to surrounding objects. The pharynx
leads behind into a short intestine which is attached to the ectoderm
high up on the side far in advance of the root of the tail. Hence
the process of shifting forward the anus and the corresponding
development of a purely muscular tail have been carried much
further in the Ascidian tadpole than in *Amphioxus*.

The nervous system in the tail is a simple neural tube ; but in
the trunk it expands into a thin walled vesicle, the so-called sense-
vesicle, which is the representative of the cerebral vesicle of *Amphioxus* and the forerunner of the brain of the higher Vertebrata.
As in the larva of *Amphioxus*, the sense-vesicle opens to the exterior,
but the spot where this occurs is involved in the invagination which
forms the stomodaeum. The tube connecting the sense-vesicle and
the stomodaeum is called the neuropore (2, Figs. 180 and 181).
Part of the side-wall of this vesicle is modified so as to form a cup-
shaped eye with a simple cuticular lens directed inwards. From the
roof hangs a ball of lime suspended by a pillar of cells ; this acts as
an otolith, and the whole forms a rudimentary ear.

Thus both in the structure of the nervous system and the
position of the anus, the Ascidian tadpole is more advanced than
the *Amphioxus*.

Although, as we have seen, both mouth and anus are present,

yet they cannot be used, for they are closed by a sheet of gelatinous matter. This is the test which is secreted by the ectoderm cells and envelopes the whole body, so that during its brief free-swimming life the Ascidian takes no food.

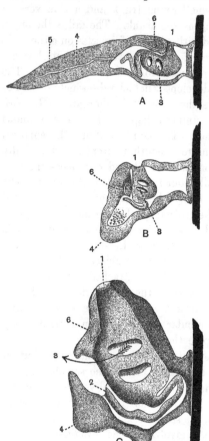

After swimming for a short time the larva fixes itself by its chin-warts to a suitable substratum and undergoes a very rapid metamorphosis. The tail shrinks and is absorbed, notochord and nerve-tube disappear : the sense-vesicle also disappears, only its hinder thickened wall persisting as the adult ganglion (6, Fig. 182). The neuropore however persists and developes into a mass of tubes underlying the ganglion, which is called the sub-neural gland. Its opening acquires a crescent form with thickened lips, and is called the dorsal tubercle. Meanwhile the chin grows enormously, so as to rotate the mouth up and away from the substratum, and thus the long axis of the pharynx becomes vertical instead of horizontal. The skin of the region where the anus becomes situated is depressed so as to form a groove. This be-

Metamor-phosis.

FIG. 182.   Diagram of the fixing and changes undergone by a larval *Ascidian*. From Lankester.

1. Mouth.   2. Anus.   3. Gill-slits.
4. In A, notochord; in B and C, vanishing tail.   5.   In A, tail.   6. Brain.

comes confluent with the outer parts of the two gill-slits, so as to form a single dorsal cavity termed the atrial-cavity, the opening of which is not far from the mouth. It must be noticed that this

cavity does not correspond to the similarly named cavity in *Amphioxus* : in
the case of the last-named animal the
atrial walls originate from the dorsal
edges of the gill-slits and meet one
another beneath the animal ; whereas
in the Urochordata they arise from the
ventral edges of the slits, and are united
with one another on the dorsal edge.
The gill-slits themselves become changed
by the growth of numerous partitions,
transverse to the axis of the pharynx,
into a series of narrow slits ; and then
by the formation of another series of
stronger bars parallel to the long axis,
into a veritable ciliated trellis-work.
All this trellis-work is supported by
horny rods like the gill-bars of *Amphioxus.* The test thickens enormously
and becomes invaded by a finger-like
outgrowth from the hinder part of the
body, which carries blood-vessels to it
and buds off cells into it which nourish
it and change its character. With these
changes the adult form is attained.
Few would see any resemblance to a
Vertebrate in the motionless sac-like
body fixed to a stone or rock and looking more like a plant than an animal
(Fig. 183).

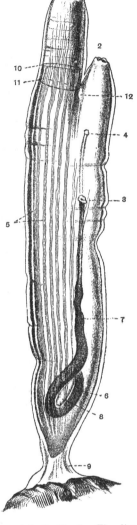

Nevertheless, the Tunicate in some
points, even when adult,
recalls the structure of
Amphioxus. Thus we
encounter a ring of delicate tentacles
a short distance inside the
mouth strikingly recalling
the velar tentacles of *Amphioxus.* As in that animal
also there is a long hypopharyngeal groove or

Structure of
adult.

FIG. 183.   *Ciona intestinalis* × 1. The live
animal seen in its test; some of the organs
can be seen, as the test is semi-transparent.

1. Mouth.   2. Atrial orifice.   3. Anus.
4. Genital pore.   5. Muscles.   6. Stomach.
7. Intestine.   8. Reproductive organs.
9. Stalk attached to a rock.   10. Tentacular
ring. 11. Peripharyngeal ring.   12. Brain.

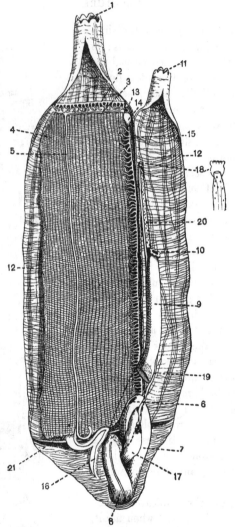

FIG. 184. View of *Ciona intestinalis* lying on its right side. Both the branchial and the atrial cavities have been opened by longitudinal incisions.

1. Mouth.   2. Tentacles.   3. Peripharyngeal groove.   4. Perforated walls of branchial sac.   5. Endostyle.   6. Oesophageal opening leading from the branchial sac to the stomach, rather diagrammatic.   7. Stomach.   8. Intestine showing typhlosole; part of it removed to show subjacent structures.   9. Rectum.   10. Anus.   11. Atrial aperture.   12. Inner surface of mantle showing longitudinal and transverse muscle fibres.   13. Dorsal tubercle.   14. Subneural gland and brain.   15. Cut edge of branchial sac.   16. Heart.   17. Ovary. 18. Pore of vas deferens.   The openings of the oviduct and the vas deferens are shown enlarged to the right.   19. Testicular tubes on intestine. 20. Oviduct.   21. Septum shutting off that part of the body-cavity which contains the heart, stomach and generative organs.

endostyle passing in front into a peripharyngeal band, and secreting a cord of mucus which is worked forward. This mucus is torn into strings by the inrushing current of water and swept backwards to the opening of the oesophagus, entangling in it food particles just as in *Amphioxus*. Instead of a hyperpharyngeal groove, there is a series of tags hanging down from the dorsal wall of the pharynx, called languets. These in life curve round so as to form a row of hooks supporting and directing the mucous strings.

The oesophagus leads into a dilated stomach which bends on itself and leads into an intestine which after one or two coils runs forward and opens into the atrial cavity. Its ventral wall is folded inwards, forming a typhlosole similar to that of an Earthworm. As usual the straight terminal portion of the intestine is called the rectum. Near the anus open the ducts of the ovary and testes, for the animals are hermaphrodite. These organs are branched clumps of tubes, the testis being spread over the surface of the stomach, the ovary forming a mass between the stomach and intestine. Oviduct and vas deferens are closely applied to one another, the vas deferens being the more superficial. The latter opens by a rosette of small pores, the ovary by a broad opening, and so the water from the gill-slits, as it passes out of the atrial cavity, sweeps away the sexual cells.

On the ventral side of the pharynx is a V-shaped heart, which is enclosed in a space called the pericardium. The heart is only a specially thickened part of a ventral blood-vessel, which lies immediately under the endostyle and communicates through a network of vessels in the gill trellis-work with the dorsal blood-vessel. Waves of contraction pass over the heart so as to drive the blood forward. After a certain interval the direction of these waves is reversed, so that the blood alternately goes to the dorsal vessel from the heart and *vice versâ*. With the exception of the heart, however, the blood-vessels do not seem to have definite walls, and are really, as in the Enteropneusta, crevices left between various organs.

The sluggish life of the Ascidian has as its only external manifestation the sudden closing of the mouth and atrial cavity by sphincters, and the consequent ejection of water—whence the popular name Sea Squirt. In consequence metabolism is at a low level and not much waste is produced. A good deal of this waste is probably got rid of by the throwing off of the mantle from time to time, but for the rest no definite excretory organ is required.

The nitrogenous excretion is stored up as crystals of insoluble uric acid in little vesicles attached to the hinder part of the intestine. These vesicles, together with the cavities of the genital organs and the pericardium, may be looked on as the remnants of the coelom, so that here a similar phenomenon has taken place to what was met with in the case of Arthropods, namely an obliteration of the coelom through the expansion of blood-vessels.

The Tunicata or Urochordata abound on every rocky shore and exhibit a surprising diversity of form. Their principal divisions are as follows.

Class I. Copelata or Larvacea.

Small forms which retain the larval condition throughout life. The gill-slits are undivided and the anus ventral. There is no atrial cavity : each of the two gill-slits opens directly to the exterior. The tail is usually carried bent forward at a sharp angle with the body. A temporary test devoid of blood-vessels is found : the animal when disturbed wriggles out of it and forms another.

Class II. The Acopa.

FIG. 185. Two groups of individuals of *Botryllus violaceus*, after Milne Edwards. Magnified.

1. Mouth opening.　　2. Common cloaca of the group.

Forms which have lost the tail with its nerves and muscles. These are divided into

　　Order I. The **Ascidiaceae**, fixed forms.

　　Order II. The **Thaliaceae**, which have secondarily acquired the power of swimming by contractions of the whole body carried out by transverse bands of muscle.

The Ascidiaceae constitute the great bulk of the Urochordata. Some of them, such as the form taken as a type in the general description given above, remain solitary throughout life, but others bud and form colonies embedded in a common test; these are called Compound Ascidians. But the group is not a natural one, since budding is carried out in different ways in different families, and has therefore probably originated several times. The commonest method is by the outgrowth of a hollow finger-shaped process of the pharynx, called a stolon, arising at the hinder end of the endostyle, which becomes divided into pieces, each forming a bud. On the other hand in *Botryllus* a different method is followed, since in this case the buds originate simply as little pockets of the atrial-wall of the parent. *Botryllus* is one of the most beautiful colonial forms; in it the buds are arranged in circles; the atrial openings of the members of a circle open into a common pit in the centre called the cloaca. *Pyrosoma* is a free floating colonial form, with the shape of a cylinder open at both ends, the atrial cavities of the constituent persons opening on the inner surface, their mouths on the outer.

The Thaliaceae are extraordinary forms. They have the shape of cylinders with the mouth at one end and the atrial opening at the other, and their body is surrounded wholly or partly with muscular hoops like the hoops encasing a barrel. The commonest

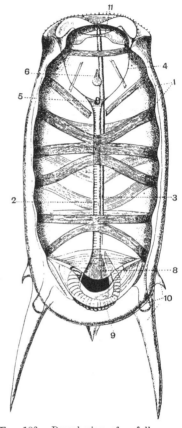

FIG. 186. Dorsal view of a fully-grown specimen of the solitary form of *Salpa democratica* × about 10. From Brooks.

1. Muscle bands.   2. Narrow band representing the dorsal wall of the pharynx, the so-called "gill."   3. Endostyle.   4. Peripharyngeal band. 5. Brain.   6. Ciliated pit.   8. "Nucleus," consisting of stomach, liver, intestine.   9. Stolon or row of young. 10. Processes of mantle.   11. Mouth.

form is *Salpa*, which at intervals may be seen in countless numbers swimming at the surface of the sea. In this anima

the test is of a glassy transparency. The two original atrial
openings or gill-slits of the larva do not become divided by
partitions, but develope into two huge vacuities in the side walls
of the pharynx, reducing its dorsal wall to a mere band, the so-
called " gill." There are two distinct forms of this animal, a sexual
and an asexual, one giving rise to the other, so that here we have
a case of " alternation of generations." In the asexual form we
find an endostyle-process or stolon which gives rise to a chain of

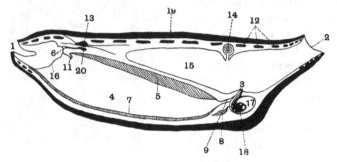

Fig. 187.   Semi-diagrammatic view of left side of *Salpa*.   From Herdman.

1. Branchial aperture.   2. Atrial aperture.   3. Anus.   4. Branchial sac.
    5. "Gill."   6. Sub-neural gland.   7. Endostyle.   8. Heart.
    9. Oesophagus.   11. Languet.   12. Muscle bands.   13. Nerve
    ganglion.   14. Embryo in ovisac.   15. Peribranchial cavity.
    16. Peripharyngeal band.   17. Stomach.   18. Testes.   19. Test.
    20. Sub-neural gland.

small sexual forms which one by one drop off.   Each sexual form
produces only one egg.   This when fertilised does not give rise to
a tailed larva, but becomes attached to the atrial wall of the
mother by a knob of maternal tissue containing blood-vessels, called
the placenta, which is embedded in a disc of embryonic tissue,
through which nourishment diffuses from mother to embryo.   In
this position it grows up into an asexual form and eventually
breaks loose and swims away.

CHAPTER XIV.

INTRODUCTION TO SUB-PHYLUM IV, CRANIATA.
THE CYCLOSTOMATA.

ALL the remaining Vertebrata are distinguished by possessing a skull and brain, and are grouped together as Craniata. The Craniata are separated by a deep gap from the lower forms: but they themselves present a fairly continuous and graded series from the lowest to the highest forms, and their comparative anatomy, especially when we take into account the fossil representatives of the sub-phylum, gives us a fairly good idea of the course which the evolution of Vertebrata has pursued; so much so indeed, that the group might be compared to the fairly reliable and complete records of a country during the historical period, whilst the Hemichordata, Cephalochordata and Urochordata represent the few scattered and scarcely decipherable documents of prehistoric epochs.

The Craniata are defined, as we have seen, by the possession of a skull and a brain, though these are only two of the many characters which distinguish them from the other Vertebrata. The skull is composed of either cartilage or bone; and even in cases where the adult skull is completely bony, in the embryo the bone is partly, at any rate, represented by cartilage. Cartilage and bone are really only two peculiar modifications of connective tissue whose fundamental characters it may be useful to recall. There is in every case a gelatinous ground substance traversed by fibres, and applied to these fibres are cells, which are connected with one another by delicate threads of protoplasm and which secrete the greater part of the ground substance and fibres contained therein. In cartilage, the ground substance becomes cheesy in consistence, the fibres being masked, and the cells are arranged by twos and threes in little pockets. In bone, on the other hand, the cells remain single while the ground substance becomes hardened by depositions of phosphate and carbonate of lime. The spaces occupied by the cells are known as lacunae, and the delicate processes which connect the cells give rise to the capillary canals known as canaliculi in the dried bone, whilst the spaces occupied by blood-vessels traversing the bone are known as Haversian canals.

In the simplest form the skull consists of two pairs of pieces of cartilage, one pair embracing the front end of the notochord and termed the parachordals. In front of these is the second pair, the trabeculae, united behind and before with each other but diverging in the middle so as to embrace between them the pituitary body, which is described with the brain. The parachordals develope ridges which wall in the sides of the brain and may form a roof over its hinder portion.

Primitive skull.

The brain is only the enlarged and modified anterior end of the neural tube, and the existence of a skull is correlated with the presence of neural arches protecting the hinder part of the nervous system. These arches consist of paired pieces of cartilage meeting above the neural tube. They have been shown to be formed as solid outgrowths of the myotomes which represent the hollow sclerotomes of *Amphioxus*, and hence it may be that the cranium itself is derived from the walls of the most anterior myotomes which early become fused with one another and otherwise modified.

Haemal arches, paired pieces of cartilages with their upper ends implanted in the sheath of the notochord and their outer ends directed downwards, are also always present, and like the neural arches are derivatives of the myotomes. In the region of the tail the haemal arches meet each other so as to form a V beneath the notochord, but in the trunk they simply project out between adjacent myotomes as transverse processes, the ends of which may become movable on the basal parts and are then known as ribs[1].

The brain of all Craniata is sharply divisible into three primary regions called fore-brain, mid-brain and hind-brain (Fig. 188). Of these the first is certainly the enlarged and highly developed representative of the sense vesicle of the Urochordata and of the cerebral vesicle of *Amphioxus*.

Brain.

---

[1] This statement applies to the transverse processes of the lower Craniata : those of the higher Craniata are secondary outgrowths from the neural arches.

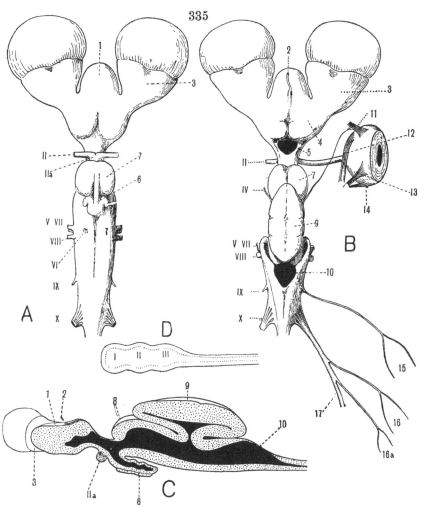

FIG. 188. *Scyllium catulus.* Dissection of the brain and of some of the cranial nerves. A. Ventral view. B. Dorsal view. C. Longitudinal median section. D. Diagram of embryonic brain showing the three primary vesicles.

1. Cerebrum. 2. Pineal stalk. 3. Olfactory lobe. 4. Cerebral hemisphere. 5. Thalamencephalon. 6. Pituitary body. 7. Optic lobes. 8. Optic lobes. 9. Cerebellum. 10. Roof of the hind-brain. 11. Superior oblique muscle. 12. Internal rectus muscle. 13. Superior rectus muscle. 14. External rectus muscle. 15. Ninth or glossopharyngeal nerve. 16. Branch of vagus nerve to second branchial cleft. 16a. Branch of vagus nerve to third branchial cleft. 17. Main trunk of vagus to fourth and fifth gill-slits, to lateral line and to viscera. II. Optic nerve. IIa. In A, optic chiasma. IV, V, VI, VII, VIII, IX and X. Roots of fourth to tenth cranial nerves. In D, I, II, III represent the first, second and third primary vesicles of the embryonic brain.

In the embryo it is a simple thin-walled vesicle, the lateral walls of which become changed into the retina or the essential sensory portion of the eye. This, as is the case in the Ascidian tadpole, has its perceptive surface turned inwards towards the brain cavity. The nerves by which the eyes are connected with the brain are really the narrowed connections of the lateral portions of the fore-brain with the central portion. The roof of the fore-brain remains thin throughout life and from it a stalk arises leading to a third median eye, the so-called pineal body, vestigial in all living forms. From the front wall of the fore-brain an outgrowth takes place, giving rise to a bilobed vesicle termed the cerebrum, each of the two lobes of which it is composed being termed a cerebral hemisphere. This in the higher Craniata is the seat of the more complex mental processes, but in the lower it appears to be intimately connected with the organ of smell. The cerebrum in these cases remains thin-roofed, but its base thickens owing to a great development of nervous matter. In order to distinguish it from the cerebrum the original fore-brain is denoted by the name thalamencephalon. This pituitary body is compounded of a downgrowth of nervous tissue from the fore-brain with a portion of tissue evaginated and constricted off from the lining of the buccal cavity. It represents the sub-neural gland of the Urochordata, and in the higher Vertebrates produces a substance which is of importance to the normal metabolism of bone and connective tissue, and recent research suggests that its secretion also influences the activity of the kidneys.

The mid-brain acquires thick lateral pouches, the so-called optic lobes : the hind-brain remains thin-roofed, except in front where a transverse nervous band, the cerebellum, is formed. The cerebellum is believed to be the portion of the brain intimately connected with the semicircular canals of the ear and to have for its function the control of the muscles so as to maintain the equilibrium of the body. The rest of the hind-brain is termed the medulla oblongata or spinal bulb; it controls the beating of the heart, the respiratory movements and other vital processes. The hinder part of the neural tube is known as the spinal cord, and it developes thick walls, so that its cavity is exceedingly small.

The essential element in the nervous system of Vertebrata, as in all other nervous systems, is a kind of cell which has been variously styled nerve-cell, ganglion-cell and neuron. This last name is undoubtedly the best, as it avoids the old misapprehension that regarded the nerve-cell and nerve-fibre as two independent structures.

The minute structure of the nervous system.On page 54 it was pointed out that the nerve-fibre is a very fine basal outgrowth of a modified ectoderm cell which is the nerve-cell. The cell, including its outgrowth, is termed the neuron. Important discoveries have recently been made on the minute structure of the nervous system of Vertebrata, and we are now able to form a simple and connected idea of the principles on which it is built up. Originating as a simple strip of ectoderm which becomes rolled up so as to form a tube, it is at first composed of cells which extend through its entire thickness and which all abut on the cavity of the tube. Some retain this position but develope branches and deposit a large amount of cuticular substance in their protoplasm: these, constituting the supporting elements of the system, are termed collectively neuroglia. Other cells retire from the cavity of the tube, becoming more or less rounded in form, but developing a number of outgrowths: these cells are the neurons. Each neuron is provided with a number of branching processes, sometimes arising from a single thick stem; these are called receptive dendrites (Gr. δένδρον, a tree), and they receive impulses. Impulses are transmitted through one long basal process, called the axis-cylinder process or axon, which ends in a tuft of processes often thickened at the tips, which are called terminal dendrites. The name axis-cylinder is suggested by the circumstance that amongst Vertebrata this process is in many cases surrounded by a fatty sheath of a conspicuous white colour, called the myelin; a process with or without this sheath making up what is known as a nerve-fibre. The tuft of dendrites in which the axon ends is found to be in close contiguity either with the receptive dendrites of another neuron, by which means the impulse is transmitted from one neuron to another, or else with the muscle-plate of a muscle-fibre, by which means the fibre is stimulated. The muscle-plate is a disc of protoplasm with several nuclei situated at the side of the muscle-fibre. The axon may give off several branches termed collaterals. These like the main stem end in tufts of dendrites; in this way an impulse may spread over several paths. The receptive dendrites of a neuron may also receive impulses from the terminal tufts of several axons, and in this way impulses are co-ordinated and combined.

As mentioned above, the skull and brain are by no means the only characters which distinguish the Craniata from other Chordata.

Perhaps the next in importance is the possession of three well-
developed pairs of sense-organs, nose, eyes and ears.
Of these the nose is the most simply constructed.
It consists merely of a pair of pits in the skin at the
most anterior portion of the body, the lining of which developes
ridges covered with sensory cells, having an olfactory function
(Fig. 189). The essential element in all sense-organs is the sense-
cell, which resembles the neuron in possessing a basal process
terminating in a tuft of dendrites by which the stimulus is transmitted as an impulse through a neuron, for in Craniata a sense-cell

Sense-organs.

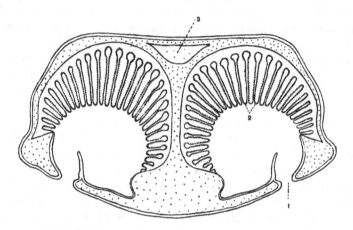

FIG. 189.   Transverse section through the snout of a Dogfish, *Scyllium canicula*,
to show the structure of the nose × 2.

1.   Opening of olfactory sac.   2.   Olfactory epithelium.   3.   Ethmoidal
region of the cartilaginous skull.

is never in direct communication with a muscle-fibre. An olfactory
sense-cell differs from a neuron in possessing one or more stiff
peripheral processes projecting from the surface of the body, by
which stimuli are received from the external world. These are
termed sense-hairs, and they are excessively delicate in structure.
Sense-cells are never combined by themselves into an epithelium:
they are always intermixed with stiff supporting cells which usually
have at the base several root-like branches. The front end of the
brain comes in direct contact with the wall of the nasal sac and the
axons of the sensory cells stretch into the brain, thus constituting
the olfactory nerve (Fig. 188).

The ears are also at first pits of the skin placed further back at
the sides of the hind-brain. In the lower forms these pits retain a
narrow connection with the exterior throughout life through a long

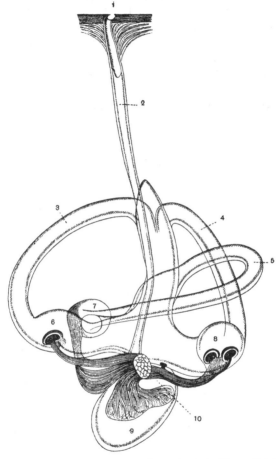

Fig. 190. Ear of *Chimaera monstrosa* L. × about 4. From Retzius. Seen
from the inner side.

1. External aperture on roof of skull. The wall of 2, the "ductus endolymphaticus," is partly removed to show that it is a tube.   3, Anterior,
4, posterior, and 5, horizontal semicircular canals.   6, Anterior,
7, external, and 8, posterior ampullae.   9. Sacculus.   10. Auditory
or 8th nerve.

tube called the ductus endolymphaticus (2, Fig. 190). In the
higher forms this tube is still recognizable but no longer opens to
the exterior. Each pit contains lymph and becomes constricted in

22—2

the middle into an upper portion, the utriculus, and a lower portion, the sacculus. With the exception of the Cyclostomata the former gives rise to three flat outgrowths placed in planes at right angles to one another (Fig. 190). These outgrowths become converted into half-rings by the meeting of their walls in the middle of each, and in this way three semicircular canals are formed, called respectively anterior, posterior and horizontal. The primary function of the whole organ, like that of the otocysts of Medusae, Crustacea and Molluscs, is to enable the animal to perceive its position. Where each semicircular canal arises from the utriculus it is swollen, and the swelling

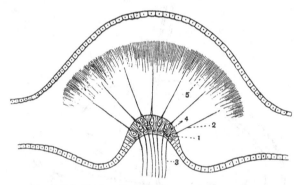

FIG. 191.   Section of an ampulla of the internal Ear.

1.   Sense-cell bearing a long hair.   2. Sense-hair.   3. Nerve termination branching round base of sense-cell (dendrites of a deeply placed neuron).   4. Interstitial cell.   5. Gelatinous cup in which the sense-hairs are embedded.

is termed an ampulla. The wall of each ampulla projects inwards, and the projection contains cells with exceedingly long sense-hairs which project into the cavity of the ampulla (Fig. 191). The free ends of these hairs are embedded in a gelatinous cup, and thus the whole organ is admirably adapted to record change of position in any direction, since any change of position can be completely analysed into movements in three planes. The lower part of the organ or sacculus has cells adapted to be stimulated by vibrations in the surrounding lymph. It often contains calcareous 'ear-stones.' In higher forms it gives off a spiral tube, the cochlea, which contains the true auditory sense-cells. These form the organ of Corti, a more complex structure than the sensory epithelium of the sacculus and ampullae, but resembling it in consisting of hair-cells which are embraced by the receptive dendrites of neurons. The grouped cells of the neurons of all these sensory structures form the several auditory ganglia. Both nose and ear have cartilaginous or bony coats which become firmly connected with the skull; these are known as the sense-capsules.

The eye is the most complicated, and in the higher Craniata by far the most important, of the sense-organs. In its origin, as we have seen, it is the lateral portion of the fore-brain which when constricted off is known as the primary optic vesicle (Fig. 192). The outer wall of this becomes modified into a sensory epithelium called the retina. This consists of a row of visual cells, their free ends directed inward towards the brain and produced into the characteristic striated rods. Beneath these sense-cells lie a number of neurons, the dendrites of which, mingling with the dendrites of the sensory cells, give rise to a comparatively thick bed of nervous tissue. Long, however, before the sense-cells are developed, the primary vesicle of the eye has completely altered its shape. The outer wall has become pushed in on the inner so as to completely reverse the shape of the sac (Fig. 192). Its cavity is reduced to a

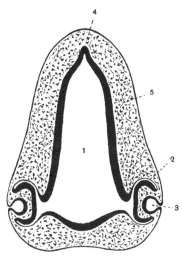

FIG. 192. Transverse section through a third day Chick to show origin of the retina from the brain and of the lens from the ectoderm. Highly magnified.

1. Cavity of brain.   2. Outer layer of retina surrounding the black, thicker layer which will form the rods and cones.   3. Lens arising as a hollow invagination.   4. Pineal body originating.   5. Embryonic connective tissue.

mere slit, and it takes on the form of a very deep double cup with its concavity directed outwards. This is the cavity of the eyeball, or so-called secondary optic vesicle, the clear gelatinous connective tissue inside which is known as the vitreous humour. The connective tissue surrounding the vesicle peripherally forms a tough fibrous or even cartilaginous capsule called the sclerotic coat, lined by a thin vascular tissue, the choroid coat. The sensitive and nervous outer layer of the primary vesicle is known as the retina, the other layer (which becomes loaded with pigment) as the pigment epithelium of the retina. If we analyse the structure of the retina, we find that it has fundamentally the same structure as the central nervous system of which, as its origin shows, it is really a part. Thus there are a number of branched and cuticularized supporting cells called fibres of Müller, extending throughout the whole thickness of the retina, and the main mass of the retina is made up of neurons. There is, however, in addition a layer of characteristic visual cells;

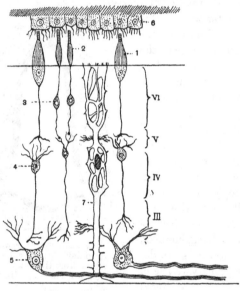

Fig. 193. Diagram to illustrate structure of a retina. The several "layers" are indicated by the numerals III, &c. in order from within (vitreous humour) outwards.

1. Cone. 2. Rod. 3. Nucleus of rod-cell. 4. Small neuron.
5. Large neuron. 6. Pigment epithelium. 7. Fibre of Müller or a supporting cell.

that is to say, of sense-cells, with a comparatively thick striated rod in place of the ordinary sense-hair. Visual rods have already been described in the eyes of Anthomedusae (p. 55) and of Arthropoda (p. 131); they occur wherever the capacity for vision is developed. In the retina of Craniata there are two varieties of visual cell, called respectively rod-cells and cone-cells. In the first, the visual rod is narrow and cylindrical, and the body of the cell beneath is filamentous with a rounded swelling for the nucleus; the basal process ends in an unbranched knob, that is to say, in a single

dendrite. In the cone-cell the rod is conical with a broad base, to
which the body of the cell containing the nucleus is immediately
applied; the basal process ends in the normal manner in a tuft of
dendrites. The basal processes of both kinds of sense-cell are in
close relation to the receptive dendrites of a layer of neurons with
small cell bodies; the axis-cylinder processes of these in turn end
close to the receptive dendrites of a layer of neurons with large
cell bodies situated close to the outer basal surface of the retina,
which give rise to the fibres constituting the optic nerve. Taking
a general view therefore we may say that the retina is a sensory
nervous epithelium consisting of a layer of sense-cells underlaid by
two layers of neurons. Before its structure was thoroughly understood, however, the appearance of the retina in transverse section
was a bewildering mass of fibres and nuclei, in which for descrip-
tive purposes different layers were distinguished. These, reckoning
them in the order proceeding from the inner side of the eyeball
towards the lens, were as follows:—(a) the layer of rods and
cones ; (b) the outer nuclear layer (VI. Fig. 193) consisting of the
bodies of the visual cells containing their nuclei; (c) the outer
molecular layer (V. Fig. 193) consisting of sections of the basal
processes of the visual cells and of the receptive dendrites of the
neurons with small cell bodies; (d) the inner nuclear layer (IV. Fig.
193) consisting of the bodies of the above neurons; (e) the inner
molecular layer (III. Fig. 193) consisting of sections of the basal
processes of the above neurons and of the receptive dendrites of the
neurons with large cell bodies; (f) the layer of nerve-cells consisting of the bodies of the last-named neurons; and finally (g) the
layer of nerve-fibres consisting of the basal processes of the last-
named neurons which constitute the optic nerve.

The remainder of the eye is to be looked on as a part of the
skin of the side of the head which has been rendered transparent
in order to allow light to reach the retina. It consists of a lens
and cornea, separated by a chamber containing the aqueous
humour. The lens is an originally hollow plug of ectoderm cells,
which breaks loose from the skin and lies in the mouth of the
secondary optic vesicle (Fig. 192). The skin outside the lens forms
the cornea, which is transparent. The cornea being joined to
the edges of the sclerotic completes the boundary of the eyeball,
as the fully-elaborated sense-organ may be termed. If the above
description has been followed it will be seen that in a Craniate
light must reach the visual cells through their basal and not

through their visual ends. As this is contrary to the almost universal rule obtaining throughout the animal kingdom, we cannot believe it to be a primitive arrangement. Rather we must believe that when the eye was being evolved the rods of the visual cells were directed towards the light, and that the epithelium of which they form a part was exposed and not rolled up into a neural tube; in a word, that the front portion of the nervous system of Vertebrata at any rate was once a plate of sensitive skin. It is most suggestive to note that in the larva of the Hemichordata we find such a plate with two eye-spots at the apex of the prae-oral lobe.

The external layer of the skin or ectoderm of Craniata is quite peculiar in the animal kingdom, in that it consists not of one, but of many layers of cells. On closer inspection, however, it is seen that the deepest layer, consisting of columnar cells alone, really repre-

Skin.   sents the ectoderm of the other phyla. This layer instead of becoming directly converted into cuticular substance, as, for example, in the Arthropoda, buds off flattened cells from its outer surface which become bodily converted into horny matter and scale off. The ectoderm rests on a specially firm bed of connective tissue called the dermis.

A very peculiar feature in the Craniata is the character of the

Nerves.   scattered sense-cells of the skin. These end in sense filaments embedded in the ectoderm, not pro-jecting beyond it. These filaments have however grown enormously, and with their growth the bodies of the cells with the nuclei have come to lie deep down in the body. Here they form segmentally arranged packets of cells lying at the side of the nerve-cord and known as the spinal ganglia. They are connected with the nerve-cord by their basal outgrowths or nerve-tails, which constitute the dorsal roots of the spinal nerves corresponding to the dorsal sensory nerves of *Amphioxus*. To each myotome a motor nerve is given off, as in *Amphioxus*, but in the Craniates the fibres of this nerve are bound up for a certain distance with the long peripheral hairs of the sense-cells constituting the spinal ganglia, so as to form a compound sensory-motor nerve, which is then said to have a dorsal sensory and a ventral motor root.

The power of transmitting and modifying impulses, characteristic of the nerve-cell, is merely one of the fundamental properties of all protoplasm, specially developed. It therefore probably resides to a small extent in all cells. In the ectoderm from its exposed con-dition this function has been largely exercised, and hence the nervous system of most animals consists of modified ectoderm cells. But the endodermic tube is likewise stimulated by the passage of food through it and it is therefore not surprising to learn that some of its cells develope nerve-tails and are even converted into small neurons. In this way a tangle or plexus of fibres with intermixed cells is formed, which is the basis of the nervous system of the gut, or 'sympathetic' system. In most groups of animals the endodermic nervous system is never developed beyond this point; but in Craniata this plexus is connected with portions of the spinal ganglia at regular intervals which early separate from the rest and are called the sympathetic ganglia. These ganglia retain their connection with the spinal cord by nerves called the rami communicantes, in which motor fibres going to the alimentary canal are included. Successive sympathetic ganglia are connected by a longitudinal commissure, and so there is a chain of sympathetic ganglia on each side of the spinal cord.

It is usual to reckon ten pairs of nerves as appertaining to the brain, but these are of very unequal value. The first or olfactory pair are really drawn-out portions of the cerebrum. In the lower Craniata these parts have the shape of swellings connected by narrow stalks with the brain, and these stalks were confused with nerves (Fig. 188). The terminal swelling comes into close contact with the epithelium of the nasal sac, and a large number of small nerves—the true olfactory nerves—connect the two. The second or optic nerve is formed by nerve-fibres growing along the stalk uniting the primary optic vesicle with the brain (Fig. 192). The nerve fibrils which run in this stalk go mainly but not entirely to the opposite side of the brain. Thus in the floor of the thalamencephalon, or primitive fore-brain, there is a crossing of fibres proceeding from the two eyes. This part of the floor becomes nipped off as a groove from the rest—and is known as the optic chiasma. The chiasma is connected with the combination of the stimuli received by the two eyes so as to produce single vision, each side of the brain receiving impulses from both eyes. The third or motor oculi, the fourth or patheticus, and the sixth or abducens nerves are motor nerves, supplying the eye muscles derived from the head cavity, the collar cavity and the first myotome respectively. The fifth or trigeminal, and seventh or facial, are most interesting nerves, being sensory as well as motor, and the sense-organs they supply in the lower Craniata are peculiar. These organs are scattered over the prae-oral part of the body or snout and the sides of the head, and are known as the mucous canals. On the snout they have the shape of deep tubes swelling out at the bottom into sacs or ampullae; and on the head, of canals communicating at intervals with the exterior by vertical tubes. Certain of the cells lining these tubes develope blunt, freely projecting sense-hairs, recalling the character of the auditory cells, whilst others secrete the mucus with which the tubes are filled and whence they derive their name. It is probable that the function of these organs is somewhat allied to that of the ear, balancing combined with hearing (or at any rate, perception of vibrations), for it has been proved that a fish deprived of its eyes is still able to guide itself along tortuous passages so long as this organ remains intact, and this is explicable only on the assumption that the reflected pulses of the water are felt by these organs. The branches of the fifth and seventh nerves which supply them are usually for some distance in close juxtaposition and are known as the ophthalmic nerves. The eighth or auditory cranial nerve goes to the ear, and arises in such close proximity to the seventh that it may be regarded as a specialized branch of it, the ear itself being very possibly a highly specialized mucous canal. The motor divisions of the fifth and seventh are distributed to the region of the mouth and to that of the first gill-slit respectively. They both fork; the upper branch of the fifth goes to the upper jaw and the lower to the lower jaw, while one branch of the seventh passes in front and the other behind the spiracle. The ninth or glosso-pharyngeal nerve is similarly forked round the first true gill-slit (Fig. 188). The tenth or vagus or pneumogastric nerve, which is certainly a compound one, gives off a branch to each of the remaining slits, to which it bears a relation similar to that borne by the ninth nerve to the second slit. The main stem of the nerve passes along the alimentary canal and sends nerves to its muscles and to those of the heart, all these muscles being develop-ments of the inner or splanchnic wall of the unsegmented coelom. The tenth nerve has also in the lower Craniata a sensory division. This separates from it soon after it leaves the brain and passes backward, supplying an immensely long mucous canal, called the lateral line, which extends from head to tail along the mid-lateral portion of the body and is provided with a series of openings to the exterior. On account of its extensive area of distribution the tenth nerve has received the name of vagus (wandering).

The alimentary canal exhibits a marked difference from the condition found in the lower Chordata. The gill-slits are reduced in number, there being as a rule not more than eight: it would indeed be more correct to speak of them as gill-pouches. In this respect Craniata agree with the Hemichordata in contrast to the Cephalochordata and Urochordata. No trace of a tongue-bar has however been found in any Craniate.

Alimentary canal.

The endostyle becomes shut off from the pharynx and thus loses entirely its original function; it branches and forms a mass called the thyroid gland. The evil results attendant on its removal or diseased condition and experiments on living animals show that it secretes into the system a substance which has a beneficial influence on metabolism, especially as regards the "tone" of the nervous system and the growth of connective tissue.

The sub-neural gland of the Urochordata, on the other hand, seems to be represented by a structure called the pituitary body. This, like the sub-neural gland, is a dorsal pocket of the stomodaeum, but it becomes cut off from all connection with the mouth and intimately associated with a downgrowth of the brain, called the infundibulum, to form an organ having an influence on the well-being of the animal (see p. 336). Since in the case of the Urochordata the sub-neural gland is fashioned out of the persistent communication of the sense vesicle with the exterior, one is tempted to regard the close connection of the infundibulum and the pituitary body as remnants of the former connection of the brain and stomodaeum in the ancestors of the Craniata. Some authors maintain that a rudiment of the infundibulum is to be seen even in the cerebral vesicle of *Amphioxus* (see Fig. 174).

Except in the lowest forms the alimentary canal is differentiated into several well-marked divisions. There is to begin with a stomodaeum lined by an epithelium consisting of many layers similar to that forming the epidermis. The first division of the endodermal tube is called the pharynx, and into this the gill-slits open. The line of demarcation between ectoderm and endoderm is entirely obliterated in the adult. Following on the pharynx is a tube of narrow diameter, termed the oesophagus or gullet, which leads into the stomach. The stomach, consists of the first of the loops into which the alimentary canal is bent in consequence of being longer than the body, it is a greatly dilated portion of the canal and in it the food is stored until a large amount of digestion is accomplished. As in other animals, the food is moved from place to place by peristaltic contractions of the visceral muscles derived from the inner wall of the coelom. There is a particularly powerful girdle of these called the pyloric sphincter, which by remaining contracted keep the distal end of the stomach, the so-called pylorus, closed until the work of digestion is accomplished, when they relax and allow the food to pass on into the next division of the canal, the intestine. The walls of the proximal part of the stomach are produced into small pouches, the lining cells of which secrete a substance called pepsin, which has the power of turning the proteid of the food into soluble peptone.

Pepsin is an example of the class of substances known as digestive ferments or enzymes : these are complex substances of unknown constitution which have the power of effecting a large amount of chemical change without themselves undergoing a permanent alteration. The object of their action on food-stuffs is to render them soluble, and therefore fitted for absorption by the wall of the canal. Pepsin is active only in an acid medium, and free hydrochloric acid is found in the contents of the stomach in small quantities, produced by special cells in the walls of the pouches just mentioned.

An organ called the liver is very conspicuous (Fig. 207). It consists of a ventral outgrowth of the gut, arising just behind the stomach, which extends forwards and branches into an immense tree-like mass of tubes welded together by connective tissue into a solid mass extending forwards and nearly obliterating the front part of the body cavity. Whether this organ really performs the same function as the so-called liver in *Amphioxus* is doubtful. It has been proved that the function of the Craniate liver is largely the elaboration of an alkaline fluid called the bile. This is partly excretory in nature, but has an important influence upon the processes of digestion and absorption in the intestine. The main stem of the liver tubes is called the bile-duct; there is often a lateral outgrowth from this which acts as a reservoir for the bile, called the gall-bladder. Besides this, the liver cells can form from the sugar brought to it from the intestine a substance called glycogen, allied to starch in composition, which acts as a reserve of carbohydrate material available for the system as needed. Among other influences which the liver exercises on the chemical processes of the body is the very important one of transforming the nitrogenous waste products into a suitable form (urea or uric acid) for excretion by the kidneys.

Another outgrowth from the intestine arises sometimes just

behind the opening of the bile-duct, sometimes from the duct itself. This outgrowth, like the liver, branches into a tree of tubes which are bound together by connective tissue to form a solid mass, though one of much smaller size than the liver. This organ is called the pancreas and it produces a secretion called pancreatic juice, by which the process of digestion is completed. This juice contains three ferments: these are amylopsin, which converts starch into soluble sugar; trypsin, which, acting only in an alkaline medium, converts proteid into peptone and simpler derivatives; and steapsin, which splits up fat into soluble fatty acids and glycerine. The fatty acids unite with the alkalis present in the mixed contents of the intestine to form soluble soaps, and these are absorbed along with the glycerine, a reconstruction into fat taking place in the intestinal epithelium.

The intestine is always somewhat longer than the body. Hence it must be to some extent looped or twisted (Fig. 207), though this may express itself only in a slight curvature. A fold projecting into it in some forms represents the typhlosole of the Urochordata and is known as the spiral valve, since it shares in the twisting. In the intestine the digested food is absorbed and transferred to the blood-vessels and lymph-canals. The last portion of the intestine is usually of larger diameter than the rest and is called, when thus distinguishable, the large intestine. In it the indigestible material is elaborated into faeces for expulsion by the anus.

The blood system of the Craniate is distinguished by the possession of a large and well-developed heart, which, like the heart of the Urochordata, is an enlargement and specialization of part of the ventral vessel. The space in which it apparently lies—really, into which it protrudes,—is called the pericardium, and is only an anterior part of the coelom shut off from the rest by the development of a transverse septum. The heart is constricted into four chambers, becoming successively more thick-walled as we proceed forwards, and named, beginning from behind, the sinus venosus, the atrium, the ventricle and the conus arteriosus (Fig. 195). It is bent into an S-shape, so that the sinus venosus is dorsal and posterior, the atrium dorsal and anterior, the ventricle ventral and posterior, and the conus arteriosus ventral and anterior. The conus arteriosus leads into the ventral aorta, which gives off the arterial arches; these are branches which ascend between the gill-sacs and ramify on their walls. From the gills the blood collects into epibranchial vessels which join

Circulatory system.

to form two longitudinal vessels on the dorsal wall of the pharynx, the roots of the dorsal aorta. These unite behind the pharynx into a single dorsal aorta, giving blood to all the hinder part of the body. The forward extensions of the two longitudinal epibranchial vessels carry blood to the head and are known as the carotid arteries. There can be little doubt that the impulse leading to the evolution of the heart came from the necessity of having a strong force to drive the blood through the capillary channels on the walls of the gill-sacs.

In the embryos of all Craniates the number of these paired connections between the ventral aorta and the roots of the dorsal aorta is six, but the two anterior pairs, viz., those traversing the wall of the pharynx parallel with those parts of its supporting skeleton known as the mandibular and hyoidean visceral arches respectively (see pp. 371 and 372), are found in adult forms only as remnants in connection with the carotid arteries. Whatever may have been the case in primitive forms, these first two arterial arches have now no part in aërating the blood, this function being performed by the succeeding four pairs of arches, along whose course only are gill-sacs developed. We shall see that the arterial system near the heart is in all groups of Craniata a modification of the six pairs of arterial arches now described.

The fore-limb is supplied by a vessel called the subclavian artery, but the origin of this differs in the several classes of the phylum. In Amphibia, Lizards and Mammalia other than Cetacea, it arises from the epibranchial artery near or behind the sixth branchial arch, while in Crocodiles, Turtles, Birds and Cetaceans its origin is from the ventral end of the third branchial arch. As in both Lizards and Cetaceans these two vessels exist side by side, but only one of them supplying the fore-limb, it is clear that the subclavian arteries are not homologous throughout the group.

Each chamber of the heart is separated from the one behind by valves, which are flaps of membrane free to move in one direction so as to open and admit blood from behind, but restrained by tendinous chords from being driven further back than so as just to meet when the chamber contracts, and thus prevent any backward movement of the blood. In the conus there may be several transverse rows of pocket valves. These valves as their name implies are loose pockets of membrane which are pressed flat against the wall of the conus during the forward movement of the blood, but which when

the conus contracts become filled with blood and swollen out so as to
meet one another and prevent the reflux of blood into the ventricle.

The development of the liver has exercised a profound influence

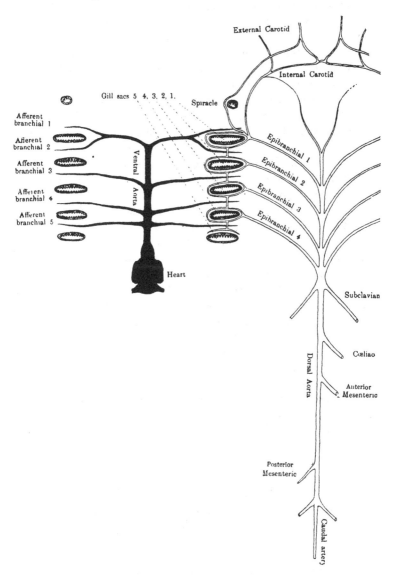

FIG. 194.   Diagram of the ventral and dorsal aortas and their branches in
Scyllium × about 1.

on the afferent part of the blood system corresponding to the hinder part of the sub-intestinal vein of *Amphioxus*. The vast mass of tubes projecting into it has broken it up into a network of capillary channels called the hepatic portal system. In front of this, where it enters the sinus venosus, it is known as the hepatic vein; behind, branches from the walls of the intestine so overshadow the original ventral trunk that this, embedded between the limbs of the spiral valve, appears as merely a small branch of the composite trunk or portal vein.

The blood from the muscles and kidneys, in a word, from the dorsal and outer parts of the coelom, collects into two longitudinal channels called the cardinal veins. These empty into the sinus venosus by transverse trunks called ductus Cuvieri. These transverse trunks divide the veins into anterior cardinals returning blood from the head, and posterior cardinals returning it from the rest of the body. In the tail the two posterior cardinals are represented by the median caudal vein, which further forward splits into two. Just as the course of the original sub-intestinal vein has been obstructed by the growth of the liver, so that of the posterior cardinal has been choked by the growth of the kidney tubes. The blood from the tail and hind limbs is forced to filter amongst these in a series of narrow channels called the renal-portal system. The front part of the vein retains the name posterior cardinal: the hinder part is called the renal-portal vein. Since the kidney tubes also receive blood from the dorsal aorta they, like the liver, have a double supply.

The blood of Craniata has in addition to the ordinary amoebocytes a much larger number of oval or round cells impregnated with haemoglobin, called red blood-corpuscles. Haemoglobin has been mentioned when describing *Lumbricus*, in which worm it is found diffused in the blood fluid. The great characteristic of haemoglobin is its power of forming a bright red, unstable compound with oxygen. This compound is formed in the respiratory organ and carried by the circulation to all parts of the body. In the capillaries it is broken up and the oxygen absorbed by the tissues. The haemoglobin having lost its oxygen changes in colour, and the impure blood which leaves the tissues is in consequence bluish. From the tissues the blood takes up carbon dioxide which, like the oxygen, is conveyed in loose chemical combination, though with the sodium of the blood instead of with the haemoglobin. The carbon dioxide is set free in the respiratory organs.

On page 128 it was pointed out that both blood and connective tissue have been derived from a jelly-like secretion such as is found in Coelenterata. This in the embryo coelomate animal fills up the interstices between ectoderm, endoderm and coelomic sacs, these interstices being collectively termed the primary body-cavity or haemacoel. It was also pointed out there, that whereas in the part of the jelly which was converted into connective tissue a large number of fibres were developed, in the portion destined to form blood, on the contrary, no fibres appeared and the jelly remained fluid, and in consequence the amoebocytes which had wandered into it from the neighbouring epithelia were able freely to move about. In Annelida, Arthropoda and Mollusca certain of the blood-spaces acquire muscular walls derived from the adjacent coelomic sacs, and thereby attain contractility which may be specially localized in a dilatation called the heart. The spaces with muscular walls are the arteries. In Craniata a further differentiation has taken place : we find not only a definite heart and arteries leading away from it, but also equally definite veins leading into it as described above, and arteries and veins are connected with one another by narrow channels called capillaries with well-marked walls. Heart, arteries, veins and capillaries are all lined by a single layer of flattened cells called an endothelium, which has been developed from the flattening out and union of a certain number of amoebocytes. The capillaries possess no other wall, but arteries and veins have outside this a wall of elastic and fibrous connective tissue in which is embedded a zone of circular muscle-fibres. These structures are all derived from the adjacent coelomic sacs. The muscles of blood vessels do not contract rhythmically and spontaneously like those of the heart, but are in a state of continued contraction called tone. This tone is under the control of the nervous system through the medium of special "vasomotor" fibres, and thus the supply of blood to an organ can be varied according to its needs.

In Craniata however, outside the definite arteries, veins and capillaries, there exists a large portion of the haemocoel in the form of irregular channels and interstices, in many cases without definite walls, an endothelium being found only in the larger trunks. This system of spaces is known as the lymphatic system. It contains a clear fluid in which amoebocytes float, but no haemo-globin-containing cells, and at one or several points the main trunks of the system open into the large veins. The finer branches of the system ramify amongst all the organs of the body. There is no circulatory current in the lymph canals except in those belonging to the viscera, but there are valves arranged so that with every contraction of neighbouring muscles some fluid can pass forwards in one direction but not backwards.

It will be seen that in Craniata, unlike Arthropoda and Mollusca, the blood, being everywhere confined to vessels with definite walls, does not directly bathe the tissues of any organ; but that materials must first diffuse through the walls of the blood-vessels into the lymph-spaces before they can reach the tissue. One explanation of the separation of the lymph-system from the blood-system is that the hae-moglobin is not diffused in the fluid of the blood, but is carried in cells which have no power of movement in them-

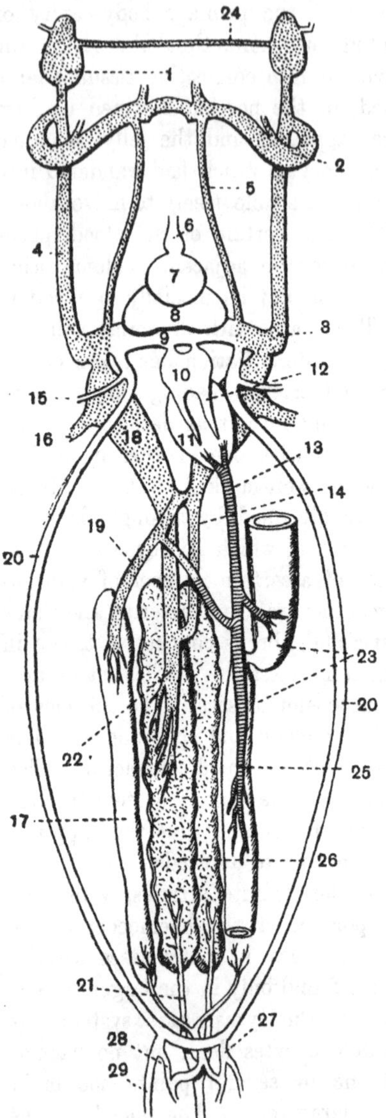

Fig. 195. Diagram of the venous system of *Mustelus antarcticus*. From T. J. Parker.

1. Orbital sinus.   2. Hyoidean vein.   3. Ductus Cuvieri. 4. Anterior cardinal vein.   5. Jugular vein.   6. Conus arteri-osus.   7. Ventricle.   8. Atrium. 9. Sinus venosus.   10. Hepatic vein.   11. Liver.   12. Hepatic vein.   13. Hepatic portal vein. 14. Left cardinal vein.   15. Bra-chial vein.   16. Sub-clavian vein.   17. Gonad.   18. Pos-terior cardinal vein.   19. Sper-matic vein.   20. Lateral vein. 21. Renal-portal veins from caudal vein to kidney.   22. Right posterior cardinal vein. 23. Alimentary canal.   24. Vein connecting orbital sinuses.   25. Sub-intestinal vein. 26. Kidney. 27. Pelvic vein.   28. Cloacal vein.   29. Femoral vein.   30. Caudal vein.

selves   Did these cells enter the lymph-system they would speedily
block its finer channels.

The supply of amoebocytes to both blood and lymph is provided
for by widely distributed actively growing nodules of cells which
bud off amoebocytes into the adjacent lymph-channels.   These
packets of cells are called lymphatic glands: the largest collection
is in the spleen, an organ having several other functions, which is
attached to the mesentery just dorsal to the posterior end of the
stomach.

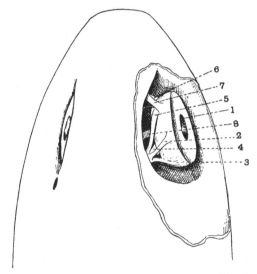

F IG. 196.   Dorsal view of head of *Scyllium canicula* × 1.   The right orbit has
     been exposed so as to show the muscles that move the eye and the second
     and fourth nerves.

1.   Lens of the eye.   2.   Superior rectus muscle of the eyeball.   3.   Ex-
     ternal (or posterior) rectus muscle.   4.   Inferior rectus muscle.   5.   In-
     ternal (or anterior) rectus muscle.   6.   Inferior oblique muscle.
     7.   Superior oblique muscle; the slender nerve entering this muscle is the
     fourth cranial.   8.   Second cranial or optic nerve, the nerve of sight.

The muscles of the Craniata like those of the Cephalochordata
are developed from the inner walls of a series of dorsal
coelomic pockets, in a word, from myotomes.   Unlike
that of the Cephalochordata the trunk coelom does not become at
first completely divided into separate sacs, the ventral portions of
which fuse later.   In the Craniata this stage is skipped in development, and the coelom appears from the first as a pair of elongated
sacs undivided below, but segmented above.   After the complete

Coelom.

separation of the dorsal portions as myotomes the ventral parts of the two sacs unite beneath the intestine, whilst above it their walls become apposed, forming the vertical sheet of tissue known as the mesentery, in which the intestine is slung.

It is necessary of course for the efficient action of the eyes that they should be movable, and this is brought about by the space around the eyeball becoming converted into a cavity called the orbit, which in the lower Craniata is continuous with the anterior cardinal vein, and thus contains blood (Figs. 195 and 196). To each eyeball six muscles are attached, two arising from the anterior part of the orbit and inserted one above and one below the eyeball, and named respectively the superior and inferior oblique; and four arising close together from the posterior corner of the orbit and inserted on the eyeball, one above and one below, the superior and inferior recti, and one antero-laterally, the internal or anterior rectus, and one postero-laterally, the external or posterior rectus.

The proboscis and collar coelomic cavities of the Hemichordata are represented in the Craniata by two cavities found in the embryo on each side, in advance of all the myotomes, termed the head-cavities. The most anterior, termed the pre-mandibular, is joined to its fellow by a narrow canal running underneath the eyes —the pair really constitute a bilobed cavity—from whose walls the inferior oblique, superior, inferior and internal recti muscles are developed.

The collar-cavities are represented by the mandibular-cavities, a pair of long, narrow cavities running down the sides of the mouth and curving up over the eye on each side. From the wall of this portion of the cavity the superior oblique muscle is derived. The external rectus muscle arises from the first myotome. The muscles derived from the anterior head-cavity are supplied by a common nerve, the third cranial; the superior oblique is supplied by the fourth, and the external rectus by the sixth cranial.

Most of the muscles which compress or expand the gill-sacs are derivatives of the wall of the unsegmented ventral portion of the coelom. From the inner wall of this part of the coelom all the muscles of the alimentary canal, which in Craniata are longitudinal as well as circular, arise, as do the muscles in the walls of the blood-vessels. From the myotomes are derived the muscles by which the locomotion of the animal as a whole is carried out. In the lower Craniata these have the same simple arrangement as was

found in the case of *Amphioxus*, but in the higher forms where the
movements are complicated by the development of limbs, these
muscles are divided into numerous bundles with a very complex
arrangement. All the muscles derived from the myotomes are
composed of striated fibres. Most of those governing the movements of the alimentary canal and blood-vessels are composed of
smooth fibres, but to this statement the muscles of the heart form
an exception.

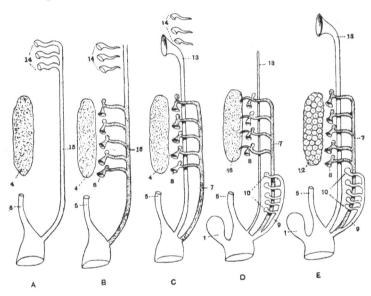

Fig. 197. Diagrams illustrating the development of the urino-genital organs
of Craniata. [For full explanation see sections on Elasmobranchii and
Amphibia.]

A. Development of pronephros and segmental duct.   B. Atrophy of pronephros, development of mesonephros.   C. Differentiation of pro- and
mesonephric ducts.   D. Development of metanephros, male type.
E. Female type.   1. Allantoic bladder.   4. Gonad.   5. Intestine.
7. Mesonephric duct.   8. Nephrostome.   9. Metanephric duct.
10. Metanephros.   12. Ovary.   13. Oviduct.   14. Pronephros.
15. Archinephric duct.   16. Testis.

The excretory and reproductive organs are closely related
                        in development, and by recent research their relation
Urino-genital   to those of the Cephalochordata has been made
organs.   tolerably plain. The unit in the excretory system
is a tube opening into the body-cavity at one end and at the other
into a longitudinal duct which opens into the proctodaeum behind.

If it opened to the exterior directly it would be essentially identical with the **nephridium** which constitutes the excretory organ of Annelida and Mollusca. Of these tubes in the Craniata there are two kinds: the first or **pronephric tubules**, called collectively the **pronephros**, develope in continuity with the duct into which they open. This is called the **archinephric duct**, or sometimes the **Wolffian duct**, after Caspar Wolff, who first saw it develope in the embryo. The pronephric tubules are situated at the upper and outer angle of the unsegmented coelom, a position exactly corresponding to that of the nephridia of *Amphioxus*, except that in the Craniata they are not developed in the shortened branchial region but immediately behind it. From the root of the mesentery opposite the inner openings of the pronephric tubule a swelling projects into the body-cavity. It is covered by a thin layer of peritoneum and is richly supplied with blood-vessels by branches from the aorta. The cells of the peritoneum appear to extract water and excreta from the blood and pour it into the coelom, whence they are excreted by the pronephric funnels. This structure is termed a **glomerulus**.

If the homology of the two sets of organs be accepted—and in view of Goodrich's researches it must be regarded as doubtful—the somewhat startling conclusion follows that the atrial cavity of *Amphioxus* must be the homologue of the archinephric duct. This conclusion, however, must not be expressed in the form that the archinephric duct is derived from the atrial cavity, but rather that both seem to be developments of a primitive groove, overhung by a longitudinal ridge which may be called the **Wolffian ridge** and which corresponds to the atrial fold in *Amphioxus*. This view is to some extent supported by recent observations on the development of the archinephric duct. It appears probable that this arises as a solid ingrowth of ectoderm, a method of development which, from the study of other cases where it occurs, may legitimately be regarded as a modified form of invagination or intucking of ectoderm. The exceptional development of the Wolffian ridge in *Amphioxus*, so as to form a veil over the gills and wall in the atrial cavity, is perhaps advantageous in a burrowing animal. Van Wijhe, to whom we owe the first clear account of the head cavities, is a strong supporter of the view of the ectodermal origin of the archinephric duct just mentioned. It ought however in justice to be mentioned that other observers do not accept this view and confidently assert that the archinephric duct is derived from the mesoderm. They regard it as

developed out of a series of secondary connections between the
nephridia just as the ureter is formed, and similar to connections
between the nephridia found in some Polychaeta.

The pronephric tubules are developed only in the larval form,
and so far as is known they persist in no adult Craniate. They
become replaced by the second kind of tubules, termed the mesonephric tubules, which, like the pronephric, open into the coelom,

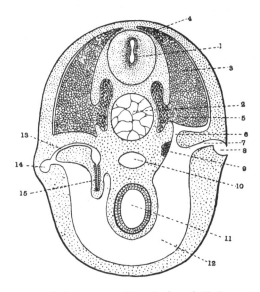

Fig. 198. Diagrammatic transverse section of a hypothetical ancestral Elasmobranch to show origin of renal and genital organs. On the left side a later
condition is shown than in the right.

1. Nerve-cord.   2. Notochord.   3. Myotome.   4. Dorsal sclerotome.
5. Ventral sclerotome.   6. Nephrotome.   7. Pronephric tubule.
8. Archinephric duct as an open groove.   9. Rudiment of genital
gland.   10. Aorta.   11. Alimentary canal.   12. Unsegmented
coelom.   13. Mesonephric tubule.   14. Archinephric duct.
15. Seminal tubule.

but unlike these swell out immediately beyond this, forming thin-
walled capsules, termed the Malpighian capsules, into which a
glomerulus, that is a thin-walled plug containing a plexus of
blood vessels, projects. Its function is similar to that of the
glomerulus opposite the pronephric tubules.

The mesonephric tubules are developed from the necks connecting the myotome with the general coelom, that is, from the lower
end of the myotome, a position corresponding to the point of origin

of the genital organs in *Amphioxus*. These necks, which may be termed nephrotomes, break loose from the myotomes, curve round and acquire openings into the archinephric duct. It is the central part of the tubule which swells out to form the Malpighian capsule. It has been suggested that the mesonephric tubules are the homologues of the genital sacs of *Amphioxus*, and this suggestion receives support from the fact that the genital organs in the lower Craniata originate from the lower ends of the nephrotomes, just where these enter the general coelom, and consequently the genital organs are at first segmented. The change of function suggested above seems at first sight a little forced, but we must remember that in primitive Vertebrata, as in Annelida, the whole coelomic wall had probably an excretory function, and hence if permanent instead of temporary pores for the discharge of genital products were formed, these pores would allow the escape of excreta thrown out by the neighbouring portions of the coelom, the excretory function of which would therefore be stimulated.

The genital rudiments soon coalesce to form continuous ridges projecting into the general coelom on either side of the root of the mesentery. In the female the sexual cells or ova, when ripe, drop into the coelom and are conveyed to the exterior by an oviduct opening into the body-cavity with a wide funnel. The oviduct arises from a groove in the dorsal coelomic wall, the edges of which meet. Posteriorly it opens into the proctodaeum. In the male, on the other hand, the sexual cells arrange themselves in tubes, the seminiferous tubules, which retain in most Craniata a permanent connection with certain of the mesonephric tubules through a set of tubes called the testicular network which like the organs they connect are derived from the nephrotomes. The archinephric duct therefore acts as vas deferens or male duct. Since through the opening of the proctodaeum not only the faeces are expelled but also the excretory products and spermatozoa from the archinephric duct, and ova from the oviduct, this aperture has received the name of cloaca, the Roman name for 'sewer.'

Division I. CYCLOSTOMATA.

The Craniata are divided into two main groups, namely, the Cyclostomata and the Gnathostomata. The former division, distinguished by the absence of true visceral arches and of jaws, includes at the present day only a few, probably degenerate, worm-like animals, with short tails like *Amphioxus*, and with naked skins. The name Cyclostomata means Round-mouthed (Gk. κύκλος, a circle ; στόμα, mouth), and alludes to the circumstances that the

Structure.

edges of the mouth are stiffened by a ring-shaped cartilage, the annular cartilage, so that the mouth cannot be closed (2, Fig. 201). There is a piston-shaped tongue supported by a lingual cartilage, and the whole is protruded by

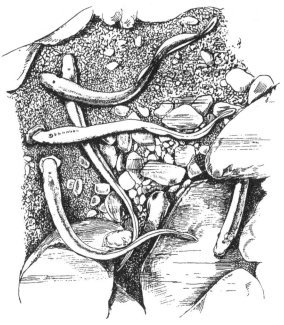

Fig. 199.   The Musk Lamprey, *Petromyzon wilderi*, in the act of spawning. From Bashford Dean and Sumner.

a muscle attached to the annular cartilage in the lips (Fig. 200). Both the tip of the tongue and the walls of the stomodaeum are beset with horny teeth, developed from the agglutinated cells of the skin. The expansion of the stomodaeum causes the mouth to act like a sucker, and the whole animal is thus enabled to adhere to some foreign body, such as a stone, or to some victim, usually a fish, in which case the rasp-like tongue works a hole in the flesh of the prey. The stomodaeum is greatly elongated and is supported in its roof by several broad cartilages, the so-called labial cartilages ; in consequence the eyes and gill-slits appear to be pushed very far back.

The condition of the sense-organs is one of the most marked characteristics of the Cyclostomata. The nose is represented by a

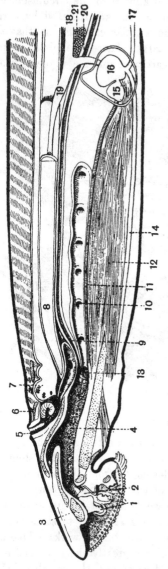

FIG. 200. The Lamprey, *Petromyzon marinus*; right half of a female specimen, the various structures being, for the most part, seen in longitudinal section (nat. size).

1. Horny teeth on inside of stomodaeum. 2. Mouth. 3. Dorsal half of annular cartilage in section. 4. Mouth cavity. 5. Nasal aperture. 6. Nasal capsule lying over nasal sac. 7. Brain. 8. Notochord. 9. Oesophagus. 10. Respiratory tube. 11. Ventral aorta. 12. Retractor of tongue. 13. Velum. 14. Inferior jugular vein. 15. Ventricle. 16. Auricle. 17. Hepatic vein. 18. Posterior cardinal vein. 19. Anterior cardinal vein. 20. Intestine. 21. Ovary.

single sac placed far back in consequence of the elongation of the stomodaeum, as above explained. This single sac is drawn out

into a long tube passing beneath the brain, and in one Order,
the Myxinidae or Hag-fishes, this opens into the roof of the
stomodaeum. The tube-like prolongation is really the pituitary
body, which in the embryo developes close to the nasal sac. The
groove connecting the two organs becomes closed so as to form a
canal, and then by the great development of the suctorial mouth
the external openings of the two organs are widely removed from
one another.

The eye developes no proper cornea or aqueous humour, the
lens remaining in connection with the skin. The ear is represented
either by two semicircular canals and a vestibule (or sacculus)
in the Lamprey, or by a single membranous tube in the Hag-fish.

The gill-slits, usually seven in number, have the form of regular

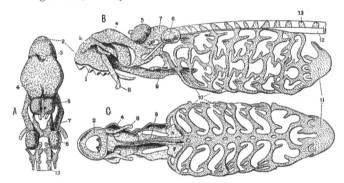

FIG. 201. A, dorsal; B, lateral, and C, ventral view of the skull of *Petromyzon
marinus* × 1 (after Parker).

1. Horny teeth.   2. Annular cartilage.   3. Anterior labial cartilage.
4. Posterior labial cartilage.   5. Nasal capsule.   6. Auditory capsule.
7. Dorsal portion of trabeculae.   8. Lateral distal labial cartilage.
9. Lingual cartilage.   10. Branchial basket.   11. Cartilaginous cup
supporting pericardium.   12. Sheath of notochord.   13. Anterior
neural arches fused together.

gill-sacs, recalling those of the Hemichordata, only without the
tongue-bars. The external opening is circular; the connection
with the gullet, on the other hand, is a vertical slit (Fig. 200).
The whole set of sacs is supported on a framework of cartilage
consisting of longitudinal dorsal and ventral bars, connecting cross-
pieces pass between the sacs and give off branches encircling their
outer openings (Fig. 201). The whole of the branchial basket,
as it is called, is a development of the dermis and has nothing to
do with the visceral arches of the Craniata, as will be shown
later.

The commencement of the true alimentary canal is marked, as in *Amphioxus,* by a velum. What corresponds to the hyper-pharyngeal groove in that animal is in many species of Cyclostomata completely constricted off from the remainder of the gullet and is known as the oesophagus, though this word is used in a different sense from that in which it is used in the case of the Gnathostomata. The lower part of the gullet which communicates with the gill-slits ends blindly behind and is called the respiratory tube.

The hinder part of the alimentary canal is a nearly straight tube, the spiral valve having a very slight deviation from a straight course. There is no dilatation of any kind in its course. The large liver empties its secretion by the bile-duct, which opens into the intestine a short distance behind the branchial region.

The skull consists of the simplest elements, viz. the trabeculae, with a wide hole for the infundibulum, and the parachordals, forming only a slender arch over the hinder part of the brain, but developing a low side wall throughout their extent with which the simple auditory capsule is fused. The nasal capsule is represented by cartilage stiffening the nasal tube. The brain is remarkable for having a thin membranous roof except just at the front end of the hind brain where a narrow band of nervous matter represents the cerebellum.

The only fins present consist of a fringe of skin similar to that found in *Amphioxus* surrounding the hinder end of the body in the vertical plane. This fringe is divided by a notch into an anterior (or dorsal) and a caudal fin. The dorsal fin is supported by cartilaginous rays situated above the neural arches which protect the spinal cord; the caudal fin has, in addition to these, rays situated below the haemal arches. A caudal fin of this description, which the notochord divides into two equal lobes, is called diphycercal.

Besides the neural arches (13, Fig. 201) and small haemal arches in the tail no other cartilage is developed in connection with the axial skeleton, the notochord with its thick fibrous sheath persisting unchanged throughout life.

The pericardium is not completely separated from the remainder of the body-cavity, and the genital organs take the form in both sexes of a single median ridge projecting into the body-cavity (21, Fig. 200). No connection of the testis tubules with the kidney tubules exists, nor is there any trace of an oviduct, both ova and sperma-tozoa being freely shed into the body-cavity and escaping by two abdominal pores or simple openings in the body-wall placed ventrally to the openings of the kidneys. Inasmuch as these latter open directly to the exterior and are quite independent of the opening of the intestine, which is placed more ventrally, we may state that no cloaca has yet been developed.

Living Cyclostomata, represented by a single class which may be called Marsipobranchii (Gr. μάρσῖπος, a pouch), are divided into two families : (i) PETROMYZONTIDAE, (ii) MYXINIDAE.

Classification.

(i) In the first family, familiarly known as the Lampreys, the pituitary body appears as a blind process from the nasal sac : each gill-sac opens directly to the exterior, and the hyperpharyngeal groove is separated from the rest of the alimentary canal as a distinct tube, the so-called oesophagus.

The Lampreys (*Petromyzon*) are conspicuous in the early spring, when they ascend small brooks to spawn. Several species inhabit the rivers of Great Britain, Canada and the United States, but the differences between them are trifling, depending mainly on the development of the horny teeth covering the tongue. One species, *Petromyzon marinus*, attaining a much larger size than the others, inhabits the sea. It may reach a length of three feet, whereas the other forms do not grow longer than from ten to twelve inches. The eggs of Lampreys develope into a most interesting larval form which stands in many respects nearer to the other Craniates than does the adult and supplies an intermediate stage between *Amphioxus* and an ordinary Craniate. This larva is called the Ammocoetes, and its mode of life resembles on the whole that of *Amphioxus*. Like that animal the Ammocoetes lives on what the currents of water, produced by the cilia inside the velum, bring. The thyroid gland, which, as we have seen, represents the endostyle, remains open, and still performs its primitive function of secreting a cord of mucus, which is carried up dorsally by a ciliated groove, the peripharyngeal band, situated just behind the velum. The hyperpharyngeal groove is represented by a dorsal strip of ciliated cells, the current produced by which sweeps the mucus backward into the alimentary canal just as it does in *Amphioxus*.

The tubular suctorial stomodaeum is represented by a hood-like upper lip and a distinct short under lip, and when the mouth is contracted the velum is produced into tentacles just as in the Urochordata and in *Amphioxus*. The lateral eyes are exceedingly rudimentary, but there is a large pineal eye, and the nasal sac has a median septum.

(ii) The MYXINIDAE are characterized by the persistent connection of the pituitary body with the stomodaeum, so that there is a tube leading from the nasal sac to the mouth. There are eight tentacles called barbels at the sides of the mouth, and there is no special oesophagus distinct from the rest of the gullet. The skin has a double series of mucous glands placed at the sides of the body, and so much mucus can be thrown out that a large amount of water can be rendered semi-solid. The intestine has no spiral valve. The Myxinidae are the animals known as Hag-fish. They adhere to fish on whose flesh they feed, but, unlike the Lampreys, they can actually burrow into their victims so that the stomodaeal region is completely buried. In connection with these habits the stomodaeal region is enormously elongated, and the eyes remain in a rudimentary condition, whilst the gill openings are pushed very far back.

The Myxinoids include two genera, *Bdellostoma* and *Myxine*. In the former, which is a genus inhabiting the southern Atlantic and Indian Oceans, the gill-sacs are seven in number on each side and open separately ; in *Myxine*, on the other hand, each external opening of the six gill-sacs is drawn out into a long tube, and the tubes of each side curve back and unite to open by a common atrial pore placed so far back that the animal can insert almost half its length into the body of its victim without interfering with its breathing. The portal vein is rhythmically contractile. *Myxine* is common on both the Atlantic and Pacific coasts of North America and on the European coast.

Division II. GNATHOSTOMATA.

The great division of the Gnathostomata includes all the remaining Craniata, and is characterised by the development of definite visceral arches, jaws and paired limbs. The visceral arches are jointed rods developed from the inner or splanchnic wall of the coelom ; they cannot therefore be considered as corresponding to the branchial basket of Cyclostomata. They are placed in the forms which retain gill-slits between these openings, and hence are often called gill-bars. The first pair of visceral arches lie in the sides of the mouth, and consist on each side of two pieces, hinged on one another and called the upper and under jaws respectively. By the motion of these on one another the mouth can be opened and closed. The nose is always represented by two sacs and the ear has three semicircular canals.

The Gnathostomata are divided into five classes. In the first three of these the temperature of the body varies with that of the surrounding medium.

Class I. PISCES.

Gnathostomata with fins supported by fin-rays and breathing chiefly by gills.

Class II. AMPHIBIA.

Gnathostomata with pentadactyle or five-fingered limbs and without fin-rays. Gills and gill-slits functional in the young but generally entirely lost in the adult. An amnion is not formed in the embryo. The skin is soft and moist.

Class III. REPTILIA.

Gnathostomata with pentadactyle limbs. The young are born similar to the adult and in embryonic life develope an amnion. Skin with horny scales.

Class IV. AVES.

Gnathostomata agreeing with Reptilia in most points, but having a constant temperature independent of that of the surrounding medium : the skin is provided with feathers instead of scales and the fore limb is used as a wing.

Class V. MAMMALIA.

Gnathostomata agreeing in many points with Reptilia, but clothed with hair instead of scales. The body, like that of Aves, has a constant temperature independent of that of the surrounding medium. The young are nourished after birth by the secretion of certain glands of the mother termed milk glands or mammary glands.

CHAPTER XV.

Sub-Phylum IV. Craniata.

Class I. Pisces.

THE class Pisces, or true Fishes, are not, as many would imagine,

Characters. characterized by their gills (since some Amphibia retain these throughout life), but by their fins. In addition to the vertical flap of skin with which we have become acquainted in the case of the Cephalochordata and the Cyclostomata, we find typically two pairs of lateral flaps, an anterior pair called the pectoral fins, and a posterior pair known as the pelvic fins (Figs. 205 and 206). Both from a study of their development and their condition in the oldest fishes, it is believed that the paired fins are derived from the division of two originally continuous lateral flaps, of which the intermediate portions have disappeared. In the embryo the remains of these ridges are known as the Wolffian ridges, which can be with some probability identified with the flaps overhanging the groove that, as is held by some, becomes converted into the primitive kidney duct. If this be so, we have representatives of the lateral fins in the walls of the atrial cavity of *Amphioxus*, of which the rudimentary folds known as the meta-pleural folds form part. If we accept this view it follows, since Cyclostomata possess a kidney duct, that they once possessed either a continuous lateral fin or the two pairs possessed by modern fishes.

While the possession of paired fins discriminates Pisces from the lower Vertebrata, the forms of these members equally sharply mark Pisces off from the class with which they are most nearly allied, namely, Amphibia. In all Pisces the limb or fin is a blade-like organ which never exhibits the slightest resemblance to the typical form familiar to all in the human limb, but Amphibia have as the representative of the paired fins limbs in which the plan of the human arm and leg can be at once recognised. The blade-like type of fin is known as the ichthyopterygium (ἰχθύς, a fish; πτερύγιον, a little wing), the other type of limb as the cheiropterygium (χείρ, the hand). Pisces therefore are defined by the possession of ichthyopterygia.

The median and the paired fins are stretched on a skeleton with a two-fold origin, (i) a series of horny or cartilaginous rods or pterygiophores which support the basal part of the fins, and (ii) a series of horny fibres or bony dermal fin-rays which support the distal part of the fins. In most Orders of fishes these are not externally discernible as they are covered by muscles and skin, but in the median fins of the Teleostomi the former series have sunk into the body, and the dermal fin-rays being covered with a thin translucent skin without scales become more or less apparent.

The class Pisces is divisible into four Orders, namely, the Elasmobranchii, the Holocephali, the Dipnoi and the Teleostomi. These Orders can also be distinguished in the case of fossil fish, though the differences become less marked as we proceed backward in time, and many indications point to the conclusion that the common ancestors of the four Orders would, if we could examine them, be classed as Elasmobranchii. Hence the Elasmobranchii may be termed the basal group of the Pisces, although modern Elasmobranchii, like all modern animals, are specialized in many respects.

Order I. Elasmobranchii.

The Elasmobranchii are distinguished (i) by not possessing any gill-cover or operculum, as it is called, each gill-sac opening separately to the surface; (ii) by the absence of an air-bladder opening into the alimentary canal; and (iii) by the absence of large bones in the skeleton. In addition to these negative characters, they are distinguished (iv) by the possession of a peculiar scale quite characteristic of the Order.

Characters.

This scale, the so-called placoid scale, consists of a little spike attached to a small plate at its inner end. The plate consists of true bone: the spike of a modification of true bone called dentine. Dentine is distinguished from bone by possessing no Haversian canals or spaces occupied by blood-vessels, nor even lacunae, since the cells of the connective tissue, out of which it is formed, remain external to the dentine. Their protoplasmic processes known as dentine fibres do however

Placoid scales.

penetrate it and give rise to canals called dentinal canals. The core of soft connective tissue is called the dentinal pulp.

The spike therefore may be described as a little wart of dermis calcified on the outside. It pushes the ectoderm before it, and it becomes encrusted with crystals of carbonate of lime forming the enamel layer (Fig. 202). These closely set crystals are secreted by the inner or basal ends of the ectoderm cells. One would naturally expect that structures like scales, which are closely arranged all over the body, would also invade the stomodaeum, which is merely a part of the skin. This we find to be the case, but here the scales are very greatly enlarged in size and changed in function; they are the well-known teeth which are used for the purpose of retaining and lacerating prey which has been seized. The spike of the tooth is usually flattened and blade-like, and provided with strongly serrated

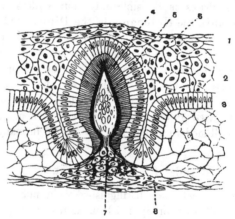

Fig. 202. Section through the skin of an Elasmobranch showing formation of a dermal spine. Highly magnified.
1. Horny layer of ectoderm.   2. Malpighian layer.   3. Columnar cells of ectoderm secreting 4.   4. Enamel. 5. Dentine (black). 6. Dentinal pulp. 7. Bony basal plate. 8. Connective tissue.

edges. Fusions of several teeth can occur. The teeth are developed in a deep fold of skin, part of the stomodaeum, situated just inside the lower jaw, and usually speaking only the outermost row are in use at one time, the skin working forward the next set as each row wears out.

The skull is much better developed than is the case in the Cyclostomes. In the cranium the parachordals and trabeculae give rise to a firm continuous plate, in which the pituitary fossa is reduced to a minute hole; there is a

Skull.

high and well-developed side wall and the roof extends a long distance forward. The sense-capsules, nasal and auditory, are well developed and firmly united with the cranium. The eyes are large and highly developed, and the side wall of the cranium is indented to make room for the spacious orbits in which the eyes move. There is a considerable part of the head in front of the brain, which usually also projects in front of the mouth. This is the rostrum or snout, and it is supported by three cartilaginous rods, one ventral and two dorsal, projecting from the front end of the cranium. These rods are the forerunners of the ethmoidal region in other forms. In most species the opening of the nasal sac is connected with the mouth by a groove called the oro-nasal groove (Fig. 207). There are usually six gill-clefts and seven visceral arches in Elasmobranchs. The first cleft, sometimes called the spiracle, is rudimentary and in some cases entirely absent. On the other hand there is one family, the Notidanidae, with two extra clefts behind, so that there are in all eight clefts and nine visceral arches in this family.

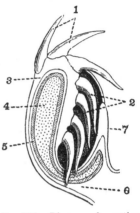

Fig. 203. Diagram of a section through the jaw of a Shark, *Odontaspis americanus*, showing the succession of teeth. From Reynolds.

1. Teeth in use.   2. Teeth in reserve.   3. Skin.   4. Cartilage of the jaw.   5. Encrusting calcification of cartilage. 6. Connective tissue.   7. Ectoderm lining the mouth.

The first pair of visceral arches form as we have seen the jaws. The upper jaw is known as the palato-pterygo-quadrate bar, a compound appellation derived from the names of the bones by which it is represented in the higher forms: the term is sometimes shortened to pterygo-quadrate. The lower jaw is called Meckel's cartilage or mandibular bar: in front a strong ligament, the so-called ethmo-palatine ligament, attaches the upper jaw to the skull. The second pair of arches is spoken of as the hyoid, and this too is divided into two portions, an upper, the hyomandibular, which is firmly connected to the cranium just below the auditory capsule, and a lower, the ceratohyal (Fig. 204). The upper jaw is connected with the cranium either directly, by articulation with the cranium in front of the auditory region, an arrangement called autostylic and prevailing among recent Elasmobranchs only in

24—2

the Notidanidae ; or the upper jaw has lost its articulating process with the cranium and is instead firmly connected or slung by ligament into the hyomandibular, which thus suspends the jaw from the skull. This arrangement, called hyostylic, is that seen in the majority of fishes. A modification, termed amphistylic, occurs in *Cestracion*, where the jaw is slung by the hyomandibular but also has acquired direct articulation with the skull behind the orbit. The remaining visceral arches have only a muscular connection with the skull and are termed the branchial arches, since to their sides

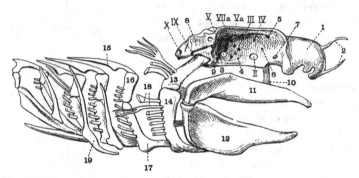

Fig. 204.   Lateral view of the skull of a Dogfish (*Scyllium canicula*) × ⅔.   From Reynolds.

1. Nasal capsule.   2. Rostrum.   3. Interorbital canal for the passage of a blood-vessel.   4. Foramen for hyoidean artery.   5. Foramen for the exit of the ophthalmic branches of Vth and VIIth nerves.   6. Foramen through which the external carotid leaves the orbit.   7. Orbito-nasal foramen which allows a blood-vessel to reach the nose.   8. Auditory capsule.   9. Foramen through which the external carotid enters the orbit. 10. Ethmo-palatine ligament.   11. Palato-pterygo-quadrate bar. 12. Meckel's cartilage.   13. Hyomandibular.   14. Cerato-hyal. 15. Pharyngo-branchial.   16. Epi-branchial.   17. Cerato-branchial. 18. Gill-rays ; nearly all have been cut off short for the sake of clearness.   19. Extra-branchial.   II. III. IV. V. Va. VIIa. IX. X. foramina for cranial nerves.

are attached the gills. The branchial arches are jointed into several pieces, which are placed in an oblique position and so arranged that when they are raised by the levatores arcuum—muscles attaching them to the skull—they diverge and expand the gill-sacs lying between them. The segments of each branchial arch are typically four in number, named respectively pharyngo-branchial, epi-branchial, cerato-branchial, and hypo-branchial. The first-named are situated in the dorsal wall of the pharynx and are horizontal in direction ; the epi- and cerato-branchial stiffen the sides of the pharynx—the cerato-branchial being the main portion of the arch, whilst the hypo-branchial pieces are found in the ventral wall of the pharynx and converge to unite in a median plate, the basi-branchial. To the cerato-branchials are attached a number of thin rods of cartilage which run outwards in the wall of the gill-sac and are called gill-rays. Lying outside the visceral arches are a varying number of cartilaginous rods. Those situated at the sides of the gape are called labial cartilages, those external to the hinder visceral arches extra-branchials (19, Fig. 204). They are equivalent to gill-rays which have become detached from the arches.

The first gill-slit, called the spiracle, is situated between the jaw and the hyoid just outside the internal ear (Fig. 208). It is a narrow tube, and its use in the more typical forms appears to be to allow vibrations to come more closely in contact with the ear, and in some cases to admit the water for breathing. The other slits are really flattened sacs, the walls of which are raised up into thin folds richly supplied with blood-vessels, which are the true gills and are supported by the gill-rays. A rudimentary gill, the pseudobranch, is sometimes developed on the front wall of the spiracle. No gill is developed on the posterior wall of the last gill-sac.

In Elasmobranchs we find, as in Cyclostomata, well-developed dorsal (or neural) and ventral (or haemal) arches, with their ends deeply embedded in the thick sheath of the notochord. This sheath has been converted into cartilage by amoebocytes wandering into the gelatinous layer secreted by the cells of the notochord, and it is divided into separate pieces called centra. Between the centra the sheath remains membranous, and in the middle of each centrum the notochord becomes very much narrowed, so that instead of being a uniform rod it is like a row of beads. The haemal arches meet beneath in the tail, but further forward they stretch out horizontally and become jointed; their outer segments are the ribs, this is the first appearance of these organs. There are usually twice as many neural arches as there are centra, and every alternate one is small and does not meet its fellow, and hence is called an intercalary piece: the haemal arches are as numerous as the centra. The cranium, visceral arches and centra are all strengthened by a calcareous deposit in the ground substance of the cartilage. This calcified cartilage is to be carefully distinguished from true bone, represented in Elasmobranchs by the bases of the scales and teeth.

The primitive tail-fin of Vertebrata, as we have seen, is a fringe

surrounding the end of the tail. Only a small and narrow remnant of this persists in Elasmobranchs, the whip-like end of the
tail being bent up; beneath it there is a well-marked fin, and this
together with the remains of the primitive caudal fin constitute a
secondary tail-fin, which is now denominated heterocercal, since
the axial skeleton does not divide it into two equal parts (Fig. 208).

The paired fins are attached to hoops of cartilage (the limb
arches), called respectively the pectoral and pelvic
girdles, the pectoral being situated just behind the
last gill-cleft, the pelvic just in front of the anus. The pectoral

Limbs.

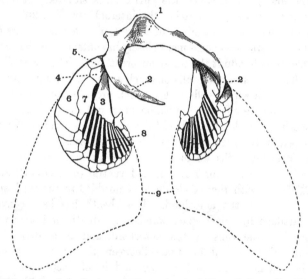

Fig. 205. Dorso-lateral view of the pectoral girdle and fins of a Dogfish,
Scyllium canicula, × ⅔. From Reynolds. The gaps between the radialia
are blackened.

1. Hollow in the midventral part of the pectoral girdle which supports the
   pericardium.   2. Dorsal (scapular portion) of pectoral girdle.   3. Metapterygium.   4. Meso-pterygium.   5. Pro-pterygium.   6. Pro-pterygial
   radial.   7. Meso-pterygial radial.   8. Meta-pterygial radial.   9. Outline of the distal part of the fin which is supported by horny fin-rays.

girdle extends a considerable distance up the side of the animal:
the pelvic is little more than a transverse bar. The fins in modern
Elasmobranchs are of what is called the uniseriate type, that is
to say, there is a thick jointed main axis with cartilaginous rays
attached only to its anterior border. Fossil Elasmobranchs show in
one case, *Pleuracanthus*, a biseriate fin with rays attached to both

borders; and in another, *Cladoselache*, a still more primitive condition, where the fin is merely a lateral flap supported by parallel bars of cartilage. By the coalescence of these at the base the axis was formed, and later by the disappearance of the rays on one side, the uniseriate fin.

In the pectoral fin the basal portions of some of the rays coalesce to form two large cartilages called propterygium and mesopterygium, whilst the axis itself is called the metapterygium. In the pelvic fin of the male the axis bears distally a grooved rod which is termed the clasper, and is used in transferring spermatozoa to the female. The axis is called the basipterygium. The distal joints of the rays in both pectoral and pelvic fins are. made up of numerous small cartilages called radialia.

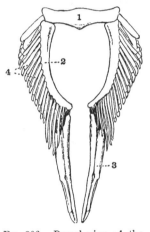

The brain of Elasmobranchs is re-
Brain.
markable for the great development of the olfactory lobes, which are in close contact with the nasal sac and are attached by a narrow stalk to the cerebrum. This is only imperfectly divided into two hemispheres and has nervous tissue on its roof as well as its floor. The cerebellum is developed into a great flap which projects back and covers the thin roof of the medulla oblongata (Fig. 188). It has also lateral outgrowths called cerebellar lobes.

Fig. 206. Dorsal view of the pelvic girdle and fins of a male Dogfish, *Scyllium canicula*. From Reynolds.

1. Pelvic girdle.   2. Basipterygium.   3. Clasper.
4. Radialia.

The alimentary canal is considerably longer than the body and is consequently folded. It has, as a matter of fact, a U-shape: the first limb and a part of the next constitute the
Alimentary Canal.
stomach, which is marked off from the intestine by a constriction and a powerful development of the circular muscles forming a sphincter or circular muscle. To the posterior aspect of the loop is attached the prominent spleen. The intestine, although outwardly straight, is probably derived from a corkscrew coil by the adhesion of successive turns: for the "spiral valve" which, as we said, is merely a ventral unfolding, has a very

strongly marked spiral course. The liver opens by the bile-duct into the beginning of the intestine, and close to its opening is situated that of the duct of the pancreas. A small gland of unknown function, the rectal gland, opens into the hinder end of the intestine.

The pericardium is almost completely separated from the rest of the coelom, communicating only by two narrow holes with it. The heart has the typical structure described in the last chapter (see p. 349). In the conus there are at least two transverse rows of pocket-valves, occasionally more. The arterial arches arising from the ventral aorta run up between successive-gill sacs and break up into capillaries on the surface of the gills: from these the blood is collected by vessels in the form of loops completely surrounding the gill-sacs. From these loops four pairs of epibranchial vessels arise and run backwards in the dorsal wall of the pharynx converging to form the single dorsal aorta, which supplies blood to all the hinder part of the body. The last gill-sac has a gill only on its anterior border; the blood from this does not reach the dorsal aorta directly but is connected by a transverse vessel with the loop surrounding the preceding gill-sac. The dorsal aorta gives off on each side a subclavian artery to the pectoral fin and then four median arteries which run down through the mesentery and supply the alimentary canal. These are named the coeliac, anterior mesenteric, lieno-gastric and posterior mesenteric arteries respectively (Fig. 194). The most anterior, the coeliac, has two important branches, (1) one supplying the liver and the proximal part of the stomach with arterial blood, and (2) the other supplying the anterior part of the intestine and the pancreas. The anterior mesenteric artery supplies the greater part of the intestine and sends branches to the reproductive organs. The lieno-gastric supplies the posterior part of the stomach and the spleen and part of the pancreas. The posterior mesenteric supplies the rectal gland. After giving branches to the genital organs, kidneys and pelvic fins, the aorta continues its course into the tail as the caudal artery. From the two most anterior branchial loops a pair of vessels arise running forward in the dorsal wall of the pharynx and at the same time converging. These are the common carotid arteries, which supply blood to the head. Each divides into two main branches, an external carotid, which pierces the floor of the orbit and supplies the eye and the jaw, and an internal carotid, which pierces the floor of the skull near the middle line and supplies the brain. The pseudo-branch on the front wall of the spiracle receives its blood from the hyoidean artery which, branching from the loop surrounding the first gill-sac, runs forward in the roof of the mouth parallel with the common carotid artery and eventually joins the internal carotid. In the venous system the anterior portion of the sub-intestinal vein is represented by a pair of hepatic veins returning the blood from the liver, opening into the sinus venosus close to the middle line, whilst the posterior portion has dwindled to a small vein embedded between the folds of the spiral valve; this however is joined by branches from the sides of the intestinal wall to form the main trunk of the portal vein. Both anterior and posterior cardinal veins are represented by wide, somewhat irregular spaces. Each anterior cardinal has an expansion called the orbital sinus which surrounds the eye. The two orbital sinuses communicate by an interorbital canal tunnelled in the base of the skull. The blood from the ventral sides of the gill-sacs and pharynx is returned to the Ductus Cuvieri by a pair of independent trunks called the jugular veins. These are each connected with the anterior cardinal vein of its side by the hyoidean vein lying in a groove on the hyomandibular cartilage (Fig. 195). The blood from the tail is returned by a median caudal vein lying beneath the caudal artery and like it enclosed between the centra and the united ventral ends of the haemal arches. At the level of the posterior end of the kidneys the caudal vein divides into the two renal portal veins lying on the outer edges of the kidneys. These veins, as has been already explained (see p. 352), are the hinder portions of the posterior cardinal veins which break up into the renal portal system of capillaries. These filter amongst the kidney tubules and reunite on the inner side of the kidney to form the spacious posterior cardinal sinuses, as the front portions of the posterior cardinals are named. These two sinuses lying ventrally to the kidneys partly coalesce. Each sinus curves forwards and outwards to join the Ductus Cuvieri and at this point it is met by the large subclavian vein returning blood from the pectoral fin. The pelvic vein receives the blood from the side of the cloaca by the cloacal vein and the blood from the pelvic fin by the femoral vein. It then opens into a longitudinal trunk, called the lateral vein, which runs along the side of the body beneath but parallel to the posterior cardinal vein. The lateral vein in front receives the brachial vein from the ventral side of the pectoral fin (not to be confounded with the subclavian

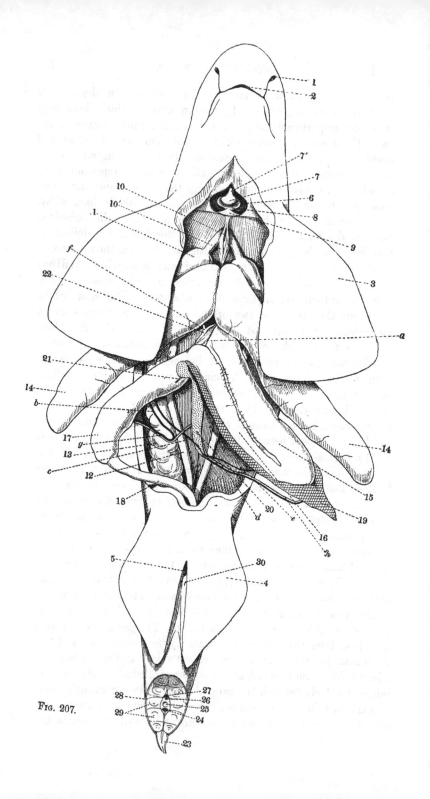

FIG. 207.

FIG. 207. *Scyllium canicula* ♀. Ventral view of viscera.

1. Left naris. 2. Mouth. 3. Pectoral fin. 4. Pelvic fin.
5. Aperture of cloaca. 6. Pericardial cavity. 7. Ventricle. 7′. Conus
arteriosus. 8. Auricle. 9. Sinus venosus. 10. Coelomic opening
of oviducts. 10′. Falciform ligament. 11. Shell-gland. 12. Oviduct. 13. Ovary reflected over to the right so as to show 12, which lies
external to the attachment of the ovary. 14. Liver. 15. Proximal
limb of stomach. 16. Distal limb of stomach. 17. Intestine.
18. Rectum. 19. Spleen. 20. Pancreas. 21. Pancreatic duct.
22. Bile-duct. 23. Dorsal fin. 24. Spinal cord. 25. Notochord in centrum of vertebra. 26. Caudal artery. 27. Caudal vein.
28. Lateral line. 29. Myotomes. 30. Abdominal pores. *a.* Hepatic artery. *b.* Intestinal branch of anterior mesenteric artery.
c. Lieno-gastric artery. *d.* Gastric branch of lieno-gastric artery
(posterior gastric artery). *e.* Splenic branch of lieno-gastric artery.
f. Portal vein. *g.* Intestinal vein. *h.* Splenic vein.

from the dorsal side of the same organ) and then opens into the
Ductus Cuvieri. The cloacal veins further give off median branches
which unite and then distribute blood to the viscera, so that some
blood from the pelvic fin may also return to the heart through a
portal system.

The ovary is a single ridge of the dorsal coelomic wall; the
oviducts are long and united far in front so as to open by a
common internal opening, situated ventral to the liver (Fig. 207).
In the middle of its length each oviduct has an enlargement caused
by a thickening of its walls due to the development of gland
cells. This is called the oviducal gland, and its function is to
secrete the pillow-shaped elastic egg-shell. In all cases a considerable amount of development takes place before the egg is laid : in
many cases development goes so far that the egg-shell is absorbed,
and the embryo takes in nutriment from the wall of the oviduct
so reaching a very large size before birth. The egg is large and
well charged with yolk. The oviducts unite posteriorly to open into
the proctodaeum or cloaca behind the anus. There are two large
testes, and these are united anteriorly and connected to the front
end of each of the kidneys, which extend along the entire length of
the abdominal coelom. The anterior region or mesonephros (for
no pronephros is developed) is narrow and its excretory function has
almost disappeared. The testis is connected with the front end of
the mesonephros by vasa efferentia uniting into a single coiled
tube or epididymis, which structures are derived from mesonephric
tubules. The archinephric duct has also lost its original function
and become a vas deferens, which lies on the ventral surface of the
kidney and conveys spermatozoa from the epididymis to the cloaca.
It enlarges at its hind end into a vesicula seminalis. The

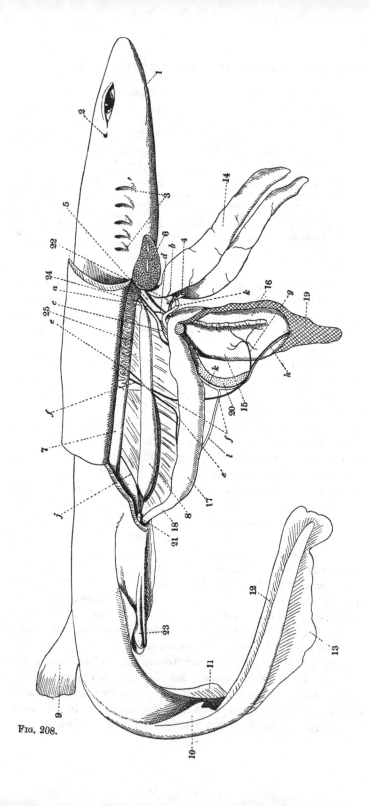

FIG. 208.

FIG. 208. *Scyllium canicula* ♂. View of viscera from the right side.

1. Mouth.　　2. Spiracle.　　3. Gill-slits.　　4. Gall-bladder.　　5. Oesophagus.
6. Pectoral fin cut off.　　　7. Vesicula seminalis lying on metaphros.
8. Testis.　　　9. Anterior dorsal fin.　　　　10. Posterior dorsal fin.
11. Median ventral fin.　　12. Dorsal lobe of caudal fin.　　　13. Ventral
lobe of caudal fin.　　14. Right lobe of liver.　　15. Proximal limb of
stomach.　16. Distal limb of stomach.　17. Intestine.　18. Rectum.
19. Spleen.　　20. Pancreas.　　21. Rectal gland.　　22. Bile-duct.
23. Claspers.　24. Ligament carrying the vasa efferentia.　25. Vas
deferens.　a. Coeliac artery.　b. Hepatic artery.　c. Anterior gastric
artery.　d. Pancreatic branch of the coeliac artery.　e. Anterior
mesenteric artery.　f. Lieno-gastric artery.　g. Posterior mesenteric
artery.　h. Splenic artery and vein.　j. Posterior mesenteric artery.
k. Portal vein.　l. Intestinal vein.

posterior and functional part of the kidney is the metanephros, and its tubules unite into about six main ducts, which converge to form a metanephric duct or ureter. There is also a blind sperm sac into whose posterior end the vesicula seminalis opens and which immediately after receives the ureter. The compound duct thus formed meets its fellow in the middle line and so there is a single urinogenital sinus which opens into the cloaca behind the anus. In the female the mesonephros is more vestigial than in the male and its duct (archinephric duct) is in front a very fine tube which lower down dilates and meets its fellow to form a median urinary sinus. This receives the ducts from the metanephros, and opens into the cloaca behind the oviduct.

Actual sexual congress or copulation takes place in the Elasmobranchs; the most posterior rays of the pelvic fins called the claspers are enlarged, and used to distend the cloaca of the female to allow of the entrance of spermatozoa (Fig. 208). This is correlated with the large size and small number of the eggs and their long retention in the oviduct. In the male the spermatozoa are stored in a swollen portion of the vas deferens, the vesicula seminalis, or in special pouches termed the sperm-sacs. It is probable that the claspers, the large eggs and the division of the kidney into two parts are specializations peculiar to modern Elasmobranchs.

The Elasmobranchs are the Sharks, Dog-fish, Skates and Rays of our seas. They are almost exclusively marine and are a group much detested by fishermen, since they are excessively voracious and their flesh is of little value.

They are divided into two sub-orders, the Selachoidei and the Batoidei. The first consists of powerful swimmers with cylindrical bodies, well-developed tail-fins and moderate pectoral fins; the latter are ground fish with broad backs and bellies and narrow sides,

whip-like tails with rudimentary tail-fin, and enormous pectoral fins
extending forward to the extreme end of the snout.

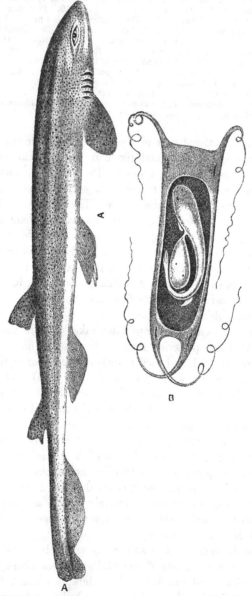

Fig. 209. A. *Scyllium canicula.* Reduced. From Day.
B. Egg-case opened to show young embryo with yolk sac.

The SELACHOIDEI are known as dog-fishes or sharks, according to
their size. The common English dog-fish, *Scyllium*
Classification. *canicula*, is about two feet long (Figs. 207, 208, and
209); another kind, the Spiny dog-fish, *Squalus acanthias*, is
distinguished by having a spine, a greatly enlarged scale, in front
of each of the two dorsal fins. This latter genus is very common
on the Atlantic coast of North America, where it is known as
the Spiny-dog. The American Smooth dog-fish, *Galeus canis*, is
distinguished from *Scyllium* by being viviparous. Amongst the
sharks the most remarkable are *Zygaena*, the Hammerhead, in
which the roofs and floors of the orbits are produced outwards,
so that the eyes are set as it were on peduncles; and *Carcharodon*,
the great White shark, which has lost its spiracles and possesses
a tail-fin with crescentic under lobe. Owing to their powerful
swimming capacities, sharks are as a rule not limited in distribution. *Carcharodon* is the dreaded man-eater of the Adriatic and
the warmer seas everywhere. *Zygaena* occasionally carries terror
into the bay of Naples, and species of both genera are found off the
American coast. The Notidanidae are a family with many interesting traits. They possess one (*Hexanchus*) or two (*Heptanchus*)
extra gill-clefts, and the upper jaw directly articulates with the skull
behind the orbit. Teeth of the same character as those borne by
living representatives of this family have been found in the Lias
shales of England. The Port Jackson shark of Australia, *Cestracion*,
is the sole surviving type of another family, representatives of which
are common in the Coal Measures. In it the snout is reduced so
that the mouth is thrust forward and the jaw is attached to the
skull in front of the orbit. The teeth are flat and pavement-like
and adapted for crushing the Molluscs on which the animal feeds.

The BATOIDEI or Rays are, as we have said, ground feeders. All
have the true gill openings on the underside of the body: the
spiracle alone opens on the dorsal side and is enlarged. It has in
fact in this group taken on the function of pumping water into
the pharynx, a duty which cannot be conveniently undertaken by
the mouth when this is burrowing in the mud at the bottom. *Raia*
is the common skate on both sides of the Atlantic: it has no caudal
fin but two dorsals. *Torpedo* is distinguished by a more elongated
body. The muscles on either side of the head are converted into
electric organs, consisting of batteries of vertical hexagonal tubes
filled with a clear gelatinous fluid, each tube representing a metamorphosed muscle-fibre. By means of these organs it can inflict

a severe shock on its enemies. *Trygon* is the sting-ray. In it the tail is long and thin and the dorsal fin-rays are practically absent, but at the spot where the tail merges into the body there is a large recurved spine, at the base of which is a poison gland, so that by a blow of the whip-like tail it can inflict a severe wound.

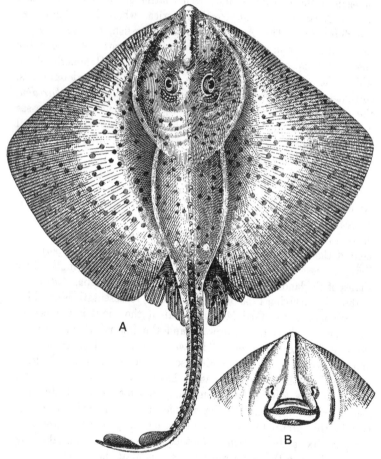

Fig. 210. *Raia maculata.* From Day.

A.   Dorsal surface, showing the spiracles just behind the eyes.   Reduced.
B.   View of mouth and olfactory pits.

The pectoral fins are joined in front of the snout. *Pristis* is the saw-fish. It has an immensely elongated rostrum, at the sides of which large pointed teeth are set; the body is elongated, but it shows all the essential features of the Batoidei. The teeth in the mouth, like those of other Batoidei, are flattened. *Pristis* is found both in the Mediterranean and Caribbean Seas and elsewhere. In some of the extinct representatives of the family the upper jaw is attached to the cranium behind the orbit. This variation in the place of attachment indicates that the connection between the two structures is secondary.

The two most interesting fossil representatives of the groups are *Cladoselache* and *Pleuracanthus* whose fins are described above (p. 375).

Order II. Holocephali.

The second Order of Pisces, the Holocephali, differ from Elasmobranchs chiefly in the skeleton; in the viscera they resemble them very closely. The Holocephali are distinguished by having the

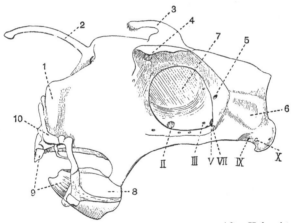

Fig. 211. Skull of a male *Chimaera monstrosa*. After Hubrecht.

1. Nasal capsule. 2. Cartilaginous appendage to the ethmoid region, representing the rostrum of Elasmobranchii. 3. Erectile appendage. 4. Foramen by which the ophthalmic nerves leave the orbit. 5. Foramen by which the ophthalmic branch of the Vth nerve enters the orbit. 6. Auditory capsule. 7. Interorbital septum. 8. Meckel's cartilage articulating with an outgrowth from the posterior part of the palato-pterygo-quadrate cartilage. 9. Teeth. 10. Labial cartilage. II. III. V. VII. IX. X. foramina for the passage of cranial nerves.

upper jaw completely confluent with the cranium, a condition called autostylic (Fig. 211): the orbits are so deeply indented that the brain is pressed back from between them, and their two cavities are only separated by a vertical plate of cartilage, called the inter-

orbital septum.   There is no spiracle and the last gill-cleft is also
closed.   A fold of skin, called the operculum, extends back over
the gill-slits.   The gills are, however, still borne on the walls of
sacs.   The snout or praeoral part of the body is much reduced in
size and supported only by a single rod of cartilage.

The scales have almost entirely disappeared and are represented
only by the great spine, the so-called icthyodorulite, which
stiffens the front edge of
the dorsal fin, by the teeth
and by the prickles on a
peculiar tentacle situated
on the snout of the male.
The teeth are confluent,
forming ridges of dentine
covered with enamel. Of
these there are a pair in
the lower jaw, called dentary plates, and two pairs
in the upper, termed vo-
merine   and   palatine
plates respectively, placed
one behind the other. Each
plate has certain areas,
where the dentine is especially   thickened,   called
tritors. The arrangement
of these tritors is used in
classifying the fossil species. The peculiar tentacle
on the head of the male
arises from a pit situated
in the middle line of the
snout, and bears sharp
tooth-like scales at its tip.

The notochordal sheath
is not broken up into
centra, but in *Chimaera* it
has developed within it a
large number of calcified
rings, three to five times
as numerous as the neural
arches.

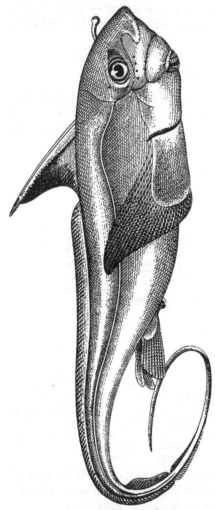

Fig. 212. *Chimaera monstrosa*, L.
Male with process on snout. Reduced.

The Holocephali were once a numerous group; now they are represented by three closely allied genera, of which the best known is *Chimaera*, sometimes called the Rabbit-fish, common to the Mediterranean and to the Atlantic coast of Europe and Africa. *C. monstrosa* is found on the East coast of N. America. On the Pacific coast *C. collei* occurs in such numbers as to be a serious nuisance to fishermen. It eats the baits off their lines. It is known as the Rat-fish, in allusion to the shape of the tail. *Callorhynchus* occurs in the temperate waters of the Southern Hemispheres. The third genus, *Harriotta*, is a deep-sea form.

Order III. Dipnoi.

The third Order of Fishes, the Dipnoi, are very interesting animals, inasmuch as they afford suggestions as to how land animals were evolved from fishes. They are distinguished by possessing true lungs, in the form of one or two sacs opening by a common tube into the ventral side of the oesophagus. The blood is supplied to

FIG. 213. *Lepidosiren paradoxa.* Male, showing the feathered pelvic fins of the breeding season. Much reduced. From Graham Kerr.

the lungs by vessels given off from the last two pairs of efferent gill-arteries (Fig. 216), and it is returned not into the general circulation, but direct to the heart, where it opens into a special section of the atrium, the left auricle, cut off from the rest by a septum. As in the Holocephali, the upper jaw is fused with the cranium. There are four or five gill-clefts and no spiracle. There is a large gill-cover which completely covers the clefts; it is strengthened in *Ceratodus* by two strong membrane-bones, the so-called squamosal or pre-operculum, and the operculum; there is also a third smaller bone called the inter-operculum. The oro-nasal groove has become closed so as to form a canal, the end of which opens within the stomodaeum as the posterior naris or choana. Owing to the forward growth of the jaws the mouth has become terminal.

The cranium is not narrowed between the orbits, and the cartilage behind is replaced by true bone, there being two exoccipital

bones at each side of the hole called the foramen magnum from
which the spinal cord issues. Above the foramen magnum the
cranium is quite obviously composed of fused neural and ventral
arches; the spines of the former and the ribs of the latter are quite
distinct. As in *Chimaera* the teeth have coalesced to form great
dentary ridges in the lower jaw and in the roof of the mouth,
the so-called vomerine and palatine plates, the first named being
anterior (Fig. 214).

It will be observed that in the Dipnoi we for the first time meet
with large bones, and it is instructive to notice under what circumstances they appear. The exoccipital bones are quite distinct from

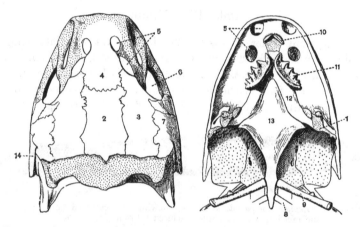

FIG. 214. Dorsal (to the left) and ventral (to the right) views of the cranium of
Ceratodus miolepis. After Günther.

1. Cartilaginous part of the quadrate with which the mandible articulates.
2, 3, 4. Roofing membrane-bones.   5. Nares.   6. Orbit.   7. Pre-
opercular (squamosal).   8. Second rib.   9. First rib.   10. Vomerine
dental plate.   11. Palatine dental plate.   12. Palato-pterygoid.
13. Parasphenoid.   14. Interopercular.

the calcifications of the cranium met with in the Elasmobranchii.
In the latter case there was hardening of the
ground substance of the cartilage; here the bone
is formed in the membrane surrounding the cartilage
and eats into and destroys the cartilage. This sort of bone is true
cartilage-bone. Besides such bones, however, we have many
which are traceable to the fusion of the small bony bases of the
typical scales which we found in Elasmobranchs. These little bony
plates fuse together to form large structures which are then termed

*Evolution of
Bone.*

membrane-bones. A first trace of this
process is seen in the fusion of the teeth
to form compound plates, as in the Holocephali.

The Dipnoi are eel-like fish with elongated whip-like paired fins, and they are
covered all over with thin rounded cycloid
scales. These scales are equivalent to the
enlarged bases of the scales of Elasmobranchs without the spikes. On the upper
side of the head these scales have joined
to form median and lateral bones; in
front near the nasal sacs there are two
smaller bones. The Dipnoi are quite
peculiar among Craniata in having an
unpaired series of large roofing bones on
the top of the head. The palatine dental
plates are supported by a bone which surrounds and replaces the cartilaginous upper
jaw, and is called the palato-pterygoid.
The roof of the mouth is sheathed by a great
plate of bone called the parasphenoid,
derived from the bases of vanished teeth
(Fig. 214). Beside these there is a
large membrane-bone outside the pectoral girdle, which is the first trace of a
collar bone or clavicle, and the two
bones already mentioned in the opercular
flap. The sheath of the notochord is
converted into cartilage, but is not divided into centra.

The paired fins are remarkable for
having a long jointed axis and two rows
of rays; they are in a word biseriate
(Fig. 215) like those of the extinct Elas-

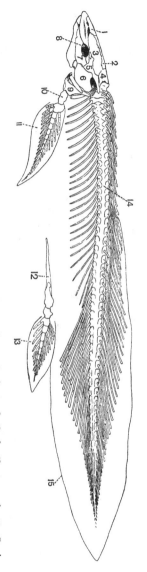

FIG. 215. Lateral view of the skeleton of *Ceratodus miolepis*. After Günther.

1, 2, 3. Roofing membrane-bones.   4. Cartilaginous posterior part of cranium.
   5. Pre-opercular (squamosal).   6. Opercular.   7. Suborbital.
   8. Orbit.   9. Pectoral girdle.   10. Proximal cartilage of pectoral fin.
   11. Pectoral fin.   12. Pelvic girdle.   13. Pelvic fin.   14. Spinal
column.   15. Caudal fin (diphycercal).

mobranch *Pleuracanthus.* A fin of this type is called an archi-
pterygium. The tail-fin is of the primitive type found in *Amphioxus* and Cyclostomata, in which the fringe of skin supported by the
fin-ray is equally developed above and below the noto-chord. It is in fact a diphy-cercal tail.

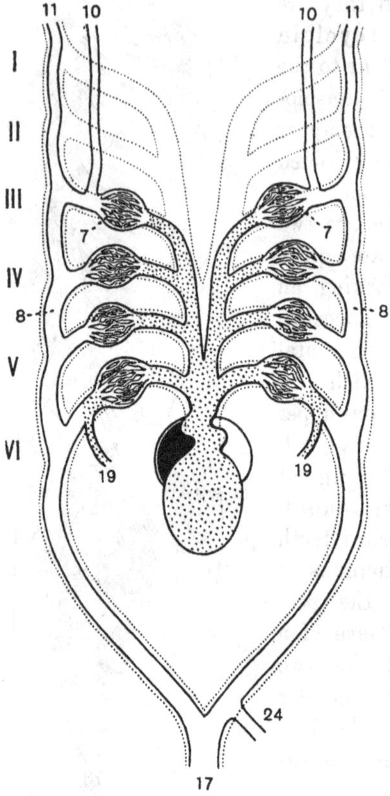

As was to be expected, the blood-system has undergone interesting modifications. The conus arteriosus is long, and as in Elasmobranchs and Holocephali has several transverse rows of pocket valves. From its anterior end four arteries are given off in a bunch on each side to the gill arteries to supply the gills: there is no ventral aorta. From the last efferent vessel on each the artery going to the lungs arises. An oblique septum divides the cavity of the conus into two in such a way as to cut off the openings of the last gill arteries from the front ones, so that the blood passing to the lungs does not mingle with that going on to the head. This is very like the arrangement found in Amphibia. The likeness to the Amphibian blood-system is increased by the presence of a median "vena cava"

Fig. 216. Diagram of the arterial arches of *Ceratodus*, viewed from the ventral side.

I. II. III. IV. V. VI. First to sixth arterial arches. 7. Gills. 8. Epibranchial. 10. Anterior carotid. 11. Posterior carotid. 17. Dorsal aorta. 19. Pulmonary. 24. Coeliac.

which returns the blood from one kidney directly to the heart. One posterior cardinal vein persists, the other has atrophied except at its origin from the kidney (Fig. 217).

The lungs are long, wide sacs, extending between the intestine and notochord, although their opening into the gullet is ventral. The gill-arches and gills are on the other hand very small, and the opening between the gill-cover and the body is narrow.

Only three species of Lung-fish are still living, but the group has very many fossil representatives. The Australian lung-fish, *Ceratodus forsteri*, has well-developed paired fins (Fig. 215). It

Classification. inhabits rivers which at certain seasons become foetid with decaying vegetation, and during this time it breathes air. The African and South American lung-fishes (*Protopterus annectens* and *Lepidosiren paradoxa*) have whip-like paired fins consisting of little more than the axis of the limb skeleton (Fig. 213). They bury themselves in mud during the dry season, a necessary precaution, since they inhabit swamps which dry up. *Lepidosiren*, the South American form, has been shown to have larvae with long feathery external gills, strikingly recalling the larvae of Amphibians. The young *Protopterus* has similar structures and retains traces of its external gills throughout life. *Ceratodus*, on the other hand, has a development practically completed within the egg-shell. The fossils referable to this order are very interesting. They occur in a great variety of forms. Some of them—referred to a Sub-order called the Arthrodira—having not only the head but also the anterior part of the trunk clothed in great bony plates. The head skeleton articulated with the trunk skeleton by ball-socket joints.

The occurrence of Dipnoi in great numbers in the rocks immediately pre-ceding in the geological series those in which the first remains of Amphibia are found is very suggestive.

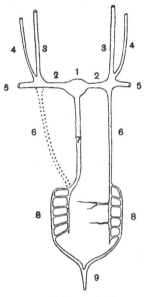

Fig. 217.　Diagram to show arrangement of the principal veins in a Dipnoan.

1. Sinus venosus — gradually disappearing in the higher forms.　2. Ductus Cuvieri = superior vena cava.　3. Internal jugular = anterior cardinal sinus.　4. External jugular = sub-branchial.　5. Subclavian.　6. Posterior cardinal, front part = venae azygos and hemiazygos.　7. Inferior vena cava.　8. Renal portal = partly hinder portion of posterior cardinal. 9. Caudal.　The hepatic portal system is omitted.

Order IV. Teleostomi.

The fourth Order of Fishes, the Teleostomi, is by far the largest
and contains the overwhelming majority of living fishes. They
differ from the Dipnoi, in that the lung or air-bladder, as it is
called, receives its blood from the dorsal aorta and returns it to the
general circulation, so that the organ is as a rule not so much
respiratory as hydrostatic. The lung is undivided and its opening
has in most cases apparently become shifted up the side of the
throat to the mid-dorsal line. Since the opening of the air-bladder
is dorsal some authorities have held that it is quite a different
organ from the lung, the opening of which is ventral. But it is
very difficult to believe that we have to deal with two totally
distinct organs in *Polypterus* and *Lepidosteus*, more especially as
in both these fishes it is probable that the air-bladder subserves
respiration. The air-bladder is never paired : if we suppose it to
represent one lung we can imagine how the opening could be
gradually shifted dorsally. The anus of Urochordata has, we know,
undergone such a shifting.

Another point of difference, distinguishing Teleostomi from the
Dipnoi, is that there are no median membrane-bones on the head,
all the bones being originally paired. Further, a set of membrane-
bones bearing teeth appears in the sides of the mouth outside the
primitive jaws, or, as we may express it, in the lips. These lip-bones
functionally replace the true jaws, and as the mouth has now
received its full armature, the name Teleostomi, or Perfect-mouthed
Fish (τέλειος, perfect; στόμα, mouth) has been given to the Order.
As in Elasmobranchs, the upper jaw is joined to the skull only by
the upper half of the hyoid arch, which is ossified as the hyomandibular. There is a strongly developed gill-cover, armed
with several bones, called the operculum. The septa between the
gill-sacs are reduced to narrow bars, so that there are gill-clefts,
not gill-pouches. The gills themselves develope into long triangular
processes freely projecting and attached only at the base. Often there
is a rudimentary gill attached to the posterior aspect of the hyoid
arch. This is called the pseudobranch or sometimes the opercular gill. This is not to be confounded with the pseudobranch
of Elasmobranchii, which is attached to the first visceral arch. The
opening of each nasal sac is divided into two by a bridge and there
is no oro-nasal groove. The cloaca is divided into two openings;

an anterior, the anus, communicating with the intestine, and a posterior, serving for the discharge of the genital products and excreta of the kidneys.

It was formerly customary to divide the fishes here grouped as Teleostomi into two Orders, the Teleostei, or completely bony fish (ὀστέον, a bone), and the Ganoidei, or fish with shining scales (γάνος, glitter, brightness). As, however, there is far more difference between different families included in the Ganoids than there is between some Ganoids and some Teleostei, this arrangement is really unnatural. The Ganoids are in fact composed of widely different families, retaining certain primitive characteristics once shared by all Teleostomi. A far more rational division of the Teleostomi is that now usually adopted into Crossopterygii and Actinopterygii.

Sub-Order A. Crossopterygii.

The Crossopterygii include only two living genera, *Polypterus* and *Calamoichthys*, which inhabit the rivers of Africa. In former geological periods, however, the Crossopterygii constituted an immense group. They are distinguished by retaining the biseriate paired fin with a shortened and broadened axis covered by scales. This scaly lobe is consequently fringed by the rays, whence the name (κροσσοί, a fringe; πτέρυξ, wing or fin). The so-called air-bladder is paired and the two halves open on the underside of the pharynx. The spiracle still persists, having a special little gill-cover strengthened by several small bones. It has recently been proved that the air-bladder is a respiratory organ and that the expired air escapes by the spiracle.

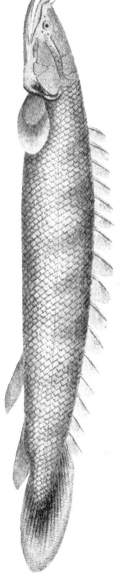

Fig. 218. *Polypterus.*

The whole body is covered with lozenge-shaped scales covered by a thick layer of shining enamel, and the dorsal fin is subdivided into a large number of finlets, each supported by a large stiff ossified fin-ray (Fig. 218). The head is covered with numerous membrane-bones arranged in pairs, replacing the cartilage of the cranium. There are not only exoccipital bones, but also a basi-occipital situated beneath the foramen magnum, and the front wall of the auditory capsule is ossified by a pro-otic bone. The notochord is narrowed in part by the formation of bony centra; these however are not, as in the Elasmobranchs, mere rings in the notochordal sheath, but are formed of the coalesced expanded bases of the bony neural and haemal arches. The term chorda-centra has been proposed for centra such as those of Elasmobranchii, formed by the segmentation of the sheath of the notochord; whilst centra such as those of the Crossopterygii are called arco-centra. The centra have concavities before and behind, and the space between two vertebrae is filled with an expanded section of the notochord, which in the middle of the vertebrae is reduced to a mere thread if not obliterated altogether. Vertebrae hollowed on each side are said to be amphicoelous (Gr. ἀμφί, both; κοῖλος, hollow).

It is obvious that an axis composed of vertebrae is a much more efficient organ of support than the flexible notochord with its loosely adherent neural and haemal arches. Hence it is not surprising to find that vertebrae have been independently developed in Fishes, Amphibia and Reptiles, and have had an independent origin even in different families of the same order—the Halecomorphi and Lepidosteidae, for example. Different investigators therefore in their endeavours to find exactly corresponding parts in the vertebral columns of various animals have arrived at discrepant results. After years of research however Gadow has been enabled to give a consistent account of the evolution of the column in all the Craniata, and his account seems on the whole the most probable.

Vertebral column.

According to Gadow, all arco-centrous vertebrae originate from four pairs of cartilaginous pieces; some of which may be entirely suppressed in some vertebrae, but all of which are represented in some part or other of the column. These are (1) basi-dorsals, the expanded bases of the neural arches; (2) basi-ventrals, the expanded bases of the haemal arches (to these when present the rib is always attached); (3) inter-dorsals, cartilaginous pieces situated between successive basi-dorsals, and derived from the segmentation of the apical ends of the haemal arches which have extended upwards round the notochord; (4) inter-ventrals, cartilaginous pieces placed beneath the notochord alternating with the basi-ventrals, and derived from the segmentation of the apical ends of the neural arches which have extended downwards round the sheath of the notochord.

In an ancestral Craniate therefore corresponding to each myotome there were on each side of the notochord four cartilaginous pieces. These pieces in different groups of animals have been variously combined with their successors and predecessors, the joint between two successive vertebrae being formed not between two pieces but by the absorption of cartilage in the middle of a piece at the place where the maximum bending occurs.

The Crossopterygii are probably in some respects more nearly related to the ancestors of Amphibia than are modern Dipnoi. This view is strengthened by the fact that the air-bladder is occasionally used as a lung. The young have one large external gill attached to the operculum. As in the Dipnoi the mouth is terminal.

Sub-Order B. **Actinopterygii.**

The Actinopterygii are distinguished by having a uniseriate fin, the base of which is not covered by scales, and by having the opening of the air-bladder dorsal.   Among the most primitive subdivision is that of the CHONDROSTEI, or Sturgeons.   In these fish there is a long shovel-shaped snout or prae-oral part of the head, which is used for shovelling up the mud at the bottom of rivers in search of prey. This old-fashioned feature is not found in any other family of the Actinopterygii, all of which have terminal mouths. The notochord of the Sturgeons has a thick sheath without any trace of centra, and the only cartilage bones in the cranium are small ossifications on its side walls, called the orbitosphenoid and the alisphenoid bones, and a pro-otic in the front wall of the auditory capsule. The hyomandibular segment of the hyoid arch is ossified by two bones—a hyomandibular where it articulates with the skull, and a symplectic below where it joins the first visceral arch.   A large part of the cranium is covered with a number of membrane-bones which pass insensibly into the great scutes or bony plates which cover the body. These latter, like the bones of the head, are derived from the fusion of scales.   Though the mouth is often toothless, there is a distinct maxillary bone in the upper lip outside the pterygoid bone which supports the pterygoquadrate cartilage. A great parasphenoid supports the base of the skull.

The common sturgeon *Acipenser* is found in the Pacific and North Atlantic, and enters the rivers of Europe and America. The ovary forms the Russian delicacy known as caviare. The spoonbill *Polyodon*, found in the Mississippi, has a very broad snout and has lost nearly all its scutes, but it retains some at the sides of the tail and a series forming a comb-like fringe along the dorsal edge of the upper lobe of the tail, called fulcra. The extinct members of this subdivision were clothed all over with ganoid scales and the membrane-bones on the head were fewer and more regular than in recent forms.

The subdivision of the HALECOMORPHI is represented at the present day by only one species, *Amia calva*, the Bow-fin of the St Lawrence and other American rivers. Like *Polypterus* this species has complete amphicoelous centra formed by the fusion of the expanded bases of the neural and haemal arches: the cranium is also largely replaced by bone, not only behind where the foramen magnum is encircled by four bones, the supra-occipital, two ex-occipitals and a basi-occipital, but at the sides and in the snout where a mesethmoid is formed. In addition there is a complete helmet of membrane-bones surrounding the cranium above and at the sides, and the lips have above a pre-maxilla and a maxilla, and below a dentary, all well provided with teeth. The whole body is covered

FIG. 219 *Acipenser sturio,* the Sturgeon. From Day.

with thin rounded scales. *Amia* is a voracious fish about two feet long.

The next family, that of the LEPIDOSTEIDAE, are the most bony of all fish. They are the Bill-fish or Gar-spikes of the Northern and Central American and Cuban lakes and rivers. The whole body is covered with scales like those of *Polypterus*, coated with a thick layer of ganoin. The jaws are long, and in both the upper and under lips there is a series of three or four bones bearing teeth, a rare condition. The skull is as bony as that of *Amia*. The vertebral centra have become opisthocoelous, that is to say, each centrum is convex in front, fitting into a concavity of the hinder surface of the one in front of it, so that the notochord has almost disappeared. The ovary, by the fusion of its free edge with the coelomic wall, has become converted into a sac, to which the funnel of the oviduct has become adherent, so that ovary and oviduct are continuous. The spiral valve on the intestine is rudimentary.

The Gar-pike are said to lie in wait among the reeds on the banks of the lakes, in order to seize small animals visiting the swamp in their long pincer-like jaws.

The remaining families of the Actinopterygii are grouped together as TELEOSTEI and have certain well-marked characters in common which distinguish them from the preceding families grouped together as the GANOIDEI. The first of these characters is the structure of the tail. In all the Actinopterygii so far mentioned the tail is heterocercal, though in some, such as *Amia*, the ventral lobe is very predominant. In the Teleostean families the ventral lobe forms the whole of the tail and is placed in a straight line with the rest of the body, the end of the notochord being sharply bent up almost at a right angle with the remainder and surrounded by a bony sheath. Such an apparently symmetrical tail-fin is called homocercal, and it is to be sharply distinguished from the really symmetrical diphycercal tail of the Dipnoi.

Another important point in the anatomy of Teleostei is the structure of the heart. The conus arteriosus with its several rows of valves has become completely absorbed into the ventricle. Only the anterior row of valves remains, separating the enlarged ventricle from the ventral aorta, at the origin of which there is a non-contractile swelling, the bulbus arteriosus.

The stomach is distinctly sac-like, the two limbs of the loop of the alimentary canal of which it is constituted showing a tendency to coalesce.

The intestine has become longer and more coiled and has completely lost its spiral valve. Close to the entrance of the bile-duct there is a series of short, blunt diverticula of the intestine, called pyloric caeca. These were supposed to be a modified true pancreas, but the true pancreas has recently been discovered in the form of a number of very delicate tubes intermixed with the pyloric caeca. In no Teleostean, so far as is known, does the pancreas form a compact gland. The ovary is a hollow organ continuous with its duct; in this point, it is true, *Lepidosteus* has Teleostean characters, but in Teleostei the testis is also continuous with its duct, which shows no relation to the kidney; so that in the male

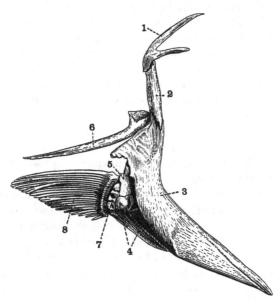

Fig. 220. The right half of the pectoral girdle and right pectoral fin of a Cod, *Gadus morrhua*, × ½.

1, 2. Supra-clavicle. 3. Clavicle. 4. Coracoid. 5. Scapula. 6. Accessory piece. 7. Ossified radialia of the fin. 8. Dermal fin-rays.

there is a complete departure from the normal Craniate arrangement. The cloaca has undergone more division than even in the primitive Actinopterygii, for the conjoined kidney ducts have an opening behind and distinct from that of the united genital ducts.

In the brain the roof of the cerebrum is a thin non-nervous membrane, but the optic lobes and cerebellum are greatly developed.

The interlacing of the fibres of the two optic nerves, which was alluded to above as the optic chiasma, has entirely disappeared, so that the optic nerve from the left side proceeds straight to the right eye, crossing but not interlocking with those from the right side to the left eye. The eyes are exceedingly large, so that in nearly all Teleostei, as in the Holocephali, an inter-orbital septum is formed, along the upper edge of which the olfactory stalks run.

In the skeleton there is hardly a feature which is not shared by some or other of the members of the more primitive families of the Actinopterygii, but in the combination of features it is characteristic, and it is of such an uniform type throughout a large number of families that it merits a special description.

The notochord is constricted by the formation of amphicoelous centra, which have developed four ridges projecting outwards in two planes at right angles to one another.

The Skeleton.

In the intervals between these ridges the ends of the neural and haemal arches are inserted, all being converted into bone. The pectoral arch is represented by a small plate of cartilage on each side, in which are two bony centres, an upper scapula and a lower coracoid (Fig. 220). Outside this there is a strong curved membrane-bone, strengthening the hinder border of the opercular slit, the clavicle. This is connected with the skull by additional bars, whilst from its lower end a pre-clavicle runs forward beneath the gills. From its hinder edge a post-clavicle often projects back into the muscles. The anterior paired fins are attached to the girdle where the scapula and caracoid join. In them the cartilaginous skeleton is almost all absorbed, the basal portions of the fin rays becoming changed into true membrane-bone and articulating directly with the pectoral girdle. The pelvic fin has a similar structure, and the pelvic girdle has quite disappeared.

The cranium is completely covered with bones. Even where cartilage persists it is covered with the bones which in other allied forms have displaced and absorbed the underlying cartilage. The cranium, as already mentioned, is so strongly compressed between the orbits that it is divided into two portions, an ethmoid region in front of the eyes and an occipital region behind them, connected by a narrow sphenoid isthmus running between them. In the cartilaginous roof there are several membranous windows or "fontanelles." The anterior fontanelle lies between the nasal sacs; the posterior fontanelles are a pair of windows situated one at each side of the supra-occipital bone.

The skull.

The supra-occipital, two exoccipitals and a basi-occipital
form the hindermost portion of the brain-case surrounding the
foramen magnum (Fig. 221). The supra-occipital, which is situated in the mid-dorsal line, has a great vertical ridge to which the
powerful longitudinal muscles of the body are attached. The
auditory capsule is completely replaced by bone; above it is

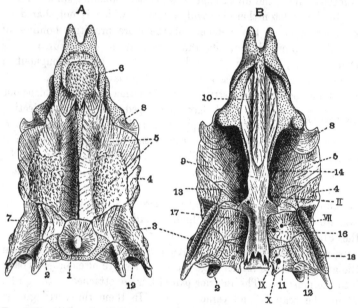

FIG. 221. A. dorsal and B. ventral view of the cranium of a Salmon, *Salmo
salar*, from which most of the membrane-bones have been removed. After
Parker. Cartilage is dotted.

1. Supra-occipital.   2. Epiotic.   3. Pterotic.   4. Sphenotic.   5. Frontal.
6. Median ethmoid.   7. Parietal.   8. Lateral ethmoid.   9. Parasphenoid.   10. Vomer.   11. Exoccipital.   12. Opisthotic.
13. Alisphenoid.   14. Orbitosphenoid.   16. Foramen for passage
of an artery.   17. Prootic.   18. Articular surface for hyomandibular.
II. VII. IX. X. foramina for the passage of cranial nerves.

covered by a pointed epiotic, in front and below by the pro-
otic, behind by the opisthotic, and on the outer side by the
pterotic with which the hyomandibular articulates. On its inner
side the capsule is only separated from the brain by membrane. In
the inter-orbital septum there is an orbito-sphenoid in front, an
alisphenoid behind and a sphenotic above partly extending into
the auditory capsule, while below and behind in the base of the

skull there is a Y-shaped bone, the basi-sphenoid, just in front of
the basi-occipital (Fig. 222). The ethmoid region in front of the
eyes is the part in which most cartilage persists; it is ossified
above by the median ethmoid and on the sides by the lateral
ethmoids. Connected with the cartilaginous upper jaw there is a
large quadrate bone developed where it articulates with the lower
jaw; an ectopterygoid bone replacing the cartilage on the outer
and an entopterygoid on the inner side; lastly a metaptery-
goid above and in front of the quadrate. The palatine bone is
situated in front of the pterygoid and articulates with the lateral
ethmoid. It ossifies around the anterior part of the originally
cartilaginous upper jaw. In the lower jaw we have a bone,
the articular, which moves on the quadrate. In the second
or "hyoid" visceral arch we find, as in all hyostylic fish,
an upper segment, the hyomandibular, articulating with the
skull, and a lower segment, the ceratohyal, which supports the
opercular flap and from which in Teleostei bony rays, called
branchiostegal rays, extend, on which the membranous part of
the membrane is stretched, as an umbrella on its ribs. The hyomandibular segment is formed, as in almost all Actinopterygii, of
two bones, an upper hyomandibular, *sensu stricto*, uniting with
the skull, and a lower symplectic joined to the quadrate. The
ceratohyal segment likewise is represented by three bones, a main
ceratohyal, a small interhyal uniting the latter to the symplectic,
and a hypohyal extending inwards from its lower end. The two
hypohyals are joined by a median bone, the glossohyal. This
supports a projection of the floor of the pharynx which is the
rudiment of the tongue. The other visceral arches are each composed of four bones, except the last, which is rudimentary. The
uppermost segments extend inwards on the roof of the pharynx
and bear teeth; they are called the superior pharyngeal bones.
Owing to the forward slope of the visceral arches these bones are
directly above the seventh or last pair of arches, which consist on
each side of a small bone bearing teeth, the inferior pharyngeal
bone. Whatever chewing is done by fish is effected by these bones,
the teeth in the front part of the mouth are chiefly used for
retaining prey. The presence of teeth on the gill-arches is not
easy to explain. Teeth, as we have seen, are structures developed
from the dermis and ectoderm, while the gill-arches support the
walls of the pharynx, which are certainly endodermal. We must
either suppose that some portion of the ectoderm has migrated

inwards through the gill-slits or that the endoderm has acquired the power of producing teeth : on the whole the first supposition seems the more probable.

The membrane-bones in the Teleostean skull are numerous and are traceable to two sources, namely, the scales of the skin and the bases of the teeth. To the former category belong the roofing bones

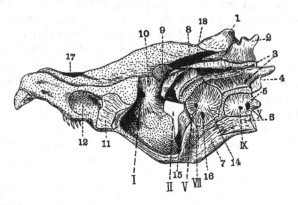

Fig. 222. Lateral view of the cartilaginous cranium of a Salmon, *Salmo salar.* After Parker. A few membrane-bones are also shown. Cartilage is dotted.

1. Supra-occipital.   2. Epiotic.   3. Pterotic.   4. Opisthotic.
5. Exoccipital.   6. Basi-occipital.   7. Parasphenoid.   8. Sphenotic.
9. Alisphenoid.   10. Orbitosphenoid.   11. Ectethmoid.
12. Olfactory pit; the vomerine teeth are seen just below.   14. Prootic.
15. Basisphenoid.   16. Foramen for the passage of an artery.   17. Anterior fontanelle.   18. Posterior fontanelle.   I. II. V. VII. IX. X. Foramina for the passage of cranial nerves.

of the skull, the **parietals** at each side of the supra-occipital, and the great **frontals** which cover the anterior fontanelle, but which also extend outwards and roof in the orbits. A pair of delicate **nasal** bones lie on the nasal sacs. On the side of the head there is a chain of bones surrounding the eye, called the **orbital ring**, and four bones stiffening the upper part of the **opercular** flap, a **pre-operculum** in front, an **operculum** above and behind, a **sub-operculum** below it, and an **inter-operculum** between the pre-operculum and the sub-operculum.

To the fusion of the bases of teeth we must ascribe the origin of the great **parasphenoid** bone, which stiffens the roof of the mouth and extends back under the basi-occipital, although in no Teleostean does it bear teeth. Similarly, in front of it there is the **vomer** made of two bones joined in the middle line, which still bear teeth in their anterior portions. The palatine and pterygoid are also traceable to the fusion of groups of teeth placed more laterally, but in this case the membrane-bone has become a cartilage bone by extending into and replacing the cartilaginous bar beneath it.

The membrane-bones in the lips, which are essentially characteristic of the Teleostomi, have probably had a double origin—scales of

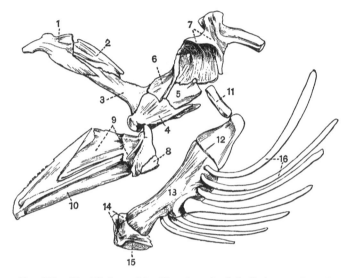

FIG. 223.   Mandibular and hyoid arches of a Cod, *Gadus morrhua* × ½.

1. Palatine.	2. Ento-pterygoid.	3. Pterygoid.	4. Quadrate.
5. Symplectic.	6. Meta-pterygoid.	7. Hyomandibular.	8. Angular.
9. Articular.	10. Dentary.	11. Interhyal.	12. Epihyal.
13. Ceratohyal.	14. Hypohyal.	15. Glossohyal.	16. Branchiostegal rays.

the outer skin united with teeth developed just inside the mouth. There are, in the upper lip, the pre-maxilla in front, the maxilla behind, and occasionally behind these a third bone, the jugal. As we have already seen, the number of these upper lip bones is greater in the Lepidosteidae. In the lower lip there are in front the dentary, behind and on the inner side the splenial, and behind and below, the angular. The branchiostegal rays mentioned above are membrane-bones, and between those of opposite sides there is a median bone called the basi-branchiostegal.

Taking a general view of the skeletons thus far studied, we see

26—2

that the replacement of cartilage by cartilage bone tends to take place first wherever there is a joint. Thus in Dipnoi we find the hinder part of the cranium replaced by exoccipitals, whereas in the Chondrostei, where all the skull and front part of the notochordal sheath form one solid mass, no such bones are developed. Again, in Teleostei, the only replacement of the lower jaw by cartilage bone

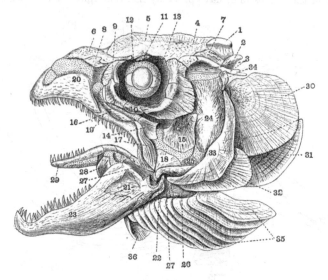

FIG. 224. Lateral view of the Skull of a Salmon, *Salmo salar.* After Parker. Cartilage is dotted.

1. Supra-occipital. 2. Epiotic. 3. Pterotic. 4. Sphenotic. 5. Frontal. 6. Median ethmoid. 7. Parietal. 8. Nasal. 9. Lachrymal. 10. Sub-orbital. 11. Supra-orbital. 12. Cartilaginous sclerotic. 13. Ossification in sclerotic. 14. Ento-pterygoid. 15. Meta-pterygoid. 16. Palatine. 17. Jugal. 18. Quadrate. 19. Maxilla. 20. Pre-maxilla. 21. Articular. 22. Angular. 23. Dentary. 24. Hyomandibular. 25. Symplectic. 26. Epihyal. 27. Ceratohyal. 28. Hypohyal. 29. Glossohyal. 30. Opercular. 31. Sub-opercular. 32. Infra-opercular. 33. Pre-opercular. 34. Supratemporal. 35. Branchiostegal rays. 36. Basi-branchiostegal.

takes place at its proximal end, where it joins the upper jaw. Then again we find the lateral ethmoid developed where the front part of the palatine articulates with the skull, and the articulation of the hyomandibular has probably had something to do with bringing about the ossification of the auditory capsule.

With the exception of the Ganoidei all the Teleostomi differ fundamentally from the Elasmobranchii and Holocephali in having small eggs. These rapidly develope into larvae with diphycercal tails and other primitive features which ally modern fish with the ancient fossil forms. The development into the adult is very slow and dangerous, and during this period an enormous number of the young perish. The habits, distribution and food of these young form one of the most important economical problems that zoologists of the present day are seeking to solve.

Eggs and larvae.

Since in the animal kingdom small eggs and a prolonged larval life constitute the more primitive arrangement, and eggs charged with yolk and cared for by the parent a secondary modification, we conclude that in this matter the Teleostei have retained primitive habits. This throws some light on a question which must often arise in the minds of all students of zoology, namely, how it is that comparatively primitive forms, like the Elasmobranchs, and highly modified forms, like the Teleostei, exist side by side? We can see that in each case there has been some great modification of the primitive arrangement, in Elasmobranchs in the egg and young, in Teleostei in the skeleton and scales. It is improbable that any living animal preserves all the characters of the common ancestor of the group to which it belongs; some traits are preserved in one case, others in another.

There are some 10,000 species of fish included amongst the Teleostei. The limits of this work forbid us to mention more than a very few, and these will be chiefly the commoner food-fish. It is possible to separate a number of families, as Physostomi, distinguished by the two ancient characters of having the air-bladder opening into the oesophagus and having the pelvic fins abdominal in position, that is to say, placed far back near the vent. All the rest are termed Physoclisti; in them the air-bladder is a closed sac having lost its connection with the oesophagus, and the pelvic fins have been in nearly every case shifted forwards and are either thoracic, that is, placed just behind the pectoral fins, or jugular, that is, placed in front of them.

Systematic.

The families of the Physostomi which we shall mention are the Cat-fish (Siluridae), the Eels (Anguillidae), the Herrings (Clupeidae), the Salmon and Trout family (Salmonidae) and the Carp family (Cyprinidae).

The Cat-fish receive their name from the peculiar appendages called barbels which hang down from their upper and under lips. Other fish may have them on the under lip, but in the Siluridae

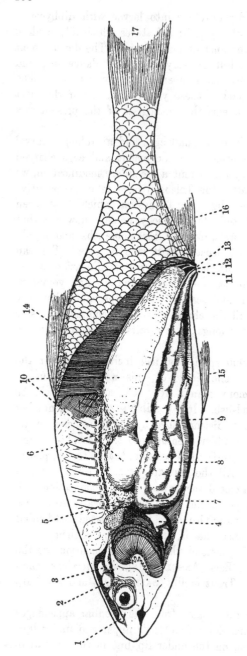

FIG. 225. A Roach, *Leuciscus rutilus*, dissected to show its brain, gills and viscera. Slightly diminished in size.

1. Olfactory lobe.   2. Right optic lobe.   3. Cerebellum.   4. Auricle of heart.
5. Lymphatic gland (the so-called head-kidney).   6. Kidney.   7. Liver.   8. Intestine.
9. Genital gland.   10. Air-bladder.   11. Anus.   12. Genital opening.   13. Opening
of ureter.   14. Dorsal fin.   15. Pectoral fin.   16. Pelvic fin.   17. Caudal fin.

the maxilla bears no teeth and forms only a support for the great barbels. The Cat-fishes are mainly a tropical group abounding in South America and Africa and nearly all are fresh-water. The forms with bony scutes are confined to South America, others are naked and without scales. These as a rule wallow in the mud at the bottom of the streams they live in. The teeth are feeble. The skull is not narrowed between the orbits; the anterior vertebrae are fused together and the sub-opercular bone is wanting. The first ray in the dorsal and in the pectoral fin is replaced by a strong spine. Several however of the naked species are found in North America, two being marine Sea-Cats. *Ictalurus* is the White Cat, an excellent food-fish ranging north to New England. *Amiurus* is common in the Eastern States, and one species, the Bull-head, *A. nebulosus*, is found in the St Lawrence.

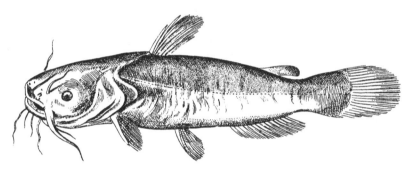

FIG. 226. A Cat-fish, *Amiurus catus*. Diminished. From Cuvier and Valenciennes.

The CYPRINIDAE, including the Carp, Gudgeon, Barbel, are allied to the Siluridae with which they agree in having the "Weberian organ." This is a modification of the anterior ribs and vertebrae to form a movable chain of bones connecting the air-bladder and the internal ear and thus enabling the animal to receive impressions of the pressure in the bladder. The Cyprinidae differ from the Siluridae in having typical scales and in wanting the maxillary barbel and in having a typical skull. *Carassius auratus*, the Gold-fish, belongs here.

The ANGUILLIDAE or Eels are long cylindrical creatures with either small deeply imbedded scales or none. The dorsal, ventral and caudal fins are contiguous, the pelvic fins being absent. The skull resembles that of the Cat-fish in not being narrowed between the orbits, and in having only one or two bones in the gill-cover. Eels spawn in the sea. *Anguilla,* the common eel, ascends streams and even crosses wet grass to get to isolated ponds; it has small scales. *Echelus (Conger),* the conger eel, is entirely marine, and devoid of scales; it attains a length of six feet. The eggs of both species develope into a peculiar ribbon-shaped larva with colourless blood and slightly developed tail called Lepto-cephalus.

The CLUPEIDAE or Herrings are distinguished by the fact that their maxillae bear teeth and form part of the edge of the jaw, and further that the maxilla is really composed of two or three pieces placed end to end, recalling the condition in the Lepidosteidae. The tail-fin is forked. The genus *Clupea,* with compressed belly edged by projecting scales, includes not only the Herrings, which come in shoals to the English and Scotch and American coasts to spawn, but also the Shad, a high-backed herring, which frequents the coasts of Canada and New England and which spawns in the rivers. The eggs of Clupeidae, as a rule, are large and heavy and sink to the bottom, unlike those of most Teleostei, which float at the surface of the sea. The Anchovy, with a projecting snout, and a rounded belly, the Pilchard and the Sprat also belong to this family. The Sardine is the young Pilchard.

The SALMONIDAE have a toothed maxilla and a jugal bone in the upper lip and a small soft fin devoid of rays behind the dorsal, called the adipose fin. In the female the oviduct has disappeared, the eggs escaping by two pores. The genus *Salmo* includes the well-known Salmon, which ascends rivers to deposit its spawn, and also the Brown River Trout of Europe, which is permanently confined to fresh-water. The brook-trout of North America belongs to a different genus, *Salvelinus,* and the great King-Salmon of British Columbia, canned in such enormous quantities, is *Onco-rhyncus tschawytscha.* From a sportsman's point of view it is distinguished from the true Salmon by the circumstance that it does not take the fly. *Coregonus,* the White Fish of the great American lakes, much esteemed for its delicate flavour, also be-longs to this group. The habit of ascending the rivers to spawn probably points to the conclusion that the whole group was origin-ally fresh-water, and that the Salmon is a river fish which has taken to the sea, whereas the Eel is a salt-water fish which has taken to fresh-water. Whether all the Herring family were ever fresh-water or not is doubtful, but it is interesting to note that the Pilchard has a small floating egg.

The great division of the Physoclisti includes five main subdivisions, the *ANACANTHINI*, the *ACANTHOPTERI*, the *PHARYNGOGNATHI*, the *LOPHOBRANCHII* and the *PLECTOGNATHI*. The last two groups are in some respects aberrant.

I.   The *ANACANTHINI* have all the fin-rays soft and flexible and the pelvic fins are shifted forward in front of the pectorals. The GADIDAE are the Cod family and include the Cod, Haddock, Whiting and Pollack. These fishes spawn out at sea, one female cod producing as many as 9,000,000 eggs. The PLEURONECTIDAE, or flat fish, also belong here. They are fish with compressed backs and bellies, and broad sides; they habitually swim on one side, and the eye belonging to the side kept downwards is twisted on to the upper side which deforms the bones of the skull. The anus is very far forward, the dorsal and ventral fins both being about equally long; the air-bladder is absent. The most valued member of this family in British water is the Sole, *Solea*, distinguished by its elongated shape. Other food-fish belonging to this family are the Flounder, the Plaice, the Turbot, the Brill and the Dab. On the coasts of Europe and of North America is found the immense Halibut, *Hippoglossus*, which may attain a length of 6 feet and a weight of 400 pounds.

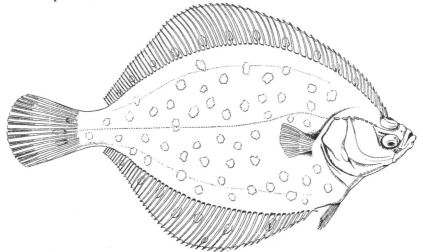

FIG. 227.   *Pleuronectes platessa*, the Plaice, found from the coast of France to Iceland.

II. The *ACANTHOPTERI* are the spiny-rayed fish in which the first rays of the dorsal and ventral fins are converted into bony dermal spines. Many families of most varied structure are included in this subdivision : the two best known are the SCOMBRIDAE, including the Mackerel and the PERCIDAE or Perches. The first are

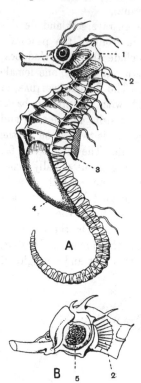

distinguished by the small dorsal fin supported by spines only, followed by a long dorsal fin the end of which is broken into finlets. In the Percidae the scales have a toothed posterior border, that is, they are ctenoid and the first spinous dorsal fin is long. The American fish called Bass belong to two families, the river-bass being one of the CENTRARCHIDAE, distinguished by high compressed body and undivided dorsal, whilst the sea-bass belongs to the SERRANIDAE, a family closely allied to the Perches.

III. The *PHARYNGOGNATHI* have the lower pharyngeal bones firmly united with one another. This division includes the LABRIDAE or Wrasses, distinguished by their thick lips and protrusible premaxillae; the TRIGLIDAE or Gurnards, which walk on the spines of the pectoral fin; the GOBIIDAE and many other small fish.

FIG. 228. A. The Sea-horse, *Hippocampus* sp. B. Head of the same, the operculum has been turned back to show the gills.

1. Branchial aperture. 2. Pectoral fin.   3. Dorsal fin. 4. Brood pouch.   5. Gills.

IV. The *LOPHOBRANCHII* are peculiar fish covered with rings of large plates; their gill processes are club-shaped instead of triangular, and are attached in tufts to the side of the clefts; there are no pelvic fins and the tail-fin is rudimentary. The jawbones, the pterygoid and maxilla, are elongated so as to form a long muzzle at the end of which is the tiny mouth. *Syngnathus* is the Pipe-fish which moves slowly amongst the long green fronds of the green sea-weed Zostera, which it resembles, picking off minute Crustacea and Molluscs. It is common both on the British and American coasts. *Hippocampus*, the Sea-horse, has the muzzle bent down at an angle with the rest of the body so as to present a whimsical resemblance to a horse's head. It anchors itself by curving its tail round weeds, and swims slowly but with dignity by means of the dorsal fin. This genus is found in the Mediterranean, on the southern parts of the American coast and elsewhere. The Lophobranchii show a peculiar mode of caring for the young ; the male has a brood pouch enclosed by two folds of skin on the underside of his body in which he carries the eggs until they are hatched.

V. The *PLECTOGNATHI* resemble the Lophobranchii in being covered with plates instead of scales and in having lost the pelvic fins. The pelvic fins are represented by spines or are entirely absent. The pre-maxilla and hyomandibular are immovably fused to the skull, while the inferior pharyngeal bones remain distinct. In *Ostracion*, the trunk-fish, the plates form as compact a cuirass as the shell of an Echinus : the only flexible spots being around the articulations of the fins and the lower jaw. In *Diodon* and *Tetrodon* the teeth have coalesced to form great transverse ridges of enamel, and the dermal plates bear spines which are usually directed backwards, but which are erected when the body is rendered tense by swallowing air into the gullet and stomach. These extraordinary fish are confined to tropical waters ; they haunt small crevices of the rocks, in which, when the tide retires, very small quantities of water are left, and it appears that the gills absorb oxygen from the air they swallow.

The Pisces are classified as follows :

Order 1. **Elasmobranchii.**

Pisces devoid of air-bladder or lung ; with placoid scales ; no bones developed except at the bases of these scales. Cartilaginous centra formed by the division of the notochordal sheath and not corresponding to the neural arches. The jaws slung to the skull by the second arch. No operculum : well-developed gill-sacs present.

Sub-order (1). **Selachoidei.**

Elasmobranchii of cylindrical form with well-developed tail-fin and pectoral fins of moderate size. Spiracle small or absent.

Ex. *Carcharodon, Scyllium, Acanthias.*

Sub-order (2). **Batoidei.**

Elasmobranchii of flattened form, the tail whip-like and the tail-fin rudimentary, the pectoral fin very large and joined to the skull. Spiracle very large and opening on the dorsal surface, openings of the other gill-slits ventral.

Ex. *Raia, Trygon.*

Order 2. Holocephali.

Pisces devoid of air-bladder or lung; the skin naked, a series of slender bony rings in the unsegmented notochordal sheath, besides this no other bones. Upper jaw completely confluent with the skull. An operculum present, well-developed gill-sacs.

Ex. *Chimaera.*

Order 3. Dipnoi.

Pisces with a large lung, sometimes divided into two, opening by a ventrally situated glottis into the oesophagus: the atrium of the heart divided into two, the left division receiving blood from the lung only. The body is covered with thin flat scales. Membrane-bones covering the skull and roof of the mouth. Cartilage bones in the upper jaw and in the hinder region of the skull, but the notochordal sheath is undivided. The upper jaw completely fused with the skull. An operculum present, the septa between the gill-sacs reduced so that they become gill-slits.

Ex. *Ceratodus, Lepidosiren, Protopterus.*

Order 4. Teleostomi.

Pisces with an air-bladder which returns blood into the cardinal veins, the atrium of the heart being undivided. The sheath of the notochord sometimes remains undivided, but when centra are present they are formed by the fusion of the expanded bases of the neural and haemal arches, not by the segmentation of the sheath of the notochord. Membrane-bones in the skull and roof of the mouth, and in addition a series in both upper and lower lips bearing teeth, situated outside and replacing functionally the original jaws. Cartilage bones in the jaws and skull. The upper jaw slung to the skull by means of the hyoid arch. A well-developed operculum present: the septa between gill-sacs are so narrowed that they form gill-slits with long branchial processes. The cloaca divided into two openings.

Sub-order 1.　**Crossopterygii.**

Teleostomi in which the pectoral and pelvic fins have the form of a lobe covered with scales fringed anteriorly and posteriorly with fin-rays. The air-bladder is bilobed and its opening is ventral. The scales are rhomboidal and covered with enamel.

Ex.　*Polypterus.*

Sub-order 2.　**Actinopterygii.**

Teleostomi in which the paired fins bear rays only on their posterior borders and in which the base of the fin is never covered with scales. The opening of the air-bladder is dorsal.

Division A.　Ganoidei.

Actinopterygii which retain the optic chiasma, several rows of valves in the conus arteriosus and a spiral valve in the intestine. A heterocercal tail.

Subdivision (1).　*Chondrostei.*

Actinopterygii clothed with large bony plates which pass uninterruptedly into the membrane-bones of the head, the notochordal sheath undivided; very few cartilage bones. A long snout projecting in front of the mouth. Teeth rudimentary.

Ex.　*Acipenser.*

Subdivision (2).　*Halecomorphi.*

Actinopterygii with thin scales; many cartilage bones and the notochordal sheath surrounded by well-formed amphicoelous vertebrae. Mouth terminal.

Ex.　*Amia.*

Subdivision (3).　*Lepidosteidae.*

Actinopterygii with rhomboidal scales covered with enamel; the skull completely ossified and the vertebrae opisthocoelous. Jaws very much elongated, each carrying a row of long teeth.

Ex.　*Lepidosteus.*

Division B. TELEOSTEI.

Actinopterygii in which the optic nerves cross without intermingling, the conus is absorbed into the ventricle, leaving one row of valves. No spiral valve in the intestine. A homocercal tail.

Section 1. **Physostomi.**

Teleostei which retain the opening of the air-bladder into the alimentary canal; the pelvic fins are abdominal in position.

Family (1). *Siluridae.*

Physostomi with a skin naked or covered with bony plates; the skull unconstricted between the orbits, maxilla without teeth and bearing a long barbel. Weberian chain present.

Ex. *Amiurus, Ictalurus.*

Family (2). *Cyprinidae.*

Physostomi with scales, the skull is constricted between the orbits, maxilla without teeth or barbel. Weberian chain present.

Ex. *Cyprinus.*

Family (3). *Anguillidae.*

Physostomi devoid of scales, skull not constricted between orbits. Weberian chain absent. No reproductive ducts. A continuous dorsal-caudal-anal fin and no pelvic fins.

Ex. *Echelus, Anguilla.*

Family (4). *Salmonidae.*

Physostomi with thin scales, the skull constricted between the orbits, the maxilla forming part of the edge of the jaw and bearing teeth. A small soft dorsal fin behind the main dorsal.

Ex. *Salmo, Salvelinus, Oncorhyncus.*

Family (5). *Clupeidae.*

Physostomi in most points resembling the last family; the maxilla consists of several pieces and there is no soft dorsal fin.

Ex. *Clupea.*

Section 2. **Physoclisti.**

The remaining sections of the Teleostei have lost the opening of the air-bladder into the alimentary canal, so that it becomes a closed vesicle.

Sub-section 1. **Anacanthini.**

Teleostei in which the fin-rays are all soft and flexible, and the pelvic fins are shifted forward anterior to the pectorals.

Family (1). *Gadidae.*

Anacanthini of symmetrical shape and not especially compressed.

Ex. *Gadus.*

Family (2). *Pleuronectidae.*

Anacanthini very much compressed laterally, which swim always on one side : the eye belonging to the lower side being rotated on to the upper side.

Ex. *Solea, Platessa, Hippoglossus.*

Sub-section 2. **Acanthopteri.**

Teleostei in which some at least of the fin-rays of the median fins are hard and unjointed.

Family (1). *Scombridae.*

Elongated Acanthopteri in which there is a short spinous dorsal fin followed by a long softer one, the hinder portion of which is broken up into finlets.

Ex. *Scomber.*

Family (2). *Percidae.*

Short stout Acanthopteri with one long dorsal fin followed by a short one and having a toothed posterior border to the scales.

Ex. *Perca.*

Family (3). *Serranidae* (Sea Bass).

Closely allied to the *Percidae*, but distinguished by having the dorsal fin undivided. Teeth large and numerous ; a large pseudobranch.

Family (3).   *Centrarchidae* (River Bass).

Acanthopteri which have a laterally compressed body and undivided dorsal fin. Teeth small; a small pseudo-branch.

Sub-section 3.   **Pharyngognathi.**

Teleostei in which the bones (rudimentary fifth gill-arches) bearing the pharyngeal teeth are firmly united together.

Ex. *Labrus.*

Sub-section 4.   **Lophobranchii.**

Teleostei covered with bony plates, the facial bones elongated so that the jaws are at the end of a tube-like proboscis. The branchial processes are arranged in tufts and thickened at their free ends. The pelvic fins absent.

Ex. *Syngnathus, Hippocampus.*

Sub-section 5.   **Plectognathi.**

Teleostei covered with bony plates, the pre-maxilla and hyomandibular immovably joined to the skull. The gills normal. Pelvic fins absent or represented by spines.

Ex. *Ostracion, Diodon, Tetrodon.*

CHAPTER XVI.

SUB-PHYLUM IV. CRANIATA.

Class II. AMPHIBIA.

THE class Amphibia includes the familiar frogs and toads, the less-known newts and salamanders, and some very curious worm-like tropical forms which burrow in the earth. The name means double life (Gr. ἀμφί, double ; βίος, manner of living), and refers to the fact that all the typical members of the class commence their lives as fish-like larvae, breathing by gills, and afterwards become converted into land animals, breathing by lungs. This strongly marked larval type of development is one of the great distinctions between the Amphibia and the only other class of Vertebrata with which they could be confounded, viz., the Reptiles. In the Reptiles, as in the Birds, a large egg abundantly provided with nutritive material is produced, and the young animal practically completes its development within the egg-shell and is born in a condition differing from the adult chiefly in size.

It might at first sight be thought that the fact that Amphibia breathe air in their later life and live on land would be sufficient to mark them off from the fish. But we have already seen that one order of fish—the Dipnoi—possesses lungs and breathes air, and on the other hand some Amphibia retain gills throughout life and rarely if ever leave the water.

The unbridged gap between true fish and Amphibia is to be found not in the breathing organ but in the structure of the limb. Fish possess fins—median and paired—which are in both cases supported by horny rays, as well as an internal skeleton ; and the paired fins have an internal skeleton which has the form of a jointed axis bearing similar rays on one or both sides (Figs. 205 and 215).

The Amphibian limb, on the other hand, is what is known as a pentadactyle limb; that is to say, it is constructed on the familiar type of the human limb, and the median fin when present has no fin rays (Figs. 229 and 231).

The pentadactyle or five-fingered limb (Gr. πέντε, five; δάκτῦλος a finger), also called the cheiropterygium (Gr. χείρ, a hand; πτερύγιον, little wing, hence an appendage), consists of three segments, a proximal, containing one long bone; a middle, containing two bones placed side by side and occasionally fused into one; and a distal, containing a series of small squarish cartilages or bones arranged in lines so as to give rise to a series of diverging rays ; the last-mentioned constitute the skeleton of the fingers and toes. In the proximal part of this lowest segment the bones are much crowded together and the rays tend to coalesce : this part has received a special name, as has also the portion where the rays although separate are embedded in the same muscular mass.

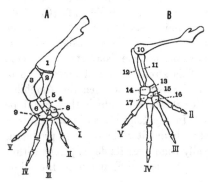

Fig. 229. A. A skeleton of a right posterior, and B of a right anterior limb of a Newt, *Molge cristata* × 1½.

1. Femur.   2. Tibia.   3. Fibula. 4. Tibiale. 5. Intermedium. 6. Fibulare. 7. Centrale of tarsus. 8. Tarsale 1. 9. Tarsalia 4 and 5 fused. I. II. III. IV. V. Digits. 10. Humerus. 11. Radius. 12. Ulna. 13. Radiale. 14. Intermedium and ulnare fused. 15. Centrale of carpus, the pointing line passes across carpale 2. 16. Carpale 3.   17. Carpale 5.

The fore-limb is called the arm, and its divisions the brachium or upper arm, the ante-brachium or fore-arm, and the manus or hand (B, Fig. 229). The hind limb is the leg, and its divisions are the femur or thigh, the crus or shank, and the pes or foot (A, Fig. 229).

The manus is divided into three regions, viz. : (*a*) the carpus or wrist where the rays tend to coalesce; (*b*) the meta-carpus or palm where the rays although separate are bound together by flesh and skin; (*c*) the digits or free ends of the rays.

The pes is similarly divided into tarsus or ankle, metatarsus or sole, and digits or toes.

The bone of the brachium is called the humerus, that of the femur bears the same name as the segment to which it belongs;

those of the ante-brachium are called radius and ulna, those of
the crus tibia and fibula (Fig. 229).

The skeletons of the pes and manus are typically exactly the
same. Situated proximally close to the middle segment of the limb
is a transverse row of three small bones, the central one being called
the intermedium in both limbs, whilst the outer and inner are
named after the bones of the middle segment of the limb adjacent
to them. Thus we find in the wrist a radiale and ulnare and
in the ankle a tibiale and fibulare. Beyond this row of bones
there is a single central bone which probably belongs to the middle
ray, and beyond it a row of five small bones corresponding to the
digits. This last row are denominated carpalia in the wrist and
tarsalia in the ankle. The individual bones are called carpale
(or tarsale) 1—5 in accordance with the digits opposite which they
are situated.

In almost every case this typical skeleton of nine bones has
undergone some modification, owing either to the absence of some
bones or the fusion of others, but in the hind-limb of the lower
Amphibia it is exactly typical. In the higher Amphibia not only
has great reduction of the elements taken place but the radius and
ulna in the fore-limb and the tibia and fibula in the hind-limb have
coalesced, a groove only being left to show their primitive distinctness.

The primitive position of the limbs with reference to the trunk
is, from the study of development, assumed to be one in which they
are stretched out at right angles to it, with the inner surface of the
hand and the sole of the foot directed ventrally and in such a
position that a line joining the tips of the fingers is parallel to the
long axis of the body. If we suppose an imaginary line or axis to
run down the centre of each limb, we shall be able to distinguish a
pre-axial from a post-axial side. In the lower Amphibia the only
change from this position that has taken place in the hind-limb is
that each segment of the limb is bent at right angles on the one
which follows it. The fore-limb is bent similarly, but it is also
rotated backwards so that its upper segment is almost parallel to
the axis of the body, and the elbow points backwards. If this
position were maintained the first digit would become external; but
the manus in most cases is at the same time twisted forwards so
that the lower end of the radius lies internal to that of the ulna,
and the radius thus crosses the ulna in its course. In the higher
Vertebrata this twisting can be undone and the hand reverted to

an untwisted position. This movement is known as supination, the reverse movement being known as pronation.

The hind-limb in the higher Amphibia and other Vertebrata is likewise rotated forward so that the knee points forward and the first digit is internal, but this does not occur in the lower Amphibia, such as *Molge*.

The pectoral girdle is not essentially different in the lower Amphibia and the more primitive Teleostomi, but the pelvic girdle is firmly joined to the transverse process of one of the vertebrae, which is called the sacral. This is one of the most distinctive features of all pentadactyle animals; it is a consequence of the adaptation of the pentadactyle limb to raise the body from the ground (Fig. 230). It is necessary for this purpose that the limb should have a firm purchase on the axial skeleton. Consequently when we find some Amphibia which never use their limbs for crawling but only for swimming, we assume that this is a secondary degenerate condition.

Next to the character of the limb one of the most distinctive features of Amphibia is the nature of the skin. Indeed the five great classes of Gnathostomata—Fishes, Amphibia, Reptiles, Birds, and Mammals—are each perfectly characterised by the nature of their skin. In a typical Amphibian the skin is soft and moist and devoid altogether of any ossifications like the scales of fishes. The skin is a most important breathing organ, since the lung alone cannot meet the demand for oxygen, and if the skin becomes dry and consequently incapable of absorbing oxygen the animal dies. The necessary moisture is supplied from a series of pockets, to form which the ectoderm is pouched inwards—or to use a more convenient term 'invaginated'—at various points, and the cells lining these pouches have the power of secreting great quantities of mucus. As the cells become

Fig. 230. Skeleton of Triton, *Molge cristata* × 1.

broken up into mucus, new cells take their place, being budded off from the underlying Malpighian layer just as the horny cells are. These pouches are known as dermal glands.

The skull and brain are very characteristic, recalling in many points those of the Dipnoi. The axis of the brain appears straight, as in fishes; in higher Vertebrates this axis is more or less folded. In contrast, however, with fishes, the cerebral hemispheres of the fore-brain are relatively large, whereas the cerebellum, usually so large in fishes, is reduced to a mere band (Fig. 240).

The skull always articulates by two pegs—the occipital condyles—with the first vertebra (Fig. 232). It is remarkable for its extremely flattened shape; the jaws are widely bent outwards so that the large eyes in no way compress the cranium, which is thus evenly cylindrical. Both membrane- and cartilage-bones are present, but the ossification is by no means complete. The exact arrangement of the bones will be given when a type is studied.

The vertebrae are either procoelous (Gr. πρό, in front; κοῖλος, hollow), or opisthocoelous (Gr. ὄπισθο-, behind), that is to say either concave in front and convex behind, or vice versâ, and the arrangement may differ in allied genera, while amphicoelous vertebrae also occur.

The vertebrae articulate with one another, not only by the centra but also by facets called zygapophyses (Gr. ζυγόν, a yoke), on the sides of the neural arches. The anterior facets, pre-zygapophyses, look upwards and are covered by the posterior facets or post-zygapophyses of the vertebra in front, which look downwards.

The circulatory system closely resembles that of the Dipnoi. The atrium is divided into two auricles, and the blood from the lungs returns direct to the left auricle by the pulmonary veins. A median vein, the inferior cava, returns the blood from the kidneys directly into the sinus venosus, receiving in its course the hepatic vein. The anterior portions of the posterior cardinals are much reduced in size and may be altogether absent.

The lungs open by a common stem, the laryngeal chamber, into the throat. The opening is called the glottis, and its sides are stiffened with cartilage.

The kidneys and reproductive organs show essentially the same arrangement as in the Elasmobranchs, the kidney being divided into a sexual part connected with the testis and a posterior non-sexual part. There is one opening for all ejecta, the cloaca.

The ventral wall of the cloaca, however, is produced outwards into a great thin-walled sac, the allantoic bladder, in which when the cloaca is closed the urine accumulates. This organ acquires immense importance in the development of the higher animals and is found in no fish.

In the larva, which is to all intents and purposes a fish, there are present those peculiar sense organs called mucous canals, supplied by the 5th and 10th nerves, but these are usually lost in the adult.

Living Amphibia are divided into three well-marked Orders, viz. the URODELA, the ANURA and the APODA. The URODELA (Gr. οὐρά, tail; δῆλος, conspicuous) have long cylindrical bodies and long flattened tails. The limbs are short and comparatively feeble, barely strong enough to lift the belly from the ground. Both pairs of limbs are about equal in size. The ANURA (Gr. ἀν-, no; οὐρά, tail) have much broader and shorter bodies; the tail is totally lost and the hind limbs are powerfully developed and adapted for jumping. The APODA (Gr. ἀ-, no, πόδα, feet) have lost both pairs of limbs and their cylindrical bodies give them a worm-like appearance; their habits heighten the resemblance since they burrow in moist earth. They have embedded in the skin small bony plates, relics of the scales which their Stegocephalous-like ancestors once possessed. The tail has in these animals almost disappeared.

In the Carboniferous rocks the remains of a large number of Amphibia have been found which have been called STEGOCEPHALA (Gr. στέγος, a roof; κεφἄλή, the head) from the circumstance that the head is covered with a compact mosaic of membrane-bones extending from the mid-dorsal line of the cranium outwards to the lips. Similar small bones or scales are found on the ventral surface. These features bear resemblances to what is found in Dipnoan or Crossopterygian fish from which Amphibia have probably descended, and the small scales of the Apoda seem to be the last remnants of this armature. Stegocephala include both long and short tailed forms, and while some of their descendants—the Labyrinthodonta—became highly specialized in the structure of their teeth and died out in the next geological period, others, in all probability, gave rise to modern Amphibia.

URODELA.

Returning to the Urodela, which are the most primitive of modern Amphibia, we find that in Great Britain they are represented by three species, all belonging to the genus *Molge* (*Triton*) and popularly known as efts or newts. *Molge cristata*, the warty eft, and *Molge vulgaris*, the common eft, are found in ponds and ditches all over the country, but *Molge palmata* is much more local. We may select *Molge cristata*, the greater or warty eft, or crested newt, as a type of the anatomy of Urodela (Fig. 231).

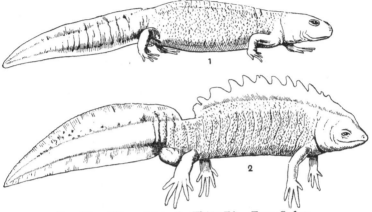

Fig. 231. *Molge cristata*, the Warty Eft. From Gadow.
1. Female. 2. Male at the breeding season with the frills well developed.

The animal is about five or six inches long, half the length being made up of the tail, which has a continuous fringe of skin, the median fin. This fin in the male extends forwards to the head dorsally and is greatly enlarged in the breeding season, but it is at all times devoid of fin rays.

The skin is clammy, owing to the secretion of the dermal glands: it is dark coloured above and yellow spotted with black below. The opening of the cloaca is placed behind the hind legs: it is a longitudinally placed oval slit which in the male has thickened lips.

The fore-limbs have only four fingers, the innermost corresponding to the human thumb being wanting, but there are five toes in the hind-limb. The animal when out of water crawls feebly along, but it swims actively in the water by means of its vertically flattened tail. The head is flattened dorso-ventrally and of somewhat oval outline, and the gape is of moderate extent. The eyes are small and project but little. The nostrils are very small and situated at the extreme front end of the snout.

If the newt be carefully watched when out of the water the skin of the underside of the head between the two sides of the lower jaw will be seen to throb at regular intervals, being alternately puffed out and drawn in. It can be further seen that the nostrils are closed when the skin is drawn in and opened when it is puffed out. These movements constitute the mechanism of breathing in the newt. As in the case of the Dipnoi, the paired nasal sacs communicate with the interior of the mouth by an opening called the choanae or internal nares, and the air passes through these from the nostril when the cavity of the mouth is enlarged. When the cavity of the mouth is compressed the nostril is closed by a flap of skin constituting a valve, and the air is forced through the open glottis into the lung, whence it is forced out again by the elastic recoil when the pressure is removed.

If the animal be laid on a board with the ventral side uppermost and skinned, a thin sheet of muscles, the mylo-hyoid, will be seen stretching between the two halves of the lower jaw. When this muscle is relaxed the floor of the mouth is arched upwards and the underside of the head consequently becomes concave. When the muscle contracts and straightens the cavity of the mouth enlarges and air is drawn in. Above the mylo-hyoid (underneath from the point of view of the dissection) are two longitudinal muscular bands, and in these are embedded the reduced remains of the visceral arches to which the gills of the larva were attached (Fig. 233). These muscles are called genio-hyoid in front of the arches, sterno-hyoid between them and the pectoral girdle, and they are continued backwards along the belly as the straight muscles of the abdomen, the recti abdominis.

These sterno-hyoid muscles can draw the visceral arches downwards and backwards and probably assist the mylo-hyoid in depressing the floor of the mouth. The genio-hyoid muscle on the contrary pulls the arches forwards and helps to restore them and the floor of the mouth with them to their old position. In this action muscles called petro-hyoid, which run from the arches to the outer surface of the auditory capsule, also take part. These muscles are representatives of the levatores arcuum of fish, and they raise the arches and consequently the floor of the mouth.

The glottis or opening into the lungs is stiffened at the sides by a pair of cartilages, which it seems probable are the remains of a hinder pair of visceral arches : and these cartilages have muscles attached to their sides which drag them apart and which belong to the same series as those which raise the arches. Hence the same muscular action which lifts the floor of the mouth opens the glottis and admits air into the lungs.

The remaining muscles of the body are not much altered from those of the fish. In the tail and the ventral part of the trunk there are V-shaped myotomes, but this arrangement is disturbed in the neighbourhood of the limbs.

Turning now to the skeleton we find that the vertebrae bear
Skeleton.
stout transverse processes with which are articulated short ribs (Fig. 230). The ribs borne by the sacral vertebra are expanded in accordance with the strain put on them by the attachment of the ilium. Of the vertebrae those of the tail are the most primitive since they are composed of all the four arcualia; but of these only the basi-dorsals and the basi-ventrals become ossified, and joining together form the bulk of the vertebra, while the inter-dorsals and inter-ventrals, although likewise fusing together, remain cartilaginous and form the inter-vertebral cartilage. This either remains continuous and owing to its flexibility acts as a joint, or it becomes more or less separated into a cup-and-ball portion. Joints in which the cup belongs to the posterior end of the vertebrae are called opisthocoelous, e.g., in *Desmognathus triton*. The basi-ventrals of the tail vertebrae form long downward haemal arches.

In the trunk the basi-ventrals occur only in early larvae; in the adult they have disappeared so that the bulk of such vertebra is formed only by the pair of basi-dorsals which alone carry the ribs, and these to compensate the loss of their capitular process have gained a new process dorsally from the tuberculum.

It is of importance to note that in many of the extinct Stegocephala, e.g., *Archegosaurus*, the caudal vertebrae were represented by four pairs of distinct arcualia, while the trunk vertebrae consisted of three separate pairs of pieces, namely the basi-dorsal, the inter-dorsal and the basi-ventral; but that in the typical Labyrinthodonts, the highest of the Stegocephala, all these constituent pieces were formed into solid vertebrae; lastly, that in some of the lowest, e.g., in *Branchiosaurus*, each vertebra consisted of a thin shell of bone surrounding the chorda, and composed of the basi-dorsals and basi-ventrals, which met each other, forming a broad-based section along the side of the vertebra, both partaking in the formation of a transverse process which carried the rib. The haemal arches of the tail are, like the ribs, outgrowths of the basi-ventral, but they do not exactly correspond to the ribs, for they are placed nearer the middle line.

In the skull the cranium is cylindrical, being quite uncompressed between the eyes. The bones of the jaws and face are

widely arched outwards, so that the whole skull has a flattened
shape. The nasal and auditory capsules form easily recognizable
buttresses projecting from the cranium.

In both the floor and roof of the cartilaginous cranium the
proper wall is largely deficient. The deficiency of the roof is the

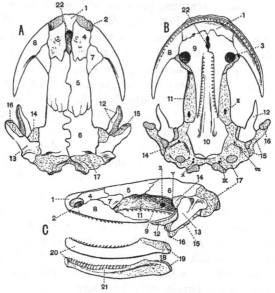

Fig. 232. A dorsal, B ventral, and C lateral views of the skull of a Newt,
Molge cristata × 2½. After Parker.

The cartilage is dotted, the cartilage-bones are marked with dots and dashes, the
membrane-bones are left white.

1. Premaxilla.   2. Anterior nares.   3. Posterior nares.   4. Nasal.
5. Frontal.   6. Parietal.   7. Prefrontal.   8. Maxilla.   9. Fused
vomer and palatine.   10. Parasphenoid.   11. Orbitosphenoid.
12. Pterygoid.   13. Squamosal.   14. Pro-otic region of fused exoccipital
and pro-otic.   15. Quadrate.   16. Calcified cartilage forming the
articular surface of the quadrate.   17. Exoccipital region of fused
exoccipital and pro-otic.   18. Articular.   19. Articular cartilage.
20. Dentary.   21. Splenial.   22. Middle narial passage, a cleft in
the cartilage of the snout filled with connective tissue.   II. V. VII. IX.
X. Foramina for the exit of cranial nerves.

anterior fontanelle, in the floor the greatly enlarged pituitary fossa.
But these deficiencies are not seen in the uninjured skull, because
the hole in the roof is closed in by two pairs of membrane-bones, the
frontals and the parietals, and that in the floor is underlaid by
a broad parasphenoid membrane-bone (Fig. 232).

Only at its extreme front and hind ends is the wall of the cranium
converted into cartilage-bone. In front there is on each side an

orbito-sphenoid bone, in the side wall, extending into the roof and
floor and ossifying also the hind wall of the nasal sac ; behind, two
exoccipital bones[1] are placed at the sides of the foramen magnum,
which they nearly encircle (Fig. 232).   These bones bear the two
condyles, so characteristic of Amphibia, for articulation with the
vertebral column.

The first visceral arch, which constitutes the cartilaginous jaws, is
almost entirely cartilaginous.   It consists of an upper part immovably attached to the skull, corresponding to the pterygo-quadrate
bar or upper jaw of Fish, and a lower part, Meckel's cartilage,
forming the basis of the lower jaw.   It will thus be seen that
Amphibia, like Holocephali and Dipnoi, are autostylic.   The same
is true of all the higher groups of the Craniata.   The upper jaw
consists of two regions, the suspensorium which is fused with the
skull and to which the lower jaw is attached, and the pterygoid
process, a spur of cartilage which runs
forward towards the nasal capsule.
Both suspensorium and the articular
end of Meckel's cartilage are slightly
calcified.   They are denominated quadrate and articular in Fig. 232, but
there is no true bone present in either
case.   The front of the auditory capsule
is ossified by a large bone, the pro-otic,
which in fully adult specimens becomes
confluent with the exoccipital.   The
hinder visceral arches in the adult are
present in a very degenerate condition.
Traces of three remain (Fig. 233).

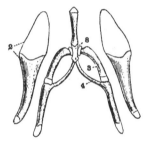

Fig. 233.   Visceral arches of
Molge cristata.   The ossified
parts are slightly shaded, the
cartilage is white.   From
Parker.

2.   Hyoid arch.   3.   First
branchial arch.   4.   Second
branchial arch.   8.   Copula,
i.e. the median piece connecting successive arches.

It is usual to speak of the hinder
visceral arches of Amphibia and higher
Vertebrata as the hyoid apparatus,
or simply as the hyoid.   The name suggests a misleading comparison with the second visceral arch of Fish; it is distinctly to be
remembered that the hyoid bone of even Man contains more than
this second arch; a good definition of the hyoid of Amphibia and
higher animals would be "the degenerate remains of the hinder
visceral arches."

Turning now to the membrane-bones of the skull, we find that it

[1] Exoccipitals as here used are equivalent to the lateral occipitals.   The
exoccipitals of Owen = the epiotics or occipitalia externa.

is roofed by three pairs, viz., the nasals, frontals and parietals. The nasals of course roof in the nasal sacs. In the palate there is one median bone, the parasphenoid, and three pairs of lateral bones, viz., the vomers in front of the posterior nares, the palatines fused with them and running along the edges of the parasphenoid, and lastly the pterygoids underlying the pterygoid process. Some of these bones are actually built up by the fusion of the bases of minute conical teeth in the larva. The vomers and palatines retain their teeth in the adult, whilst the parasphenoid loses them.

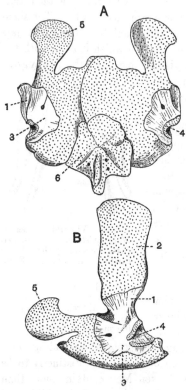

Fig. 234. A, ventral, and B, lateral view of the shoulder girdle and sternum of an old male Crested Newt, *Molge cristata* × 3. After Parker.

1. Scapula. 2. Supra-scapula. 3. Coracoid. 4. Glenoid cavity. 5. Precoracoid. 6. Sternum.

The upper lip has tooth-bearing pre-maxillary and maxillary bones developed, the lower has a dentary on the outside of Meckel's cartilage and a splenial on the inner. Above the maxilla there is a small pre-frontal bone.

If we examine the skeleton of the limbs we find that the pectoral girdle consists of two plates of cartilage which slightly overlap in the mid-ventral line. The lower half of each is forked, the forks being called precoracoid and coracoid respectively. The centre of each half of the girdle has a hollow termed the glenoid cavity for the articulation of the arm. All around the glenoid cavity the girdle is converted into bone; there is a bone termed the scapula above, and a coracoid bone below. The unossified part of the coracoid is simply termed the coracoid cartilage. The upper part of the girdle dorsal to the scapular bone is called the suprascapula. It remains cartilaginous, but is often calcified. The two coracoids are fastened behind to a small median cartilage called the sternum. The meaning of this is discussed later.

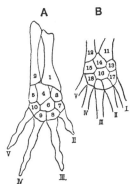

The manus has only four fingers, the thumb and the corresponding small bone in the wrist or carpus having disappeared and the ulnare and intermedium being fused, although they are distinct in the larva (Fig. 235 A). Otherwise the limb corresponds to the scheme given in the beginning of the chapter.

The pelvic girdle on each side is firmly joined to the rib of the sacral vertebra, and the two halves meet in the mid-ventral line. The upper part of the girdle above the cavity for articulation of the thigh is a bone, the ilium; below this cavity, which is termed the acetabulum, is a so-called "ischiopubic" cartilage, in the hinder part of which a small bone, the ischium, is developed. In the mid-ventral line, in front of the union of the two halves of the pelvic girdle, there is a forked piece of cartilage, the epipubis (Fig. 236).

Fig. 235. A, Right antebrachium and Manus of a larval Salamander, *Salamandra maculosa*. After Gegenbaur.

B, Right Tarsus and adjoining Bones of *Molge sp.* After Gegenbaur.

1. Radius. 2. Ulna. 3. Radiale. 4. Intermedium.
5. Ulnare. 6. Centrale.
7. Carpale 2. 8. Carpale 3.
9. Carpale 4. 10. Carpale 5. 11. Tibia. 12. Fibula.
13. Tibiale. 14. Intermedium. 15. Fibulare.
16. Centrale. 17. Tarsale 1.
18. Tarsalia 4 and 5 fused.
I. II. III. IV. V. Digits.

In the pes the only departure from the typical arrangement is the fusion of the tarsalia 4 and 5.

Viscera.

If the muscles be carefully cut through in the middle line and reflected, the body cavity and the organs contained therein will be exposed. In general the difference from the arrangement of the organs in a dog-fish is only in the relative size of the organs, in a word, in details.

The alimentary canal is thrown into a number of loops. The oesophagus is not in any way sharply marked off from the stomach, and

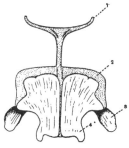

Fig. 236. Pelvic girdle of *Molge cristata* from below × 4.

1. Pre-pubic process. 2. Ilium foreshortened. 3. Pubo-ischium. 4. So-called hypo-ischium.

the latter is nearly straight, extending only a short way round the bend of the first loop. There is a well-marked large intestine or rectum, ventral to which lies the bladder. The spleen is an

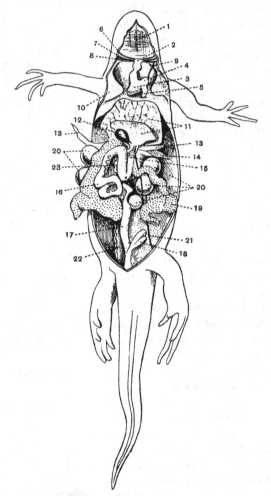

FIG. 237. A male *Molge cristata* cut open so as to expose the internal organs, about natural size.

1. Mylo-hyoid muscle with genio-hyoid underneath.   2. Conus arteriosus.
3. Ventricle.   4. Auricle.   5. Sinus venosus.   6. Carotid arch.
7. Systemic arch.   8. Pulmonary artery.   9. Anterior vena cava of left side.   10. Coracoids pulled outwards.   11. Liver.   12. Gall-bladder.
13. Lung.   14. Spleen.   15. Stomach.   16. Intestine.   17. Rectum.
18. Allantoic bladder.   19. Fat-body.   20. Testes.   21. Anterior abdominal vein, displaced.   22. Kidney with duct.   23. Pancreas.

oval red body lying at the side of the stomach and attached to the mesentery. The ducts of the pancreas and liver coalesce into a common stem before opening into the intestine.

The newt feeds on small worms and aquatic insects, which it seizes with its jaws. Both upper and lower jaws are armed with minute teeth, and there are in addition two longitudinal rows of teeth on the roof of the mouth borne by the conjoined vomer and palatine on each side. The function of these teeth is not so much to crush as to retain a hold of the prey, which is swallowed whole. The tongue is a circular cushion on the floor of the mouth, supported by the second visceral arch. Its hinder edge is partially free. The lungs are long, smooth-walled, tube-like elastic sacs, attached to the liver and other organs at their base, but their tips float freely in the body-cavity.

The heart lies far forward, between the roots of the lungs, enclosed in the pericardium. Externally all the four divisions of the piscine heart are visible, viz., sinus venosus, atrium, ventricle, conus. The venous system is essentially that of the dog-fish, only the veins are indicated by names borrowed from human anatomy. Thus the blood from the head is returned by two internal jugular veins, representing the anterior cardinals of the fish. These are joined by external jugulars from the superficial part of the throat and face and by a sub-clavian vein from each arm. The common trunk formed by the union of all three is, of course, the Ductus Cuvieri, but it is called the superior vena cava, and it receives on each side close to the middle line a posterior cardinal vein. As in fishes, this vein in its course breaks up into capillaries through the kidney, and along the outer edge of the kidney, its posterior portion, the renal portal, may be made out. The two renal portals when followed further back are found to coalesce in the caudal vein which returns the blood from the tail: each receives a sciatic vein from the dorsal side of the leg joined by a femoral from the ventral surface of the limb.

The increased importance of the hind limb has brought with it this increase in the vessels draining it, which are represented only by the small pelvic vein in fishes.

There are certain vessels, however, unrepresented in any fishes except the Dipnoi. These are: first, the pulmonary veins, which receive the blood from the lungs and open directly into the left side of the atrium, which is separated from the rest by a septum

Vascular System.

and constitutes the left auricle; secondly, the inferior vena
cava, a large trunk situated in the median dorsal line just beneath
the aorta, which receives most of the blood that has traversed the
kidneys and conveys it into the sinus venosus just between the
openings of the two superior venae cavae. The inferior cava
coalesces with the hepatic veins
returning blood from the liver :
these thus lose their independent
openings into the sinus venosus
which they had in the Dog-fish.
In its hindermost portion between
the kidneys the vena cava joins
the posterior cardinal.

So far the peculiarities of the
Newt are shared by the Dipnoi :
but there remain two veins highly
characteristic of Amphibia. The
musculo-cutaneous vein receives blood from the skin and
pours it into the subclavian ; we
have already seen that the skin
is a very important breathing
organ, and this vein returns the
blood which has been oxygenated
in the skin to the heart. The
anterior abdominal vein
arises on the ventral side of the
body near the cloaca from the
union of two forks given off by
the femoral veins ; it runs forward in the mid-ventral line,
eventually joining branches of
the portal vein and entering the
liver. This vein is found also in
the lower Reptiles and in the
embryos of Mammalia, where it is of the utmost importance in both
nutrition and respiration.

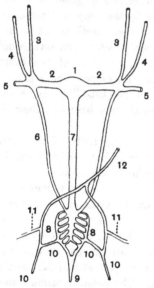

Fig. 238. Diagram to show arrangement of the principal Veins of an
Urodele.

1. Sinus venosus, gradually disappearing in the higher forms. 2. Ductus
Cuvieri = superior vena cava. 3. Internal jugular = anterior cardinal
sinus. 4. External jugular =
sub-branchial. 5. Subclavian.
6. Posterior cardinal, front part.
7. Inferior vena cava. 8. Renal
portal = hinder part of posterior cardinal. 9. Caudal. 10. Sciatic.
11. Femoral. 12. Anterior abdominal.

When the veins are cut away it is possible to follow out the
arteries. There is no ventral aorta, since on each side three arterial
arches arise in a bunch from the front end of the tubular conus.
The first of these is called the carotid arch, and is derived from

the third arterial arch of the embryo, but unlike its equivalent in Dipnoi it does not communicate with the dorsal aorta. It gives off a lingual artery to the tongue and throat and then passes up round the gullet, to which it gives off some twigs and continuing as the common carotid supplies the upper part of the head and brain. Just after giving off the lingual artery the arch swells up into a little knot, called the carotid gland. In this structure the channel of the artery is broken up into a network of fine passages and its function is believed to be that of holding back the blood from entering the head until, at the close of the contraction of the ventricle, the blood has returned from the lungs. The second arch, derived from the fourth embryonic arch, supplies most of the blood to the root of the dorsal aorta, and on this account is called the systemic arch. The fifth and sixth embryonic arches in later stages unite on each side into one trunk, which passing round the gullet joins the systemic arch. From the sixth arch is given off

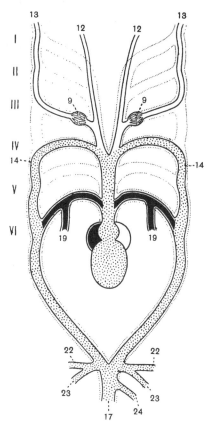

FIG. 239.   Diagram of arterial arches of *Molge*, viewed from the ventral aspect.

I. II. III. IV. V. VI. first to sixth arterial arches.   9. Carotid gland.   12. Lingual (ventral carotid).   13. Common carotid (dorsal carotid).   14. Systemic arch. 17. Dorsal aorta.   19. Pulmonary. 22. Subclavian (dorsal type).   23. Cutaneous.   24. Coeliaco-mesenteric.

the pulmonary artery which supplies the lung. On this account it is called the pulmonary arch. The systemic arch on either side gives off a subclavian artery to the fore-limb: and from its place of origin it will be seen that this subclavian is of the dorsal type

(p. 350). The subclavians originate close to the junction of the two systemic arches and each gives off a large branch to the other breathing organ, the skin, which is known as the cutaneous artery. In *Molge* the fifth arterial arch disappears, as it does in all Vertebrates above the Amphibia, but in the allied genus *Salamandra* it is retained in the adult.

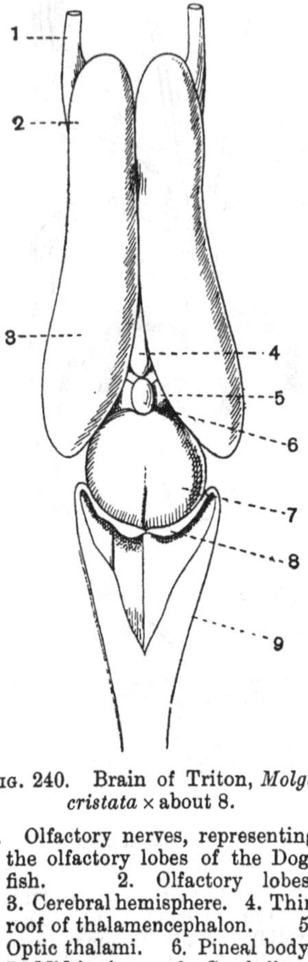

It is comparatively easy to uncover the brain and spinal cord of the newt owing to the thinness of the bones which cover them. The cerebral hemispheres are long and cylindrical, and devoid of any other connection with one another than that by way of the thalamencephalon; through the thin roof of the latter two thickenings in its floor, the optic thalami, can be clearly seen. The mid-brain is a simple smooth vesicle, and the cerebellum is a slight inconspicuous transverse band (Fig. 240).

The olfactory lobe of Amphibia differs from that of Pisces in being separated from the cerebrum only by a slight constriction. From its anterior end a brush of nerves is given off which goes to the nasal sac. The so-called olfactory nerve of the Dogfish which is the stalk connecting the olfactory lobe and the cerebrum is unrepresented in the Amphibia.

FIG. 240. Brain of Triton, *Molge cristata* × about 8.

1. Olfactory nerves, representing the olfactory lobes of the Dogfish.   2.   Olfactory lobes. 3. Cerebral hemisphere. 4. Thin roof of thalamencephalon.   5. Optic thalami.   6. Pineal body. 7. Mid-brain.   8. Cerebellum. 9. Medulla oblongata.   From Burckhardt.

The course of the cranial nerves is substantially the same as in the Dogfish; owing, however, to the loss of the gills and the mucous canals in the adult, the branches are simplified. The 9th or glossopharyngeal, as its name implies, is distributed to the pharynx and tongue. The vagus supplies the larynx and glottis, but its main stem runs on to the heart and stomach.

The first spinal nerve comes out from behind the first vertebra and is called the hypoglossal; it runs directly to the respiratory muscle, the mylo-hyoid, crossing the vagus and glossopharyngeal in its course. At the sides of the dorsal aorta the two chains of sympathetic ganglia can be made out, connected by cross branches with the spinal nerves.

To turn now to the excretory system, the kidney can be seen when the alimentary canal is removed. It is a long narrow strip on each side adjacent to the aorta. In front it tapers to the merest thread, but behind, close to the cloaca, it thickens somewhat. Along its outer edge runs the archinephric duct, and external to the archinephric duct is situated the long oviduct.

Urino-genital organs.

The tubules which compose the kidney retain throughout life the ciliated openings into the body-cavity, and if the narrow part of the kidney be cut off and mounted in a little salt solution it is possible, at least in small specimens, under a low power of the microscope, to see the funnels and to observe the whirlpools due to the currents produced by their cilia.

The genital gland in both sexes is represented by a pair of ridges suspended to the inner edges of the front parts of the kidney by slings of peritoneum similar to the mesentery suspending the gut, and on this account called mesenteries. In the female the oviduct opens

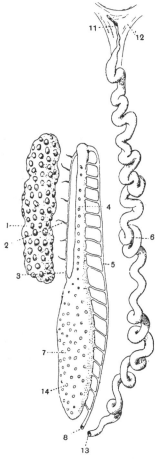

FIG. 241. Urino-genital organs of a Female *Molge cristata* × about 5.

1. Ovary.   2. Remnant of vasa efferentia.   3. Remnant of longitudinal canal connecting the vasa efferentia.   4. Sexual portion of kidney.   5. Archinephric duct.   6. Oviduct.   7. Posterior non-sexual portion of kidney.   8. Opening of archinephric duct.   11. Internal opening of oviduct.   12. Suspensory ligament.   13. External opening of oviduct.

by a ciliated funnel adjoining the root of the lung. The funnel
leads into a long convoluted tube running back to open into
the cloaca. The testis, which
takes the form of two conical
bodies with their broad ends
apposed, or sometimes a row
of three rounded lobes, communicates by a number of
vasa efferentia with the anterior part of the kidney,
which is on this account
termed the sexual portion or
mesonephros. In the male
the kidney tubules belonging to the hinder non-sexual
portion, or metanephros, are
split off from the archinephric
duct and unite into a short
common trunk, the ureter,
which joins the archinephric
duct just before the latter
enters the cloaca.

It has been stated above
that the genital glands are
a pair of ridges. In the
larva the inner portions of
the ridges degenerate, the
cells becoming largely converted into fat-bodies. In
the adult these fat-bodies
appear running parallel to the
genital organs on the inner
side. They serve as a store
of nourishment for the eggs
which develope during the
winter-sleep. The Newt, like
other Amphibia, passes the
winter buried in the mud at
the bottom of ponds and takes
no food. The conversion of

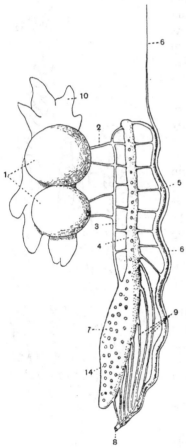

Fig. 242. Urino-genital organs of a Male
        Molge cristata × about 5.

1. Testes. 2. Vasa efferentia. 3. Longitudinal canal connecting the vasa efferentia.   4.   Sexual portion of kidney
showing nephrostomes.   5.   Wolffian
duct. 6. Rudimentary oviduct. 7. Non-
sexual portion of kidney.   8. External
opening of the archinephric duct which
has received the ureter 9 made up of
a number of ducts from the posterior
part of the kidney.   10.   Fat-body.

some of the possible eggs into fat to feed the rest is simply an example of the same principle as the sacrifice of some of the dogs in an Arctic expedition to feed the rest.

The development of *Molge* is interesting.   The male emits the spermatozoa in a bundle which the female then introduces into her cloaca, and the eggs commence their development in the body of the mother.   Soon afterwards they are laid and attached to water plants.   After some time larvae are hatched out which in many respects resemble fishes.   They are provided with three long feathery appendages on each side of the neck, in which there is a rich blood supply and active circulation. These are the external gills found only in Amphibia, Dipnoi and in *Polypterus*.   There is also a pair of curious rod-like organs in front of the gills attached to the sides of the head.   These "balancers," as they are termed, are possibly a first pair of external gills peculiarly modified.   They have mucous cells at the tip, and by means of

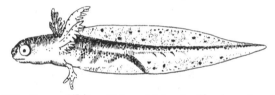

FIG. 243.   Larva of Triton, *Molge cristata* × 5.   Showing external gills.
After Rusconi.

them the young larva suspends itself for hours at a time to plants. There is a long fish-like tail, the organ of locomotion, with a fringed fin.   The fore-limbs are tiny buds.   No trace of hind-limbs exists and the gill-slits are not open.

As development proceeds the fore-limbs make their appearance provided with only two toes.   The gill-clefts, three in number, appear on each side.   After a considerable time the third finger appears and the hind legs sprout out as buds; still later the fore-limbs get all four fingers and the hind-limbs five.   The animal has now attained the appearance of the adult except in so far as the gills are concerned.   These are retained for a long time, and exceptionally, in Switzerland in high Alpine localities, the larva may become sexually ripe and never leave the water.   More usually with the closing of the gill-slits and the shrivelling of the external gills the adult state is attained.

The Urodela have for a long time been divided into two main groups, according to the presence during adult life of gill-slits and gills. Huxley thus divided them into ICHTHYOIDEA and SALA-

Classification.

MANDROIDEA. But this has been criticized as not being based upon fundamental characters. Huxley's ICHTHYOIDEA are those which retain throughout life gill-slits or external gills or both. Invariably the limbs are reduced in size, the animals rarely if ever leaving the water. In one case the hind-limbs have totally disappeared.

North America is the great head-quarters of the Ichthyoidea. *Menopoma* (*Cryptobranchus*) retains one gill-slit throughout life. This animal attains a length of 18 inches. It is fairly common on the Mississippi and its tributaries. An allied species found in Japan, and attaining a length of two feet, is the largest living Amphibian.

Amphiuma is a snake-like animal about 18 inches long, with one gill-slit. It is found in the same region as *Menopoma*. The limbs are exceedingly rudimentary, each having only two toes.

Necturus, the Mud-puppy, has small but well-developed limbs. It retains throughout life two gill-slits and three external gills on each side. *Necturus* is abundant in the shallows of the St Lawrence, wriggling in and out around the roots of aquatic plants. A somewhat similar animal, *Proteus*, with more rudimentary limbs, is found inhabiting the limestone caverns of Carniola in Austria. Lastly, there is the aberrant *Siren*, which has a horny beak en-sheathing the premaxilla and dentary; it has no hind-limbs, but is similar to *Necturus* in its gills: it is found inhabiting the swamps of the Southern United States.

Since the Ichthyoidea possess both gills and lungs it is tempting at first sight to regard them as the little modified descendants of an animal just making the transition from water-breathing to air-breathing life. There are however insuperable difficulties in the way of such an explanation. If we turn to other groups of the animal kingdom we find that the first step in fitting an animal for a land life is the covering up of the respiratory organ so as to protect it against drying up. But in hardly any fish are the respiratory organs so exposed as in *Necturus*, *Proteus* and *Siren*.

Further, it was pointed out that the great gap between fishes and Amphibia is to be found in the structure of the limb. But the Ichthyoidea do not in any way assist in bridging the gap. On the contrary their limbs are obviously degenerating, a fact which seems to show that the aquatic life has been re-acquired. Now when the similarity between say *Necturus* and the late larva of *Molge* is borne in mind, and the further fact that these larvae may abnormally become sexually ripe, the conclusion is irresistibly suggested that the Ichthyoidea are larvae in which the adult stage has been suppressed. In the case of one large American newt, *Amblystoma tigrinum*, the larva (the "Axolotl") often breeds under certain circumstances and was at one time regarded as a distinct genus (*Siredon*).

The second division of Urodela, the SALAMANDROIDEA, are in general very similar to *Molge*, both in appearance, anatomy and size.

As in Ichthyoidea, so likewise North America is very rich in Salamandroidea. These have been divided into families on grounds of differences in the skeleton which have little effect on the external appearance. The most abundant are the AMBLYSTOMATINAE represented by the genus *Amblystoma* of which there are many species, nine being found in the Eastern States and Canada. The members of this family are distinguished by having the palatine bones directed transversely, so that the vomero-palatine rows of teeth run across the roof of the mouth instead of along it, and by having amphicoelous vertebrae.. *Molge* (*Diemyctilus*) *viridescens* is the common Water-Newt of Lower Canada. It is a member of the same genus as the English Newt which has been selected for detailed description, but unlike its English congener the American species does not develope a crest in the breeding season. These Newts are representatives of the SALAMANDRINAE distinguished by having the vomero-palatine teeth in a longitudinal row and by possessing opisthocoelous vertebrae. The family DESMOGNATHINAE are closely allied to the Amblystomatinae, but differ from the latter in possessing a cluster of teeth on the parasphenoid in addition to the transverse row of vomero-palatine teeth and in having opisthocoelous vertebrae. The species of this family are common Water-Newts in the Eastern United States. *Desmognathus nigra*, the black Salamander, occurs near Montreal. The PLETHODON-TINAE includes the American Cave- and Land-Newts which rarely enter water but wriggle about actively on land. These Newts resemble the Desmognathinae in their teeth, but differ in possessing amphicoelous vertebrae. Although the most terrestrial in their habits of the New World Urodela, these animals and some of the Desmognathinae have undergone an extraordinary modification in their respiratory system. The lungs have disappeared and the septum between the auricles has become absorbed : so the animals depend for their oxygen entirely on their skin and the lining of the pharynx, the walls of which still execute active respiratory movements. This curious association of terrestrial habits with the absence of lungs suggests the idea that the lung in such Urodela as retain it may be chiefly used as a hydrostatic organ like the air-bladder of fish, for were it of prime importance as a respiratory organ it would be difficult to explain its disappearance in terrestrial forms. *Spelerpes* includes the Cave-Newts, of which there are twenty species in America and one isolated species in Italy. In these animals the tongue is long and not adherent to the floor of the mouth. It can be suddenly protruded and is used to catch insects in the same way as the tongue of the Anura. This is an exceptional action amongst Urodela, most of which seize their prey with the jaws. *Plethodon erythronotus* has the typical tongue. This is the common Land-Newt in the neighbourhood of Montreal, being found under old logs and in other damp situations.

II. ANURA.

The Anura or Batrachia are at once recognized by their broad, flattened, tailless bodies and their powerful hind-limbs.

Structure.

These limbs are not only efficient in jumping but also in swimming, and the toes are connected with one another by a thin web of skin in order to aid them in performing this function. The toes are stretched apart in the back stroke to present a large surface to the water, in the forward stroke they are folded together and offer little resistance.

Anura are much more abundant than Urodela and are found all over the world, whereas the Urodela are restricted to the Northern hemisphere. They are in fact the dominant Amphibia of the present day, but they are highly specialized, and the Urodela give a much better idea of the relation of the Amphibia to the Fishes on the one hand and the Reptiles on the other, for which reason *Molge* was selected as the type.

Besides the absence of a tail, the powerful character of the hind-limbs and the differences in the skeleton connected therewith, Anura differ from Urodela in the skull and jaws, in the pectoral girdle, in the heart and lungs, and in the kidneys, genital organs and development.

Two genera and four species of Anura occur in the British Isles. *Rana temporaria*, the common frog, and *R. esculenta*, the edible frog (the last named is thought by some not to be indigenous but to have been introduced), represent the family RANIDAE, while the BUFONIDAE or toads are represented by *Bufo vulgaris*, the common toad, and by *B. calamita*, the Natterjack, which occurs in numbers in certain restricted localities, as a rule those with a sandy soil.

As the Common Frog, *Rana temporaria*, is easily attainable, the principal points in which it differs from *Molge* will be briefly described.

The Frog.

The animal when at rest normally squats on its haunches, supporting itself slightly on its palms. Under these circumstances, the pelvic girdle makes a considerable angle with the vertebral column and the powerful iliac bones raise the skin of the back into a well-marked hump, the so-called sacral prominence.

The gape is enormous, and is caused by the lower end of the suspensorium, or part of the skull to which the lower jaw is hinged, slanting backwards instead of projecting directly downwards as in Urodela. The tongue is fixed to the floor of the mouth in front, but is free behind; it can be rapidly thrust out of the mouth by bending the posterior end forwards and it can be as rapidly retracted. It is used to whisk the insects on which the animal feeds into the mouth.

Behind the eye is a circular patch of thin, tightly stretched skin. This is the ear-drum or tympanic membrane, which closes externally the Eustachian pouch of the gullet. It is believed that this pouch or tympanum is the remains of the first gill-cleft, the spiracle of Elasmobranch fishes. Sound impinging on the ear-drum is conveyed to the wall of the ear capsule by a row of several small cartilages, the so-called columellar chain of the ear. In the Urodela sound has to find its way as best it can through the skin and muscle of the head to the auditory organ. All Anura possess Eustachian pouches and a columella auris, but all do not have a well-developed ear-drum.

The skin is most loosely attached to the muscles underneath. Large spaces containing lymph are interposed between them. These lymph spaces form a protection against the danger of drying up. There are two pairs of sacs placed, one pair just between the upper ends of the pectoral girdle, and another pair just at the sides of the rudimentary stump of a tail, which have the power of contraction and pump the surplus lymph into the veins of the neighbourhood. These are called the anterior and posterior pairs of lymph-hearts.

Turning now to the skeleton we observe many points of

difference between the Frog and the Newt. The ribs in the Frog
are indistinguishably fused with the transverse processes; in very
few Anura are they distinct and they are always rudimentary. The
vertebrae differ from those of the Urodela in the entire suppression
of the inter-ventral element so that the centrum is constructed out
of basi-dorsal, inter-dorsal and basi-ventral elements, the last
named being very rudimentary. In some Anura the basi-ventral
piece is entirely absent, and in this case, since the centrum is
constructed entirely of dorsal elements, the notochord is found for
a considerable period of development lying in a groove on its under
surface. This is the so-called epichordal type of development.

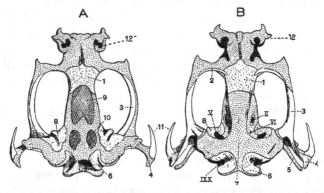

Fig. 244. A, Dorsal, and B, Ventral view of the Cranium of a Common Frog,
   Rana temporaria, from which the membrane-bones have mostly been
   removed × 2. After Parker.

1.   Sphenethmoid.   2.   Palatine.   3.   Pterygoid.   4.   Suspensorium.
5.   Columella.   6.   Exoccipital.   7.   Ventral cartilaginous wall of
cranium.   8.   Pro-otic.   9.   Anterior fontanelle.   10.   Right
posterior fontanelle.   11.   Quadratojugal.   12.   Nasal capsule.
II. V. VI. IX. X. foramina for exit of cranial nerves.

The tail vertebrae are represented by a bony style, the urostyle.
Besides it there are only nine vertebrae. The transverse processes,
or "diapophyses" of the ninth or sacral vertebra, to which is
attached the ilium, are either cylindrical as in *Rana*, or they are
more or less wide and flat as is the case in *Bufo* and *Hyla*. In
most cases to the distal end of the diapophysis is attached a nodule,
the rudimentary rib, which may either fuse with the diapophysis as
in *Rana* or remain distinguishable throughout life as in *Alytes*.
   The skull is constructed on the same plan as that of *Molge*,
but it is broader and flatter; this is due to the wide arching
out of the upper jaws, leaving a very large opening between them

and the cranium. The cause of this again is to be sought in the large protruding prominent eyes, so marked a feature of all Anura. The floor of the cartilaginous cranium is complete in the Frog, the pituitary fossa having shrunk to insignificant dimensions. The orbitosphenoids have coalesced to form a box-like bone which ossifies not only in the front part of the cranium but also in the hinder parts of the nasal sac, and is called the sphenethmoid. The parietal is fused with the frontal.

The suspensorium sends forward a pterygoid process which

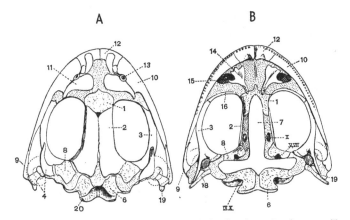

A　　　　　　　　　　B

Fig. 245. A, Dorsal, and B Ventral views of the Cranium of a Common Frog, *Rana temporaria* × 2. After Parker.

In this and the next two figures the cartilage is dotted, cartilage bones are marked with dashes, membrane-bones are left white.

1. Sphenethmoid. 　2. Fronto-parietal. 　3. Pterygoid. 　4. Squamosal.
6. Exoccipital. 　7. Parasphenoid. 　8. Pro-otic. 　9. Quadratojugal.
10. Maxilla. 　11. Nasal. 　12. Premaxilla. 　13. Anterior nares.
14. Vomer. 　15. Posterior nares. 　16. Palatine. 　18. Columella.
19. Quadrate. 　20. Occipital condyle. 　II. Optic foramen. 　V. VII.
Foramen for exit of trigeminal and facial nerves. 　IX. X. Foramen for exit of glossopharyngeal and pneumogastric nerves.

becomes attached to the skull in the nasal region. Underneath the posterior part of the pterygoid process there is a pterygoid bone which surrounds it and partly replaces it. The pterygoid sends out a fork which underlies that part of the suspensorium which forms an articulation for the lower jaw. The front part of the pterygoid process where it bends in to rejoin the skull is ossified by the palatine, which like the pterygoid has become a cartilage bone. The palatine is transverse to the axis of the skull, as in *Ambly-stoma*. Neither palatine nor pterygoid bears teeth, but the vomers

bear a little group of teeth towards their hinder edge. These
vomerine teeth are used for crushing the food.

The upper lip has a series of three bones on each side, reaching
completely to the suspensorium, an additional quadrato-jugal
being added to the two present in the Newt. The presence of this
bone suggests that the ancestors of the Anura are to be sought
amongst that highly modified group of the Stegocephala termed the
Labyrinthodonta. In them as

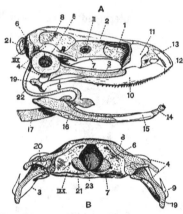

Fig. 246. A, Lateral view of the Skull,
B, Posterior view of the Cranium, of a
Common Frog, *Rana temporaria* × 2.
After Parker.

1. Sphenethmoid.   2. Fronto-parietal.
3. Pterygoid. 4. Squamosal. 5. Tympanic membrane.   6. Exoccipital.
7. Parasphenoid.   8. Pro-otic.
9. Quadratojugal.   10. Maxilla.
11. Nasal. 12. Premaxilla. 13. Anterior nares.   14. Pre-dentary.
15. Dentary. 16. Splenial. 17. Basi-
lingual plate.   19. Quadrate.
20. Columella. 21. Occipital condyle.
22. Anterior cornu of the hyoid (ceratohyal).   23. Foramen magnum. II.
IX. X. Foramina for the exit of cranial
nerves.

in the Anura the interventral
element was absent but at
any rate in the older forms
the basi-dorsals, the basi-ventrals and the inter-dorsals
were distinct pieces. In all
Anura there is a large membrane-bone of a characteristic
T-shape, known as the squamosal, lying outside the suspensorium. In the lower lip
there is a splenial and a
dentary, whilst in front the
cartilaginous lower jaw is replaced by a pre-dentary
bone. In the frog only the
premaxilla and maxilla and
vomer bear teeth. Most Anura
agree with the Frogs in this,
but the Toad, *Bufo*, and its
allies are entirely toothless.

The hinder visceral arches
are reduced to a still more
rudimentary condition than
those of *Molge*. They are
represented by a thin plate
of cartilage called the basi-lingual with short blunt processes, of
which only the last pair, which embrace the glottis, are ossified
(Fig. 247). This pair are termed the thyro-hyals. The whole
"hyoid" is thus the remains of the visceral arches.

The pectoral girdle is much more strongly developed than in
the Urodela. The coracoid and pre-coracoid processes are joined
at their inner ends by a longitudinal bar, the epicoracoid, so as to

enclose a space called the coracoid foramen. The two epicoracoids are in the frog firmly united in the middle line. In many Anura however they merely overlap (Fig. 248, B).

The upper portion of the pectoral girdle is ossified by a bone called the scapula. As in Urodela, however, the cartilage projects a long way beyond it, and this portion is called the supra-scapula and may become partially ossified. There is a distinct coracoid bone ossifying the coracoid process, and the pre-coracoid is underlain by a membrane-bone called the clavicle. In front of the pectoral girdle in the middle line lies a small rounded piece of cartilage called the episternum, followed by a bony piece, the omosternum. Behind the girdle in a similar position is a carti-

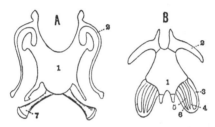

Fig. 247. Visceral arches of Amphibia. A. *Rana temporaria* adult. After Parker. B. Tadpole of *Rana*. After Martin St Ange.

In A the ossified portions are slightly shaded, while the cartilaginous portions are left white.

1. Basilingual plate. 2. Hyoid arch. 3. First branchial arch. 4. Second branchial arch. 5. Third branchial arch. 6. Fourth branchial arch. 7. Thyrohyal = fourth branchial arch.

laginous bar with a flattened end, ensheathed by a bone called the sternum; the flattened end is called the xiphisternum. The omosternum has proved to be composed of a portion budded off by the conjoined epicoracoids. The sternum is supposed to be the first sign of the breast-bone of higher Vertebrates, but as their breast-bone originates in connection with long ribs, which meet one another in the mid-ventral bone, this must be considered doubtful.

In the arm the two points to be noticed are the complete fusion of the radius and ulna into one bone, and the reduction of the carpus, in which there are only six bones, three of the distal small bones having coalesced and the centrale being absent. The first digit or pollex is rudimentary.

In the pelvic girdle there is no epipubis: the ilium is a very long cylindrical bone: the ischium ossifies most of the ischio-pubic

cartilage and is closely applied to its fellow. In the leg the tibia
and fibula are fused into one bone, which is about the same length
as the femur. The ankle is remarkably elongated, the tibiale and
the fibiale being long cylindrical bones, easily mistaken for the
middle segment of the limb. The distal bones of the tarsus have
nearly disappeared, only two or three small nodules being present
on the axial side. The longest toe is the fourth, that corresponding to the human big toe (hallux) is the shortest. It is a matter
of great interest to see on the inner side of the foot a spur
supported by a small bone which may be the vestige of a sixth

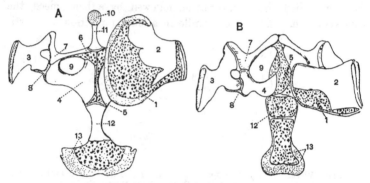

FIG. 248.   Shoulder-girdle and Sternum of

A.   An old male Common Frog, *Rana temporaria.*

B.   An adult female, *Docidophryne gigantea.* After Parker, to illustrate the
structure in Arcifera.

In both A and B the left suprascapula is removed. The parts unshaded
are ossified; those marked with small dots consist of hyaline cartilage, those
marked with large dots of calcified cartilage.

1.   Calcified cartilage of suprascapula.   2.   Ossified portion of suprascapula.
3.   Scapula.   4.   Coracoid.   5.   Epicoracoid.   6.   Precoracoid.
7.   Clavicle.   8.   Glenoid cavity.   9.   Coracoid foramen.   10.   Episternum.   11.   Omosternum.   12.   Sternum.   13.   Xiphisternum.

digit. It is a common occurrence for the number five to be
diminished, but very rare for it to be increased. It is believed that
the pentadactyle limb is derived from a fin like that of the Dipnoi
by a shortening of the main axis and a reduction in the number of
rays, and it would be not unnatural to expect to find in the lower
groups of land animals traces of extra rays.

The main differences between the circulatory system of the Frog
and that of the Newt are to be found in the arterial system. Not only
as in the Newt does the fifth arterial arch of the embryo disappear
altogether, but the sixth becomes entirely cut off from the aorta and

in addition to supplying the lung it sends a large branch to the skin, for which reason it is called the pulmo-cutaneous arch. The conus arteriosus, as in *Molge*, has two transverse rows of pocket valves, one near the heart and one near the outer end, but in the Frog there is in addition a longitudinal valve with a free ventral edge running somewhat obliquely from the one row of valves to the other. When the ventricle contracts it is, at first, full of venous blood from the right auricle. At this stage the conus is relaxed, a condition which arranges the longitudinal valve in such a way as to divert the blood almost exclusively into the pulmonary passages, whose width and shortness also favours its flow into them. As these become filled the conus contracts, and this has the effect of making the longitudinal valve lie against the openings into the pulmonary arches and so preventing any more blood entering them, while at the same time the path into the systemic arches is widely opened. By this time some of the blood has returned from the lungs to the left auricle, and so mixed blood passes to the hinder portion of the body. When the pressure in the ventricle rises to its highest point, the last blood, which is almost completely arterial,

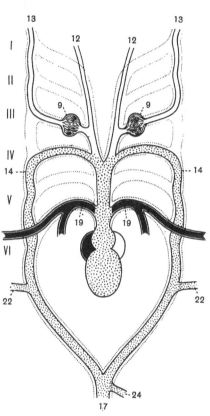

Fig. 249. Diagram of Arterial Arches of Frog viewed from the Ventral Aspect.

I. II. III. IV. V. VI. First to sixth arterial arches. 9. Carotid gland. 12. Lingual (ventral carotid). 13. Common carotid (dorsal carotid). 14. Systemic arch. 17. Dorsal aorta. 19. Pulmo-cutaneous artery. 22. Subclavian (dorsal type). 24. Coeliaco-mesenteric.

—all from the right auricle having been driven out,—is able to

overcome the resistance in the carotid gland and go to the head, which contains the organs having the greatest need for thoroughly oxygenated blood.

The posterior cardinal veins are represented only by their hinder portions, the renal portals, all the blood from the kidneys being carried by the inferior vena cava.

The brain of the Frog and of Anura in general is more highly

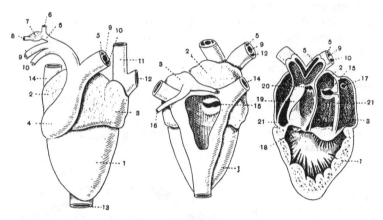

FIG. 250.

A. The heart, after removal from the body, seen from the front, the aortic arches of the left side having been removed. From Howes.

B. The same from behind, the sinus venosus having been opened up, to show the sinu-auricular valves.

C. The same, dissected from the front, the ventral wall together with one of the auriculo-ventricular valves having been removed.

1. Ventricle.   2. Right auricle.   3. Left auricle.   4. Truncus arteriosus.   5. Carotid arch.   6. Lingual artery.   7. Carotid gland.
8. Carotid artery.   9. Systemic arch.   10. Pulmocutaneous arch.
11. Innominate vein.   12. Subclavian vein.   13. Vena cava inferior.
14. Vena cava superior.   15. Opening of sinus venosus into right auricle.
16. Pulmonary vein.   17. Aperture of entry of pulmonary vein.
18. Semi-lunar valves.   19. Longitudinal valve.   20. Point of
origin of pulmocutaneous arch.   21. Rod passed from ventricle into the
truncus arteriosus, indicating the course taken by blood which flows into
the carotid and aortic arches.

developed than that of the Urodela. Thus (Fig. 251) the olfactory lobes of the cerebral hemispheres are connected together, and the optic lobes of the mid-brain are well developed.

It was pointed out (p. 368) that the limbs of Vertebrates are in all probability derived from two lateral flaps of skin—two longi-tudinal fins. The muscles in these fins were originally prolongations of the myotomes, and the nerves were of course branches of the motor nerves going to the myotomes. Now as these longitudinal flaps were converted into paired fins, and these by a continual narrowing of their bases acquired greater distinctness from the body, the portions of the myotomes supplying the musculature and the nerves in connection therewith became so to speak bunched together at the base of the limb. In adult Craniata all trace of the original metameric arrangement of the limb muscles is lost; but the metamerism of the nerves can still be seen, and the bundles of these supplying the pectoral and the pelvic limbs are known as the brachial and the sciatic plexus respectively. In the Frog, where the limbs are of far greater importance to the life of the animal than are the fins to fish, the nerves forming the brachial and the sciatic plexus are powerful trunks (Fig. 251, 2, 3, and 7—10).

The lungs are shorter than in the Newt but much wider, and their inner surface is covered with a network of low ridges which much increases their area. The kidney is a comparatively short and broad organ, very different from the long tapering organ of the Newt. The testis is connected by vasa

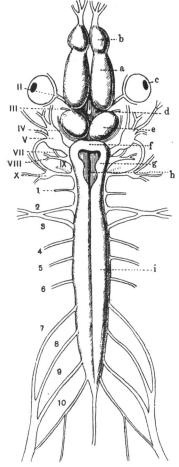

FIG. 251. Brain and Spinal Cord of a generalized Anuran. In the Phaneroglossa the 1st spinal nerve is suppressed. × about 2.

a. Cerebral hemisphere.   b. Olfactory lobe. c. Eye. d. Thalamencephalon. e. Optic lobes.   f. Cerebellum.   g. Medulla oblongata. h. Fourth ventricle. i. Spinal cord. I. Olfactory nerves. II. Optic nerve. III. Oculomotor nerve. IV. Patheticus.   V. Fifth nerve.   VII. Facial nerve. VIII. Auditory nerve. IX. Glossopharyngeal nerve.   X. Vagus nerve. 1—10. First to tenth spinal nerves. 2 and 3 unite to form the brachial, and 7, 8 and 9, to form the sciatic plexus.

efferentia with certain special tubules of the kidney. These tubules
do not open into the archinephric duct, but into a special duct
which runs along the surface of the kidney and opens into the
archinephric behind. Thus in a somewhat different way the
separation of urine and spermatozoa is carried out quite as

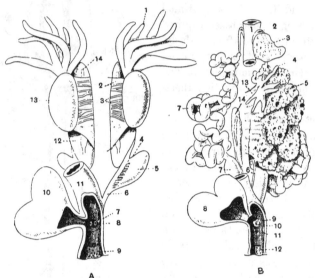

Fig. 252. The Frog.

A. The urino-genital organs of the male, dissected from the front, after
removal from the body. From Howes.

B. The urino-genital organs of the female, dealt with in the same manner as
the above, except that, in order to show the natural relations of the mouth
of the oviduct, the left lung and a portion of the oesophagus were also
removed from the body.

A. 1. Fat-body.   2. Fold of peritoneum supporting the testis.   3. Efferent
ducts of testis.   4. Ducts of vesicula seminalis.   5. Vesicula seminalis.
6. Archinephric duct.   7. Cloaca.   8. Orifice of ureter.
9. Proctodaeum.   10. Allantoic bladder.   11. Rectum.   12. Kidney.
13. Testis.   14. Adrenal body.

B. 1. Oesophagus.   2. Mouth of oviduct.   3. Left lung.   4. Corpus
adiposum.   5. Left ovary.   6. Archinephric duct.   7. Oviduct.
8. Allantoic bladder.   9. Cloaca.   10. Aperture of oviduct.
11. Aperture of archinephric duct.   12. Proctodaeum.   13. Fold of
peritoneum supporting the ovary.   14. Kidney.

efficiently as in the Newt. The archinephric duct has a number
of pouches developed on its walls which collectively form the
vesicula seminalis in which the spermatozoa are stored up. In
Bombinator the vasa efferentia apparently open directly into the
archinephric duct in front of the kidney.

Lying on the ventral surface of the kidney near its inner edge is an elongated body called the adrenal body (Fig. 252 A, 14). This organ is found under various forms in most Vertebrates; it has been recently shown to be derived from a peritoneal furrow which becomes shut off from the general coelom and loses its cavity, forming a solid rod of cells. Experiments made on higher animals and the observation of cases where it is attacked by disease, show that the adrenal bodies, like the thyroid, produce an "internal secretion." The substances poured into the blood by both these organs are essential to the proper conduct of metabolism, that of the adrenal bodies being stimulating to the muscular tissues in particular.

The eggs develope entirely outside the body, and there is a large thin-walled swelling of the oviduct in which the ripe eggs accumulate

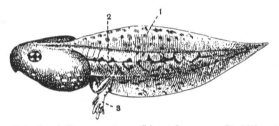

FIG. 253.   Tadpole of *Rana esculenta*, Lin. taken near St Malo × 1.   From Boulenger.

1.   Dorsal fin.   2.   Tail showing myotomes.   3.   Hind-limb.

just before being discharged. The male clasps the female round the waist and remains in this position sometimes for weeks, uttering loud croaks at intervals until the eggs are discharged. When the eggs are discharged he emits the spermatozoa on to them. The croaks are made by pumping the air from the lungs through the glottis into the pharynx and *vice versâ*. The pharynx has usually two side pouches, the vocal sacs, which become inflated with air. It is thus possible for the frog to croak when under water.

The development is in many respects different from that of Urodela. Soon after the young are hatched they acquire, it is true, three external gills on each side, but there is no trace of limbs and the gill-slits are closed, and as the mouth does not open into the alimentary canal no food is taken. Later the gill-slits appear; but a flap of skin, the gill-cover, grows back from the second visceral arch (the hyoid) and covers up the gill-slits and the external gills. The external gills then soon disappear. The two gill-covers unite with one another beneath the animal, so only one little opening to the gill-chamber remains, usually on the left side. The mouth has by this time opened into the alimentary canal, and it is provided with two horny ridges, one above and one below, besides rows of little horny prickles. The horny jaws crop the water-weeds upon which the tadpole lives.

The larva is now the well-known tadpole, with a rounded body and a long flat tail, with which it swims. The limbs gradually grow, but for a long time the front limbs are hidden beneath the gill-cover. When they finally burst through the animal sheds its horny jaws and leaves the water. For a short time the tail is retained, but absorption soon removes all trace of it and the development is complete.

The Anura are divided into two main groups according to the development of the tongue. In the **Aglossa** it is Classific-ation. entirely absent and the two Eustachian tubes have a common opening into the pharynx. This curious group only includes two genera. In one species, *Pipa americana*, the Surinam toad, the eggs are emitted from the protruded oviduct on to the back of the female, and here the young pass through the tadpole stage enclosed in deep pockets of the moist skin. This species as its popular name implies is an inhabitant of S. America. In the **Phaneroglossa**, on the other hand, the tongue is well-developed, being usually free behind, and in this case used to flick the prey, which consists of insects, into the capacious mouth. The Eustachian tubes are separate. The Phaneroglossa are divided into the **Arcifera** and the **Firmisternia**. In the first division the two epicoracoids of each side overlap (Fig. 248, B); in the second they are firmly united in the middle line (Fig. 248, A). The first division includes several families, but the two largest and most important are those of the toads or BUFONIDAE and the tree frogs or HYLIDAE.

The toads have no teeth whatever: their wrinkled skin is beset with wart-like poison glands in the upper parts, while numerous little horny spines occur superficially in the epidermis. They only enter the water at the breeding season and toads are in many respects more adapted to a land life than are frogs. Two species live in Great Britain; *Bufo vulgaris*, found everywhere, and *Bufo calamita*, the natterjack, a species with comparatively feeble hind-limbs, which crawls and does not jump. The natterjack frequents sandy places and is thus local in its distribution.

One species of Bufo (*B. americana*) is found in the north of North America. But besides the Bufonidae another family of the Arcifera, the PELOBATIDAE, which have teeth, is represented by *Scaphiopus*, a burrowing species, provided with a sharp spur on the inner side of each foot, whence the name "spade-foot" toad.

The HYLIDAE have teeth on the vomers and on the upper jaw, but their most remarkable peculiarity consists in the possession of fleshy cushions underneath the terminal joints of the digits, the bones of which are bent up and claw-like. By means of these cushions the Hylidae are able to adhere to smooth vertical surfaces, and so climb trees, in which they mostly live, only approaching the water for the purpose of laying their eggs. There is no species of this family in Great Britain and only one in Europe. In North America there are several species belonging to three genera; *Hyla*, *Chorophilus*, and *Acris*.

The *FIRMISTERNIA* have the two epicoracoids fused in the middle line and include the Frogs or RANIDAE. There is only one species, *Rana temporaria*, which is here taken as the type of the Anura, really native to Great Britain, but there exist a few colonies of the common European species, *Rana esculenta*, mostly in the Eastern Counties. The frogs of this species are most powerful croakers, and as their name implies they are used as food. It is believed that they were introduced by monks from Europe, who before the Reformation used to pay periodical visits to England to supervise their property.

In Canada and the Northern United States there are eight species of frogs. A species believed to be identical with *Rana temporaria* is found, but the two commonest are *Rana virescens*, of a green ground colour with lines of velvety black patches, and the great Bull-frog, *Rana catesbiana*, which attains three or four times the size of *Rana temporaria*, and is of brownish-yellow colour, peppered over with minute black dots.

III. APODA.

The order Apoda is, as has already been mentioned, distinguished by the entire absence of limbs and the worm-like appearance and habits of its members. In the skeleton the retention of a complete roof of bones over the space between cranium and upper lip, known as the temporal fossa, and the existence of minute bony scales embedded in the dermis, are features retained from the Stegocephala. In accordance with their retiring burrowing habits the members of this Order have very small eyes, which in some cases are rendered quite functionless by being concealed under the skin. The internal anatomy is in many respects like that of the Urodela, but the pulmonary arterial arch does not in all cases join the aorta. These animals often live at some distance from water and the larval development is passed through inside the egg-shell, but even there the embryo developes large external gills. The species of this family are restricted to the tropics; *Ichthyophis* is found in India, *Coecilia* in South America, and *Hypogeophis* in Africa. The extinct Stegocephala have been alluded to many times. Under this comprehensive head are comprised all the fossil Amphibia, remains of which are found in the Coal Measures and the Red Sandstones overlying them. It has been already pointed out that some of them, like *Branchiosaurus*, appear in the structure of the vertebral column to be the forerunners of the Urodela, while others, like the *Labyrinthodonta*, appear to lead on to the Anura. Besides these, limbless forms are also known, and there seems to be some probability that these were the ancestors of the Gymnophiona. Hence within this ancient group the beginnings of the division of the Amphibia into the three Orders by which it is now represented had already shown themselves.

The class of recent Amphibia is divided as follows:

Order 1. **URODELA.**

Amphibia retaining throughout life a long tail.

Family (1) AMPHIUMIDAE.

Both the upper and lower jaws are furnished with teeth. Fore and hind limbs small. Eyes small and devoid of lids. The gill-slits are in a vanishing state, the gills disappear in the adult.

Ex. *Amphiuma, Cryptobranchus japonicus, C. (Menopoma) alleghaniensis* the Hell-bender.

Family (2) SALAMANDRIDAE.

Both the upper and lower jaws are furnished with teeth. Eyes with movable lids. No gills or gill-slits in the adult.

Ex. *Molge, Salamandra, Desmognathus, Plethodon, Amblystoma.*

Family (3) PROTEIDAE.

Both the upper and lower jaws with teeth. Eyes without lids. Maxillary bones absent. With permanent gills.

Ex. *Proteus, Necturus.*

Family (4) SIRENIDAE.

Both jaws are toothless. The hind limbs, the maxillary bones and the eyelids are absent. With permanent external gills.

Ex. *Siren.*

Order 2. **ANURA.**

Amphibia which lose when adult all trace of tail, hind-limb much more powerful than the fore-limb and used for leaping.

Group I. **Arcifera.**

Phaneroglossa in which the epicoracoids of opposite sides overlap.

Family (1) DISCOGLOSSIDAE.

Arcifera with a round disc-shaped tongue, adherent at the whole of its base ; vertebrae opisthocoelous. Teeth in the upper jaw only.

Ex. *Discoglossus, Bombinator.*

Family (2) PELOBATIDAE.

Arcifera with a protrusible tongue, dilated sacral ribs and with teeth in the upper jaw only.

Ex. *Pelobates, Scaphiopus.*

Family (3) BUFONIDAE.

Like the previous family, but without any teeth.

Ex. *Bufo.*

Family (4) HYLIDAE.

Arcifera with dilated sacral ribs, with teeth in the upper jaw and adhesive discs on the fingers and toes.

Ex. *Hyla, Chorophilus, Acris.*

Family (5) CYSTIGNATHIDAE.

Arcifera with cylindrical sacral ribs.

Ex. *Pseudis, Ceratophrys.*

Group II. **Firmisternia.**

Phaneroglossa in which the epicoracoids are firmly united in the middle line.

Family (6) ENGYSTOMATIDAE.

Firmisternia with dilated sacral ribs.

Ex. *Engystoma.*

Family (7) RANIDAE.

Firmisternia with cylindrical sacral ribs.

Ex. *Rana.*

Order 3. **APODA.**

Amphibia of worm-like appearance, without limbs or tail and with vestigial eyes.

Ex. *Coecilia, Hypogeophis, Ichthyophis.*

CHAPTER XVII.

Sub-Phylum IV. Craniata.

Class III. Reptiles.

The name Reptile denotes literally anything that creeps (Lat.
General Char- *repo* or *repto*, to crawl). Zoologically the term
acteristics. denotes cold-blooded quadrupeds which are covered
with horny scales and which lay large eggs, inside the shells of
which the whole development is completed.

But it is not merely the size of the egg nor even the character
of the embryonic development which distinguishes Reptiles from
Amphibia. There are isolated cases of species of Amphibia in
which the development is practically completed within the egg-shell,
but in all Amphibia the whole egg becomes converted into the body
of the larva. In Reptiles on the other hand part of the egg is made
into a hood termed the amnion, which is wrapped around the body
of the embryo. This structure is cast off entirely at birth and the
wound caused by its tearing is healed. With it is also cast off a
portion of the urinary bladder—the allantois in the stricter and
original sense—which extends into the amnion and appears to
subserve respiration during embryonic life. In strictness, therefore,
only a part of the egg is converted into the body of the embryo.

This peculiar mode of development is shared by Birds and
Mammals, for which reason these two classes are often included with
the Reptiles in the term Amniota.

Next to the development perhaps one of the most characteristic
features of Reptiles is the nature of their skin. They are typically
covered with scales which are widely different from the scales of
fish. The latter are essentially areas of the dermis hardened by the
deposition of lime with sometimes the addition of a layer of crystals
from the basal ends of the ectoderm cells (enamel).

The scale of the Reptile on the contrary is nothing but an area of the horny layer of the skin where the cells are converted into horn or Keratin and are adherent to one another. In the mass of the scale the horn is rendered brown by the presence of pigment, but the outermost layer is composed of clear cells and is known as the epitrichial layer. A corresponding layer covers the embryos of Birds and Mammals, but is shed before birth. A sloughing or ecdysis of the scaly epidermis is a constant feature of the Reptilia. It may take place bit by bit, or as is the case with many Sauria the whole 'skin' is cast in one piece.

The dermal glands so characteristic of the Amphibia have almost totally disappeared, being restricted to a small area, as, for

FIG. 254.   Section through the Scale of a Lizard.
1. Epitrichial layer.   2. Heavily cornified cells forming the scale.   3. Pigment cell.   4. Ordinary cells of horny layer.   5. Innermost Malpighian layer.   6. Dermis.

instance, the front of the thigh in a lizard. It follows that a Reptile is essentially a dry-skinned animal and by no means a "slimy beast."

Besides the structure of the skin Reptiles are distinguished from Amphibia by many other points in the anatomy. Thus the skull has a larger number of cartilage bones, and includes what may be considered primitively part of the vertebral region since the hypoglossal nerve (see p. 435) is now a cranial nerve. The skull articulates with the vertebral column by one condyle. In the heart the conus arteriosus has disappeared and the ventricle is partly divided. The sexual part of the kidney is entirely disjoined from the asexual.

The lungs have to some extent acquired a spongy texture, and the mechanism for inhaling and exhaling air is usually to be found in the ribs, not in the hyoid or remains of the hinder visceral arches as in the Amphibia.

Living Reptiles are divided into five Orders, of which one consists only of one species, *Sphenodon punctatus*, found in New Zealand. This animal is the type of the Order (i) **Rhyncho- cephala**, and is especially interesting as not only being to some extent intermediate in structure between other Orders of living Reptiles, but as recalling very closely the structure of some of the oldest fossil Reptiles known to us; indeed it retains in many respects a structure which we believe was possessed by the common ancestors of the remaining four groups. These are the (ii) **Lacertilia** (Lizards), the (iii) **Ophidia** (Snakes), the (iv) **Chelonia** (Turtles and Tortoises) and lastly the (v) **Crocodilia** (Alligators and Crocodiles). Of the five Orders only the second and third are repre- sented in Great Britain and these by very few species; in North America the last four are well represented. The Lacertilia and the Ophidia are the most closely related and they are often grouped together under the same term **SAURIA**.

As type we may select the common lizard, *Lacerta vivipara*, which may be seen on very warm days disporting itself in sandy and stony places in the south of England. On the Continent it and allied species are far more abundant: in the South of Europe in summer the whole country is alive with lizards. Almost every step in the country causes two or three specimens to rush rapidly away into some retreat, either a hole under a stone or a cleft in the bark of a tree.

Structure of Lacerta.

The English Lizard has roughly the shape of a Newt, but there is a distinct neck region in front of the fore-limb, and the limbs are sufficiently powerful to completely raise the belly well above the ground and also to run at a comparatively rapid rate. Both manus and pes have five digits which end in sharp claws. The body is covered all over with minute scales (Fig. 254), of which the prevail- ing colour is reddish-brown above, and orange passing into yellow beneath. On the ventral surface and the top of the head the scales are larger and arranged in pairs. The ear-drum is situated at the bottom of a slight pit, which is the first appearance of the **outer ear**. It is not developed in all Reptiles.

The anal opening is a transverse slit at the root of the tail behind the hind pair of legs. In front of the thigh the scales are perforated by a row of pores, the openings of the only dermal glands which the lizard possesses.

Turning at once to the skeleton, we find that the vertebral

Skeleton. column consists of procoelous vertebrae. All the vertebrae articulate with one another by overlapping facets called pre- and post-zygapophyses as in Amphibia. Although externally similar to the vertebrae of Amphibia, the vertebrae of the Lizard and of Reptiles generally are formed of different elements. Thus the basi-dorsal and inter-dorsal have been suppressed, while the basi-ventral forms an intervertebral disc of cartilage which in the tail bears a pair of processes united with one another to form a **Y**-shaped chevron bone, recalling in its shape the haemal arches of fish. In the neck region the basi-ventrals each bear a bony wedge which is called the sub-vertebral wedge bone. The centrum is formed by the united pair of enlarged inter-ventrals, while the basi-ventrals are partly converted into intervertebral pads, and partly into small parts which occasionally are ossified as the so-called intercentra. The rib has shifted its position so that the capitulum is attached to the front end of the centrum (inter-ventral) behind the basi-ventral to which it belongs. The tubercular attachment is represented by ligament. There are two sacral vertebrae which have expanded transverse processes with which the ribs are fused. Behind these come the vertebrae of the tail—the caudal vertebrae. Each of these bones is made up of two halves, an anterior and a posterior, which are but loosely connected with one another. The consequence is that when a Lizard is seized by the tail this organ in many species snaps in two, one of the vertebrae breaking into an anterior and a posterior half.

All vertebrae in front of the sacrum, except the first two, have distinct ribs attached to their transverse processes. The first two are called respectively the atlas and axis vertebra. The first is as in the case of Amphibia a mere ring. It is composed of the first pair of neural arches united with the first pair of basi-ventrals, and is therefore to a certain extent homologous with one of the inter-vertebral discs. The second has a well-marked centrum, to the front of which is attached a peg-like process—the so-called odontoid process—which projects through the ring of the atlas. This odontoid process is quite unrepresented in the Amphibia, but it is characteristic of all Reptiles and Birds and Mammals. It is formed by the first pair of inter-ventrals and is therefore the first centrum. In young specimens it can be seen to be separated from the centrum of the second vertebra by an unossified disc representing the second pair of basi-ventrals.

The ribs in front of the pectoral girdle remain quite short—this region is the cervical or neck region. Immediately behind the pectoral girdle the ribs are very long and curved so as to half encircle the body like the hoops of a barrel. The foremost have attached to their lower ends cartilaginous bars—the sternal ribs—which are in turn united with a cartilaginous sternum in the middle line. This structure has the form of a lozenge-shaped plate with a hole in the middle, ending behind in two forks to which some of the posterior sternal ribs are attached. The whole sternum has arisen from the junction of the sternal ribs one with another. First those of the same side unite to form a sternal band, and then these two bands unite in front but remain separate behind. The hole in the middle also is a place where they do not unite.

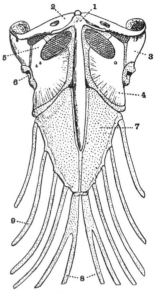

Fig. 255.   Ventral view of the Shoulder-Girdle and Sternum of a Lizard, *Loemanctus longipes* × 2.   After Parker.

1. Interclavicle.   2. Clavicle.
3. Scapula.   4. Coracoid.
5. Precoracoidal process. 6. Glenoid cavity.   7. Sternum.
8. Sternal bands not united.
9. Sternal rib.

The skull is distinguished from the Amphibian skull by many features. The jaws do not arch outwards at the sides of the cranium, as in the frog, but are bent inwards underneath it. Behind, the cartilage of the cranium is completely replaced by four bones—by the supra-occipital above the foramen magnum, the ex-occipitals at the sides of this opening, and the basi-occipital beneath. This last bone bears a single knob or condyle which articulates with the atlas vertebra. To the formation of this condyle the ex-occipitals in some degree contribute. The basi-occipital and the single condyle and the supra-occipital are highly characteristic of all Reptilia—as is also the basisphenoid bone. This is a bone replacing the cartilaginous floor of the cranium just in front of the basi-occipital. The parasphenoid so characteristic of Amphibia is reduced to a mere splint attached to the front of the basisphenoid.

The anterior part of the cranium is so compressed between the large eyes that its cavity completely disappears and it becomes replaced by a vertical sheet of membrane, the inter-orbital septum. It follows that in the dried skull the two orbits apparently open widely into one another. Almost the entire brain is pushed back

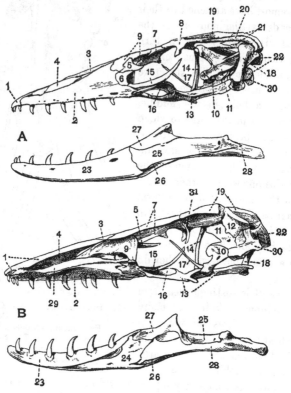

Fig. 256. A, Lateral view, and B, Longitudinal section of the Skull of a Lizard, *Varanus varius* × ⅔.

1. Premaxilla. 2. Maxilla. 3. Nasal. 4. Lateral ethmoid. 5. Supraorbital. 6. Lachrymal. 7. Frontal. 8. Postfrontal. 9. Prefrontal. 10. Basisphenoid. 11. Pro-otic. 12. Epi-otic. 13. Pterygoid. 14. Epipterygoid (columella cranii). 15. Jugal. 16. Transverse bone. 17. Parasphenoid. 18. Quadrate. 19. Parietal. 20. Squamosal. 21. Supratemporal. 22. Exoccipital. 23. Dentary. 24. Splenial. 25. Supra-angular. 26. Angular. 27. Coronoid. 28. Articular. 29. Vomer. 30. Basi-occipital. 31. Orbitosphenoid.

behind the eyes into the hinder part of the cranium. Only the olfactory stalks run through holes in the upper part of the septum. The orbitosphenoid of Urodela and the sphenethmoid of the frog are quite unrepresented, though in some of the larger lizards allied to Lacerta there is a minute orbitosphenoid bone in the upper part of the inter-orbital septum (31, Fig. 256). The inter-orbital septum is certainly a characteristic of the primitive Reptilia. It has however been lost in some of the most recent and highly modified forms.

The auditory capsule as in Teleostei is completely converted into bone, but it is ossified by three bones only, an epi-otic above, which fuses with the supraoccipital, an opisthotic behind, which joins the exoccipital, and a pro-otic which remains distinct. There is no trace of the pterotic bone so characteristic of Teleostei.

As in Amphibia the first visceral arch is represented by an upper half consisting of a suspensorium with a pterygoid process, and a lower half—Meckel's cartilage. In the upper half, however, the cartilage is completely replaced by bone. The suspensorial portion forms the quadrate bone, which is attached to the side of the auditory capsule. The pterygoid process is completely ossified by the pterygoid bone behind and the palatine in front.

A curious bone characteristic of the Lacertilia excluding the *AMPHISBAENIDAE*, and called the epipterygoid or columella bone, runs from the pterygoid vertically up to the parietal. This bone is however found only in some Lacertilia, not in Reptiles generally.

The two pterygoid bones however, instead of arching outwards converge, under the base of the cranium, and they articulate with outgrowths from the basisphenoid, called basipterygoid processes. The palatines are united in front with the floor of the nasal capsule: they bear on their inner sides slight ridges which project somewhat into the cavity of the mouth. These ridges support a flap of the lining of the mouth, the palatal flap, which is most characteristic of all Reptilia but is not found in any Amphibian (Fig. 259). It is a first trace of the process which ends in the higher animals and even in some Reptiles in the division of the mouth-cavity into an upper air-passage and a lower food-passage.

The end of Meckel's cartilage which articulates with the quadrate is converted into a bone, the articulare. Three pairs of the hinder visceral arches are preserved. These retain their rod-like form as in Urodela, the median connecting pieces (copulae) remaining small.

The membrane-bones of the skull are one of its most characteristic features. The roofing bones are the same as in the Urodela— paired nasals, frontals and parietals. On the roof of the mouth

there are two vomers and a parasphenoid. The vomers, however,
are rod-like, toothless and placed close together, and the parasphenoid is a small rudiment.

The bones of the side of the head and the upper lip form a most

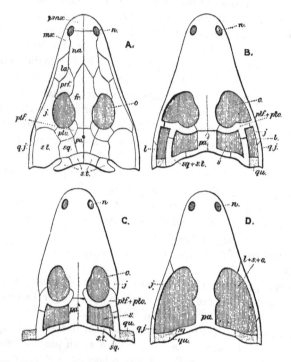

Fig. 257. Diagram of the Cranial Roof in a Stegocephalan, various types of
Reptiles, and a Bird, showing modifications in the Postero-lateral Region.
From Smith Woodward.

A. Stegocephalan (*Mastodonsaurus giganteus*, about one-fifteenth nat. size,
after E. Fraas).   B. Generalized Rhynchocephalan and Crocodilian.
C. Generalized Lacertilian, often losing even the arcade here indicated.
D. Generalized Bird.   *fr.* Frontal.   *j.* Jugal.   *l.* Lateral temporal
fossa.   *la.* Lachrymal.   *mx.* Maxilla.   *n.* Narial opening.   *na.* Nasal.
o. Orbit.   *pa.* Parietal.   *pmx.* Premaxilla.   *prf.* Prefrontal.   *ptf.* Postfrontal.   *pto.* Post-orbital.   *q.j.* Quadrato-jugal.   *qu.* Quadrate.   *s.* Supratemporal fossa.   *s.t.* Supratemporals.   *sq.* Squamosal.   Vacuities
shaded with vertical lines, cartilage bones dotted.

peculiar scaffolding which is widely separated from the cranium.
The Lizard is in an intermediate condition between the Stegocephala,
where a continuous sheet of bones extends from the cranium to
the upper lip, and modern Amphibia, where all those bones have

disappeared, leaving a large vacuity between the cranium and upper lip.

In front in the upper lip there is a premaxilla bearing teeth followed by a maxilla in which there are also teeth. The maxilla is joined to the pterygoid by an ectopterygoid or transverse bone. Between the maxilla and frontal on the side of the face are two bones known as prefrontal and lachrymal. The line of bones in the upper lip is continued by the jugal. This unites with a bone placed behind the eye termed the postfrontal, which joins both the frontal and parietal. Thus the eye is surrounded by a ring of bone.

The squamosal is a characteristic V-shaped bone. The apex of the V articulates with the upper side of the quadrate: of the two arms one is directed forwards and meets the postfrontal, thus forming a bony bar parallel to the cranium which is called the upper temporal arcade. The other limb is directed backwards and inwards and meets a crest on the parietal so that a bridge is formed extending over the hinder part of the cranium. The space in the dried skull existing between this bridge and the cranium is called the post-temporal fossa. In many reptiles, including most Lacertilia, there is a similar space between the cranium and the lateral bridge formed by the junction of the squamosal and postfrontal. This space is roofed over in Lacerta by two membrane-bones called supratemporals, but when uncovered it is known as the supratemporal fossa.

Finally, the space intervening between the quadrate and jugal on the side of the face is known as the latero-temporal fossa. In *Sphenodon, Crocodilia* and a very large number of extinct Reptiles it is bounded below by a quadrato-jugal bone which joins the jugal to the quadrate. When the quadrato-jugal is present the series of bones consisting of maxilla, jugal and quadrato-jugal is known as the lower temporal arcade. The upper temporal arcade is formed as we have seen by the postfrontal and the squamosal. The loss of the quadrato-jugal in Lacertilia is doubtless connected with the greater mobility of the jaws. In some lizards, notably in Geckos, the quadrate can move slightly on its articulation with the skull, as can also the pterygoid on the basipterygoid process. When the lower jaw is pulled downwards and backwards by its depressor muscle it tends to throw the lower end of the quadrate slightly forwards: the pterygoid slides on the basisphenoid, and pushing the ectopterygoid tilts the maxilla slightly upwards. With the maxilla all the other bones of the face move, and the membranous interorbital septum permits the ethmoidal region of the cranium to be slightly bent on the hinder portion.

The cartilaginous lower jaw is ensheathed by five distinct membrane bones. The dentary and splenial occupy the same positions as in Teleostomi and Urodela. The angular clamps the under side of the articulare, the cartilage bone replacing the upper end of the cartilaginous jaw. The supra-angular lies above the angular on the outer side of the articular. The coronoid is a small projection on the upper edge of the jaw.

The pectoral girdle is at first sight exceedingly complicated, but in reality it consists of the same parts as in the Anura. Above the cavity for articulation of the arm—the glenoid cavity—there is the cartilaginous scapula; below the girdle forks into a coracoid and precoracoid united by an epicoracoid. The cartilage bones present are the scapula, precoracoid and coracoid. The cartilage above the scapular bone is slightly calcified but is not converted into bone; this region as in Amphibia is called the supra-scapula. Along the inner edge of the supra-scapula, scapula and precoracoid runs a strong membrane-bone, the clavicle which reaches a median bone, the T-shaped interclavicle. This bone underlies the sternum. The two epicoracoid cartilages join the anterior edges of the sternum (Figs. 255 and 258).

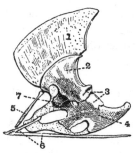

Fig. 258. Lateral view of the Shoulder-Girdle of *Varanus* × ⅗.

1. Suprascapula.   2. Scapula.
3. Glenoid cavity.   4. Coracoid.   5. Clavicle.   6. Interclavicle.   7. Precoracoidal process.

The space between the coracoid and precoracoid is called the coracoid fontanelle. Since in the Urodela it is not closed by an epicoracoid it may be regarded as a bay or indentation in the lower half of the originally simple pectoral girdle. The condition of affairs in Urodela throws considerable light on what occurs in certain other Lacertilia, such as the American Iguana. There we find that a similar deep indentation has become developed on the inner side of both scapula and coracoid, so that projections are formed to which the names mesoscapula and mesocoracoid have been given. These are not ossified by separate bones but are regions of the scapula and coracoid bones.

The fore-limb of the lizard might be taken as the type of the pentadactyle limb, since there are five fingers and the carpus has all the nine bones developed.

The pelvic girdle differs markedly from that of any Amphibian, in that in its lower portion there is a hole called the obturator foramen, corresponding to the coracoid foramen in the pectoral girdle. The girdle is ossified by three bones, viz., a vertical ilium articulating with the ribs of the sacral vertebrae, a pubis ossifying the anterior limb of the lower half of the girdle and an ischium ossifying the posterior limb. Both pubis and ischium meet their fellows in the middle line; such a union is termed a symphysis. The two obturator foramina are closed below and at the same time separated from one another by a longitudinal ligament which may have a certain amount of ossification in it. All three bones contribute to the formation of the acetabulum, the cavity for the articulation of the femur.

The presence of the obturator foramen and a distinct pubis is characteristic of all Reptiles, Birds, and Mammals, and at once distinguishes them from Amphibia.

On the hinder edge of the pubis there is a projection which is called the lateral process. In some extinct reptiles this process was extraordinarily long and ossified by a distinct bone, which has been called the post-pubis. It is the post-pubis which forms the so-called pubis of Birds and Mammals.

The most marked feature of the hind limb is the formation of a sharply-marked "ankle" joint. There is one place and one only where the foot bends on the shank; whereas in Urodela bending can occur at any place in the mosaic of small bones which forms the tarsus.

In the lizard all the three upper bones of the tarsus are joined to form a horizontal bar. The lower bones have almost entirely coalesced with the corresponding metatarsals, only the third and fourth of the series being distinguishable. Thus the lizard has what has been called an inter-tarsal joint—an arrangement which is highly characteristic of many Reptiles and of all Birds.

All trace of the division of the muscles into myotomes has disappeared, but the innermost layer of the muscles *Viscera.* of the flanks has become divided secondarily into a series of bands connecting each rib with its successor. These bands are termed the intercostal muscles and each consists of an external and an internal layer of fibres. The fibres of the external layer slope upwards and forwards and, in contracting, cause the ribs to rotate forwards; the fibres of the inner layer slope upwards and backwards and have the reverse effect. Respiration is effected by

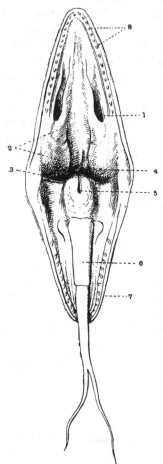

the pulling forwards and backwards of the ribs by these intercostal muscles. In their relaxed condition the ribs slant strongly backwards. When they are pulled forward by muscles attaching them to the anterior vertebrae and by the external intercostals, they rotate forwards so as to stand out at right angles to the vertebral column and thus enlarge the cavity of the chest, that is, the coelom. The diminution of pressure in this air-tight cavity at once causes an inrush of air through the glottis, the elastic lungs are expanded and their walls closely follow the chest wall. It will be noticed that the mechanism of inspiration is very different from that of Amphibians (p. 424). The network of low ridges which is found already on the inner side of the Frog's lung has in the Reptile greatly increased in complexity. The primary ridges are much higher, and between them are lower secondary and even tertiary ridges: the cavity of the lung is as it were partly filled up by a spongy mass. In all Saurians however the central cavity is easily recognized as a wide space: whilst in Crocodiles and Tortoises, still more so in Birds and Mammals, it is represented only by the bronchial tubes.

FIG. 259.   Open mouth of *Varanus indicus* × 1.

1. Posterior or internal nares. 2. Palatal folds.   3. Internal opening of Eustachian tubes. 4. Opening of oesophagus.   5. Glottis.   6. Tongue half protruded.   7. Lip of lower jaw. 8. Teeth of upper jaw.

The lungs are connected with the glottis by a comparatively long stalk, the trachea or windpipe, which is stiffened by rings of cartilage. A similar structure is found amongst Amphibia in the Gymnophiona and in a few Urodela. Immediately below the glottis the trachea is enlarged. The enlarged portion is stiffened by a large, broad, ring-shaped cartilage, the cricoid, to which are articulated two arytenoid cartilages. The whole structure consisting of the dilatation of the trachea and its cartilages is called the larynx.

The Lizard like the Frog lives principally on insects and is provided with a long mobile tongue cleft at the tip, by means of which the prey are whisked into the mouth. The tongue is free in front and attached behind, the opposite arrangement to what is found in the Frog. The teeth are simple and conical, and are implanted in a groove on the inner side of the bones bearing them. As the Lizard grows they become actually fused with the bone along the side of the groove.

When a Frog's mouth is forced open, amongst the most striking features of the roof of the mouth are the two large eyeballs shining through. When we open the mouth of a lizard nothing of the eyes can be seen. There is projecting

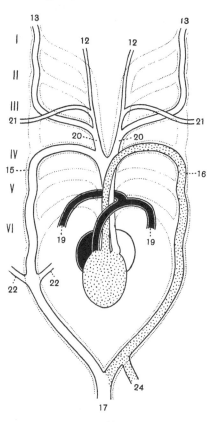

FIG. 260. Diagram of Arterial Arches of *Chamaeleo* viewed from the ventral aspect.

I. II. III. IV. V. VI. First to sixth arterial arches. 12. Tracheo-lingual (ventral carotid). 13. Common carotid (dorsal carotid). 15. Right systemic arch. 16. Left systemic arch. 17. Dorsal aorta. 19. Pulmonary. 20. Innominate. 21. Scapular (equivalent of subclavian of ventral type). 22. Subclavian (dorsal type). 24. Coeliac.

inwards from the upper lip on each side a flap, the palatal flap. This does not meet its fellow in the middle line, a cleft existing

between them. These flaps conceal the eyeballs and the inner
openings of the Eustachian tubes which lead up to the ear-drum.
Palatal flaps as already mentioned are found in all Reptilia.

Turning now to the circulatory system we find that the conus
arteriosus no longer exists as such, having been cleft into three
trunks down to its commencement
in the ventricle. One of these
trunks is ventral and slightly posterior to the others, and gives rise
to the two arterial arches, which
as pulmonary arteries supply the
lungs and have no connection
with the aorta. The other two
arterial trunks form the right and
left roots of the aorta. They cross
each other at their origin, that
which passes to the right of the
oesophagus arising from the left
of the ventricle and *vice versâ*.
The third pair of arterial arches,
corresponding to the carotid arches
of Amphibia, are well developed
in the Lizard. They have a
common stem which arises from
the right systemic arch. In some
Lizards the longitudinal epibranchial vessel of the embryo persists
between the carotid and systemic
arches on either side, so that in
this respect a Lizard may be even
more primitive than a Newt. In
others, as in all other Reptiles, this
connecting link has disappeared.

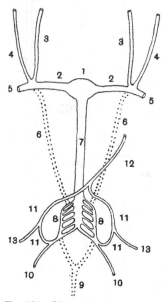

Fig. 261. Diagram to show arrangement of the principal veins in the
Anura and *Reptilia*.

1. Sinus venosus, gradually disappearing in the higher forms. 2. Ductus Cuvieri = superior vena cava.
3. Internal jugular = anterior cardinal sinus. 4. External jugular
= sub-branchial. 5. Subclavian.
6. Posterior cardinal, front part
= vena azygos. 7. Inferior
vena cava. 8. Renal portal =
hinder part of posterior cardinal.
9. Caudal. 10. Sciatic = internal
iliac. 11. Pelvic. 12. Anterior
abdominal. 13. Femoral =
external iliac.

The ventricle has projecting
into its cavity two imperfect partitions or septa. One is the continuation of the division between
the two auricles, the other is a ridge which arises from the ventral
side and tends to separate the opening of the pulmonary arteries
from that of the right and left aortic arches. When the ventricle
at first begins to contract it is full of venous blood from the right

auricle : by the time arterial blood has commenced to enter it from the left auricle, the ventral septum mentioned above has been driven against the opposite wall, so as to shut off the pulmonary trunk from the rest of the ventricle and prevent its receiving any more blood. The left aortic arch, which arises on the right, receives mostly venous blood from the right auricle, the right aortic arch arterial blood from the left auricle, and it is from this arch, as mentioned above, that the carotid arteries arise. Hence the head receives comparatively arterial blood, and all the rest of the body mixed blood. The lingual artery of Amphibia is represented in Reptiles by a vessel (tracheo-lingual or "ventral carotid") which arises from the carotid arch near the middle line and supplies the tongue, trachea and muscles of the neck and shoulder.

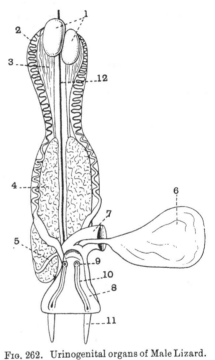

The vessels supplying the fore-limb arise together from the right systemic arch in the case of Lizards instead of as in Anura from both right and left arches, they are subclavians of the dorsal type (see p. 350), but in Chelonians and Crocodiles the subclavians are ventral in origin, coming off from the carotid trunk on either side close to its division into ventral and dorsal carotids. In

Fig. 262. Urinogenital organs of Male Lizard.
1. Testis.   2. Vas deferens = archinephric duct.   3. Epididymis = (mesonephros). 4. Kidney = metanephros.   5. Ureter. 6. Bladder.   7. Rectum cut and turned back.   8. Cloaca laid open.   9. Opening of Vas deferens.   10. Groove leading to opening of penis. 11. Penis. 12. Dorsal aorta.

Lizards this "ventral subclavian" is represented by the scapular artery which runs to the shoulder region.

The veins, on the whole, closely resemble those of *Molge*. There is however no large cutaneous vein, and the anterior part of the posterior cardinal, now called the vena azygos, is found only on the right side, where it receives the numerous intercostal veins, returning the blood from the muscles connecting the ribs. The renal-portal, sciatic, femoral and anterior abdominal veins have the same arrangement as in the Urodela.

The brain is distinguished by the comparatively large size of the cerebral hemispheres, which overlap the thalamencephalon above and at the sides. They end in front in large pear-shaped olfactory lobes. The cerebellum is a high vertical ridge and is thus much more prominent than in any Amphibian. The remainder of the hind brain, the medulla oblongata, includes a longer portion of the spinal cord than it does in Amphibia, for the hypoglossal nerve arises from its side and escapes through an aperture in the exoccipital bone. This nerve is reckoned the twelfth cranial, not the eleventh, for there is a trunk called the spinal accessory or eleventh cranial. This arises by several roots from the side of the medulla oblongata, joins the vagus in a ganglion, and then leaving the skull supplies some of the neck muscles. In the Ophidia this nerve is not distinguishable from the vagus.

In the genital organs, the Lizards and Reptiles generally are distinguished from Amphibia by the complete separation of the mesonephros from the metanephros or functional part of the kidney. The persisting part of the mesonephros, now known as the epididymis, is only developed in the male, where it is closely connected with the testis. As in the Newt it receives the vasa efferentia. In the female the oviduct is shorter and has a wider internal funnel than in the Amphibia, and it is also placed further back so as to be rather nearer to the ovary. This is an arrangement suited to the large size of the eggs, which are too heavy to be drawn any distance by the current produced by the cilia of the oviduct.

The egg is fertilized whilst still in the oviduct. The male lizard has two organs called copulatory sacs or penes, situated, one on each side, on the hinder wall of the cloaca. These, when not in use, are hollow pouches opening into the cloaca. When in use they are turned inside out, and are then seen to have grooves leading to the openings of the vasa deferentia or archinephric ducts.

Most lizards lay their eggs in crevices amongst stones and allow them to be hatched by the heat of the sun. In all cases a considerable amount of development goes on before they are laid. In the English species *Lacerta vivipara* the young burst through the egg-shell and use up all the yolk whilst they are still in the oviduct, so that in common parlance they are born alive, that is, as little lizards and not as eggs.

Order I. Rhynchocephala.

As mentioned above, the order Rhynchocephala is represented
by the single species, *Sphenodon punctatus,* found only in New
Zealand. This is a very Lizard-like animal. The back is covered
with small scales which in the middle line form a comb-like crest:

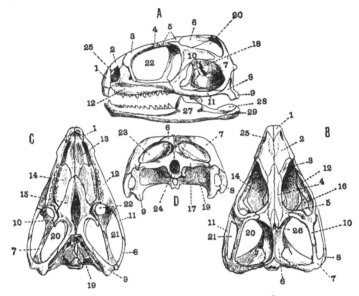

F<small>IG</small>. 263.   Skull of *Sphenodon punctatus* × 1.

A. Lateral.   B. Dorsal.   C. Ventral.   D. Posterior.   After von Zittel.

1. Premaxilla.   2. Nasal.   3. Prefrontal.   4. Frontal.   5. Postfrontal.   6. Parietal.   7. Squamosal. 8. Quadratojugal. 9. Quadrate.   10. Postorbital.   11. Jugal.   12. Maxilla.   13. Vomer.
14. Palatine.   15. Pterygoid.   16. Ectopterygoid or transverse bone.
17. Exoccipital.   18. Epipterygoid.   19. Basisphenoid.   20. Supratemporal fossa.   21. Lateral temporal fossa.   22. Orbit.   23. Post-
temporal fossa.   24. Foramen magnum.   25. Anterior nares.
26. Interparietal foramen.   27. Dentary.   28. Supra-angular.
29. Articular.

the belly is covered with large square scales. In the skeleton and
male genital organs, however, *Sphenodon* is widely different from
the Lizard. The quadrate in the skull is quite immovable, being
firmly clamped by the squamosal and quadratojugal. The latero-
temporal fossa is thus completely bounded below and the supratemporal fossa is uncovered. Between the parietals is a gap called

the interparietal foramen: in this is situated the tip of the
pineal body which has here all the characters of a simple eye.

The animal has teeth when young, but they become worn away,
while the edges of the maxilla and premaxilla become converted
into cutting edges.

The vertebrae are amphicoelous, and the basi-ventrals, represented by the sub-vertebral wedge bones and chevrons, are placed
beneath the interspaces between the vertebrae throughout the neck,
trunk and tail, and not as in Lizards in the neck and tail regions
only.

The ribs have three divisions, there being a small intermediate
piece intercalated between the dorsal and sternal rib. From the
dorsal rib a hook-like outgrowth, the uncinate process, projects
backwards, which overlaps the next rib as in Crocodiles and Birds.

Behind the sternum there is a long series of rod-like bones, the
so-called abdominal ribs, embedded in the muscles of the belly.
They are placed parallel to the direction of the sternal ribs, that is,
they slope obliquely forwards and inwards. They are regarded as
membrane bones and supposed to correspond to the ventral bony
scales of the Stegocephala.

All these peculiarities of the skeleton are found in many of the
oldest fossil reptiles.

There is no proper copulatory organ : the cloaca is used for this
purpose as in the Urodela.

From a condition in many respects represented at the present
day by *Sphenodon*, the ancestors of living reptiles appear to have
diverged in two directions.

On the one hand, the original stock gave rise to descendants
with long flexible bodies and extensible jaws—this latter feature
involving of course a movable quadrate. The cloacal opening
became converted into a transverse slit and copulatory organs
became developed behind it. This stock includes the Snakes and
Lizards which are often included in the one comprehensive Order
the Sauria.

On the other hand, the descendants of the common ancestral
form diverged in the direction of heavily armoured forms, in which
membrane bones underlying the scales were developed and in which
the jaws are very powerful, the quadrate remaining immovably
clamped by the quadratojugal. The cloacal opening became a
longitudinal slit and developed the single median copulatory organ
on its front wall. This stock includes the Turtles and Crocodiles.

SAURIA.

We now turn to the SAURIA. Most people would imagine that the task of distinguishing a lizard from a snake was an easy one. But if we were to collect together all the limbless species of Reptiles we should find not only that they differ very much from one another in the structure of the skull and in other points, but that they are more nearly related to different families of lizards than to one another. There is no doubt that the snake-like forms have been derived from four-limbed reptiles like lizards, for some of

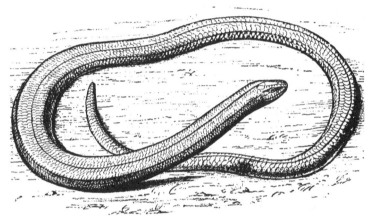

FIG. 264. A limbless Lizard, *Anguis fragilis*, the blind-worm, slightly reduced.

them have rudimentary vestiges of limbs. It is evident then that there must be an advantage in certain situations in getting rid of limbs, and it is further evident that the effect of this advantage has been that not only in one but in many families of lizards some species have lost their limbs. The kind of life to which a snake-like form is suited is a lurking one amongst crevices in stones, or thick vegetation, or in the soil, where movement is best effected by wriggling and limbs would be in the way.

Under these circumstances, we must either class together all limbless Sauria as snakes, and thus give up the idea that the members of an Order must necessarily be descended from the same ancestral species, or else we must select one group as the true Snakes (Ophidia), the members of which have many other characters in common besides the negative one of having no limbs. This latter course is that which has been adopted by Huxley, who defines true snakes somewhat arbitrarily as those forms which have lost all trace of the pectoral girdle and of the urinary bladder, although they may retain traces of hind-limbs.

Order II. Lacertilia.

The Lacertilia then include all species of Sauria which have the right and left halves of the mandibles connected by a sutural symphysis and which retain a urinary bladder and some trace of the pectoral girdle. In all other characters they are a very diversified group. Most of them possess well developed limbs, movable eyelids and movable quadrate bones, but a good many species belonging to specialised burrowing families have no limbs and scarcely a trace of the pectoral girdle, while the eyes are concealed beneath the skin and the quadrate has become more or less immovable. Some, e.g., *Draco volans*, have the hinder ribs expanded so as to press out two expansions of skin and form a parachute-like expansion on each side, by means of which they are supported as they flit from tree to tree in great leaps. Most feed on insects, worms, &c. like the English lizards; some are large enough to seize mice and birds and frogs. The limbless forms are represented in England by the Blind- or Slow-worm, *Anguis fragilis*, and in North America by the allied Glass-snake, *Ophisaurus ventralis*. These animals have skulls like that of *Lacerta* and rudiments of pectoral girdles. Besides the Blind-worm, the Common Lizard, *Lacerta vivipara*, and the Sand-Lizard, *Lacerta agilis*, are British.

In North America four families of Lizards are represented, one being that of the limbless. ANGUIDAE, while the most remarkable of the others is that of the IGUANIDAE. These animals have short thick tongues and overlapping scales which form a crest of spines on the head and back and round the throat. *Phrynosoma douglasi*, the horned " toad," is found all through the Central States and even penetrates into Ontario; it is the sole lizard found in Eastern Canada.

Order III. Ophidia.

The Ophidia, or true snakes according to definition, have the right and left halves of the mandible connected by an elastic band; they are also devoid of a urinary bladder and of any trace of a

pectoral girdle. Besides this however they have a large number
of other characters which severally are shared by some families
of Lizards but which collectively are found only in the Ophidia.

The vertebrae in addition to the zygapophyses on the sides of
the neural arch have median bosses and pits by which they fit into
one another, called respectively zygantra and zygosphenes (Gr.
ἄντρον, a cave or hollow; σφήν, a wedge). There are no sternal ribs

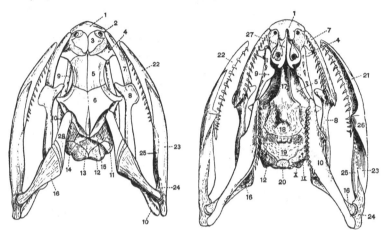

Fig. 265.　Dorsal (to the left) and ventral (to the right) views of the skull of the
Common Snake, *Tropidonotus natrix*. After Parker.

1. Premaxillae (fused). 2. Anterior nares. 3. Nasal. 4. Prefrontal.
5. Frontal. 6. Parietal. 7. Maxilla. 8. Transverse bone.
9. Palatine. 10. Pterygoid. 11. Pro-otic. 12. Exoccipital.
13. Supra-occipital. 14. Opisthotic. 15. Epi-otic. 16. Quadrate.
17. Parasphenoid. 18. Basisphenoid. 19. Basi-occipital. 20. Occipital condyle. 21. Splenial. 22. Dentary. 23. Angular. 24. Articular.
25. Supra-angular. 26. Coronoid. 27. Vomer. 28. Squamosal.
IX. X. Foramina for the ninth and tenth cranial nerves.

or sternum, but the dorsal ribs are elongated and curved ventrally,
and a snake literally walks on the ends of them; it is in a sense a
vertebrate centipede.

In the skull the chief point to be noticed is the extreme mobility
of the jaws. The jugal as well as the quadratojugal have disappeared, the pterygoids no longer articulate with the base of the
skull, and the quadrate itself is pushed away from the cranium by
the squamosal, which is a rod-like bone (Fig. 265). Some authorities
hold that this bone is not the representative of the squamosal, but
represents the supra-temporal of *Lacerta*. The result of this
arrangement is that when the lower jaw is pulled down, the

quadrate is quite free to thrust the pterygoid forward and push up the maxilla by means of the transverse bone; that is to say there is the same mechanism as was described in the lizard, only more easily set in motion and capable of much more movement. The halves of the mandible, or lower jaw, are connected by elastic fibres, and thus they can be widely separated. The result of this is, that a snake has an enormous gape and can swallow prey almost as large as itself. Snakes of quite moderate size dispose of frogs, birds, &c. The large Pythons of India can crush an animal larger than a half-grown sheep into a shapeless mass by coiling themselves around it, and they then swallow it whole.

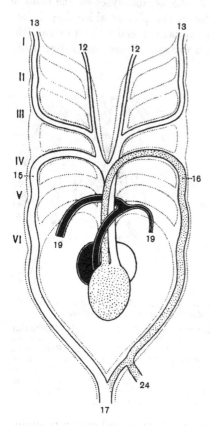

The hyoid, including under that name the remains of all the hinder visceral arches, is vestigial, consisting of a single bone on each side. This permits of the pulling of the glottis far forward between the halves of the mandible when the animal is engaged in swallowing its prey, this shifting of position being necessary to prevent choking.

FIG. 266. Diagram of Arterial Arches of Snake viewed from the ventral aspect.

I. II. III. IV. V. VI. First to sixth arterial arches.   12. Tracheal (ventral carotid). 13. Common carotid (dorsal carotid). 15. Right systemic arch.   16. Left systemic arch.   17. Dorsal aorta.   19. Pulmonary.   24. Coeliac.

In the skull the brain extends forwards between the eyes and there is consequently no interorbital septum. That this is a secondary and not a primary state of affairs is shown by the fact that the front part of the brain is protected at the sides by downward extensions of the frontal and parietal bones, whereas in animals such as the Urodela and Mammalia, where an interorbital septum has never been formed, the side-walls of the cranium are constituted by the orbitosphenoid and alisphenoid bones. It is curious to find this absence of an interorbital septum in a family of limbless lizards, the AMPHISBAENIDAE. What relation, if any, it has to the snake-like habits it is hard to guess.

The two eyelids have coalesced to form an extra guard in front of the eye, but there is a transparent portion in the lower one through which the animal can see. The outer covering of scales is shed periodically, half-a-dozen times every year or oftener, and replaced by a new set formed by the activity of the ectoderm, and during this process, since the covering of the eye is affected, the snake is blind.

One lung is small, and the other (the right) greatly elongated, the hinder part being quite smooth.

The heart resembles that of Lizards both in structure and the mode of distributing the arterial and venous blood. The differences between the vascular systems of a Snake and a Lizard depend chiefly on the absence of limbs and the correlated great development of the vertebral column, ribs and their musculature as organs of locomotion in the Snake. Thus the subclavian arteries are absent from the right systemic arch, while the vertebral and caudal arteries and veins are well developed. Another difference is that the left pulmonary artery is very slightly developed, in connection with the reduced condition of the left lung.

Snakes are divided into many families, of which two are represented in Great Britain and three in the temperate parts of North America. A rough classification would divide them according to their habits into: (a) those which poison their prey, (b) those which crush their prey, and (c) those which swallow their prey directly.

Those which crush their prey are confined to the tropics ; those which swallow their prey directly are the non-venomous snakes, and are represented in both England and North America by the family COLUBRIDAE. In this family the maxilla is long and bears numerous teeth, as do also the pterygoid and the lower jaw. The head is much broader behind than at the muzzle. There are about thirty species belonging to eighteen genera in North America, of which *Tropidonotus sirtalis*, the garter-snake frequently met with in Canada, is one of the commonest ; and in England the family is represented by two species, the smooth-snake, *Coronella laevis*, and the grass- or ring-snake, *Tropidonotus natrix*.

The venomous snakes in America belong to two families. In the first, the ELAPIDAE, the maxilla is a long bone and bears in front two large teeth which are grooved, to allow the secretion of glands in the lip to trickle down into the wound which they make. The teeth behind are not grooved. The American Harlequin Snake, *Elaps fulvius*, belongs to this family. This snake receives its name from its brilliant colours; it has seventeen crimson rings

FIG. 267. The Texas Rattlesnake, *Crotalus atrox*, reduced. From Stejneger after Baird and Girard.

bordered with yellow. Another family is that of the VIPERIDAE. The maxilla is much shortened and bears one enormous fang, which when the mouth is closed lies against the roof of the mouth : when the mouth is opened the maxilla is rotated by means of the ecto-pterygoid, so as to erect the tooth. The typical Rattlesnake— *Crotalus horridus* or *C. atrox*—derives its name from an appendage of about 8 to 9 loosely connected horny rings which it bears at the end of its tail, the shaking of which makes a noise like a rattle. This is one of the most deadly snakes known : it is found all over the United States in mountainous places and enters Canada. Like all CROTALINAE or Pit-vipers it has a sensory pit between eye and nose. The English Adder, *Vipera berus*, is, like all the Old World Viperinae, devoid of such pits.

Order IV. Chelonia.

The Chelonia or Turtles are the most peculiar order of the Reptilia. In some respects they are nearest to the Amphibia, but they are highly specialized. Their leading peculiarity is the possession of two great shields, a dorsal the carapace and a ventral the plastron, composed of bones firmly connected together, so that most of the organs of the body are enclosed in a box. The horny scales which cover in this box are very large and form what is known as tortoise-shell. The carapace is formed of a central row of neural plates which are expansions of the spines of the dorsal vertebrae with a nuchal plate in front of these and a pygal behind, the two last-named being of dermal origin.

On each side there are costal plates; this name is given to broad expansions of the outer surfaces of the ribs (Fig. 268). The ribs curve inwards to join the centrum, and since this, as in all Reptiles, is formed by the interventral, each rib is nearly opposite the interspace between two centra and sometimes unites with them both. The transverse process is represented by the expansion of the neural plate which meets the costal plate. The almost horizontally directed outer ends of the ribs are received into a series of dermal bones called marginals, which form the edge of the carapace.

The plastron is formed of one unpaired and several paired bones (Fig. 269). The median bone, called the entoplastron, is believed to correspond to the interclavicle of other Reptiles. The first pair are called epiplastra and probably represent the clavicles of other forms. The posterior pairs are called hyoplastra, hypoplastra and xiphiplastra respectively; they are firmly joined to the marginals.

In front and behind the plastron and carapace are separated by soft flexible skin; their edges project so as to form roof and floor to cavities into which the head and neck and arms in front and the

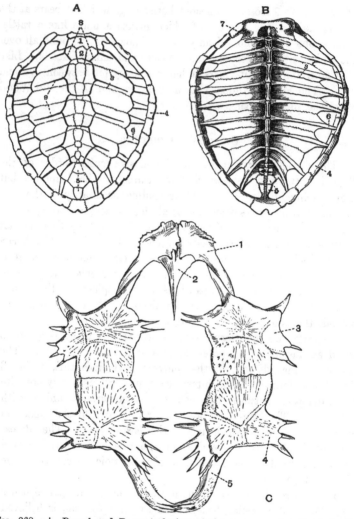

FIG. 268. A, Dorsal and B, ventral view of the carapace of a Loggerhead
Turtle, *Thalassochelys caretta*. After Owen. In A the outlines of the
superficial horny scales constituting the tortoise-shell are indicated by
heavy lines, whilst the outlines of the bones of which the carapace is
composed are represented by lighter lines.

1. Nuchal plate.   2. First neural plate.   3. Second costal plate.
    4. Marginal plate.   5. Pygal plate.   6. Rib.   7. Thoracic vertebra.
    8. First vertebral shield.   9. Costal shield.

   C. The Plastron of a Green Turtle, *Chelone mydas* × ⅓. (Camb. Mus.)

1. Epiplastron (clavicle).   2. Entoplastron (interclavicle).   3. Hyoplastron
    (cleithrum).   4. Hypoplastron.   5. Xiphiplastron.

legs and tail behind can be withdrawn. A study of the development of modern Chelonia and of the anatomy of fossil species makes
it plain that the ancestors of the present forms were provided with
a carapace composed entirely of dermal bones underlying the horny
scales, just as is the case with Crocodilia. This dermal carapace
however was gradually replaced by the development of bony expansions of the ribs and neural arches; though remnants of it
persist in the nuchal, pygal and marginal plates.

There is no trace of sternal ribs or sternum; but the pectoral
and pelvic girdles occupy the peculiar position of being within
instead of outside the ribs, a consequence of the almost horizontal
direction of these. The girdles are in fact converted into pillars or
struts which keep the plastron and carapace apart. In front the
scapula forms a vertical pillar which has a ventral process—the
acromion—projecting inwards beyond the articulation with the
coracoid. This process is unique amongst recent Reptilia but existed
in the Plesiosauria. The coracoid slopes backwards and inwards.
The ilium and pubis serve to support the carapace posteriorly. The
pelvic girdle is similar to that of a Lizard but the pectoral girdle
has no epicoracoid. The limbs are essentially similar to those of
the Lizard but the toes are shorter and blunter. The neck is extraordinarily flexible; the vertebrae composing it fit one another by
cup and ball joints, one is amphicoelous, another is biconvex. The
dorsal vertebrae have flat faces.

The skull is devoid of teeth and the premaxilla and maxilla are
short. Both they and the dentary have sharp cutting edges ensheathed in horn so as to form a beak. In all species the orbit is
encircled with a bony ring and the ectopterygoid or transverse bone
is wanting. The palatal crest on the palatine is hardly perceptible.
The squamosal does not usually join either the postfrontal or
parietal, hence the upper temporal arcade is absent and there is
no distinction between the supra-temporal and latero-temporal
fossae. In the marine *Chelone* and its allies however the postfrontal, squamosal, parietal and quadratojugal coalesce to form a
sheet of bone from the crest of the skull to the lip, roofing over a
cavity lying at the side of the cranium and containing muscles.

Breathing is performed as in Amphibia, by a mylohyoid muscle
and other muscles causing movements of the hinder visceral arches,
of which there are three pairs.

The heart in structure and mode of action resembles that of
Lizards and Snakes, the left-hand systemic arch conveying blood

chiefly venous to the viscera, while the right-hand one supplies the head, trunk and limbs with blood which is much more arterialised than that in the other arch (p. 470). The fore-limbs are however supplied by a different vessel from the subclavian of the Lizard.

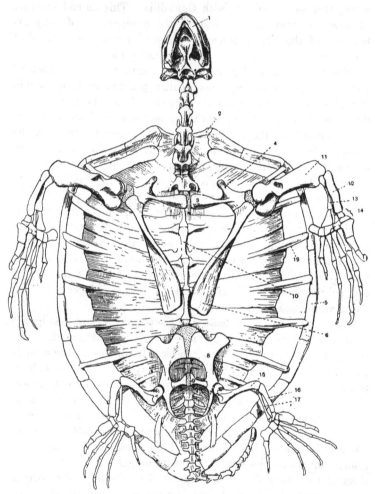

Fig. 269. Ventral view of the Skeleton of *Chelone mydas*, the Green Turtle × about ⅛. The plastron has been removed.

1. Lower jaw or mandible. 2. Nuchal plate. 3. Ventral process of scapula, the acromion. 4. Scapula (much foreshortened). 5. Marginal bone. 6. Coracoid. 7. Ilium. 8. Pubis. 9. Ischium. 10. Centrum of vertebra. 11. Humerus. 12. Radius. 13. Ulna. 14. Carpus. 15. Femur. 16. Tibia. 17. Fibula.

We have already seen (page 350) that in some vertebrates the artery to the fore-limb arises from the systemic arch on its dorsal course to join its fellow, while in others the fore-limb receives its blood from an artery given off from the ventral end or commencement of the systemic arch or else from the ventral end of the third arch near its division into dorsal and ventral carotids. As the vessel to the fore-limb is always called a subclavian artery it is convenient to express the fact that this vessel is not homologous throughout the

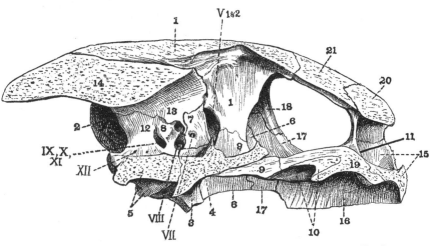

Fig. 270. Longitudinal vertical section through the Cranium of a Green Turtle, *Chelone mydas* × ⅔.

1. Parietal.   2. Squamosal.   3. Quadrate.   4. Basisphenoid.   5. Basi-occipital.   6. Quadratojugal.   7. Pro-otic.   8. Opisthotic.   9. Pterygoid. 10. Palatine.   11. Rod passed into narial passage.   12. Exoccipital. 13. Epi-otic fused to supra-occipital.   14. Supra-occipital.   15. Pre-maxilla.   16. Maxilla.   17. Jugal.   18. Postfrontal.   19. Vomer. 20. Prefrontal.   21. Frontal.   V 1 & 2, VII, VIII, IX, X, XI, XII, foramina for the exit of cranial nerves.

vertebrate groups by the terms "dorsal subclavian" and "ventral subclavian." In Amphibians and Lizards the subclavian is of the dorsal type, but in Chelonians and, as we shall see, in Crocodiles also the arm is supplied by a ventral subclavian, a vessel which is homologous with the "scapular" artery to the shoulder muscles in a Lizard. The venous system in all chief respects is like that already described in the Lizard.

The copulatory organ is a grooved rod attached to the front wall of the cloaca. The groove leads to the openings of the male ducts, the vasa deferentia.

The members of the order Chelonia have very various habits and modes of life. Some are vegetable feeders, others purely animal. None are found in Great Britain, but the representatives of six groups are found in temperate North America. These are

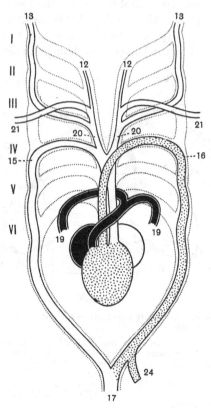

FIG. 271. Diagram of Arterial Arches of a Turtle viewed from ventral surface.

I. II. III. IV. V. VI. First to sixth arterial arches. 12. Tracheal (ventral carotid). 13. Common carotid (dorsal carotid). 15. Right systemic arch. 16. Left systemic arch. 17. Dorsal aorta. 19. Pulmonary. 20. Innominate. 21. Subclavian (ventral type). 24. Coeliac.

(1) The TESTUDINIDAE or Land Tortoises.

(2) The EMYDIDAE or Pond Turtles.

(3) The CINOSTERNIDAE or Box Turtles.

(4) The CHELYDRIDAE or Snapping Turtles.

(5) The TRIONYCHIDAE or Mud Turtles.

(6) The CHELONIDAE or Marine Turtles.

The TESTUDINIDAE have a very arched carapace and short club-like limbs in which the toes are tightly bound together by skin. Only a few species, *Testudo polyphemus*, the burrowing Gopher, and *Cistudo carolina*, the Box Tortoise, are known in temperate North America.

The EMYDIDAE are represented by many species. In this family the carapace has a wide horizontal margin and the toes are connected by a web. Most of the species are aquatic, a few however are almost as terrestrial as the TESTUDINIDAE. *Chrysemys picta*, the painted Pond Turtle, ranges north into the St Lawrence.

The CINOSTERNIDAE have a long and narrow carapace with the margins produced downwards; it is highest behind. The front part, and sometimes the hind part, of the plastron move like a hinge on the rest and close in the head and tail, whence the name Box Turtle. Sole genus *Cinosternum*, e.g., *pennsylvanicum*.

The CHELYDRIDAE are the so-called Alligator- or Snapping-Turtles. The head, neck and tail are all large and cannot be completely protected between the carapace and plastron. The carapace is highest in front. The jaws are hooked and powerful and the animals are very vicious. *Chelydra serpentina*, the "Snapper," is one of the commonest of American turtles. It is found everywhere from Canada to the tropics.

The TRIONYCHIDAE or Mud Turtles have no horny scales; both carapace and plastron are covered with leathery skin. There is a soft pig-like snout; only the three centre toes have claws. They seek their food by burrowing in the bottom of ponds.

The CHELONIDAE are distinguished by their peculiar skull and the absence of many or all of the nails. Their extremities have become flattened and form very efficient paddles.

Order V. Crocodilia.

The last and highest order of the Reptilia is the Crocodilia. These animals agree with the Chelonia in having a series of bony plates underlying the horny scales of the skin, also in having an immovable quadrate and a single median copulatory organ.

The Crocodiles are of large size and are decidedly Lizard-like in their general appearance, the chief observable external difference between them and the Lacertilia being in the jaws, which are exceedingly long in comparison with the rest of the skull, so that the gape is very wide.

The dermal plates form rings on the tail, but on the body, as in Chelonia, they form a dorsal and a ventral shield separated by intervening softer skin. In many Crocodiles the ventral shield is very rudimentary.

In the general arrangement of the bones and the temporal fossae the skull resembles that of *Sphenodon* : but there are great differences in the jaws and palate. The maxilla is very long and is armed with conical teeth which are implanted in distinct sockets or alveoli, the bone having grown up round their bases.

The two palatal folds have met so as to completely divide the upper air passage from the lower food passage : both the palatines and the pterygoids being completely united in the middle line

(Fig. 274). The choanae or posterior nares are therefore situated very far back directly over the glottis, whilst the external nostril is at the tip of the snout.

In consequence of this position of the external nostril the

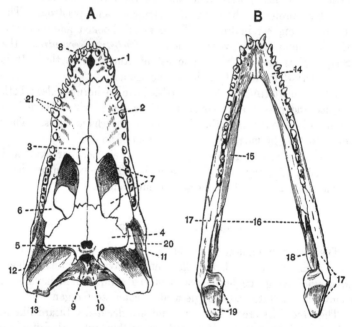

FIG. 272. Palatal aspect, A, of the Cranium, B, of the Mandible of an Alligator, *Caiman latirostris* × ⅓.

1. Premaxilla. 2. Maxilla. 3. Palatine. 4. Pterygoid. 5. Posterior nares. 6. Transverse bone. 7. Posterior palatine vacuity. 8. Anterior palatine vacuity. 9. Basi-occipital. 10. Opening of median Eustachian canal. 11. Jugal. 12. Quadratojugal. 13. Quadrate. 14. Dentary. 15. Splenial. 16. Coronoid. 17. Supra-angular. 18. Angular. 19. Articular. 20. Lateral temporal fossa. 21. Openings for the passage of blood-vessels supplying the alveoli of the teeth.

crocodile can lie for hours hidden under the water with only the tip of the snout exposed, and so surprise any unwary animal coming to the water to drink.

All the cervical and trunk vertebrae and some even of the caudal vertebrae bear ribs. The manner in which these ribs are articulated to the atlas and axis vertebrae throws much light on the relation of these peculiar vertebrae to the rest. Thus we observe that the first pair of ribs are articulated with their heads to the lower part of the atlas, showing that this represents a basi-ventral homologous with the intervertebral cartilaginous pads of the rest of the column. The heads of the second pair of ribs are united to an intervertebral cartilage separating the odontoid process from the centrum of the second vertebra. This cartilage is therefore the second basiventral, and the odontoid process is the first interventral, homologous with the centra of all succeeding vertebrae. The tubercle of each of the second pair of ribs has also an attachment to the odontoid process lying obliquely above and behind the capitular attachment and hence the centrum of the axis vertebra has no rib attached to it. The third pair of ribs has shifted its capitular attachment back on the centrum of the third vertebra.

This backward shifting of the capitular attachment has taken place in all suc-ceeding vertebrae, and the head of the rib is attached directly under its tubercle. In the trunk, as we proceed backwards, the capitular attachment to the centrum is gradually raised till it reaches the transverse pro-cess and is confounded with the tubercular attachment, and the hindermost ver-tebrae are single headed.

There are abdominal ribs, as in *Sphenodon*; they are arranged in transverse rows, each row on each side consisting of three or four bones (Fig. 274).

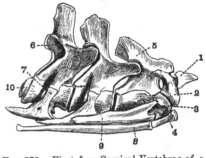

Fig. 273. First four Cervical Vertebrae of a Crocodile, *C. vulgaris*. Partly after von Zittel.

1. Neural spine of atlas. 2. Lateral portion of atlas. 3. Odontoid process. 4. Ventral portion of atlas. 5. Neural spine of axis. 6. Postzygapophysis of fourth vertebra. 7. Tubercular portion of fourth cervical rib. 8. First cervical rib. 9. Second cervical rib. 10. Convex posterior surface of centrum of fourth vertebra.

The pectoral girdle consists of simply a scapula and coracoid, the latter reaching the sternum, which is cartilaginous but protected ventrally by an interclavicle (Fig. 275). In the fore-limb the carpus has retained three bones in the proximal row, but the distal row consists of a block of cartilage representing the first and second carpalia and a bone representing the remaining three. There is consequently an intercarpal wrist-joint corresponding to the inter-tarsal joint common to Reptiles.

The pelvic girdle is very peculiar. The ilium is broad and

rounded above and joins the two sacral vertebrae. The true pubis is a small round bone inserted and fused with the anterior edge of the ischium and ilium, but what is ordinarily called the pubis, or better the epipubis, is a bone directed forwards which does not meet its fellow, nor does it form any part of the socket for the femur or acetabulum (Fig. 275). This so-called pubis or epipubis can be compared only to that of Urodela and Marsupials. The tarsus, like the carpus, is much reduced and modified. It consists of a proximal row of two bones, one of which, the fibulare or calcaneum, forms a distinct heel. The distal row consists of two bones, one representing the first, second and third tarsalia, the other the fourth and fifth.

The heart of the Crocodile is remarkable for the fact that the septum in the ventricle has grown forwards so as to completely divide it into two halves, the right and left ventricles. The left root of the aorta arises from the right ventricle and crosses the right root, which arises from the left ventricle and gives off the two carotids. The left root therefore receives venous blood from the right auricle and the blood sent to the trunk is mixed. In

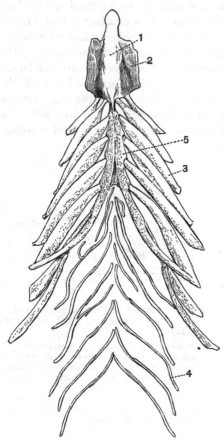

Fig. 274.   Sternum and associated Membrane-bones of a Crocodile, *C. palustris* × ⅓.

The last pair of abdominal ribs which are united with the epipubes by a plate of cartilage have been omitted.

1. Interclavicle.   2. Sternum.   3. Sternal rib.   4. Abdominal splint rib.   5. Sternal band.

addition there is a small passage, the foramen of Panizza, joining the two trunks where they cross, so that the blood leaving the right arch to go to the carotid is also somewhat mixed. The right common or dorsal carotid is very reduced, the left-hand vessel supplying both sides of the head. The fore-limb receives blood by a subclavian of the ventral type, as in Chelonians. The lung is no longer a simple sac, but has thick spongy walls and the central passage is reduced to a narrow tube. In the brain the cerebellum is large and cylindrical.

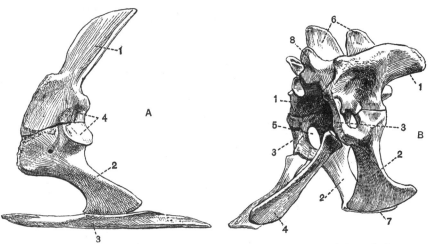

Fig. 275. A. Left half of the Pectoral Girdle of an Alligator, *Caiman lati-rostris* × ⅔.

1. Scapula.   2. Coracoid.   3. Interclavicle.   4. Glenoid cavity.

B. Pelvis and Sacrum of an Alligator, *Caiman latirostris* × ½.

1. Ilium.   2. Ischium.   3. True pubis.   4. Epipubis (so-called pubis).
5. Acetabular foramen.   6. Neural spines of sacral vertebrae.   7. Union of the two ischial bones.   8. Process bearing prezygapophysis.

All these peculiarities of the internal organs may be termed foreshadowings of what is found in Birds and Mammals, and hence Crocodiles are styled rightly the highest of the Reptiles.

Crocodiles are inhabitants of rivers and swamps and spend most of their life in the water. The best known and classical example is the Crocodile of the Nile, *Crocodilus niloticus*. There is but one species in the southern states of North America, the Alligator, *Alligator mississipiensis*, which has a much shorter and broader snout than the Crocodile. This animal lies for hours absolutely motionless at the surface of the water so as to greatly resemble a log, and thus entrap any unwary animal which may venture near. The Gavial, *Gavialis gangeticus,* in India is remarkable for its excessively long and narrow jaws.

So far as we can learn from fossils the Reptiles seem to have been the dominating type of land animals in the ages which intervened between the close of the Coal epoch and the end of the Chalk period when the white limestone which constitutes the Southern cliffs of Britain was deposited as a sediment in the quiet waters which covered what is now Western Europe. A rough sketch of the history of the Class as deduced from fossils may be given here.

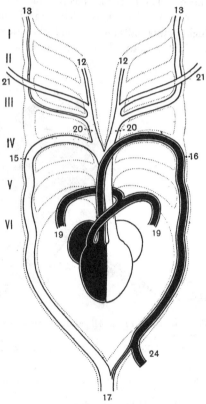

Fig. 276.   Diagram of Arterial Arches of Crocodile viewed from the ventral aspect.

I. II. III. IV. V. VI. First to sixth arterial arches.   12. Tracheal (ventral carotid). 13. Common carotid (dorsal carotid) [right side nearly atrophied].   15. Right systemic arch.   16. Left systemic arch. 17. Dorsal aorta.   19. Pulmonary.   20. Innominate.   21. Subclavian (ventral type). 24. Coeliac.

The Reptilia seem to have arisen from the Stegocephala.   At least included in the latter group are some forms like *Eryops* and *Cricotus* in which the bones flanking the notochord, which have not yet united so as to form vertebrae, are represented by basidorsals, basiventrals, and interventrals, the interdorsal as in all Reptilia being suppressed, while in the skull the basi-occipital region is ossified. In the Sandstones lying above the Coal indubitable Reptiles with fully formed vertebrae make their appearance. Some of these, termed the Pariasauria, still recall the Stegocephala in possessing a complete roof of dermal bones covering the skull.   At the same time allied forms termed Theromorpha showed gaps in this armour of dermal bone corresponding to the supratemporal and laterotemporal fossae, the latter being exceedingly small and often not present. The limbs in all these early Reptiles were short and stout, the fore and hind limbs being of nearly the same size. Some of the Theromorpha in possessing teeth divided into incisors, canines and molars, and in having the lower jaw partly supported by the squamosal, approached the characters of Mammalia. Another allied group were the Dicynodontia, which agreed with the Theromorpha in having a large supratemporal fossa but differed in having the teeth reduced to two tusks in front or completely absent. It is believed that these were the forerunners of the Chelonia.

In the same period we meet a large number of forms with supratemporal and laterotemporal fossae equally developed, these were the Rhynchocephala of which *Sphenodon* now is the sole survivor. From this group in the following age were developed (*a*) Water-Reptiles—the Plesiosauria—with long swan-like necks and limbs transformed into flippers by the shortening of the bones of the arm and leg, and (*b*) Land-Reptiles—the Dinosauria—with greatly developed limbs ; in some cases the whole weight being borne by the hind-limbs, the fore-limbs being short and used for prehensile purposes only. In a still later period from the less specialized Dinosauria were developed (1) the *Crocodilia*, which reverted to the water but retained limbs fit for progression, and (2) *Pterosauria*, possibly flying reptiles in which the "wing" was a flap of skin supported by the greatly elongated 5th finger. The forerunners of modern Sauria are found only in the Chalk period in the form of long-bodied aquatic Reptiles with however the characteristic loss of the quadrato-jugal. There remain to be mentioned the whale-like Ichthyosauria, which since they possess only a supratemporal fore-arm seem to be sprung from the Theromorpha. The limbs were converted into flippers more thoroughly than in the case of the Plesiosauria, the limb-bones being reduced to round nodules and the fingers increased in number by forking. There was no neck and no sternum but a series of "abdominal ribs" as in Rhynchocephala.

The class Reptilia is classified as follows :

Order 1. **Rhynchocephala.**

Reptilia devoid of special copulatory organs and with an immovable quadrate.

Ex. *Sphenodon.*

SAURIA.

Order 2. Lacertilia.

Reptilia with two copulatory sacs on the posterior
wall of the cloaca : with a movable quadrate, with a
pectoral girdle, and with the rami of the lower jaw united
by a symphysis.

Ex. *Lacerta, Anguis.*

Order 3. Ophidia.

Reptilia with two copulatory sacs on the posterior wall
of the cloaca : with a movable quadrate but with no trace
of a pectoral girdle ; with the rami of the lower jaw
united by ligament.

Ex. *Crotalus, Vipera, Tropidonotus.*

Order 4. Chelonia.

Reptilia with one median copulatory organ on the anterior
wall of the cloaca and an immovable quadrate. No sternum
and the ribs expanded horizontally to form a dorsal shield :
a ventral shield of dermal bone. No teeth.

Ex. *Testudo, Chelone.*

Order 5. Crocodilia.

Reptilia with one median copulatory organ on the anterior
wall of the cloaca and an immovable quadrate. A well-developed
sternum, joined by the ribs. With many alveolar teeth.

Ex. *Crocodilus, Alligator, Gavialis.*

CHAPTER XVIII.

Sub-Phylum IV. Craniata.

Class IV. Aves.

It is probable that if the first child one met were asked to describe a bird, he would say that birds were animals General Characteristics. which were covered with feathers and had wings to fly with. Though it often happens that the marks by which the ordinary person distinguishes one animal from another are not those which seem most important to a Zoologist, yet in this case the Zoologist could not find more important features to serve as the basis of a definition of the class Aves or Birds.

Birds then are vertebrate animals in which the fore-limb is modified into a wing or flying organ and in which the body is covered with feathers. Bats likewise have the fore-limb converted into a wing, but they are covered with hair, not feathers, and their wing is not constructed on the same plan as that of the bird.

Birds are sometimes classed along with the Reptiles as Sauropsida, since they have a good many features in common with them, and are thus contrasted with the Mammalia, or ordinary quadrupeds. This, however, gives a wrong view of the relationships of the three groups. Both Birds and Mammals are believed to be descended from Reptilian-like ancestors, and it is an open question whether the changes which Birds have undergone are not at least as important as those which have taken place in Mammals in the process of their evolution from ancestors which, had they lived now, would have been termed Reptiles.

Birds agree with Reptiles in that they lay large eggs from which the young are hatched in a form closely resembling the parent; they are like Reptiles also in the structure of their jaws—the lower jaw consisting of five bones and articulating with a quadrate bone—and in the structure of the hinder part of their skulls, of their breast-bones and of their ankle-joints. The vertebrae very rarely have epiphyses like those of Mammals, though these are found in Parrots. As in Reptiles, the number of neck vertebrae is variable. Like Reptiles, Birds have nuclei in the red corpuscles of the blood, and the sole remaining complete systemic arch goes to the right (Fig. 285), like the principal arch in Reptiles. On the other hand, they are "warm-blooded," that is to say, the temperature of the body remains practically the same whether the surrounding air gets hot or cold ; it is in fact higher than that of any mammal : the ventricle of the heart is completely divided into two, and in addition to the wings and feathers, the structure of the leg and hip-bones and of the brain, distinguishes them from any living Reptile.

Strange as the statement may appear, it is true, nevertheless, that the feathers are really scales like those found in lizards, immensely developed and with the edges frayed out. Like scales, they are epidermal, that is, developments of the outer or horny layer of skin. The area which is about to form the feather becomes raised into a little finger-shaped knob by the growth of the deep layer of the skin or dermis which carries the blood-vessels. The little knob thus formed is in turn sunk in a pit called the follicle, the skin immediately surrounding it being depressed. Thus the lowest part of the feather is a little hollow tube of horny cells formed round the knob of dermis, but the upper part, like the scale of a lizard, is formed only on one side of the knob, and this part as it is pushed away by the growth of the deeper parts becomes frayed out so as to form the vane of the feather. In the latter we can distinguish a central stem or rhachis, and two rows of lateral branches or barbs, which are kept in position by a number of secondary processes or barbules. The barbules bear little hooks which interlock with one another. Down consists of small feathers growing between the bases of the larger ones. In these the barbules are absent, so that the barbs are not held together but float freely about, forming a kind of fluff. When a bird is plucked it is seen that the feathers are confined to certain tracts (pterylae) separated by others called apteria devoid of feathers or covered only with down feathers. Thus in most birds the mid-ventral and mid-dorsal lines are apteria. The colour of the feathers is partly due to coloured substances or pigments in the epidermal cells and partly to minute structural detail which causes interference of the light reflected from them.

The wing is the foreleg of the bird. One can easily recognise the parts corresponding to upper arm, fore-arm and hand, but the latter is highly modified and specialised for the important function of carrying the long primaries or hand quills. When the wing is at rest the upper arm extends backwards, the fore-arm is sharply

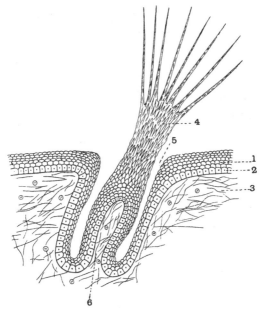

FIG. 277.   Section through the Skin of a Bird showing a developing Feather.
Highly magnified.

1. Epidermis.   2. Malpighian layer of the epidermis.   3. Dermis.
4. Young feather.   5. Follicle round base of feather.   6. Dermal
papilla which developes blood vessels and is the organ of nutrition of
the feather.

bent up on this, while the wrist is sharply bent down. When the wing is expanded these are partially, but never entirely, straightened out, so that a bird begins the down-stroke of the wing with the arm bent in a very similar way to that in which a swimmer's arm is bent when he strikes back with it. In the hand we find as a rule three digits, the first, second and third. These have their first joints, the metacarpal bones, closely united

together.   In Man the metacarpals of the various fingers are united
by skin and flesh which constitute the palm, but they are movable on one another, whereas in the bird the metacarpals of the
second and third digits are firmly joined at both ends.   The index
has in addition to the metacarpal, in most birds, three other small
bones called phalanges, of which the end one sometimes carries
a claw: the third digit has only one bone or phalanx besides the
metacarpal.   The metacarpal of the first digit or thumb is very
small, but is likewise completely fused
with the other metacarpals.   Besides this
the thumb has two joints and often a
claw.

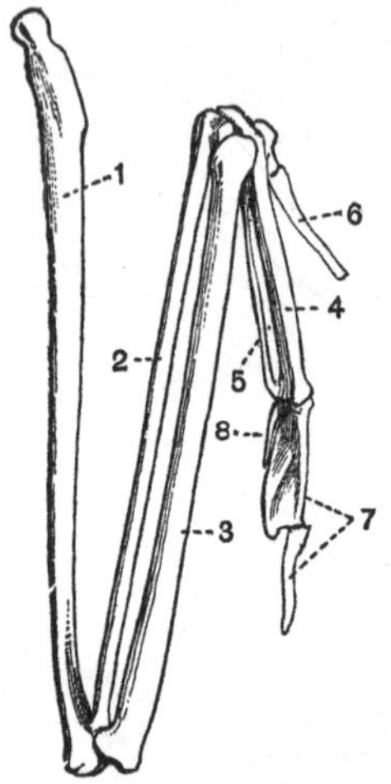

Compared with the arm or fore-leg of
other animals the arm of a bird strikes one
as having very little flesh.   This is because the muscles, especially those on the
fore-arm, have comparatively short bellies
but very long tendons, in correlation with
the often very much lengthened bones,
one of which, the ulna, serves as support
of the secondaries or arm-quills.

The movements which constitute flying, namely, the powerful down-stroke of
the whole arm and the slower up-stroke,
are carried out by the immensely developed
pectoral muscles, great fleshy masses which
cover the breast-bone or sternum.   This
bone has a more or less pear-shaped outline, rounded in front and pointed behind,
the ribs ending in its sides (Fig. 279).
In accordance with the tendency in all
birds to develope the body into a long
neck and a rounded trunk, we find evidence
that the number of ribs encircling the body
and joining the sternum has been reduced.

FIG. 278. Bones of the right
Wing of a Gannet, *Sula
alba* × ½.

1. Humerus.   2. Radius.
3. Ulna.   4. Second
metacarpal.   5. Third
metacarpal.   6. Thumb
or pollex.   7. Second
digit.   8. Third digit.
The distal phalanges of
the thumb and second
digit were wanting in the
specimen from which this
figure was drawn.

Not only do we find small free ribs connected with the hinder
cervical vertebrae, but attached to the sternum are outgrowths
called costal and xiphoid processes (Fig. 279) which are regarded
as the remains of sternal ribs the dorsal halves of which are
vestigial or lost.   If we picture to ourselves the pectoral girdle
being thrust backwards and the pelvic girdle forwards so as to crowd

the viscera into a small space we shall realize the meaning of the
differences between the skeleton of the trunk of a Reptile and that
of a Bird.   From the middle line of the sternum projects a great
vertical crest stretching outwards, called the carina or keel, and it
is from the sides of this mainly that the pectoral muscles take their
origin.   There are two main muscles on each side.   First the pectoralis major on the surface, which passes into a tendon attached
to the upper end of the humerus.   The contraction of this muscle
brings about the down-stroke of the wing, the effective stroke in
flying.   Underneath the pectoralis major is situated the pectoralis
minor, a much smaller muscle.   Its tendon passes underneath

the arch formed by the clavicle
and the coracoid bone, the latter of
which, as in Reptiles, connects the
shoulder-blade firmly with the sternum.   Having passed through this
arch the tendon is attached to the
back of the humerus, so that the
contraction of the muscle pulls the
humerus and thus the wing upwards
and backwards and not downwards,
the upper end of the coracoid acting
as a pulley round which it passes.

Returning to the wing, we must
now notice how the feathers are
arranged.   The great quill feathers
are attached chiefly to the upper
and posterior edge of the hand, but
there are also a large number which
are implanted in the posterior surface of the ulna.   These two groups
of feathers are pushed one over the
other when the wing is folded, just

Fig. 279. Shoulder-girdle and Sternum of Peacock, *Pavo cristatus*
× ⅜.

1. Carina of the sternum.   2. Coracoid.   3. Scapula.   4. Clavicle.
5. Costal process.   6. Surfaces
for articulation with the sternal
ribs.   7. Posterior (xiphoid)
and oblique processes.

like the silk of a closed umbrella, but when the wing is stretched
out they only overlap very slightly, and thus a coherent and
practically air-tight surface is formed.   Those feathers which are
attached to the hand are called primaries (6, Fig. 280, C), those
arising from the ulna, secondaries (8, Fig. 280, C); a few arising
from the upper arm are called tertiaries; any air which might
escape between the bases of the long feathers is stopped by an
upper layer of shorter feathers, called coverts (1, 2, 3, 5 and 7,

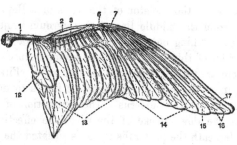

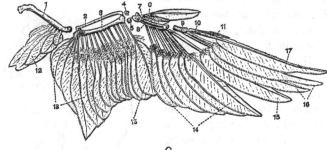

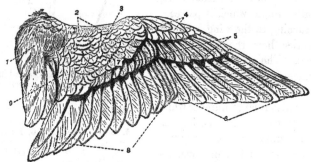

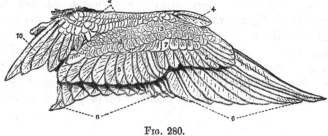

Fig. 280.

Fig. 280. Wing of a Wild Duck, *Anas boschas* × ⅓.

A. Right wing seen from the dorsal side, with the coverts removed. B. Left
wing disarticulated and seen from the ventral side, with the coverts
removed.   C. The dorsal side of a right wing.   D. The ventral side
of a left wing. From Wray.

In A and B.   1. Humerus.   2. Radius.   3. Ulna.   4. Radial carpal.
5. Ulna carpal.   6. First phalanx of first digit.   7. Second metacarpal.
8. Third metacarpal.   9. First phalanx of second digit.   10. Second
phalanx of second digit.   11. Vestigial quill.   12. Tertiaries.
13. Secondaries.   14—17. Primaries.

In C and D. 1, 2, 3, 5, and 7. Coverts.   4. Bastard wing.   6. Primaries.
8. Secondaries.   9, 10. Tertiaries.

Fig. 280, C and D).   Air is prevented from escaping in front by the
hand, which is stretched out in a vertical plane, and by two folds of
skin, one in the angle between fore-arm and upper arm, the other
between the upper arm and the body. The name bastard wing
is given to a tuft of feathers borne by the thumb (4 Fig. 280, C
and D).

The full mechanical explanation how the down-stroke of the
wing not only prevents a bird from falling but urges
it onwards is not completely understood, and much
of what is generally accepted is too complicated for an elementary
text-book, but the broad principles involved may be simply set
forth. A bird when it is in the air, like any other heavy body,
is continually falling : the blow of the wing has therefore not only
to effect a forward impulse, but also an upward one sufficient
to compensate for the distance the bird has fallen between two
strokes. These impulses are derived from the elastic reaction of
the air compressed by the down-stroke of the wing. When the
wing is expanded, it is slightly convex above and concave beneath.
This arises from the fact that the quill feathers are attached to
the upper edge of the webbed limb and project gently downwards
and backwards, so that there is a space left which is bounded
behind by the quills and in front by the bones and web of the
limb. Now if this space had a symmetrical shape the air would
be compressed in such a way that the resultant impulse would be
directly upwards ; but it is not symmetrical, for its roof has a very
steep slope in front and a very gentle one behind, and the air is
compressed in such a way that an oblique reaction results, a
reaction which we can resolve by the parallelogram of forces into
an upward and an onward one. So much for the flight of a bird in
still air. The air is, however, very rarely still, and the currents
which exist are never quite horizontal, but generally inclined

Flight.

slightly upwards, since the lowest layer of air is checked by friction
against the ground, and birds which are good flyers can, by in-
clining their wings at the proper angle, obtain quite sufficient
support from the play of the current against the wing without
exerting themselves to any great extent. This is called soaring,
and can be seen beautifully in the flight of the Gannet. In this
manœuvre birds are assisted by the tail, which is really a fan-
shaped row of strong feathers attached to the coccyx, that very
small vestige of a true tail or portion of the vertebral column
extending behind the anus, which modern birds possess (Fig. 282).
In this region the vertebrae are thin discs, several of which may be
soldered together so as to form a bone called the pygostyle.

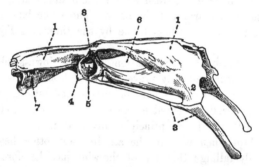

Fig. 281.   Lateral view of the Pelvis and Sacrum of a Duck, *Anas boschas* × $\frac{2}{3}$.

1. Ilium.   2. Ischium.   3. Pubis.   4. Pectineal process, the rudiment
   of the prepubis corresponding to the pubis of the Lizard.   5. Acetabulum.   6. Ilio-ischiatic foramen.   7. Fused vertebrae.   8. Facet
   on which the projection on the femur, the trochanter, plays.

The legs of birds can be shown to be constructed on essentially
the same type as those of Reptiles, but modified so as to
enable them to support the body in an upright position.
The arrangements to effect this are very interesting,
as they differ markedly from those found in the human skeleton.

Upright
position.

In the pelvic girdle the ilia are lengthened so as to be
attached to a considerable number of vertebrae, six or more, and
so a firm attachment of the limb to the main skeleton is effected.
In Reptiles only two vertebrae are joined to the ilium, but in
their case the weight of the body is supported on all four limbs,
whereas in a Bird the whole vertebral column has to be balanced
about two points of support, and hence the ilium must be quite

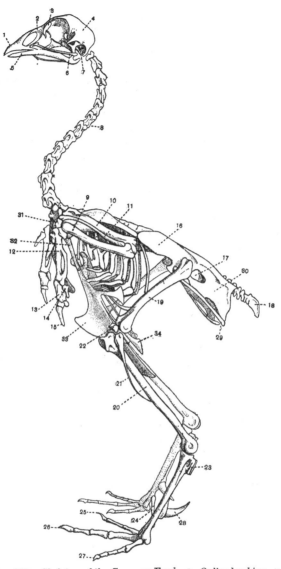

Fig. 282. Skeleton of the Common Fowl, ♂, *Gallus bankiva* × ½.

1. Premaxilla. 2. Nasal. 3. Lachrymal. 4. Frontal. 5. Mandible. 6. Lower temporal arcade in region of quadratojugal. 7. Tympanic cavity. 8. Cervical vertebrae. 9. Ulna. 10. Humerus. 11. Radius. 12. Carpo-metacarpus. 13. First phalanx of second digit. 14. Third digit. 15. Second digit. 16. Ilium. 17. Ilio-ischiatic foramen. 18. Pygostyle. 19. Femur. 20. Tibio-tarsus. 21. Fibula. 22. Patella. 23. Tarso-metatarsus. 24. First toe. 25. Second toe. 26. Third toe. 27. Fourth toe. 28. Spur. 29. Pubis. 30. Ischium. 31. Clavicle. 32. Coracoid. 33. Keel of sternum. 34. Xiphoid process. The forked bone just in front of 7 is the Quadrate.

immovably strapped to the vertebral column. The result of this
has been atrophy of some of the hinder ribs, and the ventral
halves of some of these form the xiphoid processes of the sternum.
The ischium is directed backwards parallel to the hinder part of
the ilium, and often fused with it so as to surround a space
called the ilio-ischiatic foramen. The pubis is a very slender
bone which is also directed backwards. It is in fact a postpubis
corresponding to the lateral process on the pubis of the lizard (see
p. 467). Except in the Ostrich the two pubes never unite with
one another ventrally to the cloaca, as they do in Reptiles and
Mammals, the absence of a pubic symphysis facilitating the laying
of the egg, which is very large relatively to the size of the animal.
The thigh is bent sharply forwards and the shank backwards, and
the ankle is raised to a considerable height above the ground by
the great length and upward direction of the bones of the sole or
metatarsals (Fig. 282). Thus a Bird walks on its toes and like
Reptiles possesses an intertarsal ankle-joint. In Birds however, in
order to give firmness to the leg, the metatarsals are closely united
together and the small bones of the tarsus have entirely disappeared, the proximal row having been incorporated with the tibia,
while the distal bones have fused with the metatarsals. Thus in
an adult Bird the ankle-joint is a simple hinge between two
compact bones, the upper being a tibio-tarsus, the lower a tarso-
metatarsus. There are usually four toes, but the first, corresponding to the human great toe, is sometimes, like the fifth,
absent, while its metatarsal remains distinct from the other three.
This toe, except in Steganopodes, is generally directed backwards.
The raised sole of the foot really constitutes the visible "leg" of
most birds, the thigh being altogether, and the shank mostly, buried
in the feathers. In many birds the sole is plated by scales which
are raised horny plates of skin, similar to the scales of Reptiles.

The most characteristic features about a bird, next to the limbs
and feathers, are certainly the head and neck. The
skull is high and arched behind in order to make
room for the comparatively large brain; in front it
slopes gradually downwards to the pointed beak, which is encased
in a hard horny sheath. The bones which underlie this beak are
(above) the premaxilla and (below) the dentary bone of the lower
jaw. No modern bird possesses teeth, and the maxilla, which
usually carries most of the teeth in animals which have them, is
very small and confined to the cheek behind the gape, whereas the

Head and
Neck.

premaxilla is very large. Behind the maxilla two other slender bones, the jugal and quadratojugal, complete the lower temporal arcade as in Chelonia and Crocodilia, but the jugal never sends up a process behind the orbit and the post-orbital is a mere process of the frontal bone, so that the orbit and the temporal fossa open into one another. The eyes are of great size: a bird has little or no sense of smell, and governs its life mainly by the sense of sight: in correspondence with this the orbits or eye-sockets are so enlarged that the skull between them is reduced to a thin vertical plate, the interorbital septum, in which there is no brain cavity. This great development of the eye-sockets and the obliteration of the brain cavity between them is not, however, confined to Birds: it is found as already mentioned in many Reptiles also, and is indeed one of the several points in which a bird's skull may be said to be Reptilian. It is however characteristic of Birds, as opposed to Reptiles, that this interorbital septum is largely converted into bone. In its hinder and upper portions it is composed of orbitosphenoid bones, like those found in Teleostean fishes, which support the exit of the optic nerve, but in its lower part it is composed of a vertically compressed presphenoid bone corresponding to that which ossifies the front part of the floor of the cranium in Mammalia. In front the interorbital septum is continuous with the internasal or ethmoid septum: this latter is ossified by a mesethmoid bone, which unites, but not quite immovably, with the presphenoid. The hinder part of the floor of the cranium is ossified by the basioccipital and basisphenoid bones, and the front of the latter is drawn out into a long spur called the basisphenoidal rostrum. Underlying the basisphenoid there is a membrane-bone called the basitemporal, a relic of the parasphenoid of Fishes and Amphibians. In some Reptiles traces of the front part of this bone remain, but never any of the hinder portion, and this is an indication that Birds are descended from a type of Reptile more primitive in some respects than any now existing. Other points of resemblance to Reptiles are that the lower jaw is made up of no less than five distinct bones interlocking with each other; and that instead of there being a direct hinging or articulation of the lower jaw to the skull, a bone called the quadrate is interposed, as in Reptiles, which articulates on the one hand with the lower jaw and on the other with the skull. This quadrate bone is movable, and to it in front are jointed the bones of the palate, the pterygoids and palatines, which slide on, but are not fixed to, the base of the skull. Hence

when the lower jaw is opened, i.e., pulled down, these bones are
pushed forward, and the upper beak, to which they are fastened in
front, is slightly tilted up, thus increasing the width of the gape.
In parrots the front part of the skull, including the bones of the
face, has an actual joint with the hinder part of the skull. Thus it
follows that in spite of the presence of a quadratojugal the quadrate
is movable. It is to be remembered however that the quadrato-
jugal is here a small flexible bone, very unlike the great bony bar
of Chelonia and Crocodilia. The hyoid apparatus consists of the
second and third pairs of visceral arches. The second pair, which

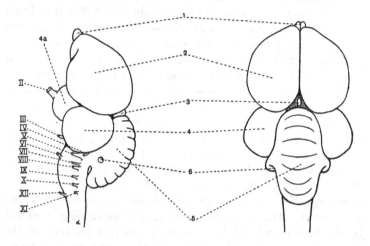

Fig. 283. Brain of Pigeon, *Columba livia* × about 2.

1. Olfactory lobes.   2. Cerebral hemispheres.   3. Pineal gland.   4. Optic
lobes.   4 a. Optic chiasma.   5. Cerebellum.   6. Lateral lobe of
cerebellum.   II. Optic nerves.   III. Motor oculi.   IV. Patheticus.
V. Trigeminal.   VI. Abducens.   VII. Facial.   VIII. Auditory.
IX. Glossopharyngeal.   X. Vagus.   XI. Spinal accessory.
XII. Hypoglossal.

correspond to the hyoid of Fishes, are very short and consist mainly
of the median piece or glosso-hyal which is closely connected to
the median piece of the third pair. The latter are elongated rods
to which are attached the muscles which protrude the tongue. As
in the Reptilia, the skull has one central knob or condyle for
articulation with the backbone, not two, as is the case with the
Amphibia and Mammalia.

The features peculiar to the Bird are, firstly, the great elongation of the premaxilla carrying the beak—this causes the nostrils to
be placed at the base of the snout instead of at the tip, as is the case
with Reptiles; secondly, the enormous size of the orbit and the
absence of any bony bar to separate it from the temporal fossa, the
hollow on the side of the skull, in which are situated the muscles that
close the jaws; and thirdly, the height and arched character of
the hinder part of the skull, which lodges the brain. The bones of
the skull are usually indistinguishably united in the adult, are
hollow and contain air, and are in consequence very light, as befits
an animal which flies. Similar air spaces also exist in the larger
bones of the trunk and limbs. The insects, which also have taken
to the air, have somewhat analogous air reservoirs. Like insects,
birds are represented
by a large number of
species which all exhibit
great uniformity of
structure.

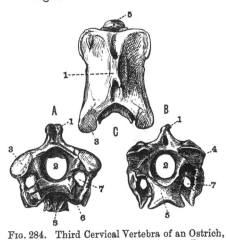

When the brain is
examined, the meaning
The Brain. of many of
the peculi-
arities of the skull is
seen. What we might
perhaps, with a little
looseness, call the organs
of thought, the hemi-
spheres of the fore-
brain, are greatly en-
larged, being high and
rounded. The parts of
the brain supplying the
nose, the olfactory
lobes, are on the other
hand very small and

FIG. 284. Third Cervical Vertebra of an Ostrich,
Struthio camelus × 1. A, anterior, B, posterior, C, dorsal view. A and B after Mivart.
1. Neural spine. 2. Neural canal. 3. Prezygapophysis. 4. Postzygapophysis. 5. Posterior articular surface of centrum. 6. Anterior
articular surface of centrum. 7. Canal
between the capitulum and tuberculum of the
rudimentary rib. 8. Hypopophysis, a
median ventral outgrowth of centrum.

poorly developed, in accordance with the feebly-developed nasal sacs,
the sense of smell being but slight, as mentioned above (Fig. 283).
The brain is bent sharply on itself, so that the optic lobes of the
mid-brain—portions connected largely with vision—are pressed downwards and the hemispheres are brought close to the cerebellum,

which, in contradistinction to what is the case in most reptiles, is large and transversely wrinkled. Evidence is accumulating that an important function of the cerebellum is to coordinate motor impulses proceeding from higher parts of the brain to the skeletal muscles.

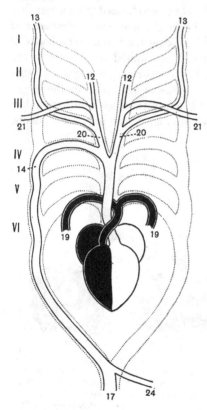

FIG. 285. Diagram of Arterial Arches of a Bird viewed from the ventral aspect.

I. II. III. IV. V. VI. First to sixth arterial arches. 12. Tracheal (ventral carotid). 13. Common carotid (dorsal carotid). 14. Systemic arch. 17. Dorsal aorta. 19. Pulmonary. 20. Innominate. 21. Subclavian (ventral type). 24. Coeliac.

All birds have comparatively long necks (Fig. 282), and the vertebrae which form the support of this part of the body have the surfaces with which they articulate with one another shaped like saddles, being concave from side to side and convex from above downwards in front and exactly the opposite curvatures behind (Fig. 284). This arrangement allows great freedom of movement, and, as all know, a bird is able to twist the head completely round and look straight backwards. In doing so of course it squeezes the skin of one side of the neck and stretches that of the other, and so the great jugular vein, which carries blood from the head, is liable to be blocked on one side (Fig. 286). To obviate this difficulty, the two jugulars are connected by a cross piece just under the head, so that the blood from both sides can always have a free passage. The carotid arteries, which take blood to the head, come close together at the base of the neck and run up just under the vertebrae. As they are placed close to the axis of rotation and are further protected by curved

Vascular System.

rods growing out from the vertebrae and forming arches over them, they are never compressed, however much the bird twists its neck.

Turning now to the consideration of the internal organs, we have first to notice the structure of the heart. In Birds the ventricle is completely divided into two, a condition found only in the Crocodiles among reptiles, and even there the great trunks leaving the two parts of the ventricle communicate. In Birds only one systemic arch remains complete; this passes round to the right, coming off from the left half of the ventricle; in Reptiles, it will be recollected, the left fellow of this one was still present. From the systemic arch there arises an innominate artery for either side, which splits up into a ventral carotid, reduced as compared with that of reptiles but, as in their case, supplying the trachea, and into a dorsal or common carotid to the head and a subclavian to the breast and wing. The subclavian artery which arises from the ventral carotid divides into a brachial artery of moderate size for the wing and a very much larger pectoral artery which supplies the pectoral muscles. These, as we have seen, are the real seat of the activities of the wing. The subclavian of Birds corresponds in origin with that of Chelonians and Crocodiles and so is the ventral type of subclavian, as opposed to the dorsal type found in Lizards and Amphibians. The arteries supplying the lungs, the pulmonaries, which, as in the Reptiles, have no longer any connection with the systemic arch, come off from the right side of the heart; one passes to each side to reach the lungs. The arteries of the hinder part of the trunk agree in their general arrangement with those of Reptilia and Amphibia. In the venous system the

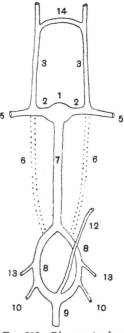

FIG. 286. Diagram to show arrangement of the principal Veins of a Bird.

1. Sinus venosus—gradually disappearing in the higher forms. 2. Ductus Cuvieri = superior vena cava. 3. Internal jugular = anterior cardinal vein. 5. Subclavian. 6. Posterior cardinal, front part. 7. Inferior vena cava. 8. Renal portal = hinder part of posterior cardinal. 9. Caudal. 10. Sciatic. 12. Coccygeo-mesenteric. 13. Femoral. 14. Anastomosis of jugulars.

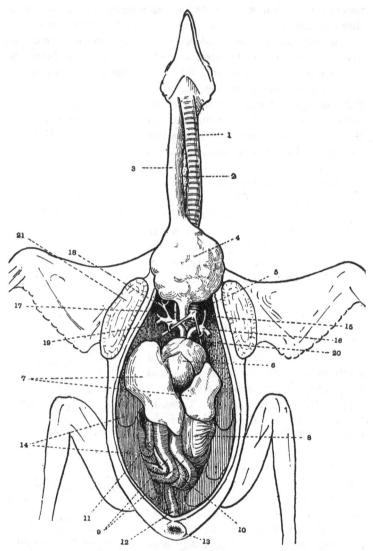

FIG. 287. The chief Viscera of the Pigeon, *Columba livia* × ⅔.

1. Trachea.   2. Thymus gland.   3. Oesophagus.   4. Crop.   5. Syrinx.
6. Heart.   7. Liver.   8. Gizzard.   9. Duodenum.   10. Pancreas.
11. Small intestine.   12. Rectum.   13. Cloaca.   14. Air-sacs.
15. Left carotid.   16. Left subclavian.   17. Right carotid.   18. Brachial
artery.   19. Right subclavian.   20. Muscles of syrinx.   21. Pectoralis major muscle cut across.

connection of the two jugulars has been already referred to. The jugular joins a large subclavian vein to form the superior vena cava. The largest part of the subclavian vein, like that of the corresponding artery, is made up of a pectoral vein returning blood from the pectoral muscles. The front parts of the posterior cardinal veins have disappeared : but their hinder parts remain as the renal-portal veins which as usual arise by the bifurcation of the caudal vein and receive on each side a femoral and a sciatic vein from the leg. The renal-portal pours its blood into the inferior vena cava, not as in Amphibia and Reptiles through a system of capillaries, but directly by a single vessel channelled through the substance of the kidney. Hence in Birds the kidney tubules receive blood only from the aorta and do not, as in the lower Craniata, receive a double supply. From the point where the caudal vein divides into the two renal-portals a vein is given off which descends into the mesentery and opens into the posterior mesenteric branch of the portal vein, thus establishing a connection between the portal and cardinal systems of veins. This vein is called the coccygeo-mesenteric (12, Fig. 286), and is quite peculiar to Birds.

The lungs are firmly fitted in against the ribs; they do not, as in most Reptiles or as in ourselves, hang freely in a cavity; their most remarkable feature is the possession of great thin-walled bladder-shaped outgrowths, the air-sacs, of which the prolongations extend even into the bones.

Respiratory System.

There are nine of these great air-sacs, one placed at the base of the neck, and the other eight situated in pairs at the sides of the body cavity under the ribs (Fig. 287). When the ribs are in their normal position, the air-sacs are expanded, but when the ribs are pulled backwards so as to compress the air-sacs, air is driven out; when the ribs and wall of the body behind come into their natural position again, the air-sacs are expanded and air rushes in, filling the lungs on its way. Breathing out or expiration is therefore the active function, drawing in air is an elastic reaction, the opposite to what is the case in man and other mammals. The windpipe or trachea is long, and the hoops of cartilage which stiffen it form complete rings, so that it is not easily compressed (Fig. 287). Like most other land vertebrates, birds have a larynx or organ of voice at the top of the trachea formed in the usual manner by the enlargement of some of these rings of cartilage, and the stretching of a thin membrane between them and two special cartilages, the arytenoids, which lie at the opening of the windpipe into the gullet.

The effective organ of voice in Birds, the syrinx, is found much deeper down, at the spot, namely, where the windpipe splits into two tubes, the bronchi, which lead to the lungs. The last rings surrounding the trachea just before it bifurcates are more or less fused with one another so as to form a box with stiff walls called the tympanum. The inner walls of the bronchi, just where they join one another, are thin and membranous, and constitute a membrana tympaniformis interna. From the fork a flexible valve, termed the membrana semilunaris, projects up into the tympanum, and as here the cartilage rings have the form of half-hoops, which are drawn together by special muscles, the width of the opening of the bronchus into the windpipe is small. When air is forcibly expelled the valve above mentioned is set vibrating like the reed in an organ-pipe, and by this mechanism the song is produced. The muscles which connect the half-rings together (intrinsic muscles) and two which connect the syrinx with the sternum (extrinsic muscles) by altering the tension of the sides of the trachea, and consequently the rate at which it vibrates, change the pitch of the note produced. A syrinx such as we have described is found in the vast majority of birds. It is termed a broncho-tracheal syrinx because both bronchi and trachea are concerned in its formation. In a few North American birds a tracheal syrinx is found in which the organ of voice is constituted by a portion of the trachea where the rings are thin and delicate, so that the sides are flexible. In a few birds allied to the Cuckoo there is a bronchial syrinx, a thin flexible membrane being formed about the middle of each bronchus by the incompleteness of some of the rings.

The alimentary canal commences with the buccal cavity or stomodaeum, partially divided by the palatal flaps into an upper air-passage, and a lower food-passage. The tongue, which is pointed and horny, ensheaths the glosso-hyal bone; it is protruded by the action of muscles which pull the enlarged third visceral arch forwards. Behind the tongue open the ducts of the sub-maxillary glands; at the corners of the gape the parotid glands pour their secretion into the mouth, whilst at the sides of the tongue the sub-lingual glands open. All these glands are pouch-like outgrowths of the

Digestive
Systems.

ectoderm of the stomodaeum and secrete a mucus which assists in
swallowing the food, and occasionally (as in Woodpeckers) in
causing the prey to adhere to the tongue. The names indicate the
position of the glands, as for instance, parotid (Gr. παρά, beside,
οὖς, ὠτός, the ear). Following on the buccal cavity and indistinguishably fused with it is the endodermal pharynx into which
the glottis opens, and also the persistent remains of the first pair
of gill-sacs, the Eustachian tubes. The pharynx leads into a long
gullet lying dorsal to the trachea, which eventually passes into the
stomach. The gullet in the Pigeon and many other birds developes
a large thin-walled outgrowth on the ventral side called the crop.
This is used as a storehouse for the food, and in the Pigeon
may be found full of unaltered seeds. The stomach has a most
characteristic form in Birds; it is sharply divided into two regions,
an anterior egg-shaped one called the proventriculus, and a large
posterior flattened one called the gizzard. In the walls of the
proventriculus are found the pepsin-forming glands, while on the
other hand the endoderm of the gizzard developes a horny lining
which is thin in Birds that live on an animal diet, but very thick in a
grain-eating Bird like the Pigeon, where it forms upper and lower
hardened plates. When by the contraction of the greatly thickened
visceral muscles of this part of the alimentary canal the upper and
lower plates are brought together, a crushing-mill is produced by
which the food is broken up. The action of this mill is assisted by
the habit which many Birds possess of swallowing fragments of
stone. A collection of these, sometimes including fragments of
glass, may be found on opening the gizzard of a Pigeon. It is a
great development of this habit which has earned for the Ostrich its
reputation of flourishing on a diet of nails, penknives and match-
boxes. The liver in Birds is remarkable for possessing two ducts,
one opening as usual close to the pyloric end of the stomach and
one into the distal end of the first loop of the intestine. The
pancreas of Birds has from one to three ducts. The intestine is
folded into four or five loops, the arrangement of which has been
made use of as a basis for classification. It ends by passing into a
short rectum or large intestine, which is marked by a pair of outgrowths, the intestinal caeca. Their size varies much, from long
and wide blind sacs, as for instance in the common fowl, Ducks,
Geese and other herbivorous birds, to quite small vestiges as in the
Pigeon and in fish- and flesh-eating birds. The rectum ends in an
enlargement termed the urodaeum, the upper part of which receives

ducts of the kidneys and reproductive organs, while into the dorsal
wall of the lower and outer part a glandular pouch of unknown
function, called the bursa Fabricii (12, Fig. 288), opens. This
becomes smaller and sometimes entirely disappears in the adult Bird.

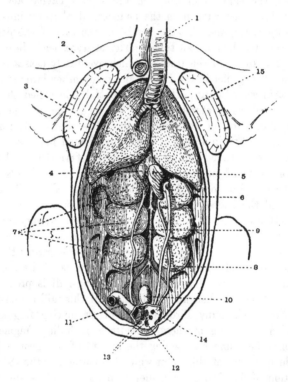

FIG. 288. The Lungs, Kidneys and Gonads of a Pigeon, *Columba livia* × ⅔.

1. Trachea.   2. Bronchus.   3. Lung.   4. Suprarenal body.   5. Ovary.
6. Oviduct.   7. Lobes of kidney.   8. Ureter.   9. Aorta.   10. Bursa
Fabricii.   11. Rectum.   12. Opening of bursa Fabricii.   13. Openings
of ureters.   14. Opening of oviduct.   15. Cut pectoral muscle.

The structure of the kidneys and reproductive organs is essentially the same as in the Reptilia. The meta-
Urinogenital   nephros in both sexes is distinctly divided into
Organs.   lobes. The mesonephros is represented by a small
lobed epididymis closely adherent to the testes. The suprarenal
body (4, Fig. 288) is homologous with the adrenal of Amphibia.

In most Birds there is no special organ for copulation, the whole end of the cloaca being turned inside out for this purpose, just as in Amphibia and Rhynchocephala. That this however is a secondary and not a primary state of affairs is suggested by the existence in Ostriches and some other Birds of a long penis on the dorsal wall of the cloaca similar in structure to one of the penes or copulatory sacs of the Lizard.

There is only one functional ovary, the left; an instance of the economy one observes throughout animated nature, for there is always a tendency when organs become expensive, that is, so large as to be a serious tax on the system, to reduce their number, and the production of eggs of the size of a Bird's is a great drain on the organism. There are two oviducts, but the right is small and useless. It must be remembered that the true egg formed by the ovary is the yolk; the white and the shell are additions derived from the oviduct.

The nests which Birds build and their care for the nestlings,

Habits.

whom they in some cases feed at short intervals for about seventeen hours out of the twenty-four, are well-known to all. Most also are well aware that many birds migrate to other lands as winter sets in. It is less well known that quite as many migrate to lands further north on the approach of spring. Few imagine the enormous distances which are covered by birds on the wing. They constantly pass from the Bermuda Islands to the Bahamas, 600 miles, without a rest. Many species which have their home in North Africa go every spring to North Siberia to build their nests. They fly, when migrating, at such heights in the air as to be quite invisible and attain a pace which seems hardly credible. There is no doubt, however, that very large numbers perish in crossing the sea.

Remains of fossil birds earlier than the tertiary period are very rare, but a few exceedingly interesting specimens have, however, been obtained. The principal of these is *Archaeopteryx*, represented by two specimens from the quarries in lithographic stone at Solenhofen in Germany. This remarkable bird had a long tail like that of a lizard, to each vertebra of which a pair of feathers was attached; the fingers of the wing bore claws and the bones of the palm (metacarpals) were free from one another. In the skull the premaxilla was as usual ensheathed in a horny beak but the maxilla bore teeth.

In all these points *Archaeopteryx* may be said to retain reptilian characters. Two other fossil birds (*Hesperornis* and *Ichthyornis*) had teeth in the maxillae but in other respects their structure was like that of modern birds.

The classification of Birds presents great difficulties. They used to be divided into six Orders, viz., *CURSORES*, or running birds; *NATATORES*, or swimming birds; *GRALLATORES*, or wading birds; *SCANSORES*, or climbing birds; *RAPTORES*, or birds of prey; and finally, the *INSESSORES*, or perching birds, a division which includes all our songsters, besides crows, rooks, magpies, sparrows and many others. Such a classification is, however, no longer accepted, because it gives fundamentally wrong ideas about the true relationships of birds. The aquatic birds, for instance, include totally different types, such as the Gull and the Duck, which have been derived from quite distinct families of land birds. All are agreed, however, in separating the Ostriches and their allies from all the rest as a great main division, the *RATITAE* (Lat. *ratis*, a raft) : but it is doubtful if this is a 'natural' group. They are distinguished by possessing a dwindled wing, which is useless for flight. This leads to the dwindling of the great wing-muscles, the pectorals, and the consequent disappearance of the crest on the breast-bone to which they are attached, and of the collar-bone, or clavicle. The breast-bone, having no longer a keel, has been compared to a raft, whence the name was suggested. The palate is also more like that of a Reptile than is the case with other Birds, for the pterygoid bones are firmly united to the basi-sphenoid by means of a basi-pterygoid process which springs from the last-named bone and articulates with the posterior end of the pterygoid. The upper jaw therefore cannot be bent up. The vomers also are very large and only united in front. Behind they join the basi-sphenoid and prevent the palatines from touching that bone.

These birds have enormously powerful legs and can run at a tremendous pace. The feathers are no longer required to form a compact surface, and hence the barbs have no barbules ; this is the cause of the soft downy character which is so characteristic of them and makes them so prized for ornament. The true Ostrich, *Struthio*, is found in Africa, *Rhea* in South America, the Cassowary, *Casuarius*, in New Guinea, the Emeu, *Dromaeus*, in Australia, and the smallest Ratite bird, the Kiwi, *Apteryx*, in New Zealand. In historical times, however, New Zealand had the largest of all Ratitae, the Moas, *Dinornis*, which had thigh bones larger and thicker than those of a horse.

Classification.

All other birds are called *CARINATAE* (Lat. *carina*, a keel), and are distinguished by having a keel or crest on the sternum and well-developed flying muscles. It is impossible to define with any certainty any large main groups into which they can be divided, but the number of smaller natural groups is very large. Here we can only briefly allude to two main points in which birds differ from one another and which are relied on in classification.

These are :—(1) The form of the palate: this resembles that of the Ostriches only in one family, the Tinamous. This type of palate is called dromaeognathous. In all other Carinatae the palatines slide on the base of the skull. The maxillae send plates inwards towards the middle line dorsal to the palatines, called the maxillo-palatines. When these are united and the vomers small or absent, or when they unite with the vomers, as in the Duck and Hawk, the bird is said to have a desmognathous palate ; when the maxillo-palatines remain separate, as in the Fowl, and the vomers form a conspicuous rod pointed in front, the palate is schizognathous ; finally when the maxillo-palatines remain separate and the vomers form a truncated wedge in front but diverge behind, aegithognathous is the term used. (2) Secondly, the manner the young are cared for is an important feature. Nearly all birds sit on or incubate their eggs. The Megapodes, or Mound-birds of Australia, however, merely collect a heap of grass and leaves, and leave the eggs to be hatched by the heat developed in fermentation. The young, e.g., of the Pheasant, Game-birds, Ducks, when hatched are able to run about and feed themselves (nidifugae) or they leave the egg in a helpless condition, blind, and have to be fed by the parents (nidicolae), for instance those of the Passeres or Songsters, which require constant care and attention for a long time. These last are considered by leading ornithologists to be the most highly developed of all birds, both as to their intellect and their structure, so that it is hardly too much to say that the increasing sacrifice of the parents on behalf of the young has had its reward, in the improvement it has brought about in all the faculties of the race.

The class Aves is classified as follows :

Sub-class I. ARCHAEORNITHES.

The three fingers and their metacarpals remain separate, each with a claw. Both jaws with alveolar teeth ; tail without pygostyle ; wings with well-developed remiges.

Only example, *Archaeopteryx*.

Sub-class II. Neornithes.

Metacarpals fused.

Division I. Ratitae.

Flightless; without a keel on the sternum; without a pygostyle. Coracoid and scapula fused.

> Ex. *Struthio*, African Ostrich; *Rhea*, American Ostrich; *Dromaeus*, Emeu ; *Casuarius*, Cassowary ; *Apteryx*, Kiwi.

Division II. Odontolcae.

Marine, flightless, without sternal keel ; teeth in furrows.

> Ex. *Hesperornis*.

Division III. Carinatae.

Without teeth, with a keeled sternum.

Family (1) Colymbiformes (Divers and Grebes). Plantigrade, nidifugous, aquatic, all toes webbed.

> Ex. *Colymbus*, Diver ; *Podiceps*, Grebe.

Family (2) Sphenisciformes (Penguins). Nidicolous; wings transformed into rowing paddles ; feathers small and scale-like.

> Ex. *Spheniscus*, Penguin.

Family (3) Procellariiformes (Petrels). Nidicolous, well flying, pelagic ; sheath of bill compound.

> Ex. *Procellaria*, Petrel ; *Puffinus*, Puffin ; *Diomedea*, Albatross.

Family (4) Ciconiiformes. Swimmers or waders ; desmognathous with basipterygoid processes ; without copulatory organ.

> Ex. *Sula*, Gannet; *Pelecanus*, Pelican ; *Ardea*, Heron; *Ciconia*, Stork ; *Phoenicopterus*, Flamingo.

Family (5) Anseriformes (Ducks and Geese). Nidifugous ; desmognathous with basipterygoid processes ; with copulatory organ ; palate bearing hard, horny, parallel ridges.

> Ex. *Anas*, Duck ; *Anser*, Goose ; *Cygnus*, Swan.

Family (6) Falconiformes (Birds of Prey). Nidicolous ; desmognathous ; beak powerful with decurved tip.

> Ex. *Falco*, Falcon ; *Aquila*, Eagle ; *Cathartes*, Turkey Buzzard.

Family (7)   Tinamiformes (Tinamous).   Nidifugous; without pygostyle; palate dromaeognathous.

> Ex.   *Tinamus*, Tinamon.

Family (8)   Galliformes (Game birds).   Nidifugous; schizognathous.

> Ex.   *Gallus*, Common Fowl; *Phasianus*, Pheasant; *Tetrao*, Grouse.

Family (9)   Gruiformes (Cranes and Rails).   Waders, nidifugous; schizognathous.

> Ex.   *Rallus*, Rail; *Fulica*, Coot; *Grus*, Crane.

Family (10)   Charadriiformes (Plovers, Gulls and Pigeons).   Schizognathous.

> Ex.   *Charadrius*, Plover; *Larus*, Gull; *Pterocles*, Sandgrouse; *Columba*, Pigeon.

Family (11)   Cuculiformes (Cuckoos and Parrots).   Desmognathous.

> Ex.   *Cuculus*, Cuckoo; *Psittacus*, Parrot.

Family (12)   Coraciiformes.   Nidicolous.

> Ex.   *Coracias*, Roller; *Upupa*, Hoopoe; *Alcedo*, Kingfisher; *Strix*, Barn-owl; *Caprimulgus*, Nightjar; *Cypselus*, Swift; *Picus*, Woodpecker.

Family (13)   Passeriformes.   Nidicolous; aegithognathous.

> Ex.   *Passer*, Sparrow; *Turdus*, Thrush; *Hirundo*, Swallow; *Alauda*, Lark; *Corvus*, Crow.

CHAPTER XIX.

Sub-Phylum IV. Craniata.

Class V. Mammalia.

THE class Mammalia (Lat. *mammae*, breasts), the last division
General Character-istics. of the phylum Vertebrata, includes those animals
which suckle their young. Like the Birds, their
temperature is constant and they have the ventricle
of the heart completely divided into two halves. But they differ
from Birds in never possessing feathers; only in one Order is the
fore-arm converted into a wing, and even in this case the arrangement of the parts is quite different from that in the Bird's wing.

Besides these characters however there are a large number of
others in which, while Mammals differ from both Birds and Reptiles,
the last-named two groups agree with one another, so that for a
long time the opinion was held that Mammals were vastly further
removed from Reptiles than were Birds; and indeed if only modern
Reptiles were considered this could not well be denied. If however
we examine the remains of the Reptiles which have existed on the
earth in past time, we come to the conclusion that the better way
to state the difference would be to say that, whereas Birds might be
traced back to Reptiles not very unlike modern lizards, Mammals
are derived from a type which has died out, leaving no modern
representatives. Thus Mammals are in all probability an older
group than Birds since they are presumably derived from the
Theromorpha and birds from some Rhynchocephalan ancestor.

Just as feathers constitute an indubitable mark of a Bird, so
true hairs are equally characteristic of Mammals. It is true that
the word hair is loosely used, being often applied for instance to the
delicate flexible spines of caterpillars, which are constructed on a
totally different plan to the hairs of Mammals. A hair in the
zoological sense is a rod composed of closely packed cells converted

into horn, and under a microscope the outline of these cells can be seen like a mosaic on the surface of the hair, the outermost ones overlapping each other like slates on a roof with the same function of letting the water run off.

We saw that a feather originated as a little knob, the outside of which was composed of horny cells, while the interior consisted of soft living tissue supplied with blood-vessels; a hair on the other hand makes its appearance as a cylinder of horny cells growing down from the epidermis into the dermis underneath. This cylinder then becomes split into an outer sheath and an inner core, the latter of which elongates and forms the hair, while the former remains

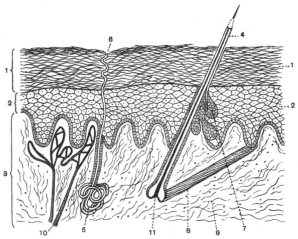

Fig. 289.   Section through the Skin of a Mammal.   Highly magnified.
Diagrammatic.

1. Outer layer of dead horny cells which are rubbed off from time to time, *Stratum corneum.*   2.   Deeper layer of cells retaining their protoplasm, *Stratum Malpighii.*   1 and 2 form the epidermis and are ectodermal in origin.   3.   Dermis or *Corium.*   4.   A hair.   5.   Sweat-gland. 6.   Opening of the duct of the sweat-gland.   7.   Sebaceous or fat gland. 8.   Erector muscle of the hair.   9.   Connective tissue fibres of the dermis. 10.   Blood-vessel.   11.   Vascular papilla at base of the hair follicle.

stationary and constitutes the follicle of the hair. The growth of the hair is rendered possible by a little plug of dermis carrying blood-vessels, which is pushed up into the lower end of the hair. In consequence of the rich supply of food brought by these vessels to the deep cells of the ectoderm lying above them, these cells bud off horny cells with great rapidity and persistence, and in this way a column of horny cells is formed which pushes out the older part of the hair and causes the whole structure to assume a great length, sometimes equalling that of the body. The plug of dermis is called the papilla of the hair. In a few cases hairs may be aggregated so as to form overlapping scales, and practically all Mammals have nails or claws on the fingers and toes which resemble essentially the corneous reptilian scale.

There is one respect in which Mammals and Birds agree with each other and differ from all other kinds of animals, and this is that their body temperature is considerably higher than that of their usual surroundings and is capable of varying with safety to the extent of only a few degrees. This condition of a constant temperature is known as the homoiothermal (so-called "warm-blooded") condition and differs strikingly from the poikilothermal (so-called "cold-blooded") one of other animals, in which the body temperature varies with that of the surroundings and is usually only one or two degrees above the latter. The temperature of a Bird or Mammal is maintained constant by regulation both of the loss of heat by radiation at the surface and of the manufacture of heat by tissue oxidation.

Perspiration or sweat is also characteristic of Mammals. This consists of a fluid secreted by certain cells of the epidermis which remain soft and are not converted into horn like most of the outer cells. The cells which manufacture the perspiration are arranged to form long tubes called sweat-glands, which penetrate far below the epidermis into the dermis underneath (Fig. 289). The production of sweat is a factor in the regulation of the body temperature and by it also certain excreta leave the body. The fluid poured out carries off a certain amount of heat and by its evaporation cools the skin. Besides the sweat-glands there are other tubes which are invaginations of the epidermis and consisting of a special kind of cell. These tubes, sebaceous glands, open into the hair follicles. They secrete the fatty substance or sebum which gives the natural gloss to the hair (Fig. 289).

Mammals, as we have seen, feed their young after they are born by suckling them, that is providing them with milk. This milk is a peculiar fluid produced by the mammary glands, consisting of epidermal tubes crowded together over certain areas of the ventral surface. They open at certain spots, raised above the general level, which constitute the nipple or teat. Many zoologists believe that the mammary glands are really only modifications of the ordinary sweat-glands and sebaceous glands, for in the lowest Mammals there is no special raised skin area which can be called a teat.

As regards their internal structure the great differences between Mammals on the one hand and Reptiles and Birds on the other, are to be found in the skull, the brain and the limbs and, to a lesser extent, in the heart and the arrangement of the great arteries and veins.

Turning first to the skull, we find that in a Mammal instead of having only one knob or condyle to fit into a cup on the first vertebra, as is the case with Birds and Reptiles, the skull has two, which are projections of the exoccipital bones that wall in the sides of the foramen magnum, whereas in Birds the single condyle is an outgrowth of the basi-occipital bone that forms the floor of the foramen magnum (Fig. 290). In Reptiles, more especially the Chelonia, the so-called single condyle is really trifid, the lateral parts being formed by the exoccipitals and the basal one by the basi-occipital. From this condition it is easy to see how the conditions in Birds

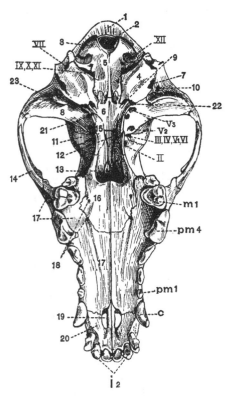

Fig. 290.   Ventral view of the Cranium of a Dog, *Canis familiaris* × ⅔.

1. Supra-occipital.   2. Foramen magnum. 3. Occipital condyle.   4. Tympanic bulla. 5. Basi-occipital. 6. Basi-sphenoid. 7. External auditory meatus.   8. Glenoid fossa. 9. Foramen lacerum medium, aperture through which the internal carotid passes to the brain. 10. Postglenoid foramen.   11. Alisphenoid. 12. Presphenoid.   13. Vomer.   14. Jugal. 15. Pterygoid.   16. Palatal process of palatine.   17. Palatal process of maxilla. 18. Posterior palatine foramen.   19. Anterior palatine foramen.   20. Palatal process of premaxilla.   21. Opening of tube in alisphenoid bone through which the carotid artery passes   22. Hole for passage of Eustachian tube.   23. Process of squamosal to act as a stay for condyle of lower jaw.

II—XII. Exits of cranial nerves.   i 2. Second incisors.   C. Canine.   pm 1, pm 4. First and fourth premolar.   m 1. First molar.

and higher Reptiles on the one hand and in Mammals on the other may have been derived. Then the brain instead of lying behind the eyes extends forward between and above them; there is consequently no interorbital septum, and the side walls of the brain-case are thoroughly and firmly ossified, not merely represented by a vertical plate imperfectly ossified, as in a Bird, or nearly entirely membranous, as in some Reptiles. These walls are in fact constituted of alisphenoid bones behind and orbito-sphenoid bones in front, whilst a strong mesethmoid bone is developed in the internasal septum. This septum is prolonged beyond the bones of the face by a cartilaginous plate forming the support of a flexible nose or muzzle; this is a feature quite peculiar to Mammals. The base of the cranium is completely ossified, not only behind by the basi-occipital and basi-sphenoid bones but in front by the pre-sphenoid bones. To the last-named the conjoined vomers are

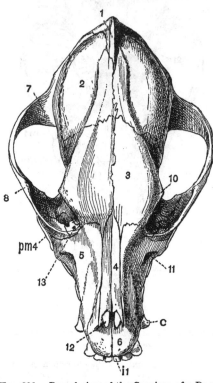

FIG. 291. Dorsal view of the Cranium of a Dog, *Canis familiaris* × ⅔.

1. Supra-occipital.   2. Parietal.   3. Frontal.
4. Nasal.   5. Maxilla (facial portion).
6. Premaxilla.   7. Squamosal.   8. Jugal.
10. Postorbital process of frontal.   11. Infra-orbital foramen.   12. Anterior palatine foramen.   13. Lachrymal foramen.   i 1. First incisor.   c. Canine.   pm 4. Fourth pre-molar.

attached forming a wedge-shaped mass which projects downwards and divides the air-space above the palatal folds into two. The name (Lat. *vomer*, ploughshare) is derived from the shape of the bone in Mammals; it is inappropriate as a description of its shape in other Craniata. The pterygoid bones take the form of thin vertical plates ; they are attached throughout their whole length to the side wall of the cranium in the region of the alisphenoid. As in Crocodilia and desmognathous Birds the palatal folds are united in the middle line ; the bones supporting them are processes of the premaxillary, maxillary and palatine bones. Between the pterygoid bones however the palatal folds form a purely muscular bridge, called the soft palate, which ends posteriorly in a projecting lobe, called the uvula, lying close to the glottis. The processes of the palatine bones always meet so as to form a bony bridge, called the hard palate; those of the premaxilla and maxilla do so to a certain extent, leaving however vacuities known as the anterior palatine foramina (19, Fig. 290). (The posterior palatine foramina are small holes in the palatine bones for the passage of blood-vessels.) As in Chelonia, there is only a lower temporal arcade, which is formed mainly by the cheek-bone or jugal. There is however no quadratojugal, and the jugal joins a process of the squamosal, which is a large bone covering the side of the skull and almost concealing the conjoined bones of the auditory capsule from view. It is characteristic of Mammalia that these bones, which in the embryo are distinct from one another, unite to form a single bone, called the periotic, which is fused to the squamosal. In Reptiles, on the other hand, the epiotic joins the supraoccipital and the opisthotic the exoccipital, while the prootic remains distinct. The outer ear, the funnel-shaped passage leading into the tympanum, which is termed the meatus auditorius externus, is surrounded by a bone called the tympanic, often swollen into a rounded form and then termed the tympanic bulla. There is often a tube-like prolongation of this bone into the base of the ear-flap or pinna.

There is no quadrate recognizable as such, the lower jaw consisting of a single dentary bone on each side, which articulates with a smooth cup-shaped facet on the squamosal, called the glenoid cavity. Occupying the position of the pre-frontal bone in Reptiles is a small bone called the lachrymal. This bone derives its name from the fact that it is pierced by a hole called the lachrymal foramen (13, Fig. 291) which permits of the passage of a duct leading from the orbit to the cavity of the nose. This duct carries off the excess of tears (Lat. *lacrima*, a tear), the secretion of the lachrymal gland, a development of the epidermis between the eyelid and the eye. The lip-bones, the pre-maxilla and maxilla, are well developed and like the dentary normally bear teeth, all of which are implanted in distinct sockets formed by the upgrowth of the bone which bears them. Many of these teeth are rooted, that is to say, after a certain time the dermal papilla on which the tooth is moulded becomes constricted at the base, so as to be connected by only a narrow neck with the adjacent connective tissue, this appearing in the dried tooth as a small hole through which a blood-vessel passes. The term root is applied to the dentine surrounding the narrow neck. When it is formed, growth of the tooth ceases.

The teeth of Mammalia are amongst their most characteristic

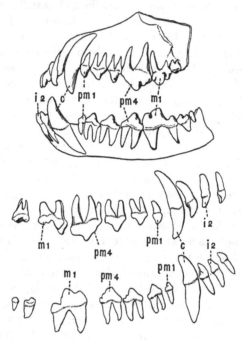

FIG. 292.　Dentition of a Dog, *Canis familiaris* × ½.

i 2.　Second incisor.　　c. Canine.　　pm 1, pm 4.　First and fourth premolars.
　　　　　　　　　　　　　　　　　m 1.　First molar.

organs; they are more differentiated than those of other Craniata, and their peculiar structure enables us to identify many fossil remains as mammalian.

They are typically differentiated into four kinds, viz. incisors or cutting teeth, canines or stabbing teeth, premolars and molars, which taken together are termed cheek-teeth or back-teeth (Fig. 292). The incisors are borne by the premaxilla and have sharp,

straight edges adapted for cutting morsels of convenient size from
the food. The canines and hinder teeth are borne by the maxilla.
The canines, popularly known as the eye-teeth, are pointed teeth
used for the purpose of killing prey or for defence against enemies,
or in the fights which occur among males for the possession of
females. The premolars have at least one cutting edge, often two
or more parallel to one another; they are used to cut up the
morsels which have been taken into the mouth. Finally the molars
have broad surfaces with which the food is sufficiently broken up
to permit of its being swallowed. The teeth of the lower jaw are
of course all borne by the dentary, and they are divided into the
same varieties as those in the upper jaw. In Elasmobranchii,
as we have seen, the teeth are enlarged scales developed on a
fold of skin which is invaginated within the lip, and as one row
of teeth becomes worn out another takes its place, the skin bearing
the old teeth slipping forward over the lip. In the higher Craniata
this fold is represented by a solid wedge of ectoderm, called the
enamel organ, but in Amphibia and Reptilia it only bears one
row of teeth. In Mammalia it normally produces two, the first of
which lasts only for a short time during the youth of the animal,
and is known as the milk dentition; the teeth belonging to this
row are pushed out of the gum by those of the second row, or
permanent dentition, which last throughout the life of the
animal. In the milk dentition there are only incisors, canines, and
molars; the milk molars are succeeded by the premolars of the
permanent dentition, while the permanent molars have no predecessors and are regarded as belated members of the first dentition.
The teeth of Mammalia have undergone profound modifications in
accordance with the different habits assumed by different members
of the class, and are one of the principal features on which its
division into Orders is based.

From a study of the dentition of living Mammals the conclusion is arrived at that the typical number of teeth, that is to
say, the number which the common ancestral form possessed, may
be estimated at 44, i.e., 11 on each side of each jaw, made up of three
incisors, one canine, four premolars, and three molars. This fact is
expressed by the formula $\frac{3 \cdot 1 \cdot 4 \cdot 3}{3 \cdot 1 \cdot 4 \cdot 3}$, where the upper line denotes
the teeth on each side of the upper jaw and the lower line those on
each side of the lower jaw.

The complete absence of the quadrate bone in the upper and of the articular in the lower jaw has given rise to much speculation as to what has become of these elements, which are so constantly present in Aves and Reptilia and are distinctly represented by cartilage even in Amphibia. For a long time the favourite theory was that they had been metamorphosed into the so-called ossicula auditûs or bones of hearing. In Anura, Reptilia, and Aves sound is conveyed from the ear-drum or tympanic membrane to the wall of the auditory capsule by a single rod, called the columella auris. In Mammalia however the connection is effected by a chain of three small bones called the malleus (Lat. hammer), incus (Lat. anvil) and stapes (Lat. stirrup) respectively, the last named being apposed to a membranous spot in the auditory capsule, called the fenestra ovalis, while the malleus is in contact with the ear-drum.

Now since in the embryo both malleus and incus are represented by blocks of cartilage which are for some time in continuity with Meckel's cartilage (v. p. 356), it was natural to suppose that they were representatives of the articular and quadrate bones which in lower Craniata are portions of the first visceral arch. The study however of the extinct Reptiles which show the closest approximation to Mammalia has rendered another view more probable. In the Theromorpha, as these are called, the quadrate is very small, and is enveloped in the huge squamosal which extends downwards and backwards and forms part of the articulation for the lower jaw. This tendency to the covering of the quadrate is also observable to a slight extent in living Chelonia; and in this group the quadrate is observed to be bent around the anterior wall of the outer ear or meatus auditorius externus and the Eustachian tube. Taking everything into consideration it appears probable that the quadrate is represented in Mammalia by the tympanic bone which like the quadrate of Chelonia encircles the outer ear, and that the ossicula auditûs owe their origin to the segmentation of the columella auris, their intimate connection with Meckel's cartilage being paralleled by the fusion of the columella and the Meckel's cartilage in Crocodilia. This tendency to fusion of the columella and lower jaw is interpreted as meaning that the columella is a relic of the hyomandibular bone of fishes, and that in the ancestors of terrestrial Vertebrata the auto-stylic articulation was derived from the hyostylic.

The articular bone has probably entirely disappeared in Mammalia; when it was no longer used as an articulation it would first cease to ossify and then become indistinguishable from the rest of Meckel's cartilage. Doubtless a similar fate would have overtaken the quadrate but for its secondary use as a protection to the tympanic cavity and as a frame for the tympanic membrane.

The chief peculiarity of the brain as compared with Reptiles is

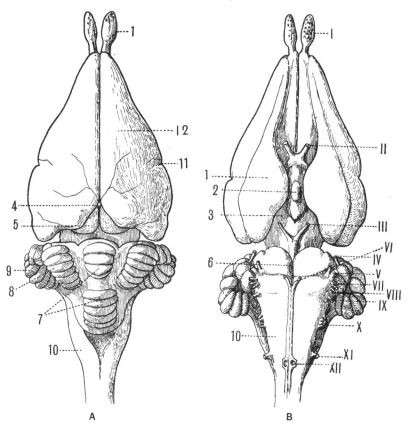

Fig. 293.   Brain of Rabbit, *Lepus cuniculus* × 2.

A.   Dorsal aspect.   B.   Ventral aspect.

1.   Olfactory lobe.   2.   Pituitary body.   3.   Crura cerebri.   4.   Pineal gland.   5.   Anterior pair of corpora quadrigemina.   6.   Pons Varolii. 7.   Cerebellum.   8.   Lateral lobe of cerebellum.   9.   Floccular lobe of cerebellum.   10.   Medulla oblongata.   11.   Sylvian fissure separating the frontal lobe 12 from the temporal lobe behind.   I.   Origin of first or olfactory nerves.   II.   Optic or second nerves arising from the optic chiasma.   III.   Third or motor oculi nerve.   IV.   Fourth or patheticus nerve.   V.   Fifth or trigeminal nerve.   VI.   Sixth or abducens nerve. VII.   Seventh or facial nerve.   VIII.   Eighth or auditory nerve. IX.   Ninth or glossopharyngeal nerve.   X.   Tenth or vagus nerve. XI.   Eleventh or spinal accessory nerve.   XII.   Twelfth or hypoglossal nerve.

the greater development of the cerebral hemispheres, in proportion to the hind brain or cerebellum. The former overlap
completely and conceal the thalamencephalon and the mid-brain,
and they are connected with one another by a great transverse band
of nerve-fibres, called the corpus callosum. It is customary to
map out the surface of the hemispheres into regions, in order to
facilitate description in delimiting the areas concerned with the
development of specific sensations and with the control of specific
movements. These regions are called frontal, parietal, occipital,
and temporal lobes. The temporal lobe is separated from the
frontal by a deep groove, called the Sylvian fissure (11, Fig.
293, A). How well the increased size of the cerebrum is reflected in the shape of the cranium will be seen when it is
recollected that the frontals and parietals, which represent merely

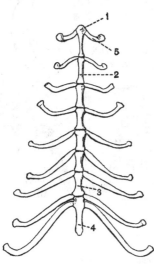

FIG. 294. Sternum and sternal Ribs
of a Dog, *Canis familiaris* × ½.

1. Presternum.   2. First sternebra of mesosternum.   3. Last
sternebra   of   mesosternum.
4. Xiphisternum. The flattened
cartilaginous plate terminating
the xiphisternum is not shown.
5. First sternal rib.

the membrane covering the anterior
fontanelle, not only form the roof
of the cranium but a large part of
its domed side wall; and further
that the orbitosphenoid and alisphenoid, which are portions of the
cartilaginous brain-case, are restricted to the base of the skull. The
cerebrum has in fact protruded
through the anterior fontanelle,
pushing the membrane before it.
The same condition is observable
in Birds, but not in Reptiles or
Amphibia. The cerebellum however is also well developed, just as
in Birds, having indeed in addition
to the lateral lobes an outer pair
of lateral projections, called floc-
culi, embedded in a hollow of the
bone that covers the inner ear (Fig.
293). The two halves of the cerebellum are connected with one
another by a conspicuous band of
fibres in the floor of the brain, called
the pons Varolii.

The nose, except in aquatic Mammalia, is a highly developed
sense-organ. The epithelium lining it is produced into scroll-like folds

which are supported by thin plates of bone arising from the mesethmoid, and called ethmoturbinals. Above, where the mesethmoid joins the orbitosphenoid, so numerous are the apertures in it to allow the bundles of nerve-fibres from the olfactory cells to pass to the brain, that this part of the bone is reduced to a sieve, whence it has received the name of cribriform plate. From the maxilla, which forms the outer wall of the lower part of the nasal tube, a similar scroll-like bone, the maxillo-turbinal, arises, which supports a corresponding fold of epithelium. This fold however is supplied only by the second division of the fifth nerve, and is not believed to have any olfactory function, but merely to act as a filter to free the inrushing air from grosser particles before it reaches the delicate olfactory epithelium.

The neck region of Mammals (with rare exceptions) always consists of seven vertebrae, and thus whereas a long-necked bird like a swan has numerous short vertebrae in this region, in a long-necked Mammal like the giraffe the same region consists of seven immensely long vertebrae. The sternum of Mammals also is peculiar, consisting of distinct pieces or sternebrae. The first of these is called the presternum, and bears a crest for the attachment of the pectoral muscles; the last ends in a spade-like xiphoid cartilage, and is called the xiphisternum. The intervening segments constitute the sternebrae of the mesosternum. The lower ends of a pair of ribs are attached opposite the junction of two sternebrae (Fig. 294).

In Mammalia as in other Amniota the centrum is formed from the interventrals, and the head of the rib is articulated between two vertebrae; the articulation is not shifted on to the vertebra as in Crocodilia. The vertebrae have occasionally in the neck region cup and ball articulations like those of Amphibia, Reptilia and Aves, but elsewhere the thick intervertebral cartilage allows of sufficient bending, and the centra have flat ends which ossify late and for some time form separable discs of bone called epiphyses. These do not represent any new elements in the centrum, but only a method of ossification found also in the limb-bones. This method consists in the ensheathing membrane or periosteum first forming a tube of bone round the centre of the cartilage, the ends of which remain soft and capable of further growth, being only replaced by bone when growth is diminishing, and united with the main ossification when growth has ceased.

In the fore-limbs of Mammals the chief point to be noticed is the reduction in size of the pectoral girdle to which the fore-limbs are attached. The lower part of this, the coracoid, which in Birds and Reptiles is a large, strong bone meeting the sternum,

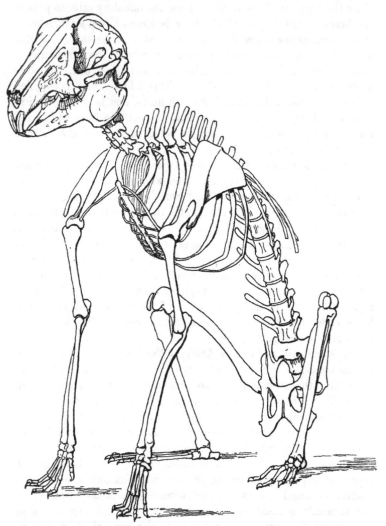

Fig. 295. Skeleton of Rabbit, *Lepus cuniculus*. × ⅔. In sitting position.

is here, with the exception of a few primitive forms, a small hook with no connection at all with the sternum. Hence the pectoral girdle is much more movable than is elsewhere the case, and takes part in the movements of the limb. It is therefore not surprising to find that the upper portion of the girdle, the shoulder-blade or scapula, is broad, affording a large surface for the attachment of muscles; and that its surface is still further increased by the presence of a sharp vertical ridge rising up along its middle line (Fig. 295). To the end of this spine, as it is called, the collar-bone or clavicle is attached; this bone extends inwards to the sternum and is loosely connected with it. In some Mammals the clavicle is absent.

The general form of the pelvic girdle to which the hinder limbs are attached is not very unlike that of Birds; but there are two important differences. First, the ilia or hip-bones are attached only for a very short distance to the backbone; and secondly, the lower bones of the girdle, the pubis and ischium, meet their fellows of the opposite side in front of the belly beneath the anus, whereas in Birds they do not even approach each other in this place, though in *Rhea* the ischia do meet dorsal to the anus. It follows that this part of the body is extensible in Birds and can be stretched as the large egg passes out through the oviduct.

The leg of Mammals differs from that of Birds and Reptiles in that the ankle-joint is situated between the bones of the shank (the tibia and the fibula) and the small bones of the ankle, instead of in the middle of these small bones (Fig. 295). The heel-bone or calcaneum is one of the uppermost tier of ankle-bones and corresponds to the bone called fibulare in the general scheme of the pentadactyle limb. It is prolonged into the heel, to which the great gastrocnemial muscles which form the calf of the leg and which raise the heel are attached.

Turning now to the blood system of Mammals we find that the red blood corpuscles which give the colour to the blood are unlike those of other Vertebrates. They have no nuclei and are bicon-cave, while they are also much smaller and (except in Camels and Llamas) round, not oval, as in all lower Vertebrates. Like Birds, but unlike most Reptiles, the Mammals have a four-chambered heart; the main blood-vessel, the aorta, is supplied by the left systemic arch alone, the right one being cut off from connection with it and being represented by the common trunk of the right carotid and subclavian arteries, the so-called innominate artery (Fig. 296), this being exactly the converse of the arrangement in Birds.

The ventral carotid arteries, which we have seen are reduced in
Birds as compared with Reptiles and Amphibians, are usually

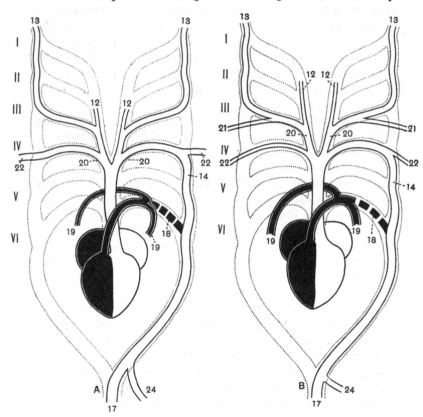

Fɪɢ. 296. Diagram of Arterial Arches of Mammals, viewed from the ventral
aspect.

A. Of all Mammals except Cetaceans.

I. II. III. IV. V. VI. First to sixth arterial arches.   12. Ventral carotid
    (small or absent).   13. Common carotid (dorsal carotid).   14. Syst-
    emic arch.   17. Dorsal aorta.   18. Ductus arteriosus.   19. Pulmonary.
    20. Innominate.   22. Subclavian (dorsal type).   24. Coeliac.

B. Of Narwhal, representing Cetaceans.

I. II. III. IV. V. VI. First to sixth arterial arches.   12. Ventral carotid
    (small).   13. Common carotid (dorsal carotid).   14. Systemic arch.
    17. Dorsal aorta.   18. Ductus arteriosus.   19. Pulmonary.   20 Innomin-
    ate.   21. Subclavian (ventral type).   22. Intercostal (equivalent to
    subclavian of dorsal type).   24. Coeliac.

absent in Mammals, though they may exist as quite small vessels,
the trachea and other structures in the neck receiving blood from

the common or dorsal carotid on its way to the head. The sub-clavian artery is of the dorsal type as in Lizards and Amphibians in all Mammals except Cetaceans, in which Order the fore-limb or paddle obtains blood by an artery corresponding in origin with that to the fore-limb of Chelonians, Crocodiles and Birds, i.e., a "ventral subclavian." The "dorsal subclavian" possessed by most Mammals is also found in Cetaceans, but it is distributed to the ribs and their muscles as the "intercostal artery."

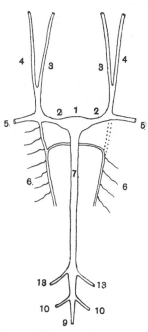

It is an interesting feature of the arterial system of a Mammal that in consequence of the embryo receiving its oxygen from the ma-ternal blood, the connection between the systemic and pulmonary arches persists in an unreduced condition until birth. This connection is known as the ductus arteriosus, and through it the blood from the right ventricle is passed direct to the dorsal aorta. The pulmonary arteries are quite small during this period, and by these arrangements circulation of blood through the as yet functionless lungs is avoided. At birth the ductus arteriosus shrinks and is rapidly reduced to a solid cord, while the enlarging pulmonary vessels provide for the deviation of the venous blood to the now expanded lungs.

Fig. 297. Diagram to show arrange-ment of the principal Veins in a Mammal.

1. Sinus venosus—gradually dis-appearing in the higher forms. 2. Ductus Cuvieri = superior vena cava.   3. Internal jugular = an-terior cardinal sinus.   4. Ex-ternal jugular = sub-branchial. 5. Subclavian.   6. Posterior cardinal front part = venae azygos and hemiazygos. 7. Inferior vena cava.   9. Caudal.   10. Sciatic = internal iliac.   13. Femoral = external iliac.

In the venous system the blood from the head is returned by ex-ternal and internal jugular veins, the former being much the larger. The hinder portions of the posterior cardinal veins have quite disappeared and the caudal vein is continued directly into the inferior vena cava, so that there is no longer even the outward appearance of a renal-portal system. The anterior portions of the posterior cardinals, however, persist as the venae azygos and hemiazygos, which on each side receive the veins from the spaces between the ribs—the intercostal veins. Only that on the right—the vena azygos—reaches the ductus Cuvieri (superior vena cava); the left cardinal or vena hemiazygos developes a transverse branch through which its blood joins that of the right cardinal, and the veins from the legs, instead of traversing the kidneys, empty at once into a great vena cava inferior situated in the middle line of the back, which is joined higher up by two short veins from the kidneys (Fig. 297).

One of the most interesting peculiarities of Mammals is their breathing mechanism. It will be remembered that whereas the Amphibia simply swallow air, in the Reptiles the size of the chest cavity is enlarged by pulling the ribs forward and then separating them, and as the lungs are closely attached to the wall of the chest, they are likewise enlarged and air rushes into them. In Mammals this same mechanism exists, but in addition there is a totally independent means of pumping air into the lung. This is rendered possible by the existence of a diaphragm, a partition convex in front which separates the coelom of the chest from the rest of the body cavity. This partition is partly muscular, and when the muscle contracts the whole membrane is tightened and necessarily flattens, with the result that the chest cavity is enlarged and air enters the lungs. The action of the diaphragm in fact is precisely similar to that of the muscular floor of the mantle cavity of the snail. The diaphragm is attached ventrally to the xiphoid carti-lage, dorsally to the vertebral column, and laterally to the ventral edges of the hinder ribs which do not reach the sternum. By it the coelom of the mammal is separated into a thoracic division in front and an abdominal one behind. Since therefore all the vertebrae which bear recognisable ribs which reach the sternum belong to the thoracic region, they are termed thoracic vertebrae, while the ribless vertebrae of the abdominal region are denominated lumbar.

In the digestive system the principal peculiarity of Mammals is the high state of development of the salivary glands. These glands are much branched tubular outgrowths of the ectoderm of the mouth-cavity or stomodaeum; they secrete a fluid which moistens the food and is swallowed with it, thus helping digestion. They are foreshadowed by small glands in frogs and snakes, but in Mammals they form three large masses, viz. the sub-lingual, underneath the tongue, the sub-maxillary, under the angle of the jaw, and the parotid, just under the ear. Glands in similar positions are found in some Birds, but those of Mammalia secrete in addition to mucus a ferment, called ptyalin, which turns starch into sugar, so that the secretion which is called saliva is a true digestive juice. The development of the large intestine withdraws water from the undigested residue of the food, thus reducing it to a semi-solid mass or faeces, which is also a characteristic of the Mammalian alimentary canal.

Mammals are divided into three great primary divisions or sub-classes according to the structure of the ovary and oviduct and to the stage of development attained by the young at birth. The lowest forms have large eggs like those of birds: in the higher forms the egg is at first small but has the power of absorbing nourishment from the wall of the oviduct, which is here enlarged to form a womb or uterus. In the highest division a special organ for the nourishment of the embryo, the placenta, is developed, as an enlargement of the embryonic bladder.

Classification.

The sub-classes are called :

    I. PROTOTHERIA, or primitive mammals.

   II. METATHERIA, or modified mammals.

  III. EUTHERIA, or perfect mammals.

Sub-class I. PROTOTHERIA.

The Prototheria include two extraordinary animals, the *Ornithorhynchus* (*Platypus*), or duck-billed mole, and the *Echidna*, or spiny ant-eater, which are found only in Australia, New Guinea and Tasmania. In these animals large eggs with a firm shell are laid in a nest and incubated by the mother, and in harmony with this arrangement the two oviducts are large throughout the whole of their length, and do not join each other at any point but open along with the intestine into a common vent or cloaca, as is the case with Birds and Reptiles. The ureters do not open into the bladder as they do in all other Mammalia, but they and the bladder open separately into the cloaca. In the male a copulatory organ or penis is present opening into the cloaca behind, and in front protruded from the cloaca. After they are hatched the young receive

Fig. 298.   The Duckbill, *Ornithorhynchus anatinus.*

milk from the mother.   There is no teat, but the fluid from the milk glands seems to soak into the hair and thence is sucked by

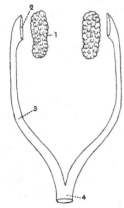

the young.   Before birth the young receive no nourishment at all from the mother, but subsist on the abundant yolk of the egg.

The skeleton of the Prototheria presents many interesting features of agreement with the Reptiles; thus the vertebrae have no epiphyses, and there is not only a complete coracoid articulating with the sternum, but also two precoracoids which overlap.   Underneath these there are two clavicles and a T-shaped interclavicle, so that the shoulder-girdle recalls the complicated one of the lizard.

Fig. 299.   Diagram to illustrate the arrangement of the female genital ducts in the Prototheria.

1. Ovary.   2. Oviducal funnel.   3. Oviduct.
4. Opening into cloaca.

Ornithorhynchus has webbed feet and lives in the water, feeding on worms and insects, which it digs out of the mud by its broad, shovel-like snout, whence the name Duck-bill (Fig. 298).   It crushes its prey by means of horny plates, which are really patches of the hardened gum : when it is young, however, it has true calcareous teeth, two or three on each side of each jaw, but these it loses when it grows older. These teeth are covered by several rows of small points or tubercles. Similar teeth are found amongst the oldest remains of Mammals which are known, the so-called Multituberculata.

Echidna lives on ants and other insects, which it ensnares by putting out its tongue covered with sticky saliva. Like other ant-eaters it has a long snout and no teeth. It is covered with stiff spines like a porcupine.

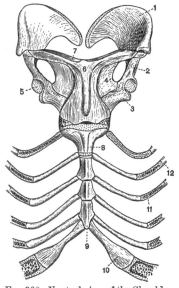

Sub-class II. METATHERIA.

The division Metatheria includes the curious pouched Mammals of Australia and the neighbouring islands and the Opossums of America. In these animals the egg is exceedingly small, and the egg-tube is divided into an upper part of correspondingly narrow diameter, called the F a l - l o p i a n t u b e, and a lower, wider part, called the u t e r u s. In this latter the small egg lies for a while and rapidly grows and developes, absorbing food from the uterus through the thin egg-membrane, since there is never any egg-shell. Beneath the uterus comes the lowest part of the egg-tube, the so-called

FIG. 300. Ventral view of the Shoulder-girdle and Sternum of a Duckbill, *Ornithorhynchus paradoxus* × ¾. After Parker.

1 and 2. Scapula.   3.   Coracoid. 4. Precoracoid.   5.   Glenoid cavity. 6. Interclavicle. 7. Clavicle. 8.   Presternum.   9.   Third seg-ment of mesosternum.   10. Sternal rib. 11. Intermediate rib. 12. Vertebral rib.

vagina. The two vaginae come into close contact with each other above and then diverge, both opening below apparently into the lowest part of the bladder, as do the vasa deferentia in the male. What seems to be the lowest part of the bladder is really the front portion of the cloaca, which has become separated from the part behind that receives the opening of the intestine. This common

vestibule for excretory and reproductive ducts is called the urino-
genital sinus, and its opening is distinct from that of the intestine
or anus, although the two openings are still surrounded by a common
muscle. From the spot where the vaginae meet above a pouch
called the median vagina is often developed. This ends blindly
in the young female, but in the mature female it acquires an
opening into the urinogenital sinus and through this opening the
embryo is born, the lateral vaginae serving merely to admit the
spermatozoa from the male.

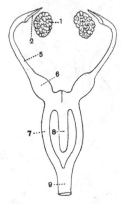

Fig. 301. Diagram to illustrate the arrangement in
the female genital ducts of
the Metatheria.

1. Ovary.   2. Oviducal
funnel.   5. Fallopian
tube.   6. Uterus.
7. Vagina.   8. Median
vaginal pouch.   9. Uri-
nogenital vestibule.

When the young are born they appear
not as eggs but as little mammals, which
are however exceedingly small in size.
They are then placed by the mother, who
is said to transfer them with her lips, in
a pouch made by a fold of skin on the
lower part of her body, whence the name
Marsupials (Lat. *marsupium*, a pouch),
often given to these animals. A pair of
sesamoid or epipubic bones run forward
from the pubes. They are ossifications
of a tendon of the external oblique muscle.
Similar structures are found in Prototheria, in Crocodiles and in Urodela.
The young are quite incapable of feeding
themselves, and therefore the mother by
compressing the muscles of the belly
squeezes the milk-gland and forces milk
down their throats. In order to allow
the young one to breathe at the same
time, the back of the soft palate is
wrapped round the upper end of the
windpipe, which projects into the throat so that the air passes from
the nose straight down the windpipe whilst milk flows down at the
sides of the air-passage into the stomach.

In the mandible the angle, that is to say the lower and posterior
end, is as a rule prolonged inwards as a horizontal shelf of bone.
By this feature fossil skulls are recognized as belonging to the
Metatheria.

The living Metatheria are divided into two great orders, of which
the first is mainly carnivorous and the second herbivorous, though
some members of both are insectivorous. The first order is termed

POLYPROTODONTIA ;   the animals composing it have at least four
incisors on each side of the upper jaw and three on each side of the
lower, whence the name (Gr. πολὺ, many; πρῶτος, foremost; ὀδόντες,

Fɪɢ. 302. *Petrogale xanthopus.* The Rock Wallaby with young in pouch.
After Vogt and Specht.

teeth).   The *DIPROTODONTIA*, as the second Order is called, derive
their name from the circumstance that in the lower jaw there is
one large pointed inciser on each side, the others being rudimentary

or absent, so that only two prominent teeth are observable (Gr. δι-, two). The Polyprotodontia are represented in America by the family of the Opossums, DIDELPHYIDAE, which is confined to that continent. It includes 24 species, most of which are found in Mexico, Central America, and Brazil, but one, the Virginian opossum, *Didelphys virginiana*, ranges north as far as the south bank of the Hudson river. In all the Didelphyidae the great toe is large and can be separated from the other toes so as with them to grasp a support; thus it is said to be 'prehensile.' In Australia the Polyprotodontia are represented by three families, viz. the DASYURIDAE, the PERAMELIDAE and the NOTORYCTIDAE. The first family includes the animals known as native cats, which resemble the American opossums, but are distinguished from them by the smaller number of incisor teeth and by having a rudimentary first digit in both fore- and hind-feet, whereas in the

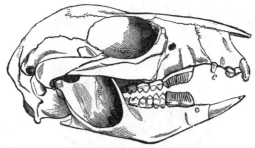

FIG. 303. Skull of Lesueur's Kangaroo-rat, *Bettongia lesueuri*. To exhibit Diprotodont type of dentition.

Didelphyidae, as we have seen, this digit is long and prehensile. The largest member of the family is *Thylacinus cynocephalus*, the Tasmanian wolf, now confined to the wilder parts of Tasmania: it has a skull which strikingly resembles that of a dog; in its habits it resembles a wolf and is very destructive to sheep. The Banded Ant-eater, *Myrmecobius fasciatus*, is an aberrant member of the same family which lives on insects, capturing them with its long tongue. The insects are made to adhere to this organ by the viscid saliva. The teeth, though rudimentary, are distinct. There is no pouch: the young when first born cling to the teats and conceal themselves in the long hair of the mother's abdomen. The PERAMELIDAE or Bandicoots are small animals somewhat resembling Rabbits and Hares in their appearance but with

pointed muzzles ; they are remarkable in possessing a type of foot
characteristic of the Diprotodontia.   The NOTORYCTIDAE include the
single genus *Notoryctes*, which in habits and appearance resembles
the Mole, a similar mode of life having brought about similar
modifications of structure.

The Order Diprotodontia includes a number of species confined,
with one exception, to Australia and the neighbouring islands.
One species, the only living representative of the family EPANOR-
THIDAE, has been recently found in South America.   This animal,
which has received the name *Caenolestes uliginosus*, has feet like

FIG. 304.   Banded Ant-eater, *Myrmecobius fasciatus* × ¼.

the DIDELPHYIDAE, and this circumstance renders it possible that
it has been independently evolved from that family, whereas
the other members of the Order seem to have been derived
from forms like the PERAMELIDAE.   The typical Diprotodontia
have the second, third, fourth and fifth toes of the hind-foot
united by a web of skin.   The fourth is the strongest toe,
the fifth is a little shorter, but usually nearly as stout as the
fourth ; the second and third, though as long as the fourth, are

much more slender, while the great toe is often rudimentary. Exclusive of the EPANORTHIDAE there are three families in the Sub-order. The first family, the PHASCOLOMYIDAE, consist of one genus, *Phascolomys*, the Wombat, represented by three species. The PHASCOLOMYIDAE are distinguished by possessing only one incisor on each side of the upper jaw, and as both upper and lower incisors are chisel-shaped the dentition resembles that of a Beaver or Rat. Wombats are heavy animals with a shuffling gait, about the size and appearance of a Badger.

The second family, the PHALANGERIDAE, or Australian opossums, have normally three incisors on each side of the upper jaw; the fore- and hind-limbs are of about the same size and the great toe is prehensile. These are small animals which like squirrels live in trees, and several species possess a parachute-like membrane extending from fore- to hind-limb, by the aid of which they sustain themselves in the air during their great leaps from tree to tree. *Phascolarctus*, the so-called native Bear, is a clumsy tailless Phalanger, in which the prehensile great toe is specially well developed. The MACROPODIDAE, or Kangaroos, are the most peculiar family of Diprotodontia, and indeed of the Metatheria. They resemble the Phalangeridae in having three upper incisors on each side, but differ totally in the structure of the limbs. The fore-limbs are so small as to be used only for grasping, and locomotion is effected by a series of leaps carried out by the hind-limbs aided by the powerful tail. The sole of the hind-foot is excessively narrow, the second and third digits being represented by bones so slender that they take no part in supporting the body. *Macropus giganteus*, the gray Kangaroo or "Old Man," may obtain a height of from 4 to 5 feet. The fourth toe of the hind-foot has a powerful claw with which when the animal is brought to bay it has been known to rip open a dog. The allied genus *Petrogale* includes smaller species, called Rock Wallabies, with only a short claw on the hind-foot. As their name implies they frequent rocky regions. The so-called Kangaroo-rats, *Bettongia* and others, are nocturnal animals of small size, which live on leaves, grass, and roots, the last of which they dig up with their fore-paws.

Sub-class III. EUTHERIA.

The highest division of the Mammalia, the Eutheria, includes all the most familiar animals, hedgehogs, rats, rabbits, cats, dogs,

lions, tigers, horses, oxen, whales, elephants, monkeys, up to and
including man himself. In them as in the Metatheria the egg is
exceedingly small, in Man and the domestic animals for instance,
it varies from $\frac{1}{120}$ to $\frac{1}{200}$ inch in diameter. The upper part of the
oviduct, the Fallopian tube, is consequently narrow; the uterus is
however enlarged, for the egg not only lies there a long time—
called the period of gestation or pregnancy—but as it is
developing into the young mammal a special organ called the
placenta is developed, which grows out and becomes interlocked

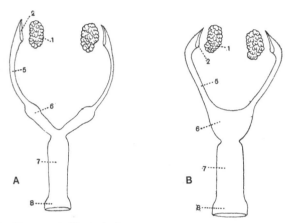

FIG. 305.   Diagrams to illustrate the arrangement of the female genital ducts in
an Eutherian Mammal.   A.   Rabbit.   B.   Man.

1.   Ovary.   2.   Oviducal funnel.   5.   Fallopian tube.   6.   Uterus.
7.   Vagina.   8.   Urino-genital sinus.

with folds in the wall of the uterus. This organ is nothing but an
enormous development of the bladder of the embryo, which is
called the allantois, and which is also developed in the embryo
of Birds and Reptiles, where it subserves respiration and lies above
the embryo beneath the egg-shell. In Eutheria the surface of the
allantois is covered with vascular outgrowths called villi, which fit
into pits on the wall of the uterus. Both the membrane covering
the allantois and the lining of the uterus degenerate, allowing the
blood-vessels of mother and embryo to come into close contact.
The placenta becomes gorged with blood driven into it by the heart
of the developing embryo, and at the same time the uterus becomes
congested and loses its epithelium, so that the blood of the mother
and that of the young approach very closely to each other. They

are separated only by the thin outer wall of the placenta, so that nourishment diffuses from one to the other, and the blood of the embryo is oxygenated and its carbon dioxide removed by the maternal blood. So close is the connection, that when the embryo is born and passes out of the uterus, carrying with it the placenta, the latter in most cases tears open the vessels in the wall of the uterus and the mother loses a considerable quantity of blood. The lowest parts of the two oviducts are completely joined and pass into a single passage, the vagina, while the middle portions, or uteri, are sometimes quite separate as in the rabbit (A, Fig. 305), sometimes partly united as in the cat, rarely completely joined as in monkeys and man (B, Fig. 305). In one or two Metatheria a placenta such as has been described has been recently discovered, but it is of very small extent. These facts lead us to believe that Metatheria are degenerate descendants of early Eutheria, and we may take as a further mark of degeneracy the almost complete disappearance of the milk set of teeth.

Order I. Edentata.

When we take a general survey of the orders or main divisions into which the Eutheria are divided we find that we have three or

Fig. 306.   Tamandua Ant-eater, *Tamandua tetradactyla*.   From Proc. Zool. Soc. 1871.

four strange groups, the relations of which to the others are most difficult to decide. These include the curious Edentata of South America, comprising three families, the BRADYPODIDAE or Sloths, the MYRMECOPHAGIDAE or American Ant-eaters, and the DASYPODIDAE

or Armadillos. With these the South African forms, included in
the families MANIDAE or Scaly Ant-eaters and ORYCTEROPODIDAE

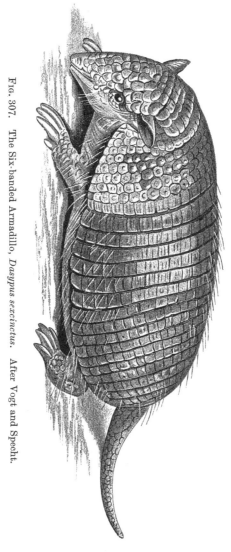

FIG. 307. The Six-banded Armadillo, *Dasypus sexcinctus*. After Vogt and Specht.

or Cape Ant-eaters, are usually grouped, though their relationship
is a matter of doubt. The name means "toothless," and was
given to them by the early naturalists because they supposed them

35—2

to be devoid of teeth. This is only the case with one small family,
the Ant-eaters, or MYRMECOPHAGIDAE, which, like Echidna, have
lost their teeth through disuse. In the rest there are teeth,
but front teeth are always wanting. In the adult none of the
teeth have enamel and all are similar to each other. The
hands and feet are armed with great curved claws, adapted
for holding on to supports, not for grasping or attacking, and
incapable of being retracted or pulled back. Consequently the
hands and feet are like hooks, on which the animals walk
clumsily, bending the fingers under them. The apparent want

FIG. 308.   White-bellied Pangolin, *Manis tricuspis.*

of utility is however explained when the animals are looked at in
their natural surroundings. It is then seen that one family, the
Sloths (BRADYPODIDAE), spend all their time climbing about on trees,
on the leaves of which they feed. There is a remarkable adaptation
which probably helps them to escape detection by their enemies.
The surface of the hairs is grooved and affords a resting-place for a
unicellular Alga which causes the animal to have a greenish appearance so as to be almost invisible amidst the foliage. The second
family include the true Ant-eaters or MYRMECOPHAGIDAE; in these
the strong claws are used for pulling down and digging up ant-hills.
The muzzle is long and toothless. There is a very long tongue,

and enormous salivary glands, the sticky secretion of which entraps the ants. The Tamandua Ant-eater, *Tamandua tetradactyla*, of Central and South America, is arboreal in its habits and lives in the dense primeval forests of the New World: it uses its strong claws for climbing and has a prehensile tail. The third family, the Armadillos or DASYPODIDAE, can dig with such rapidity that a comparatively large animal will scoop out a burrow for itself in a few minutes. These Armadillos are also very remarkable as being the only Mammals in which the dermis or deeper skin developes into hard bony plates such as we find in Turtles and Crocodiles, whilst the hair on the upper part of the body is replaced by horny scales like those of snakes and lizards, covering the bony plates.

It is thought that in comparatively recent times, geologically speaking, South America was an island, and just as Australia has preserved some curious animals which could never have held their ground against the powerful lions and tigers and wolves of the Old World, so in South America evolution seems to have run a course of its own.

In Africa there are found two other genera of Ant-eaters, *Manis*, the scaly Ant-eater, also found in Eastern Asia, and *Orycteropus*, the Cape Ant-eater, both of which are usually classed under the Edentata. *Manis* has the hair agglutinated to form overlapping scales, but has no dermal plates and no teeth. *M. tricuspis* is arboreal in its habits. *Orycteropus* has peculiar folded teeth and scanty hair. It is termed by the Boers the Aard-vark or Earth-pig and is nocturnal in its habits, sleeping during the day in burrows which are usually found in the neighbourhood of the large ant-mounds so common on the veldt. Neither genus is believed now to have any close affinity with true Edentata, their reproductive organs being markedly different from those of the S. American forms, and they are provisionally grouped together in an Order termed the *EFFODIENTIA*.

MARINE MAMMALS.

The second and third strange groups, the relations of which to the rest of the Mammals are unknown, are the two groups inhabiting the sea, viz., the Whales or CETACEA, and the Sea-cows or SIRENIA. Both of these have some peculiarities in common, due to their having adapted themselves to special and—for a Mammal—unnatural conditions. Thus both Whales and Sea-cows have lost all outward trace of hind-limbs,

_{Marine Mammals.}

although a pair of small bones representing them are found embedded in the body. In both the tail has become flattened, developing flukes or fins at the sides, and, as in Fishes, it is by strokes of the tail that these Mammals principally move. But whereas in Fishes the tail moves from side to side, here, in accordance with the necessity for coming to the surface to breathe, it moves up and down. Both Whales and Sea-cows have lost nearly all hair ; Whales retain only one or two traces about the lips, while in Sea-cows there are scanty bristles all over the body and the lips are thickly covered. The other Mammals which have taken to the sea, the Seals and their allies, are provided with a thick coating of hair (see p. 538).

We saw that the nose in Fishes was a sense-organ, adapted for stimulation by gases and other odoriferous substances dissolved in the water. When land-animals were evolved, the air needed for breathing was drawn in past the nose ; the oro-nasal groove leading from the nose to the mouth, which was at first an open gutter, became changed into a closed channel, the nasal passage, and the nose became modified for stimulation by vapours mixed with air. In the marine Mammals air is still drawn in through the nasal passage, but they are concerned no longer with perceiving things which emit vapours into the air, but rather like fishes, with substances dissolved in the surrounding water. Since the true nose, the sense-organ, owing to its new connection with breathing, cannot be used for perceiving substances in the water, it ceases to be of use, and becomes vestigial : the nasal bones shrink away into small remnants and the nasal opening is placed far back on the snout, near what appears to be top of the animal's head.

Order II.　Cetacea.

The Order of Whales is distinguished by the great rounded cranium and by the elongation of the bones of the face and jaws. These support an immense prow-like snout formed chiefly of fat, which is an admirable buttress of defence for the animal's skull. The supra-occipital bone is of great size and forms the posterior surface of the cranial dome, interposing between the small parietals and meeting the frontal. The frontals develope great orbital plates flanking the face, beneath which is the small orbit bounded below by the slender jugal. In Whales also the teats are situated far back, as they are in cows, and the mother forces the milk down the young one's throat ; for in the Whale, as in the

young marsupial, the windpipe and nose are directly connected ;
only here the connection lasts through life and allows a Whale to
swim through the water with its mouth open whilst it breathes at
the same time.

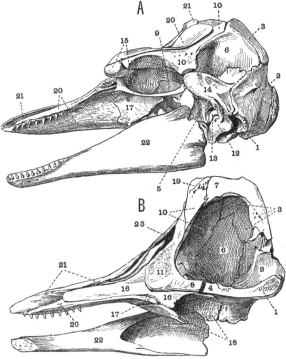

FIG. 309.   A, Lateral view, and B, Longitudinal section of the Skull of a young
            Ca'ing Whale, *Globicephalus melas* × ⅙.

1 Basi-occipital.   2 Exoccipital.   3. Supra-occipital.   4. Basisphenoid.
    5. Alisphenoid.   6. Parietal.   7. Interparietal fused with 3.
    8. Presphenoid.   9. Orbitosphenoid.   10. Frontal.   11. Mesethmoid.
    12. Tympanic.   13. Periotic.   14. Squamosal.   15. Jugal.
    16. Vomer.   17. Palatine.   18. Pterygoid.   19. Nasal.
    20. Maxilla.   21. Premaxilla.   22. Mandible.   23. Anterior
    nares.

Whales are divided into Sub-orders, the whalebone Whales or
MYSTACOCETI and the toothed Whales or *ODONTOCETI*.   In the
latter there are numerous teeth, but they are all alike and simple
(Fig. 309), and the maxilla develops a great crest which conceals the
orbital plate of the frontal.   The great Sperm-whale, *Physeter
macrocephalus*, of the Southern Seas, has teeth only in the lower jaw
and feeds on cuttle-fish and fishes, gripping the long flexible arms
of the former by pressing them against the upper jaw.   Spermaceti

oil is the melted-down fat of this monster. The Ca'ing or Pilot Whale (*Globicephalus melas*), which also feeds chiefly on cuttle-fish, has teeth on both upper and lower jaws (Fig. 309). Pilot-whales are social in disposition, and the herds are occasionally driven into bays or fiords in the North Atlantic and captured. Smaller Toothed Whales are found round the coast of Britain which have teeth in both jaws. Others are known as the Porpoise, *Phocaena*, the Dolphin, *Delphinus*, and the Grampus, *Orca*. The common Porpoise, *Ph. communis*, is the most abundant and best known of British Cetaceans. It is not more than six feet long and is often cast ashore. In the Gulf of St Lawrence the White Whale, *Delphinapterus leucas*, is fairly common. It attains a length of twelve feet.

The whalebone Whales, *Mystacoceti*, have no teeth. The orbital plate of the frontal is uncovered and there is a small ethmo-turbinal covered with olfactory epithelium. They are all large animals, although they feed on the smallest prey, such as minute pelagic mollusca, jelly-fish and crustacea. The "whalebone" or baleen consists of a large number of horny plates hanging down like curtains from the palate into the cavity of the mouth. These are placed in pairs, one on each side of the mouth, one pair behind the other, and the fellows of a pair nearly meet in the middle. The lower edges of these plates are frayed out so as to form a fringe or strainer. After the whale has taken water into its mouth it raises its tongue against the edges of the plates and allows the water to trickle out through the strainer described above; all the small animals taken in the water are thus retained and then swallowed. The best quality of whalebone is obtained from the Right Whale, *Balaena mysticetus*, an animal about fifty feet long, found only in the Arctic regions. The great Rorqual Whale, *Balaenoptera sibbaldi*, has a fin in the middle of its back, and attains a length of from 60—80 feet; it is the largest animal now found on the globe and is very abundant. The lesser Rorqual, *Balaenoptera rostrata*, is a smaller animal some 30 feet in length. On two occasions at least the animal has strayed up the St Lawrence as far as Montreal where it has been starved to death in fresh water. The head of *Balaenoptera* is much shorter than *Balaena* and the whale-bone is shorter and coarser.

Order III. Sirenia.

Sea-cows differ from whales in so many respects that they cannot have any close relation with them. They are vegetable-feeders, and

browse on sea-weeds and other water-plants. As these habits necessitate their staying under the water for some considerable time,
the bones are heavy and solid, quite different in structure from the

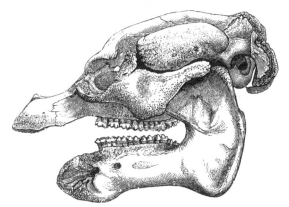

Fɪɢ. 310.　Skull of African Manatee, *Manatus senegalensis* × ⅙.

bones of whales, which are much more spongy in texture. The skull
is long, not rounded, and the face bones are only moderately
developed. The parietals are not pushed aside by the development
of the supra-occipital; the supra-orbital plate of the frontal is small,
while the orbit is large and bounded below by a very powerful jugal.

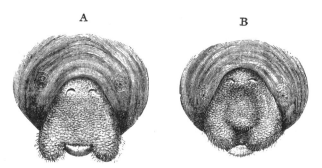

A　　　　　　　　　　　B

Fɪɢ. 311.　Front view of head of American Manatee, *Manatus americanus*,
showing the eyes, nostrils and mouth. A, with the lobes of the upper lip
divaricated. B, with the lip contracted. From Murie.

The teeth are broad and crushing, and front teeth sometimes are
found developed as tusks. There is no such snout as is found in
Whales, but there are large movable lips by means of which food
is seized (Fig. 311). The teats are placed on the breast as in Bats,

Monkeys and Man, and the mother holds the young under the arm, which is quite flexible and not a mere fin as in whales. It is supposed that the legends of mermaids have been suggested to sailors by the sight of these strange parents holding their young above water. There are two genera, each represented by a single species; the Manatee, *Manatus*, found on the warmer parts of the coasts of the Atlantic and in the estuaries of its rivers, both in America and Africa, and the Dugong, *Halicore*, found all around the coasts of the Indian Ocean and round Australia, where it is fished for and eaten. Until 1768 a third species, *Rhytina stelleri*, of great size—20 to 25 feet long—inhabited some islands in the Behring Sea. It had no teeth, horny plates on the gum supplying their place. This species was exterminated by Russian seal-hunters.

Leaving aside these curious groups of animals, we find that the relations to one another of the remaining Mammals are more easily understood. We have first of all to deal with the Insectivora.

Order IV. Insectivora.

This is a group of small animals which, as their name implies, feed chiefly on insects. They have three or four sharp pointed cusps on each of their back teeth, adapted for piercing the armour of insects, while their front teeth in both jaws are directed outwards so that they act like a pair of pincers in seizing the prey. The Insectivora are plantigrade, that is, they place the whole palm and the whole sole on the ground when they walk (Figs. 312, 313); in nearly every case they have the full number (five) of fingers and toes; they have long flexible snouts projecting beyond the mouth and their brains are of a low and simple structure, the surface of the cerebral hemispheres being smooth, while they leave the cerebellum uncovered. In many cases there is a shallow cloaca surrounded by a sphincter into which both anus and urinogenital passage open. They possess an allantoic placenta, but this covers only a small portion of the surface of the uterus, and indeed in this respect they are hardly more advanced than those Metatheria which retain an allantoic placenta. The Insectivora, as may be seen from the description, are a very primitive group, and like other primitive groups consist of a number of families widely differing from one another in structure. Taking a broad view we may say that the tropical families exhibit the highest grade of structure. Thus the GALEO-PITHECIDAE, or flying shrews, represented by the genus *Galeopithecus*, have a parachute-like expansion of skin extending from neck to hand, forming a web including the fingers. A similar expansion of skin reaches from wrist to foot, forming a web between the toes, and there is a piece of skin connecting the two legs behind. There is a ring of bone round the orbit, and the symphysis pubis is long and strong. The TUPAIIDAE, or Tree-shrews, have likewise the orbit encircled by bone and a strong symphysis pubis, but they are devoid of any parachute-like extension of skin. They are small animals with large eyes and long furry tails ; both these groups are confined to the Malay archipelago and India and both inhabit trees.

FIG. 312. African Jumping-shrew, *Macroscelides tetradactylus* × ½. From Peters.

The MACROSCELIDAE have no bony ring round the orbit but they possess a strong symphysis pubis. Their most marked characteristic is an elongated foot (see fig. 312) which enables them to make great springs. Hence the name Jumping-shrews. They are represented by 14 species distributed over Africa.

The three families which represent the Insectivora in Great Britain are all of a lower type. Not only is the orbit never surrounded by bone but the zygomatic arch is slender and sometimes

even absent. The brain cavity is very small and the symphysis
pubis is very short; sometimes the pubes are united only by ligament.

The first of these families is the ERINACEIDAE, or Hedgehogs, distinguished by the slender zygomatic arch, and by the tympanic being
in the form of a ring. The well-known Hedgehog, *Erinaceus
europaeus*, is intermediate in size between a rat and a rabbit. It has
the fur intermixed with spines, and when alarmed can roll itself
into a ball, tucking in head, limbs and tail, and in this condition can
bid defiance to its enemies. All Erinaceidae are not of this character;

FIG. 313.   Russian Desman, *Myogale moschata*.

the rat-like *Gymnura* from India and the Malay peninsula is
without spines.

The other two families are the Shrew-mice (SORICIDAE) and
the Moles (TALPIDAE); these are represented in both Great Britain
and North America, but the latter country is without Hedgehogs.
The Soricidae have lost the zygomatic arch altogether, the pubes
are disconnected and the tympanic is ring-like. As the popular
name implies these are mouse-like animals covered with fur.
There are three British species, *Sorex vulgaris*, about the size of
an ordinary mouse, *Sorex pygmaeus*, one of the smallest Mammals

known, and *Crossopus fodiens*, the Water-shrew, distinguished by having the feet frayed with stiff hairs to aid in swimming. The North American *Blarina* has the aspect of a Mole with its small eyes and rudimentary outer ears. It is called the Mole-shrew, but its normal arms and hands at once distinguish it from the true Moles. The true Moles, TALPIDAE, are above all characterised by the greatly enlarged hands and powerful though short arms by which they are adapted for a burrowing life. To make room for the large hands in narrow burrows the front segment of the sternum is greatly elongated, thus carrying the pectoral girdle and limbs forward on to the neck, where there is room for them. The clavicles are short, almost square bones, and the humerus of the arm is short and stout. The zygomatic arch is present and the tympanic is a bulla. The TALPIDAE are represented in Great Britain by *Talpa europaea*, the common Mole, which feeds on earthworms, constructing a complicated system of underground passages through which it hunts its prey. In North America the commonest is perhaps *Condylura cristata*, the Star-nosed Mole, the snout of which is encircled by a ring of fleshy outgrowths.

The Russian Desman, *Myogale moschata*, once extended as far west as Britain. It lives in burrows by the water-side and feeds chiefly on fresh-water insects and their larvae. In correspondence with its mode of life the hind-feet are webbed and the tail large and compressed, forming an efficient swimming organ. It is hunted for its fur.

There still remain four families to be mentioned, each of which however is represented by a few species. These are interesting because, (1) They have a more primitive type of molar tooth than any other living Mammals; (2) In their distribution, like the ancient genus *Peripatus*, they belong to the southern hemisphere, only over-stepping it when they go into the West Indies. The type of tooth is the tri-tubercular, which is found in the oldest remains of Mammals known: it is distinguished by the reduction of the characteristic cusps of the insectivoran tooth to three which form the points of a triangle. Of these primitive families the CHRYSOCHLORIDAE are the Golden Moles of the Cape, so-called from the iridescent sheen of the fur. They have the reduced eyes and enlarged hands and arms of the ordinary Mole, but these hands and arms are placed not at the sides of the neck but at the sides of the thorax, the ribs of which are bent inwards to create hollows for their reception. The zygomatic arch is present and the tympanic is a bulla. The remaining families have lost the zygomatic arch and the tympanic is a mere ring. These are (1) POTAMOGALIDAE, represented by a single species, Water-shrews from Central Africa with a flattened tail, short limbs and no clavicles, (2) SOLENODONTIDAE, and (3) CENTETIDAE, two closely allied families of small hog-like animals with stout limbs, the first from Cuba and Hayti and the second from Madagascar.

The most interesting circumstance about the Insectivora is the fact that when by means of fossils we trace back the higher groups of mammals they seem all to merge imperceptibly into forms which from their teeth and general organisation we should class as Insectivora. There is therefore really good ground for supposing that the living Insectivora, though modified in special details, nevertheless represent, so far as their general organisation is concerned, the earliest type of Eutheria which appeared on the globe. From these original Insectivores advance seems to have taken place along five lines :—I., some Insectivora took to attacking larger prey, including their own less fortunate relatives, and gradually developed into the Carnivora or flesh-eating mammals : II., some became vegetable feeders and gave rise to the great group of hoofed animals, relying either on their swiftness, size or strength for defence : III., some took to burrowing and developed into gnawers or Rodents, relying chiefly on their burrows for safety : IV., some took to the air, the fore-limb becoming changed into a wing; these are the Bats : V., the remainder took to escaping into trees when hard pressed, and eventually gave rise to the great group of the Primates which includes Monkey and Man.

Order V. Carnivora.

The Carnivora are distinguished above all by their teeth (Fig. 292). They have small insignificant front teeth or incisors, but the eye-teeth or canines, situated in the maxilla just where it meets the premaxilla, are large and pointed. With these the animal seizes and kills its prey. The premolars have cutting edges, consisting typically of a large central cusp and two smaller ones, one in front and one behind. The molars with the exception noted below are broad and crushing (Figs. 292 and 314). The last premolar in the upper jaw and the first molar in the lower jaw constitute what are called the carnassial teeth. These are very large blade-like teeth which bite on one another like a pair of scissors. The upper one has enlarged central and posterior cusps, the anterior cusp being small or wanting; the lower carnassial has an anterior blade-like portion consisting of two cusps and a posterior flattened portion or heel. The nails are sharp curved claws.

The most familiar examples of this class of animals are our Dogs and Cats. The wild ancestors of the domesticated pets are unknown, though the dog's ancestors were no doubt allied to the wolf, whereas the cat is probably descended from some species belonging to the

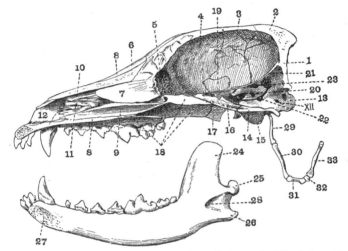

Fig. 314. Vertical longitudinal section taken a little to the left of the middle line through the Skull of a Dog, *Canis familiaris* × $\frac{3}{5}$.

1. Supra-occipital. 2. Interparietal. 3. Parietal. 4. Frontal. 5. Cribriform plate. 6. Nasal. 7. Mesethmoid. 8. Maxilla. 9. Vomer. 10. Ethmo-turbinal. 11. Maxillo-turbinal. 12. Premaxilla. 13. Occipital condyle. 14. Basi-occipital. 15. Tympanic bulla. 16. Basi-sphenoid. 17. Pterygoid. 18. Palatine. 19. Ali-sphenoid. 20. Internal auditory meatus, the passage for the eighth nerve to the internal ear. 21. Tentorium, a fold of calcified connective tissue projecting into the cranial cavity and separating the cerebrum from the cerebellum. 22. Foramen lacerum posterius, the passage for the tenth nerve. 23. Floccular fossa, the cavity in which the floccular lobe of the cerebellum is lodged. 24. Coronoid process. 25. Condyle. 26. Angle. 27. Mandibular symphysis. 28. Inferior dental foramen. 29—31. Segments of the second visceral arch. 29. Stylo-hyal. 30. Epi-hyal. 31. Ceratohyal. 32. Basi-hyal. 33. Thyro-hyal, the third visceral arch. XII. Condylar foramen, the aperture through which the twelfth cranial nerve leaves the skull.

East, allied to but distinct from the Wild Cat, *Felis catus*, still found in remote parts of Scotland and possibly in the mountains of North Wales. Possibly the domestic cat has originated from the Caffre Cat, *F. caffra*, which extends throughout Africa and was considered sacred by the ancient Egyptians, who embalmed their bodies in such amazing numbers that their mummies have been exported from Egypt and used as manure.

In the Dog, *Canis familiaris,* and the other members of the family CANIDAE, the muzzle is long and the teeth numerous. Their arrangement can be expressed by the dental formula i. $\frac{3}{3}$, c. $\frac{1}{1}$, pm. $\frac{4}{4}$ m. $\frac{2}{3} = 42$, where the upper line shows the teeth in the upper jaw, the under line those in the lower. The first figure denotes incisors, the second canines, the third premolars and the last molars. The hindermost back teeth, or molars, are still broad. The fore-legs cannot be used for grasping. The claws are comparatively blunt and cannot be retracted.

In the domesticated Cat on the other hand the muzzle is short, and the teeth reduced in number, the formula being i. $\frac{3}{3}$, c. $\frac{1}{1}$, pm. $\frac{3}{2}$, m. $\frac{1}{1} = 30$, whilst the fore-limbs can be used for seizing. The claws are very sharp, and can, when not in use, be completely retracted or rather raised, so as not to wear the points. In all these respects Cats are more perfectly adapted for a carnivorous life than Dogs, these latter still retaining traces of their descent from a different kind of mammal. Just as the Wolf, *C. lupus,* the Jackal, *C. aureus,* and the Fox, *C. vulpes*—the last-named the only wild species of *Canis* found in Britain—are species of dogs distinguished from each other by size and slight peculiarities of hair, etc., so the Lion, *F. leo,* the Tiger, *F. tigris,* the Leopard or Panther, *F. pardus,* the Lynx, *F. lynx,* and the Puma, *F. concolor* (frequently called a "Panther" in America, where it is found from Canada to Patagonia), are all Cats. The differences in the colour of the skin which help to distinguish them are in all probability due to the fact that the colours are protective, enabling the animals when in their natural surroundings to escape the notice of their prey. Thus Lions, which as a rule live in dry and rather open places, are of dun colour; the stripes of the tiger's skin deceptively resemble the alternating shadows and sunlit strips of ground found amongst the reeds in which they live; the spots of the leopard are undiscoverable amidst the alternating patches of light and shade caused by the sunlight struggling through the interstices of the foliage of a forest.

The Bears, URSIDAE, represent a third type of Carnivora. They are plantigrade, placing the whole sole of the foot on the ground; the molars are blunter than those of the Cats and Dogs and very broad, the carnassials are broad and the premolars very small and often fall out; the upper carnassial is a comparatively small tooth and the heel of the lower carnassial is larger than the blade; these peculiarities are connected with the fact that the Bears are not merely flesh feeders but can live partly on a vegetable diet. The Brown Bear of Europe, *Ursus arctos*, which used to be abundant in Britain, is so nearly allied to the Grizzly Bear, *U. horribilis*, of the Rocky Mountains, that the latter is by some authorities placed in the former species. In Eastern Canada, especially in the Province of Quebec, the Black Bear, *Ursus americanus*, is very abundant and is trapped for its fur. It is usually an inoffensive animal, feeding on berries and bark, but occasionally, especially when it has cubs, it will attack man.

The Stoats, Weasels, Martens, Minks, Polecats, Otters, Badgers and Skunks, forming the family MUSTELIDAE, are sometimes supposed to be allied to the Bears, but are really very distinct. They have very long necks, slender, flexible bodies and short limbs, and their habits are exceedingly bloodthirsty and ferocious. The chief resemblances to Bears are found in the skull and teeth, but the contrast in general build and in gait—the Mustelidae are digitigrade—is very striking. Six species of Mustelidae are found in Great Britain: (1) The Otter, *Lutra vulgaris*, an animal which has webbed toes and a long, somewhat flattened tail. It lives on fish, passing much of its time in the water. (2) The Badger, *Meles taxus*, a heavy, somewhat clumsy animal with blunt claws and short limbs, leading a nocturnal, burrowing life and feeding on mice, reptiles, insects, fruit, acorns and roots. (3) The Pine Marten, *Mustela martes*. (4) The Polecat, *Putorius foetidus*, which feeds on small mammals, birds, reptiles and eggs, and has a disagreable odour. The Ferret is a domesticated variety of the Polecat. (5) The Weasel, *Putorius vulgaris*. In cold regions the Weasel turns white in winter. (6) The Stoat, *Putorius ermineus*, which also turns white in cold climates except the tip of its tail, which remains black. Its fur is much prized. These last four are closely related species with long, slender bodies, sharp curved claws and ferocious habits.

In North America there is an interesting family, the PRO-CYONIDAE, intermediate between the Ursidae and Mustelidae. The members of this family have sharp muzzles but clumsy bodies and short necks; the Raccoon, *Procyon lotor*, is the most familiar. It is omnivorous. The Mustelidae are represented by otters, martens and a remarkable form, the Skunk, *Mephitis mephitica* (Fig. 315),

which produces a secretion of such repulsive odour as to make it
avoided by other animals and a terror to man. It is strikingly
marked and affords a much-quoted example of warning coloration ;
its conspicuous colour enabling would-be enemies to distinguish its
possessor from less offensive prey. The VIVERRIDAE should be
mentioned although they are a tropical group. In general shape
they resemble the Mustelidae, but in the shape of the carnassial
teeth and in the division of the auditory bulla by a septum they
agree with the Felidae. The best-known members of the family are
the African and Indian Civet Cats (*Viverra civetta* and *V. zibetha*)

FIG. 315. The Common Skunk, *Mephitis mephitica*.

from whose perineal glands the civet of commerce is obtained.
Fossil remains connect the Viverridae and Mustelidae and one would
not be far astray in calling them "primitive cats."

The Carnivora mentioned hitherto are often grouped together as
the *CARNIVORA VERA* or *FISSIPEDIA*. The second group of recent
Carnivora is represented by the seals and is termed the *PINNIPEDIA*.
The name is derived from the fact that fingers and toes are united
by webs of skin. The Seals are almost as purely marine animals
as the Whales and Sea-cows, but they have become adapted to
their surroundings in quite a different way. Thus their fur is

close and thick, and they are protected against the cold of the
water by it, instead of being covered all over by a thick layer of
fat as are the Whales. The tail is short and insignificant, but
they make a powerful stern oar by directing the feet backwards
parallel to the body so that the soles are turned up. Thus the feet
act just in the same way as the tail does in a whale, making up and
down strokes and driving the animal forward. The whole upper
part of the limb is buried in the body. In one group, the true Eared
or Seal-skin Seals, OTARIIDAE (the fur of some species of which is
used for making jackets), the feet can be turned forward when the
animal comes on land. There are also some traces of an external

Fig. 316.   The Patagonian Sea-Lion, *Otaria jubata*.   From Sclater.

ear, whence comes the name OTARIIDAE or Eared Seals which is
given to them (Fig. 316). They are confined to the Pacific coast of
America. The Walrus, *Trichechus rosmarus*, of the Arctic seas, is
the representative of a second family, the TRICHECHIDAE. No
external ear is present but here also the feet can be turned forward.
The canine teeth of the upper jaw are very long and give the animal
a fierce appearance. They are however chiefly used for digging up
bivalves from the mud and for climbing on the blocks of ice in the
Arctic regions where the animal is found. The name "Old Man"
sometimes given to it by whalers is suggested by the tufts of grey
hair on the sides of the face. The common Greenland Seal, *Phoca*

vitulina, and the Gray Seal, *Halichoerus grypus*, are the two species
of PHOCIDAE, the third and last family of the Pinnipedia, regularly
found round about the British coast in out-of-the-way places.
The members of this family have harsh fur and no trace of an
ear-flap and are unable to turn their feet forward, so that when
they come on land they shuffle along entirely by the aid of their
fore-limbs. They are in fact the most thoroughly adapted for
aquatic existence of all the Pinnipeds. *Phoca vitulina* is common
on the eastern shores of Canada and New England.

Order VI. Ungulata.

The great group of the Ungulata or hoofed animals represents
the second line of evolution from the primitive Insectivores. Here
we find that all power of grasping with the limbs is absent and the
feet are purely adapted for running, the toes being encased in hard
blunt nails which are called hoofs. At the present time the
Ungulata include a number of very diverse forms. But it must
be remembered that a large proportion of the group is extinct,
and that to some extent the fossil forms serve to connect the very
heterogeneous members of the group that still exist.

Sub-Order I. Sub-ungulata.

In former times there existed a great assemblage of big and
often clumsy animals belonging to the Ungulata in which the toes
were all nearly equal in length and the bones of the wrist arranged
in parallel longitudinal series. The Sub-ungulata at one time
spread over the earth and in South America, which became isolated
in early times, they gave rise to a great variety of forms. Some of
these mimicked the descendants of the Ungulata, and formed one of
the most striking examples of parallel evolution. Only two families
of the *SUB-UNGULATA*, as these animals are called, survive till
the present day. These are the Elephant family, PROBOSCIDEAE,
and the family of the *Hyrax*, HYRACIDAE.

HYRACIDAE.

The *Hyrax* (*Procavia*) is the coney mentioned in the Bible.
The Hyracidae are small, not unlike rabbits in appearance, but their hind-feet closely resemble those
of the Rhinoceros. Their front teeth are, it is true,
somewhat chisel-shaped, as in the Rodentia, but there are four of

*Family
Hyracidae.*

these below and two above, which is quite unlike the arrangement in the rabbit. It is possible however that the two teeth reckoned as lower posterior incisors may really be canines, since they do not, like the other incisors and like those of the Rabbits, grow throughout life (Fig. 317). These animals are found throughout Africa except

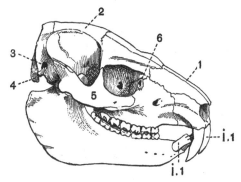

FIG. 317.   Skull of *Hyrax* (*Procavia*) *dorsalis* × ⅔.

1.   Nasal.   2.   Parietal.   3.   External auditory meatus.   4.   Process of the exoccipital.   5.   Jugal.   6.   Lachrymal foramen.   i 1.   First incisor.

in the north and also in Arabia and Syria. Only one genus is now recognized, *Hyrax* (*Procavia*), with several species. Most of these live amongst rocks, in mountains and in stony places, but some frequent the trunks and large branches of trees and sleep in holes.

PROBOSCIDEAE.

The Elephant is too well known to need much description, but it may be pointed out that the trunk is really a long flex-

Family Proboscideae.

ible snout, an excessive exaggeration of what is found in Insectivores, and that the tusks are front teeth, only those in the upper jaw being developed (though in some extinct elephants, as for instance in some species of *Mastodon*, both upper and lower incisors were present while in *Dinotherium* the tusks were developed in the lower jaw only), and finally that the upper parts of the arms and legs are quite free from the body, instead of being, as is usually the case with mammals, buried inside the general contour of the body. There are only two living species, the African Elephant, *Elephas africanus*, inhabiting the forest region of tropical Africa and hunted for its tusks, and the Indian Elephant, *Elephas indicus*, inhabiting the jungles of India, Further India, Ceylon and Sumatra,

which is frequently domesticated. The canines are lost and have
left no traces. The molars succeed one another in a horizontal row,
never more than two being at any one time functional (Fig. 318).
The ridges on these teeth when worn present the appearance
of parallel bands in the Indian Elephant, but in the African they
form diamond-shaped lozenges. The ears of the latter are very
large and the trunk ends in two nearly equal prehensile "lips"
attached to its lower margin. In the Indian Elephant the ears
are smaller; there is but one finger-like "lip" at the end of the

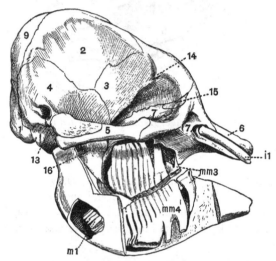

Fig. 318. Skull of a young Indian Elephant, *Elephas indicus*, seen from the
right side, the roots of the teeth have been exposed × ⅛.

1. Ex-occipital.   2. Parietal.   3. Frontal.   4. Squamosal.   5. Jugal.
6. Premaxilla.   7. Maxilla.   9. Supra-occipital.   13. Basi-occipital.
14. Postorbital process of the frontal.   15. Lachrymal.   16. Pterygoid
process of the ali-sphenoid.   i 1. Incisor.   mm 3, mm 4. Third and
fourth milk molars.   m 1. First molar.

trunk and this is attached to the upper edge of the end of the
trunk. The skull is very massive, but the exterior gives an
erroneous impression of the size of the brain-case because the bones
are enormously thickened and contain large air-spaces, especially
in older specimens, where the frontals may attain a thickness of
one foot. Till recent times (geologically speaking), an extinct
elephant, *Mastodon*, inhabited North America, Europe and parts
of Asia. Its remains are being constantly dug up from the bottom
of gravel-pits and marshes. Some species of *Mastodon* had tusks in

both jaws; in most the molar teeth were covered with tubercles like those of the pig, instead of ridges of enamel. Another fossil species allied to the Indian Elephant but covered with thick fur, the Mammoth, *Elephas primigenius*, had formerly an extensive range around the North Pole and at one time was common in Britain.

Sub-Order II. **Ungulata vera.**

All the rest of the Ungulata have the thigh and the upper arm more or less buried in the body, whilst the heel and the wrist are raised in walking so that the creature goes along on the tips of its

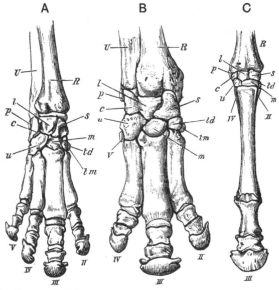

Fig. 319. Bones of right Fore-foot of existing *Perissodactyles*. A, Tapir, *Tapirus indicus* × ¼. B, Rhinoceros, *Rhinoceros sumatrensis* × ⅕. C, Horse, *Equus caballus* × ⅛.

c. Cuneiform (ulnare).　　　*l*. Lunar (inter-medium).　　　*m*. Magnum (third distal carpal).　　*p*. Pisiform.　　*R*. Radius.　　*s*. Scaphoid (radiale). *td*. Trapezoid (second distal carpal).　　*tm*. Trapezium (first distal carpal). *U*. Ulna.　　*u*. Unciform (conjoined 4th and 5th distal carpals). II—V, second to fifth digit. From Flower.

toes. The bones of the wrist are arranged in transverse rows, the members of two adjacent rows alternating with one another. The first digit in both fore- and hind-limbs is entirely absent. These true Ungulates, *UNGULATA VERA*, as they are called, are divided into two great groups: (1) the *PERISSODACTYLA*, in which there is an odd number of toes and in which the true central axis of both arm and leg runs down through the centre of the third finger or toe (Fig. 319), and (2) the *ARTIODACTYLA*, in which there is an even number of toes, and in which the axis of the limb passes down between the third and fourth toes (Fig. 322).

Division I. *PERISSODACTYLA.*

The Perissodactyla were formerly a numerous class of animals, but now three families alone survive, the tapirs, TAPIRIDAE; the various species of rhinoceros, RHINOCEROTIDAE; and the horse and numerous species of ass, EQUIDAE.

Of these the oldest and most primitive are the TAPIRIDAE. They

Tapiridae.

still have four toes on the fore-feet, which is an even number; but as they have only three on the hind-feet and in both fore- and hind-feet the axis of the limb runs through the third toe, there is no doubt that they are to be classed with the Perissodactyla (Fig. 319). The snout is long and flexible, longer than the snout of the Insectivores but not so long as the snout of the Elephant. A most interesting feature in the natural history of the Tapirs is that they are now found only in two widely separated parts of the world, viz., the north of South America and in the Malay Peninsula with the neighbouring islands of Borneo and Sumatra. We need not however suppose that there was at one time a land bridge across the Pacific, for in recent rocks we find remains of Tapirs all over Europe, Asia and America, so that the present species are to be regarded as two separated remnants of a great race of animals which once had a very wide distribution. Their present range affords an often quoted example of what is known as "discontinuous distribution."

The RHINOCEROTIDAE are represented at the present day by the

Rhinocero-
tidae.

genus *Rhinoceros*. The Rhinoceros is a heavier and clumsier animal than the Tapir; it has three toes on both fore- and hind-feet and no projecting snout. Its chief peculiarity however is the horn which it carries so to speak on the bridge of its nose. The horn has no bony core, and as it is entirely composed of horny matter may be said to be a mass of hairs stuck together. There are several species found in Asia and in Africa; the best known is perhaps the Indian, *R. unicornis* (Fig. 320); the Javan, *R. sondaicus*, is smaller. Both these species have but one horn. Two-horned rhinoceroses (the two horns stand-

ing one behind the other) are now found in the Malay Peninsula,
Borneo and in Sumatra (*R. sumatrensis*), while in Africa there are
several species; the commonest, *R. bicornis*, is frequently shewn
in menageries. It is supposed that the idea of the unicorn

FIG. 320.　Indian Rhinoceros, *Rhinoceros unicornis*.　From Wolf.

was derived from the one-horned rhinoceros, but if this be so the
imagination must have played a powerful part in evolving the
graceful animal which figures in the royal arms out of the clumsy
rhinoceros.

The general appearance of the horse, *Equus caballus*, is
sufficiently well known, but the structure of its
feet, which, next to the wings are the most highly
specialized organs of locomotion in the animal kingdom, demand
careful attention.

Equidae.

The apparent "knees" of the horse correspond to the joints of
the wrist and the ankle, the true elbow and knees are concealed in
the body of the animal, although the motion of these joints can be
clearly seen if a running horse be watched. A horse walks on the
very points of its finger and toe-nails, and it possesses only one finger
on each hand and one toe on each foot (C, Fig. 319), the fingers and
toes corresponding to the outer fingers, the toes of the Rhinoceros
being represented merely by bones entirely concealed beneath the skin
and applied like splints to the great middle finger and toe respectively. Thus the whole limb instead of being a loosely jointed
flexible organ for grasping, becomes a firmly jointed lever bending
only in one plane and suitable for quick locomotion.

The Horse, as we know it, has been domesticated and bred by man for thousands of years and is doubtless very unlike its wild ancestor. The wild animals at present existing which are called Wild Horses are all more like donkeys, with longer ears and without the peculiar wisp-like tail of the Horse; they are also all more or less striped. The Zebras and Wild Donkeys are found in Africa on the great plains in the south, in the deserts of Syria and Persia and in the central plains of India. Another African form, the Quagga, has become extinct in recent times. In America when discovered there were no horses, although the horse has since run wild there; but in the most recent geological period the horse abounded in America and why it should have died out in a country which afterwards proved to be well suited for it is a mystery. In the same country in the deposits formed at the bottom of great lakes are found the remains of a series of animals which form a complete chain from a true horse which appears in the newest deposits to animals which not only are more primitive than Tapirs but which must even be reckoned as Sub-ungulata, for they have five fingers and five toes but had the bones of the wrist and ankle in longitudinal series. This series of forms is one of the most complete evidences of evolution known to geologists.

Division II. *ARTIODACTYLA.*

Unlike the Perissodactyla, the Artiodactyla or even-toed Ungulates constitute an immense assemblage of animals, and until the invention of modern fire-arms were the dominant animals on the great plains of Africa and also of North America. The Artiodactyla may be divided into a higher and a lower section.

The lowest section may broadly be called the Pigs, *SUINAE.* Suinae. They retain four toes on fore- and hind-feet, have a snout ending in a round flat surface and are all gross feeders, eating not only roots of various kinds but also small animals if they come in their way. Their teeth are covered with tubercles a good deal blunter than the cusps on the teeth of an Insectivore but still of the same essential nature. Such teeth are termed bunodont, whence the name *BUNODONTIA* (Gr. βουνός, a hill or mound) has sometimes been applied to this division. The Hippopotamus, the sole representative of the family HIPPOPOTAMIDAE, is nothing but an enormous Pig; it differs from the ordinary Pig in having all its toes of equal length, whereas in the true Pig the outer (second and fifth) toes are small and do not reach the ground. The Hippopotamus spends most of its time in rivers and swamps feeding on the reedy vegetation of such places. It has exceedingly powerful jaws and when wounded has been known to crush a canoe between them. The true Pig belongs to the Family SUIDAE and is a domesticated variety of the wild boar, *Sus scrofa*, which, as is well known, survived in England until the middle ages and still exists in Europe. In the male the canines, or eye-teeth, are powerfully developed, those of the lower jaw projecting upwards outside the mouth. In the Babirusa, *Babirusa alfurus*, of Celebes, the upper canines do not enter the mouth but are bent upwards and pass through special holes in the skin, curving back over the head like horns. They grow persistently, their roots being kept open. The Pigs are not strictly vegetable feeders but are really scavengers, eating every vegetable or animal substance they encounter, the food they seek especially consisting of roots. A very interesting genus, the Peccary, is represented by two species, *Dicotyles tajaçu* and *D. labiatus*, which inhabit the American continent. The former ranges from Patagonia to the Red River of Arkansas, the latter between Paraguay and British Honduras. The name means 'two navels' and was suggested by the presence of a large gland in the middle of the back resembling a navel. On the hind-foot the fifth toe is wanting, so that there are only three toes; but the position of the axis of symmetry is still between the third and fourth toes. The Peccaries go in droves and are most dangerous antagonists; climbing a tree is the only chance of safety to a hunter who meets a herd.

When we leave the Pigs we have to deal with the higher section of the Artiodactyla, the *SELENODONTIA* or *RUMINANTIA*, which include most of our domestic animals, the cow, sheep, goat, camel, etc., as well as all the deer and antelopes. The latter name is derived from the habit of ruminating, that is of bringing the food back from the stomach into the mouth after it has been swallowed and chewing it again. Corresponding to this habit we find that the stomach has acquired a complicated structure. Just where the gullet opens into it we find a large pouch projecting laterally with the walls covered with little projections or papillae; this is called the paunch or rumen. Just below the oesophagus is another smaller pouch divided by a constriction from the first (Fig. 321). The second pouch is called the reticulum because its walls are raised into intersecting folds producing cavities like the cells of

a honeycomb. The food mixed with saliva is swallowed without
chewing, and after traversing the oesophagus it is driven from rumen
to reticulum and back by the action of the muscles and well soaked
with gastric juice. After some time it is pressed up again into the
mouth and thoroughly ground up by the great broad premolar and
molar teeth. When swallowed for the second time it is nearly
fluid. It now passes down a groove or channel in the side of the
gullet enclosed between two ridges. Reaching the spot where the
gullet opens into the stomach, the grooves are continued along
the upper wall of the stomach and the fluid food is led away from

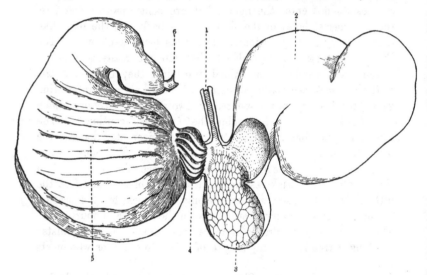

Fig. 321.   Stomach of a Sheep, cut open to show the various chambers.

1. Oesophagus.   2. Rumen.   3. Reticulum.   4. Psalterium.
                    5. Abomasum.   6. Duodenum.

the paunch into the third division of the stomach, the manyplies
or psalterium, which has numerous folds of membrane projecting
into its cavity; by means of these the food is completely filtered
from all the solid matter it contains. It then passes on into the
fourth and last compartment, the abomasum, whose walls are raised
into but a few ridges and which is lined with an epithelium containing numerous gastric glands. This leads into the duodenum
or first part of the intestine, in which the digestion is completed.
The teeth of the Ruminants have no distinct tubercles like those
of the Pig, since these projections have become confluent so as

to form hard curved ridges of enamel; and as the jaws shift on each other sideways, the upper and lower back teeth produce a grinding action just as two millstones do. The name *SELENODONTIA* (Gr. σελήνη, the moon) has been given to the Ruminantia on

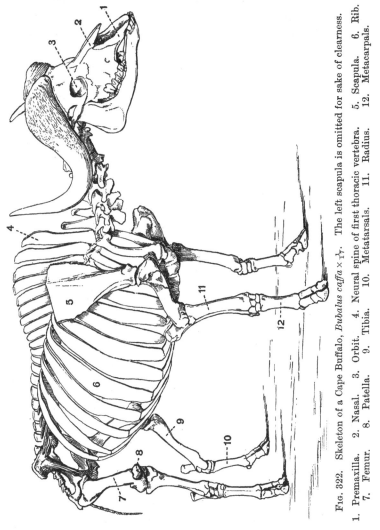

Fig. 322. Skeleton of a Cape Buffalo, *Bubalus caffa* × ¹⁄₁₇. The left scapula is omitted for sake of clearness.

1. Premaxilla.   2. Nasal.   3. Orbit.   4. Neural spine of first thoracic vertebra.   5. Scapula.   6. Rib.
7. Femur.   8. Patella.   9. Tibia.   10. Metatarsals.   11. Radius.   12. Metacarpals.

account of the crescentic ridges on their teeth, which are termed selenodont. It is interesting to note as evidence of the more advanced structure of the Ruminantia as compared with the Suina, that the selenodont teeth always pass through a bunodont stage in their development. The canines in the upper jaw are long in the male *Moschus*, who no doubt uses them in his fight for the possession of the female. The lower canines however are usually placed close to the lower front teeth and are indistinguishable from them. There are—with few exceptions—no front teeth in the upper jaw, and the grass is bitten off by pressing the lower front teeth against a patch of hardened gum.

The feet of the Ruminants are organs beautifully formed for quick motion; the ideal which Nature has so to speak striven to attain being the same in their case as in that of the horse, though she has had to start from a different basis. As in the case of the horse, the end in view has been a firm jointed lever moving only in one plane; but in Ruminantia this has been attained by keeping two fingers and two toes and so to speak glueing them together except in the bones of the hoof. Ruminants, like Horses, walk on the points of their finger- and toe-nails; the metacarpals of the third and fourth digits are fused together, while of the outer fingers and toes only vestiges remain which hardly ever reach the ground, and often do not appear externally. The "cloven hoof" is therefore formed by the nails of two fingers or two toes (Fig. 322).

The families composing the division Ruminantia are the TRAGU-LIDAE or Chevrotains; the CAMELIDAE or Camels (Tylopoda); the CERVIDAE or Deer; the GIRAFFIDAE or Giraffes; the ANTILOCAPRIDAE which has but one species, the Prong-buck, *Antilocapra americana*; and the BOVIDAE or hollow-horned cattle (Cavicornia) including antelopes, goats, sheep and oxen.

Classification.

The TRAGULIDAE comprise some small animals found in Africa and India, in which the foot is intermediate in structure between that of the pigs and that of the higher Ruminants: the outer toes are complete although very slender, and the two inner imperfectly joined with one another. The stomach and teeth however are like those of a Ruminant, except that there is no third compartment or psalterium. The African Chevrotain from the West Coast is larger than its Asiatic allies (Fig. 323). It frequents water-courses and is said to have the habits of a pig.

The Camels are familiar to all as far as their general appearance is concerned. The humps—of which the Arabian camel, *Camelus dromedarius*, has one, and the Bactrian or Asiatic camel, *C. bactrianus*, two—are masses of fat, reserve material on which the animal supports its life when deprived of food. In the foot the main

weight rests on a pad behind the hoofs; these latter are separated
from each other, so that the animal has a broader support than a
cow or a deer. A camel does not walk on its finger- and toe-nails,
but on the last joints of the fingers and toes. The stomach has no
psalterium, but both the rumen and reticulum have a large number
of water-cells, that is deep pouch-like outgrowths in which a quite
undrinkable fluid is stored. It will be noted that all the peculiarities of the structure of the Camel which have just been mentioned
are directly related to the exigencies of a life on arid, sandy wastes.

FIG. 323.　The African Water-Chevrotain, *Dorcatherium aquaticum.*

Thus the diverging toes and leathery pad on the foot enable them
to secure a broader surface of the yielding sand on which to support
the animal's weight: the humps are a provision of food and the
water-cells in the stomach contain a supply of fluid to serve the
animal in its long wanderings from oasis to oasis over the desert.
The Arabian Camel is only known in the domesticated state, but
the Bactrian Camel ranges wild over some of the more inaccessible
regions of Central Asia.

　　It is a remarkable and interesting fact that we find some
members of the Camel tribe in South America. These animals, the

Llama, *Auchenia glama*; the Vicuna, *A. vicugna*; the Alpaca, *A. pacos*; and the Huanaco, *A. huanacos*; live in the Andes. They have no humps but possess long fleeces which are used for making cloth. The skeleton of one of these animals is almost indistinguishable from that of a camel, and they have the same stupid, stubborn ways as their relatives in the Old World. It is curious to see in the stomach the same provision as is found in the camel, although water is, as a rule, plentiful enough where the Llama lives.

The higher Ruminants are divided into two main groups according to the character of their horns. In the CERVIDAE or true Deer the horns are bony outgrowths of the frontal bones. The horns are shed every year and are nearly always branched. They may be termed antlers to distinguish them from the true horns of the Bovidae. The antlers are usually confined to the male, but in the Reindeer, *Rangifer tarandus*, which is called the Caribou in Canada, they also occur in the female. When the antler has attained its full growth the blood supply ceases and the skin peels off. In a rim round the base called the "fur" absorption takes place, so that the greater part is easily detached. In the Cavicornia or BOVIDAE, the core of the horn is an unbranched bony outgrowth into which air spaces continuous with the cavity called the frontal sinus of the skull often extend. This core is permanent and is covered by a hard horny sheath made of compacted hairs. Two small families occupy an intermediate position, these are the GIRAFFIDAE, represented by the Giraffe, *Giraffa camelopardalis*, and the Okapi, *Okapia johnstonii*, and the ANTILOCAPRIDAE represented by the Prongbuck, *Antilocapra americana*. The Giraffe is now confined to the Ethiopian region; it is a conspicuous inmate of zoological gardens, on account of its extraordinarily long neck, in which however there are as usual only seven vertebrae. The Giraffe has two short horns, unbranched and covered throughout with soft fur. The Okapi is a forest giraffe with a comparatively short neck. In the Prongbuck, which is found in the prairies of North America, the horn bears a small lateral branch and is covered with a horny sheath, and this sheath, but not the horn itself, is shed once a year.

In the family BOVIDAE are included everything from an Antelope to an Ox, and, strange as it may appear, we have practically a complete series of links filling up the gap between the graceful light-limbed Gazelle and the thick-necked Buffalo, so that we cannot say very precisely where Antelopes end and Oxen begin. The Musk-ox, *Ovibos moschatus* (Fig. 324), which ranges over the

Arctic wastes of Canada in large herds, is intermediate in some respects between the Sheep and the Goats on the one hand and the Oxen on the other, but is more closely allied to the former series.

At present in Britain there are but two indigenous species of deer found wild, the Red-deer of Scotland, *Cervus elaphus*, and the Roe-deer, *Capreolus capraea*; the Fallow-deer, *C. dama*, is probably an introduced species, and at present is only represented in Britain by semi-domesticated animals. In Roman times there were wild Oxen, and some suppose that a breed of wild Oxen kept at Chillingham in Northumberland and in one or two other large parks are descended from these ancestors.

Fig. 324.   The Musk-Ox, *Ovibos moschatus*.

In Canada a large variety of the Red-deer, the Wapiti, *Cervus canadensis*, is found, also the Reindeer or Caribou, *Rangifer tarandus*, and the Elk or Moose, *Alces machlis*, with short bull-like neck and broad fan-like horns. Throughout the whole of Eastern America the so-called "red-deer," *Cariacus virginianus*, is found in the mountains. The Bovidae are represented by the Musk-ox, *Ovibos moschatus*, with horns curved like a ram, and by the Rocky Mountain Goat, *Haploceros montanus*. Until recently the American Bison, the so-called "Buffalo," *Bison americanus*, ranged in enormous herds over the Western plains of North America; but before 1883,—with the exception of a few scattered stragglers which are "protected,"—this magnificent animal had been exterminated.

Order VII. Rodentia.

The Rodentia or Gnawers (Lat. *rodo*, to gnaw) are another of the
main divisions of the Mammalia and include our rabbits, hares,
squirrels, rats and mice, besides the porcupine, beaver, guinea-pig
and many other foreign species. These are all sharply marked off
from other mammals by the structure of their teeth. The incisors,
of which there are typically only one pair in each jaw, are chisel-

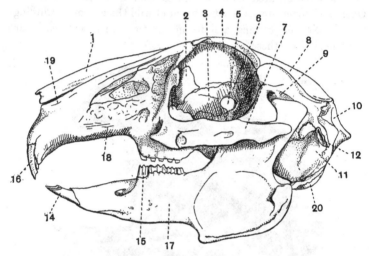

Fig. 325. Side view of the Skull of the Rabbit, *Lepus cuniculus*.

1. Nasal bone.   2. Lachrymal bone.   3. Orbito-sphenoid.   4. Frontal.
5. Optic foramen.   6. Orbital groove for ophthalmic division of trigeminal
nerve.   7. Zygomatic process of squamosal.   8. Parietal.   9. Squamosal.
10. Supra-occipital.   11. Tympanic bone.   12. External auditory
meatus.   14. Lower incisor.   15. Anterior premolar tooth.   16. Anterior
upper incisor.   17. Mandible.   18. Maxilla.   19. Premaxilla.
20. Occipital condyle.

shaped and covered with hard enamel on their outer sides only.
They constantly grow and are only kept down to proper size by
continual gnawing and rubbing against each other (Fig. 325). If
one of the teeth is destroyed the opposite one grows until it may
pierce the other jaw, prevent the mouth from being opened, and thus
starve the animal to death. There are no canines, so that there
is a great space or diastema between the front teeth and back
teeth. The claws are always blunt and nail-like, and walking is
done on the last joints of fingers and toes, not as in the case of the

Ungulates on the points of the nails (Fig. 295). Our English Rodents are the Hares and Rabbits, LEPORIDAE ; the Squirrels, SCIURIDAE ; the Voles, Rats and Mice, MURIDAE; and the Dormouse, the sole British representative of the family MYOXIDAE. In North America there are allied species, and in addition the Ground-squirrels, or Chipmunks, *Tamias*, and three species of Woodchuck or Marmot, *Arctomys*; also Porcupines, represented by the common Canadian Porcupine, *Erethizon dorsatus*, the Beaver, *Castor canadensis*, and many others. *Hystrix cristata* is the Porcupine of Southern Europe and Northern Africa. The guinea-pig is probably a domesticated variety of the South American species *Cavia cutleri*. These various species are very like each other in their general anatomy, but differing in the character of their molars, in their fur and in their tails.

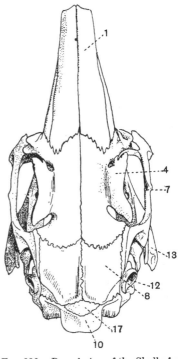

FIG. 326. Dorsal view of the Skull of a Rabbit, *Lepus cuniculus*.

1. Nasal bone. 4. Frontal. 7. Process of squamosal supporting the jugal. 8. Parietal. 10. Supra-occipital. 12. External auditory meatus. 13. Angle of lower jaw. 17. Interparietal.

Hares and Rabbits constitute the *DUPLICIDENTATA*, one of the two Sub-orders into which the Order is divided. The name is derived from the possession of an extra pair of upper incisors, which however are so small as to be useless. The tail is short and the cusps of the premolar and molar are joined so as to form ridges or folds running across the tooth. The Common Hare, *Lepus timidus*, and the Mountain Hare, *L. variabilis*, are both British; they have longer legs than the third British form, the Rabbit, *L. cuniculus*, and have fewer young at a time. In the temperate part of North America there are at least six species of Duplicidentata all referable to the genus *Lepus*. Of these the most interesting are *Lepus americanus* and *Lepus campestris*. The fur of both these species turns white at the tips in winter, enabling the animals to escape observation on the snow-covered ground.

L. americanus, the Northern Hare, is abundant in New England and Eastern Canada : its summer fur has a cinnamon colour. *L. campestris* is the famous "Jack-Rabbit" of the western prairies, which has fur of a yellowish-grey colour in summer. It can run with great swiftness.

The remaining Rodentia are called *SIMPLICIDENTATA*, and possess only two incisors above, one on each side.

The Squirrels, SCIURIDAE, are distinguished by their bushy tail, their large hind limbs and the fact that the cusps on their back teeth are distinct. *Sciurus vulgaris* is the common British Squirrel; it extends from Ireland to Japan. Two species are very common in Canada and New England, viz., *Sciurus hudsonicus*, the Red Squirrel, and *S. carolinensis*, the Grey Squirrel. These lively little animals can be seen in autumn disporting themselves in the trees lining the avenues of the suburbs of Montreal. *Sciuropterus volans* is the Flying Squirrel; this animal is provided with a furry expansion of the skin of its sides joining the elbow and knee. This expansion forms a parachute-like membrane which supports it in its great leaps from tree to tree. In these manœuvres it is assisted by the broad flattened tail. The Flying Squirrel is common in the temperate part of the United States. A similar but larger species (*S. sabrinus*) may be seen at dusk leaping from tree to tree on the Mountain of Montreal. *Anomalurus*, found in West and Central Africa, is also a Flying Squirrel, the skin of whose sides is prolonged into a parachute-like membrane (Fig. 327). It differs from *Sciuropterus* however in having a round tail provided with horny scales underneath, which assist in climbing, and in having its "parachute" supported by a cartilaginous rod arising from the elbow.

The Mice and Rats, MURIDAE, have naked tails with scales underneath. The ordinary Rat is the brown Norway Rat, *Mus decumanus*, which was introduced some time ago into England and has almost everywhere driven out the old English Black Rat, *M. rattus*. The Common British Mouse is *M. musculus* : the Wood-mouse, *M. sylvaticus*, and the Field-mouse, *M. minutus*, also occur in Britain. The Water-rat or Vole, *Arvicola*, is distinguished from the true Rat by the fact that the cusps on its back teeth, instead of being rounded as in the true Rat, are angular. *A. amphibius*, the Water-vole, *A. agrestis*, the Field-vole, which often does much damage to crops, and *A. glareolus*, the Bank-vole, represent the genus in Britain. The Dormouse, *Muscardinus avellanarius*, which like the Squirrel passes the winter in a hole in a tree, has a long bushy tail, and, in outward appearance at any rate more resembles a tiny squirrel than a rat. In its skull it resembles the MURIDAE, but it differs from both Squirrels and Rats in not

FIG. 327.   The African Flying Squirrel, *Anomalurus fulgens*.

possessing a caecum on the intestine. On this account it, along with five or six allied species from Europe and Africa, has been separated as a distinct family, the MYOXIDAE.

Amongst the most interesting American Rodents are the Beaver, the Porcupine, the Ground-squirrel, the Marmots and the Musquash. The Beaver, *Castor canadensis*, has a broad flat tail, suited for swimming, which is covered with horny scales. The Beaver, by means of its sharp incisors, cuts down trees growing on the banks of streams, so that they fall across streams thus damming them up and raising the level of the water so as to cover the entrance to their burrows. By this means large tracts of country have been converted into swamp. The Porcupines (HYSTRICIDAE) have some of their hairs developed into sharp spines which make them awkward objects to

handle. In the Canadian Porcupine, *Erethizon dorsatus*, the spines
are concealed by the fur. The commonest Ground-squirrel of
North America is the Chipmunk, *Tamias*, an active little animal
with large eyes and a short hairy tail. The Prairie Marmot,
Cynomys, the so-called Prairie-dog, lives in communities, burrowing
in the ground. Its home is often shared by a small burrowing
Owl, *Athene cunicularia*, and by a Rattlesnake, which probably
eats the young Marmots. The Musquash or Musk-Rat, *Fiber*

FIG. 328.   The Musquash, *Fiber zibethicus.*

zibethicus, one of the MURIDAE, is peculiar to North America,
and very widely distributed in suitable places (Fig. 328). It is
aquatic, living on roots and water-plants and is most active at night.
It constructs burrows in the banks of streams, the openings of which
are under water. Its fur is valuable.

Order VIII. Cheiroptera.

The Cheiroptera (Gr. χείρ, a hand ; πτερόν, wing), or Bats, have
not in their general organization, in teeth or brain or stomach,
departed far from the Insectivora ; their great distinguishing
feature is the modification of the arm into a wing. As in Birds, the
fore-arm is bent up on the upper arm, the wrist bent down on the
fore-arm; but unlike Birds' wings the flying membrane is of skin, the

greater part of which is stretched between the fingers of the five-fingered hand, only the smaller part extending, as in Birds, between the elbow and the side of the body.   The hand is enormous, the little finger being, as a rule, very greatly developed and as long as the rest, while the thumb alone is small and is not included in the

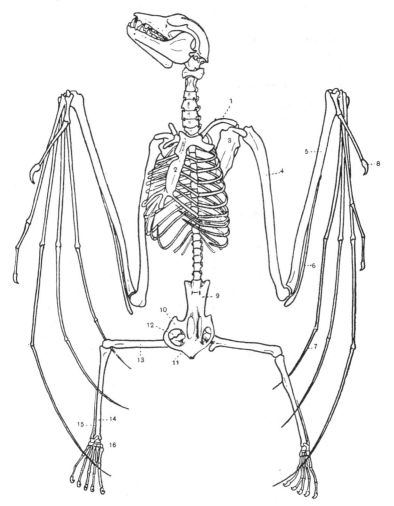

Fɪɢ. 329.   Skeleton of *Pteropus medius*, a fruit-eating Bat × about ½.

1. Clavicle.   2. Keeled sternum.   3. Scapula.   4. Humerus.   5. Radius.
6. Ulna.   7. Little finger.   8. Thumb.   9. Ilium.   10. Pubis.
11. Ischium.   12. Obturator foramen.   13. Femur.   14. Tibia.
15. Fibula.   16. Tarsus.

membrane but ends in a hook-like nail (Fig. 329). Part of the
membrane extends down the thighs, and in some even the tail is
involved. The knees are turned outwards and backwards, a most
extraordinary position which would mean dislocation if the hip-joint
of any other mammal were forced into it but which is rendered
possible in Bats owing to the fact that in them the pubes are not
directed inwards so as to meet one another in a symphysis but slope
outwards and are consequently widely separated from one another
(Fig. 329). When the Bat crawls it hooks itself along with its
thumb-nail and pushes itself awkwardly with its hind feet. It has
a most awkward gait and the animal is consequently very helpless
when not flying.

One of the most extraordinary things about Bats is the
development of sensitive patches of skin on the face for the
purpose of perceiving faint disturbances in the air. It has been
shown that the eyes of Bats, although apparently normal, are
really degenerate, that in fact the layer of visual rods in the retina,
which is the special organ of light-perception, is most imperfectly
developed. To compensate for this we find, in some species, the
outer ear, in others the skin around the nostrils, in others
again the skin on the lips and chin, developed into curious outgrowths richly supplied with nerves. By means of these sense-
organs Bats are enabled to avoid obstacles, and a blind Bat, let
loose in a room across which numerous strings have been stretched,
will fly about without touching one. Owing to their powers of
flight Bats are exceedingly widely distributed and extend to small
oceanic islands where there are no other mammals. The true blood-
sucking Bat or Vampire, *Desmodus rufus*, is found in Central and
in South America. Its back teeth are rudimentary, but its front
teeth are razor-edged. *Pteropus,* the so-called Flying Foxes or
Fox-bats, of India and Madagascar, belonging to the family PTERO-
PIDAE, are the largest Bats known; they feed exclusively on fruit, and
the cusps on their teeth are blunter than is usual amongst Bats.
The African *Xantharpyia,* one species of which frequents the
interior of the Pyramids and other dark ruins in Egypt (Fig. 330)
belongs to the same family. In Great Britain there are some fifteen
species of Bat divided amongst five genera. Of these the Long-eared
Bat, *Plecotus auritus* ; the Whiskered Bat, *Vespertilio mystacinus* ;
the Horse-shoe Bats, *Rhinolophus hipposiderus* and *R. ferrumequi-
num* ; the Barbastelle, *Synotus barbastellus* ; and the Pipistrelle,
Vesperugo pipistrellus ; represent the genera. Besides the species

just mentioned there are three more species of *Vesperugo* and three more of *Vespertilio* in Britain. South America has a large fauna of peculiar Bats, but the North American forms are allied to the

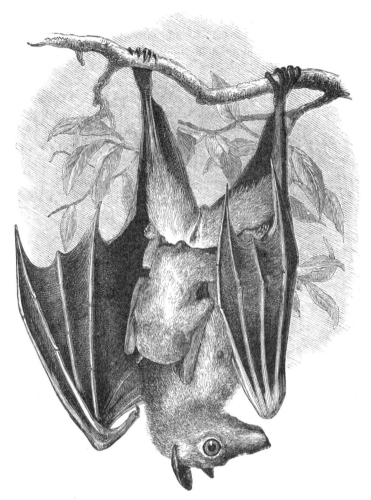

Fig. 330. Female and young of *Xantharpyia collaris*. From Sclater.

British, although the species are distinct. The Serotine, *Vesperugo serotinus*, is the only species common to the two regions. *Scotophilus humeralis* is one of the most familiar species peculiar to North America.

Order IX. Primates.

The last Order of the Mammalia is that of the Primates, which include Lemurs, Monkeys and Man. As was mentioned before, this order is characteristically arboreal, that is to say they live among trees, climbing from branch to branch. This circumstance may explain why they retain certain primitive characteristics found else-where only amongst the Insectivora. Thus the thigh and upper arm are quite free from the body and the whole sole and palm are placed on the ground when walking ; and there are five fingers and five toes. On the other hand the eyes are pushed round to the front of the skull instead of being placed at the sides of the head, and the jugal joins the post-orbital process of the frontal, so that

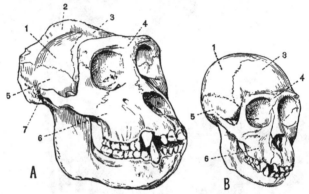

Fig. 331.   Half front view of the skulls, A, of an old, B, of a young Gorilla,
                    Gorilla savagei × ¼.

1.   Parietal.   2.   Sagittal crest.   3.   Frontal.   4.   Supra-orbital ridge.
      5.   Squamosal.   6.   Maxilla.   7.   External auditory meatus.

the orbit is surrounded by a bony ring (Fig. 331). Some at least of the toes have flat nails. The big toe is shorter than the rest, and, except in Man, can be separated from them so as to be used for grasping. In most but not in all Monkeys the thumb can be used similarly, so that monkeys are said to have four hands. There are two large mammae or nipples situated on the breast. Other mammae when present are vestigial and situated behind the func-tional ones.

There are two great divisions of the Primates, the *LEMUROIDEA* and the *ANTHROPOIDEA*. The first of these includes some curious little animals, of which the majority are found in the Island of

Madagascar, the rest in Africa, India and the Malay Archipelago. Many of the species are nocturnal, move silently and have large eyes, whence the name Lemur (Lat. *lemures,* goblins, spectres). These animals have heads recalling those of rats, with no suggestion of the human face, and in their brains and some other points they are far below the Monkeys. The cerebral hemispheres do not cover the cerebellum; the placenta of the embryo is spread evenly all over the surface of the uterus, and there are occasionally additional

FIG. 332.   The Ring-tailed Lemur, *Lemur catta.*

mammae on the abdomen. Their incisor teeth are separated in the middle line, but, as in all Primates, there are never more than two on each side. The Ring-tailed Lemur, *Lemur catta* (Fig. 332), is said to be an exception to the rule that the group is arboreal and to live amongst rocks and bushes, but other authorities say that it lives in troops amongst the forests of Madagascar. It is a gentle, graceful creature with a plaintive cry.

MAMMALIA.

[CHAP.

The *ANTHROPOIDEA*, including the true monkeys and man, are distinguished by the fact that the bony ring surrounding the orbit sends inwards a plate of bone, which completely separates the orbit from the temporal fossa. Further, the cerebral hemispheres conceal the cerebellum when the brain is viewed from above; the placenta is highly developed and concentrated on one part of the wall of the uterus, and there are never more than two mammae. This sub-order of Primates is divided into five families, viz., HAPALIDAE, CEBIDAE, CERCOPITHECIDAE, SIMIIDAE, and HOMINIDAE, the last being constituted of the single species, *Homo sapiens*, man.

The HAPALIDAE and CEBIDAE are confined to South and Central America, and are sometimes grouped together as *PLATYRRHINI* (Gr. πλατύς, broad ; ῥίς, ῥινός, nose). The animals belonging to this section have a broad internasal septum and three pairs of premolar teeth. The tympanic bone is without a tube-like prolongation. The HAPALIDAE or Marmosets are small, furry animals inhabiting the forests of Brazil and Columbia; they have the least ape-like feet of any of the Anthropoidea. The great toe is small and it alone has a flat nail ; all the other toes and all the fingers bear curved claws. There are only two pairs of molars. The CEBIDAE have flat nails on their fingers and toes and three pairs of molars, making with the premolars six cheek-teeth on each side of each jaw, the largest number found amongst Anthropoidea. The Cebidae have prehensile tails which assist them in climbing. The genus *Ateles* includes the Spider-monkeys, in which this function of the tail is prominent, the under side of this organ being naked and scaly so as to allow the animal to obtain a hold. The genus *Cebus* has the tail hairy all round ; several species of this genus are often seen in captivity.

The CERCOPITHECIDAE and SIMIIDAE are confined to the Old World. They constitute the section *CATARRHINI*, characterised by the possession of a narrow internasal septum, a spout-like prolongation of the tympanic bone extending into the base of the ear-flap, and the reduction of the number of premolar teeth to two pairs, whilst there are always three pairs of molars. The CER-COPITHECIDAE have the legs as long as the arms, or longer, and go habitually on all-fours. There are always bare patches of thick callous skin on the buttocks forming the so-called ischial callosities, on which the animals rest when they assume a sitting posture, and there is in almost every case a well-developed tail. This family include the Indian and African monkeys, among them the Bandar-log of Kipling's Jungle Tales. One species, *Macacus inuus*, the Barbary ape, is found on the Rock of Gibraltar, and this is the only species which enters Europe. It is remarkable for being completely tailless. *Semnopithecus entellus* is the sacred Langur of India, and owing to its immunity from persecution has become very abundant.

Fig. 333.   The Orang-utan, *Simia satyrus*, sitting in its nest. From a specimen in the Cambridge Museum.

The SIMIIDAE include those monkeys which in structure and appearance most resemble man. In this family the tail is completely absent, the arms are longer than the legs, and the gait might be described as that of a baby learning to walk. They never go completely on all-fours, but usually shuffle along unsteadily on

their two feet, which like those of a baby show a tendency to
turn inwards under them ; they usually steady themselves by
bending forward so that their knuckles touch the ground.   Four
genera are included in this section, viz. *Hylobates, Simia, Gorilla*
and *Anthropopithecus.*   *Hylobates* include several species known
as Gibbons, inhabiting South-Eastern Asia and the Malay Archipelago.   These are apes with exceedingly long arms ; they assume
a completely upright position when on the ground and run along
holding up their long arms in the air as if they were balancing
poles.   In their power of supporting themselves without any help
from the arms they approach man ; but in other respects they
depart widely from him, as for instance in the brain, where the
cerebellum is not completely covered by the cerebrum.   *Simia* is
represented by a single species, *S. satyrus,* the Orang-utan, a large
animal about 4½ feet high, which is found in the islands of Borneo
and Sumatra.   This animal walks on two feet supporting itself on
its knuckles.   It lives however almost entirely in trees, constructing
a sort of nest for itself out of branches (Fig. 333).   It is remarkable
for its high rounded cranium enclosing the large brain, which
presents the closest approximation to the human brain of all the
brains of apes.   The cranium is however still overshadowed by the
bones of the face and lower jaw.   There exists only a single species
of *Gorilla,* viz., *G. savagei,* confined to a limited region of Equatorial
Africa.   This is the largest of all the apes, reaching a height of 5½
feet.   It is distinguished from *Simia* by its shorter arms and more
receding forehead.   The skull of the young Gorilla strikingly resembles a child's skull, but in the adult it is deformed by the
development of great bony ridges which give attachment to the
muscles of the face (Fig. 331).   *Anthropopithecus* is represented
only by *A. troglodytes,* the Chimpanzee, which lives in Western
Africa in the same region as the Gorilla but has a wider distribution.
It is distinguished by its shorter arms, which do not reach below
the knee, and by its smoother and rounder skull.   It does not
reach a height of more than 5 feet and is on the whole the most
Man-like of all the Simiidae, though each of the other species of
the family approaches more closely to the human standard in some
particular feature.

Man is distinguished above all by the great size of the brain,
which is double the size of that of the highest monkey, and by the
modification of the leg so as entirely to support the body, in
consequence of which the big toe is no longer used for grasping.

Some hold that it was this latter modification which brought about the great development of the intelligence of Man, arguing that when once the hand was entirely at the service of the brain the varied uses to which it could be put would give the opportunity for the use of the mind. This seems probable, but the great factor which has stimulated the mental development of Man is his habit of living together in societies and undertaking concerted enterprises for the benefit of the community. To this power of combination not only intellect but also language and morals may eventually be traced back. Man did not make society, it was society that made Man.

Men are divided into three great races which are as distinct from one another as are many groups of allied species amongst other animals. All races of Men are however mutually fertile but a mixed race shows a tendency to revert to one or other of the parent races—this is true at any rate of mixtures which have taken place within the historical period. Many authorities explain the peculiarities of some island populations on the assumption that they are due to an early crossing of two of the principal races at a time when perhaps their leading features were less fixed than they are now. The three races alluded to are characterized as follows :—

(i) **Ulotrichi.** Woolly hair characterized by numerous close, often interlocking, spirals, 1—9 mm. in diameter. The hair of the head is usually long in the Melanesians and very short in the Negritos and Bushmen. It is almost invariably black. Ellipsoidal in transverse section.

Ex. Bushmen, Negrillos, Negritos, Negros and Bantus, Papuans and Melanesians.

(ii) **Cymotrichi.** Wavy hair ; undulating or it may form a curve or imperfect spiral from one end to the other (as in the Indonesians), sometimes the extremity forms long curls (as in some Europeans and Todas), in another variety the hair is rolled spirally to form clustering rings or curls a centimetre or more in diameter (as in Australians, some Dravidians and Ethiopians). The hair of non-European peoples is generally black, with often a brownish or reddish tinge. Oval in transverse section.

Ex. Dravidians, Australians, Ethiopians, Semites, Mediterraneans, Europeans, Indo-Afghans, Indonesians, Polynesians, Armenians.

(iii) **Leiotrichi.** Straight or smooth hair. Lank hair that usually falls straight down, occasionally with a tendency to become wavy (as in the Finns, some Amerinds). It is almost invariably black. Circular in transverse section.

Ex. Lapps, Ugrians, Turko-Tartars, Northern and Southern Mongols, Northern, Central and Southern Amerinds, Patagonians, Eskimo.

The fossil representatives of the class Mammalia are exceedingly numerous. It would lead us too far to give even such a general account of them as was given of fossil Reptilia, but a few hints as to the light thrown by them on the ancestry of existing groups may be given here. Mammalia seem to have been derived from the early Reptilia of the Sandstones overlying the Coal Measures. One group of these, the Theromorpha, in showing the division of the teeth into three kinds, and in the envelopment of the quadrate by the squamosal, might almost be regarded as the direct ancestors. Unfortunately the succeeding rocks are mostly of marine origin, and in them few and fragmentary remains of Mammalia are preserved. Some of these show small molars covered with many cusps similar to the teeth of *Ornithorhynchus* and these teeth are classified as the remains of an Order *Multituberculata*, the members of which are supposed, like *Ornithorhyncus*, to have had a reptilian arrangement of the genital organs. The remains are principally lower jaws, but in one case a scapula with a facette for a coracoid and an interclavicle have been found which bear out the conclusions founded on the jaws. At the same time other jaws have been found which show teeth of a different kind. These have molar teeth of the tritubercular pattern, but the angle of the jaw is inflected and these have been referred to the Metatheria. As however the latter group owe some of their peculiarities to degeneracy it would be better to regard these jaws as remains of the direct forerunners of Eutheria from which the Metatheria represent a side line.

When we come to the sands and clays lying above the Chalk which constitute the Tertiary "rocks," we find in many localities a rich assemblage of remains of undoubted Mammalia of the Eutherian type. The oldest horizon shows remains of animals called Condylarthra and Creodonta. Both groups are small plantigrade animals, with 44 teeth, but in the first group the cusps of the tritubercular molars are blunt, and in the second sharp and pointed. In this small distinction the beginning of the cleft which widens into the chasm now separating Ungulata and Carnivora is seen. Modern Insectivora are the little modified descendants of the Creodonta. In the next horizon traces of the Primates appear as Lemuroidea, the marks discriminating them from Creodonta being the enlargement of the orbit and its surrounding leg bone, while the molar teeth have a fourth tubercle. At the same time the Condylarthra show horse-like forms (*Phenacodus*), still with five fingers and five toes and of the sub-ungulate type, but true Ungulata now appear with the bones of wrist and ankle in transverse rows and reduced number of toes. The earliest of these, the Lophiodontidae, were Perissodactyla, and in the shape of the face some recall the horse, others the rhinoceros, though the limbs were like those of tapirs. The cusps on the teeth were four in number, and were commencing to coalesce into ridges. Rodentia also make their appearance as Tillodontia, animals with one pair of large incisors in each jaw, but with the other incisors and the canines present, easily derivable from the Creodonta. Passing further on, the origin of the Artiodactyla becomes apparent in the next horizon, a host of small pig-like animals making their appearance which in higher formations gradually differentiate themselves into the families of Artiodactyla. The ancestors of the South American Edentata, which at the previous horizon were not separable from Creodonta except by the fact that the tritubercular molars lost their enamel late in life, become at this period distinguished by the restriction of the enamel to bands and the reduction of the incisors. Still higher in the series bats (Cheiroptera) make their appearance, little different from what they are at present.

In the horizon above this the ancestors of whales are found, as the Archaeoceti with well-developed nasal bones, the nostrils placed about the middle of the snout, and with double-rooted serrated molar teeth, derivable from the tritubercular type by the development of additional cusps, all like the original three being in the same line. True Carnivora distinguished by the carnassials have likewise been by this time developed from the Creodonta ; in the Ungulata the earliest forms of Camels and of Tragulidae have appeared.

Once formed Carnivora rapidly become differentiated, for in the next period Felidae and Viverridae had already appeared, and contemporaneously with them the first deer (Protoceratidae) and the earliest Sirenia with visible hind-limbs (*Halitherium*). Still higher the Elephants (Proboscideae) appear represented at first by forms with both lower and upper tusks or even lower alone

(*Mastodon* and *Dinotherium*). At the same time the deer first appear with antlers and the rhinoceros acquires a horn, and the family of bears (Ursidae) is commencing to be distinct from the primitive dog-like Carnivora, the gradual reduction in size of the premolars, and the carnassial marking the change. True Apes (Anthropoidea) here succeed the Lemuroidea.

In the next period the Giraffe (*Samotherium*), *Hyrax* and *Orycteropus* appear and so practically the whole group of Mammalia has made its appearance, the remaining changes consisting chiefly in the extinction of many forms either completely, or partially, so that their representatives are now restricted to limited areas. It will be noted how completely the geological evidence bears out the idea of the central position of the group Insectivora among Mammalia.

The Class Mammalia is divided as follows :

Sub-class 1. PROTOTHERIA.

Mammalia which lay large eggs and in which the two oviducts are completely separated, and there is a persistent cloaca. No placenta.

Ex. *Ornithorhynchus, Echidna.*

Sub-class 2. METATHERIA.

Mammalia in which the young are born in a most imperfect condition, and are carried by the mother in a pouch on the abdomen. The oviducts are differentiated into vagina, uterus and Fallopian tube, the two vaginae partially united. The cloaca is divided into an anus and a urinogenital aperture. An allantoic placenta may or may not be developed but when present is more or less vestigial. In all cases there is an adhesion between the yolk-sac of the embryo and the uterus.

Order I. Polyprotodontia.

Metatheria with four or five incisors on each side of the upper jaw and with at least three pairs of incisors of approximately equal size in the lower jaw.

Family (1) DIDELPHYIDAE.

Polyprotodontia with a large opposable great toe, the

other digits of the hind-foot being subequal in size. American.

Ex. *Didelphys.*

Family (2) DASYURIDAE.

Polyprotodontia with a rudimentary great toe, the other digits of the hind-foot subequal in size. Australian.

Ex. *Thylacinus.*

Family (3) PERAMELIDAE.

Polyprotodontia with a rudimentary great toe, the other digits of the hind-foot united by a web of skin, the second and third being excessively slender: the muzzle long and pointed. Australian.

Ex. *Perameles.*

Family (4) NOTORYCTIDAE.

Polyprotodontia with rudimentary eyes, an enlarged manus and burrowing habits. Australian.

Ex. *Notoryctes.*

Order II. **Diprotodontia.**

Metatheria with not more than three incisors on each side of the upper jaw, and with, as a rule, one pair of large chisel-shaped incisors in the lower jaw, the other lower incisors being vestigial or absent.

Family (1) EPANORTHIDAE.

Diprotodontia with all the toes of the hind-foot free from one another and subequal. American.

Ex. *Coenolestes.*

Family (2) PHASCOLOMYIDAE.

Diprotodontia with the toes of the hind-foot united by a web of skin: only one pair of chisel-shaped incisors in upper jaw: limbs subequal. Australian.

Ex. *Phascolomys.*

Family (3) PHALANGERIDAE.

Diprotodontia in which the toes of the hind-foot are united by a web of skin, the great toe being well-developed,

38—2

free from the web, and opposable to the rest: limbs sub-equal: three incisors on each side of the upper jaw. Australian and Papuan.

Ex. *Phalanger.*

Family (4) MACROPODIDAE.

Diprotodontia in which the toes of the hind-foot are united in a web of skin and the great toe is rudimentary: the fore-limbs very short and suited only for grasping: three incisors on each side of the upper jaw. Australian and Papuan.

Ex. *Macropus, Bettongia, Petrogale.*

Sub-class 3. EUTHERIA.

Mammalia in which the young are born able to suck and in which there is no pouch. The two vaginae are always completely confluent. The cloaca is divided into an anus and a urino-genital aperture. An allantoic placenta always present and greatly developed.

Order I. Edentata.

Eutheria devoid of enamel on the teeth and without median teeth; the limbs are, as a rule, provided with heavy hook-like claws: uterus simple and globular: placenta dome-shaped.

Family (1) BRADYPODIDAE.

Limbs long and the fore-limbs greatly longer than the hind-limbs: face short: arboreal in habit. South American.

Ex. *Bradypus.*

Family (2) MYRMECOPHAGIDAE.

Limbs short and stout: muzzle exceedingly long: no teeth. South American.

Ex. *Myrmecophaga.*

Family (3) DASYPODIDAE.

Limbs short and stout: muzzle long with numerous teeth: a shield of dermal bones covered by horny scales. South American.

Ex. *Dasypus.*

Family (4) MANIDAE.

Covered externally with large, overlapping horny scales: no teeth: long protractile tongue. Asian and African.

Ex. *Manis.*

Family (5) ORYCTEROPODIDAE.

Covering of bristly hairs: teeth numerous and heterodont: no thumb on anterior limb: femur with a third trochanter. African.

Ex. *Orycteropus.*

Order II. Effodientia.

Eutheria resembling Edentata in teeth and claws but with bicornuate uterus and zonary or diffused placenta.

Order III. Cetacea.

Large aquatic Eutheria which have lost the hind-limbs and have developed horizontal flukes on the tail. The fore-limb is a paddle: the cranium is globular and the teats are posterior.

Sub-order 1. Mystacoceti.

Cetacea devoid of teeth in the adult and with plates of whalebone in the mouth.

Ex. *Balaena, Balaenoptera.*

Sub-order 2. Odontoceti.

Cetacea with teeth at any rate on the lower jaw and no whalebone.

Ex. *Physeter, Globicephalus, Delphinapterus, Phocaena.*

Order IV. Sirenia.

Aquatic Eutheria, with limbs and tail as in the Cetacea: the cranium is cylindrical and the teats pectoral.

Ex. *Manatus, Halicore.*

Order V. Insectivora.

Small plantigrade Eutheria, with pointed cusps on the molar teeth: the brain of low type: a flexible snout often present. The more familiar families are

Family (1) ERINACEIDAE.

Insectivora with the body covered with harsh spines: limbs subequal.

Ex. *Erinaceus.*

Family (2) SORICIDAE.

Small mouse-like Insectivora with soft fur.

Ex. *Sorex, Blarina.*

Family (3) TALPIDAE.

Mouse-like Insectivora with rudimentary eyes and large hands adapted to burrowing.

Ex. *Talpa, Condylura, Myogale.*

Order VI. **Carnivora.**

Eutheria with sharp recurved claws and powerful canine teeth : the premolars adapted for clipping flesh : the incisors small.

Sub-order 1. **Fissipedia.**

Carnivora with separated digits : a distinct carnassial tooth and one or more broad molars.

Family (1) FELIDAE.

Fissipedia with short face and a reduced number of premolar and molar teeth : with retractile claws.

Ex. *Felis.*

Family (2) CANIDAE.

Fissipedia with long face and full number of premolar teeth : claws non-retractile.

Ex. *Canis.*

Family (3) URSIDAE.

Fissipedia with long face : teeth blunt and partially adapted for a vegetable diet : plantigrade in gait.

Ex. *Ursus.*

Family (4) PROCYONIDAE.

Fissipedia with a sharp pointed muzzle and reduced number of teeth, otherwise like *Ursus.*

Ex. *Procyon.*

Family (5) MUSTELIDAE.

Fissipedia with long necks and exceedingly flexible bodies : a reduced number of teeth : in the skull and in the shape of the carnassial tooth they resemble Ursidae but they are digitigrade in gait.

Ex. *Lutra, Meles, Mustelus, Mephitis.*

Sub-order 2. **Pinnipedia.**

Aquatic Carnivora with the toes united by a web of skin: the tail is rudimentary, but the two hind-limbs are turned backwards and closely apposed so as to form a paddle: no distinct carnassial tooth and no broad molars.

Family (1) OTARIIDAE.

Pinnipedia still retaining a trace of the external ear, and capable of turning the hinder-limbs forward so as to walk on land.

Ex. *Otaria.*

Family (2) TRICHECHIDAE.

Pinnipedia devoid of external ear, but capable of walking on land: the upper canines form long tusks.

Ex. *Trichechus.*

Family (3) PHOCIDAE.

Pinnipedia devoid of external ear, and incapable of turning the feet forward, so that when on land they can only wriggle along with the help of their anterior limbs: the canines not specially enlarged.

Ex. *Phoca.*

Order VII. **Ungulata.**

Eutheria with limbs adapted entirely for progression, the terminal phalanx of each functional digit is enclosed in a short blunt nail.

Sub-order 1. **Sub-ungulata.**

Ungulata with short sub-equal toes, and with the bones of the carpus and tarsus arranged in parallel longitudinal series.

Family (1) HYRACIDAE.

Small Sub-ungulata with a very short snout: a pair of chisel-like incisors in each jaw.

Ex. *Hyrax (Procavia).*

Family (2) PROBOSCIDEAE.

Large Sub-ungulata with a very long flexible snout (trunk) used for prehension : incisors long and curved, forming tusks : molars very broad, only one pair in use at a time.

Ex. *Elephas.*

Sub-order 2. **Ungulata vera.**

Ungulata in which the bones of the carpus and tarsus are arranged in transverse rows, the members of successive rows alternating with one another. The first digit is lost.

Division I.   PERISSODACTYLA.

Ungulata in which there is, with rare exceptions, an uneven number of digits in each limb, and in which the axis of symmetry passes through the third digit.

Family (1) TAPIRIDAE.

Perissodactyla with four digits in the fore-limb and three in the hind-limb : a short flexible snout.

Ex. *Tapirus.*

Family (2) RHINOCEROTIDAE.

Perissodactyla with three subequal digits in each limb : one or two median horns without bony cores carried on the nasal bones.

Ex. *Rhinoceros.*

Family (3) EQUIDAE.

Perissodactyla with only one complete digit in both fore- and hind-limbs.

Ex. *Equus.*

DIVISION II.   ARTIODACTYLA.

Ungulata in which there is almost always an even number of digits, and in which the axis of symmetry passes between the third and fourth digits, these digits being flattened against each other so as to form two symmetrical halves of a cylinder.

Section A. **Bunodontia.**

Artiodactyla with comparatively simple stomachs : the cusps on the molar teeth are separate.

Family (1) HIPPOPOTAMIDAE.

Large Bunodontia with four subequal toes in both fore- and hind-limbs.

Ex. *Hippopotamus.*

Family (2) SUIDAE.

Bunodontia of moderate size, in which the two outer toes though complete are shorter than the others.

Ex. *Sus, Babirusa, Dicotyles.*

Section B. **Selenodontia.**

Artiodactyla with complex stomachs adapted for ruminating : the cusps on the molars coalesce so as to form crescents.

Family (1) TRAGULIDAE.

Small Selenodontia without horns, and with only three compartments in the stomach : the outer toes although excessively slender are still complete.

Ex. *Tragulus.*

Family (2) CAMELIDAE.

Selenodontia without horns, with only three compartments in the stomach : the outer toes entirely absent, the inner toes slightly diverging below, the weight resting on a pad behind them.

Ex. *Camelus, Auchenia.*

Family (3) CERVIDAE.

Selenodontia with antlers in the form of bony outgrowths of the frontal bone shed annually : four compartments in the stomach : the second and fifth digits incomplete.

Ex. *Cervus, Cariacus, Capreolus, Rangifer, Alces.*

Family (4) BOVIDAE.

Selenodontia with horns which are outgrowths of the frontal, never shed, and covered with a thick horny sheath :

four compartments in the stomach : the second and fifth
toes rudimentary.

Ex. *Bos, Ovis, Ovibos, Haploceros.*

Family (5) GIRAFFIDAE.

Selenodontia with short horns which are outgrowths of
the frontal, never shed, and permanently covered with soft
fur : immensely elongated neck and very long limbs.

Ex. *Giraffa.*

Family (6) ANTILOCAPRIDAE.

Selenodontia with branched horns which are outgrowths
of the frontal covered with a horny sheath. This sheath is
shed annually.

Ex. *Antilocapra.*

Order VIII. **Rodentia.**

Eutheria with one large pair of chisel-shaped incisors in
each jaw growing throughout life and no canines. The
Rodentia walk on the whole surface of the last joint of the
digit, not on the extreme tip as do the Ungulata : the nails are
blunt but not usually hoof-like.

Sub-order 1. **Duplicidentata.**

Rodentia in which there is a second pair of rudimentary
incisors in the upper jaw.

Ex. *Lepus.*

Sub-order 2. **Simplicidentata.**

Rodentia in which there is only one pair of incisors in the
upper jaw.

Ex. *Sciurus, Tamias, Mus, Fiber, Arvicola, Muscardinus,
Castor, Erethizon, Hystrix, Cavia.*

Order IX. **Cheiroptera.**

Eutheria in which the fore-limb is converted into a wing,
the hand being greatly enlarged and the fingers elongated in
order to support the wing-membrane ; the leg small and the
knee-joint rotated backwards : teeth and brain resembling
those of the Insectivora.

Ex. *Vespertilio, Vesperugo, Rhinolophus, Xantharpyia.*

Order X. **Primates.**

Eutheria with long limbs, the brachium and femur not being buried in the body: five digits in each limb, some of them having flat nails: the great toe or thumb or both are opposable to the other digits. The orbits are rotated on to the anterior aspect of the skull and are completely surrounded by bone : the brain is large.

Sub-order 1. **Lemuroidea.**

Primates in which the orbit is merely surrounded by a bony ring: front teeth separated by a space in the middle line.

Ex. *Lemur.*

Sub-order 2. **Anthropoidea.**

Primates in which the orbit is completely separated from the temporal fossa by an inwardly projecting sheet of bone: front teeth in contact in the middle line.

Section (1) **Platyrrhini.**

Anthropoidea with a broad internasal septum, three pairs of premolar teeth and a simple tympanic bone : the great toe opposable to the other toes: the thumb imperfectly or not at all opposable to the other fingers.

Family (1) Hapalidae.

Small thickly furred Platyrrhini with a flat nail on the great toe only, claws on all the other digits : two molar teeth on each side.

Ex. *Hapale, Midas.*

Family (2) Cebidae.

Platyrrhini with flat nails on all toes : three molar teeth on each side.

Ex. *Ateles, Cebus.*

Section (2) **Catarrhini.**

Anthropoidea with a narrow internasal septum, two pairs of premolar teeth and three pairs of molars in each jaw. The tympanic bone has a tube-like prolongation. The great toe is opposable to the other toes, the thumb imperfectly opposable to the other fingers.

Family (1) CERCOPITHECIDAE.

Catarrhini with arms not longer than their legs: bare patches on the buttocks: with rare exceptions a well-developed tail.

Ex. *Macacus, Semnopithecus.*

Family (2) SIMIIDAE.

Catarrhini with arms much longer than legs and a semi-erect gait: no tail.

Ex. *Gorilla, Hylobates, Simia, Anthropopithecus.*

Section 3. **Hominidae.**

Anthropoidea with arms of moderate length and long legs: the foot entirely adapted to support the body, the great toe not opposable to the other toes: the thumb completely opposable to the other fingers: the upright attitude habitual: no tail: brain very large.

Ex. *Homo.*

CHAPTER XX.

PHYLUM PLATYHELMINTHES.

FROM the Earthworm up to Man we have been considering animals which either in the embryo or in the adult exhibit the coelom in a characteristic and unmistakable form. Such a space is indicated even in the Actinozoa, where the endoderm lining the lateral compartments of the coelenteron gives rise to the muscular bands and the generative organs, performs the excretory functions and is probably the homologue of the mesoderm of higher forms. If this be so the space in the Actinozoa surrounded by this " endoderm " is equivalent to a coelom, but one not yet shut off from the digestive tube.

Introduction.

Thus from the Coelenterata to Man we have traced a series of organisms all of which possess in some form or other this particular organ. Since Vertebrates include Man and are among the most highly organized animals at present living on the earth, we have placed them last in the series of Coelomata, but this must not be taken to indicate that there is any kind of progression through *all* the series of lower animals up to Man. The vertebrate ancestor of Man probably separated whilst still of exceedingly simple structure from the ancestors of other animals, and there has been independent progress along many different lines, culminating for instance in an Insect, a Cuttlefish and a Sea-urchin.

Leaving now the Coelomata we must consider a few phyla which we cannot definitely assert to be Coelomata. All of these groups possess between the ectoderm and endoderm a mass of various tissues, muscular, connective, excretory and generative, hollowed out by spaces or traversed by systems of tubes; but it has not yet been shown in the case of any one of them that this mass of tissues has had in the embryo the form of sacs lying at the sides of the alimentary canal from the walls of which the said tissues have been differentiated.

In the case of some of the following phyla it is reasonable to expect that fuller knowledge will show that they are coelomate, but at present this has not been definitely proved, and thus it seems more logical to consider them apart from the organisms which undoubtedly possess a coelom.

The Phyla that follow are all Metazoa and since they possess no notochord are Invertebrates.

The Platyhelminthes consist of three large classes, (i) TURBEL-LARIA, (ii) TREMATODES and (iii) CESTODA. These three groups contain animals which are bilaterally symmetrical, each half of the body being a reflection of the other. The alimentary canal may be entirely lost, but when present it has only one aperture, which serves both as mouth and anus, as in the case of the Coelenterata. A separate anus is never found, and there is no evidence from the study of development that the ancestors of Platyhelminthes ever possessed such an opening to the alimentary canal. The alimentary cavity is practically the only cavity in the animal, as there is no space between the skin and the intestine which could be compared to the body-cavity of other animals, and except for the narrow cavities of the excretory system and the genital ducts the bodies of these animals are solid. The excretory system, often termed the water-vascular system, the function of which is the ridding the body of the waste nitrogenous materials which, as explained before, result from the catabolism of living protoplasm, is in its structure eminently characteristic of the Phylum. It consists of a series of narrow tubules permeating the body in every direction; these on the one hand communicate with larger tubes which open on to the surface of the body, and on the other receive a large number of still smaller tubules each of which ends in a cell with a single cilium hanging into the end of the tubule. The constant flickering of this cilium is thought to keep the fluid contents of the tubule in motion. Such a cell is termed a flame-cell from the fancied resemblance of the motion of the cilium to the flickering of a flame. The Platyhelminthes usually contain both male and female reproductive organs in the same animal, and it is characteristic of them to have a special portion of the ovary called the yolk-gland or vitellarium, set apart to produce small yolk-filled cells, which serve as food for the perfect ova during the early stages of development.

Class I. Turbellaria.

The Turbellaria are free-living animals, and as a rule swim about in the sea or in fresh-water ponds or streams. A few, however, have taken to living amongst moist earth, and some species, e.g., *Bipalium kewense*, are occasionally met with in hot-houses all over the world, being probably imported with the roots of some tropical orchid or fern. Other species of land Turbellaria are common in the Tropics.

Turbellaria are all very soft animals and capable of considerable change of outline. In their native habitat they are not easy to see, many of them having colours which imitate the sea-weeds, etc., amongst which they live, and many appear only at night from their hiding-places. If, however, a bunch of red sea-weed be shaken out in some clean sea-water in a white china dish, as a rule many of these animals can be seen swimming with an undulating motion like a Sole or clinging to the sides of the dish.

One of the commonest species in the fresh-water ponds of Great Britain is *Mesostoma ehrenbergii*, a flat leaf-like organism, perhaps half an inch long, the transparency of whose tissues permits at times the examination of some of the internal organs. The whole of the outer layer of cells—the ectoderm—bears innumerable cilia, by whose action the animal glides slowly along when it does not swim by the undulations of its whole body. Within this ectoderm are certain circular and longitudinal muscle-fibres, and these surround a mass of cells called the parenchyma. This consists of cells of a stellate shape, united with one another by their out-growths, the interstices between the cells being filled by a semi-fluid jelly-like substance. The parenchyma may be regarded as a primitive form of connective-tissue in which the nervous system, alimentary canal and excretory and reproductive systems are embedded.

The ectoderm cells secrete a great deal of mucus, mingled with which are a number of little rod-like bodies called rhabdites. The exact use of these is not clearly known; in their formation they recall the nematocysts of the Coelenterata. Like the nemato-cysts they are extruded on irritation. The mucus forms a bed over which the animal moves and in which the cilia work, so as to propel the animal.

The mouth of the *Mesostoma* is, as its name indicates, near the centre of the body, on the ventral surface. It leads at once into a pharynx with very muscular walls which can act like a sucker. This pharynx can be withdrawn into the body or pushed a little

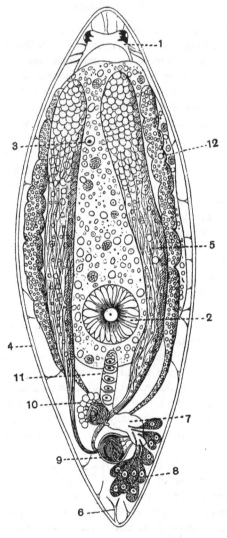

FIG. 334. *Mesostoma splendidum*, drawn from a compressed individual; the
cilia and rhabdites are omitted. Magnified. From von Graff.

1. Brain, showing two eyes.   2. Pharynx surrounding the mouth.   3. Food
vacuole in an endoderm cell.   4. Ectoderm.   5. Testis lying above
alimentary canal, showing developing spermatozoa.   6. Muscle fibres.
7. Genital atrium.   8. Glands of the genital atrium.   9. Penis.
10.   Shell-glands surrounding the spermatheca.   11.   Germarium.
12. Vitellarium.

way out of the mouth. The chamber in which it lies and which gives it room to play in and out, is an ectodermal pouch called the **pharynx-sheath**. This is small in *Mesostoma*, but in other Turbellarians it may be much larger, and the pharynx is consequently capable of stretching out a long distance. This is the case, for instance, with the fresh-water genus *Planaria* (Fig. 335).

Certain glands, called salivary glands, open into the cavity of the alimentary canal at the inner end of the pharynx, and then the cavity opens into the stomach, which is a sac-like structure with no other opening than the mouth lying in the centre of the body. The cells lining this cavity are, like the endoderm cells of *Hydra*, amoeboid, and they take up particles of food into themselves in the same way that an *Amoeba* does. This primitive form of digestion has been lost in most of the Coelomata, where the digestive cells pour out solvent fluids into the cavity of the alimentary canal, and the food is rendered soluble in this cavity before being absorbed. In the Turbellaria, as in the Coelenterata, this secondary method of digestion coexists with the amoeboid method.

Mesostoma is carnivorous and eats small worms, minute crustacea and insect larvae. It uses its mucus to ensnare and entangle its prey. Its method of devouring them recalls the habits of the starfish. It holds them fast by means of the pharynx, using this as a sucker. The so-called salivary glands secrete a strong digestive ferment which rapidly dissolves the flesh of the victim, reducing it partly to a fluid condition and partly to a disintegrated mass of particles. There is no vascular system to distribute the digested food to the different parts of the body, so that these products must be passed from cell to cell through the solid body until they arrive where they are needed. The undigested parts are passed out through the mouth.

The two main ducts of the excretory or water-vascular system open near the sides of the mouth, each then passes upwards towards the dorsal surface and divides into two longitudinal vessels, one running towards the head, the other towards the tail. These four longitudinal branches give off innumerable finer ones, which subdivide until each branchlet ends in a flame-cell. These latter are very minute and require a high power of the microscope and very careful focussing to see.

The nervous system consists of a large ganglion called the brain, divided by a shallow depression into two lobes. It is situated in front of the mouth near the anterior end of the animal, embedded

in the parenchyma. It gives off a pair of nerves which run forward
to the tip of the body, and another pair of rather stout nerves
which run back, one on each side of the pharynx, to the tail. The
nerves give off fine branches which are distributed all over the body.
A pair of eyes of a simple structure lie on the upper surface of the
brain.

The male organs consist of two long sac-like testes which lie
above the alimentary canal and are directly continuous with their
short ducts, the vasa deferentia. These ducts unite to form the
muscular penis which communicates with the genital atrium through
which it can be protruded. The proximal portion of the penis is
swollen up to form a bulb called the vesicula seminalis, in which
the spermatozoa are stored up before being transferred to another
individual. The female organs consist of a large ovary on each side,
divided by constrictions into numerous lobes; these are not well
marked in the species represented in Fig. 334. The whole of one
ovary and the greater part of the other produce only yolk-cells and
are therefore to be regarded as yolk-glands or vitellaria. The
basal lobe however of one of the ovaries (11, Fig. 334) produces ova
capable of development; this is the ovary (*sensu stricto*) or germarium. The two oviducts, or as they are generally styled, the
vitellarian ducts, are directly continuous with the yolk-glands and
lead directly into the genital atrium. Near their common opening
a thick muscular pouch opens into the atrium. This is the spermatheca which receives the spermatozoa from another individual, and
emits them on to the ova as they pass its opening. Around the
spermatheca are certain glands called the shell-glands, which also
open into the atrium. The secretion of these glands forms the
egg-cases in which one egg and many yolk-cells are enclosed. As
the egg-cases are formed they pass into two great sac-like diverticula of the atrium, one situated on each side of the body, called
the uteri. In these they are carried about by the animal for some
time, but are eventually laid, and become attached to water-plants
by the stickiness of their outside layer. There are two kinds of
these egg-cases in *Mesostoma*, one thin-walled, called "summer
eggs," and the other thick-walled, called "winter eggs." The former
are believed to contain ova fertilized by the spermatozoa of the same
individual; these develope rapidly, devouring the surrounding yolk-
cells and the resulting young hatch out in April and May. These
when they arrive at maturity cross-fertilize one another, and as a
result the thick-walled capsules termed "winter-eggs" are produced,

which lie dormant during the winter, whilst the parent turns opaque, sinks to the bottom of the water and dies. In the spring young are hatched from the winter eggs, which produce when mature summer eggs, and in some cases are supposed after laying these to live on and produce the winter eggs of the next season; but in this respect the various species probably differ from one another.

The Turbellaria are a large group, and fall naturally into two main divisions, viz., the **Rhabdocoelida** with a rod-like gut (Gr. ῥάβδος, a staff) and the **Dendrocoelida** with a branched one (Gr. δένδρον, a tree). Each of these divisions is again subdivided; thus the Order Rhabdocoelida includes the Sub-orders *ACOELA*, *ALLOIOCOELA*, and *RHABDOCOELA*, whilst the Dendrocoelida are divided into *POLYCLADA* and *TRICLADA*.

To turn first to the divisions of the **Rhabdocoelida**, the Sub-order *ACOELA* includes extraordinary forms in which there is no digestive cavity; the alimentary canal is represented by a porous mass of endoderm cells, amongst the interstices of which the digested food soaks. The endoderm completely fills the space surrounded by the ectoderm and muscle layers; there is no parenchyma. In almost every case there is a muscular pharynx by which the animal adheres to its prey, and through which a pseudopodium-like mass of endoderm is protruded. This protrusion secretes a solvent which disintegrates the victim, and it then engulfs the product after the manner of an *Amoeba*. In some cases, however (*Convoluta*), the endoderm is infested with small green Algae, and the animal lives largely on the compounds formed by these, needing only a scanty diet of Protozoa and Diatoms to supplement its internal provision.

The genital organs are simple; the ovary is not divided into vitellarium and germarium and no yolk-cells are produced. It has been suggested that the Acoela are the most primitive of all the phylum, and that they have been directly derived from large multinucleate Protozoa, but their development makes it possible that their peculiarities are due to degeneracy.

The *ALLOIOCOELA* have a parenchyma and a hollow alimentary canal which has slightly developed lateral lobes. The testes are represented by scattered masses of cells without distinct ducts; and the spermatozoa apparently find their way to the main vasa deferentia by passing through the interstices of the parenchyma. The germaria are two in number and have long ducts opening into the genital atrium distinct from those of the vitellaria.

The *RHABDOCOELA* are the most highly developed forms of the
Rhabdocoelida; their alimentary canal is cylindrical and surrounded
by a mass of extremely watery parenchyma which simulates a body
cavity. The arrangement of the genital organs has been described
above.

The first division of the **Dendrocoelida**, the *TRICLADA* (Gr.
τρι-, triple, κλάδος, a branch), derive their name from the circumstance that there are three main branches of the alimentary canal,
one in the middle line running forward from the inner end of the
pharynx, and one running backwards at each side of the pharynx.
There is a pair of germaria formed from the most anterior branches
of the great lobed ovaries; they discharge into the same ducts as
the vitellaria. The uterus is an unpaired sac. The group includes
marine, freshwater and terrestrial forms. *Planaria (Dendrocoelum)*
is a common form in the streams of both Britain and Canada.

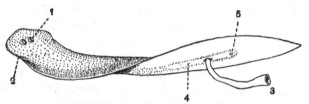

FIG. 335. *Planaria polychroa* × about 4.

1. Eye.   2. Ciliated slit at side of head.   3. Mouth of proboscis.   4. Outline of the pharynx sheath into which the pharynx can be withdrawn.
5. Reproductive pore.

The *POLYCLADA* are a marine group and are thought by some
authorities to be the most primitive of all Turbellaria. Their
name is suggested by the fact that the alimentary canal consists of
many branches radiating from a central stomach into which the
large pharynx opens. The ovary is a lobed organ not divided into
vitellarium and germarium. The eggs are laid in plate-like masses
bound together by slime. They develope into free-swimming young
known as Müller's larvae. These are little oval organisms provided
with a ciliated band drawn out into eight longitudinal loops, and
on these the cilia are arranged in transverse rows fused at the base
so as to resemble the combs of the Ctenophora. The resemblance
of these ciliated loops to the "ribs" of the Ctenophora suggested
to Lang the idea that Turbellaria were Ctenophora which had
become adapted to a creeping life, in which a marked bilateral

symmetry had replaced the generally radial arrangement of the organs of the normal Ctenophora, though traces of the latter arrangement remain in the Polyclada. The brain on this view would be the apical plate which had shifted forward; the stomach with its radiating branches would correspond to the funnel of the Ctenophora and the canals in connection therewith; the pharynx sheath would represent the stomodaeum, the so-called "stomach" of the Ctenophora; but the eversible pharynx, the copulatory organ, and above all the excretory system must be regarded as new acquisitions. This view, which at first was not received with much favour, has received strong support by the investigation of a marine organism known as *Ctenoplana*. This is a flattened animal resembling a Polyclade in shape and in the circumstance that the ventral surface is covered with cilia with which it creeps. It possesses however an apical plate of thickened nervous ectoderm supporting a mass of otoliths on bars of fused cilia, and there are eight short "ribs" radiating from this plate. These ribs as in Ctenophora are thickened bands of ectoderm bearing combs of cilia fused at the base. The funnel and its canals are represented by a lobed alimentary canal, continued on each side into a tentacle canal, from the end of which springs a long retractile tentacle. The genital organs have their independent ducts opening directly to the exterior. In all these respects therefore *Ctenoplana* is intermediate between the Polyclada and the Ctenophora. If we accept Lang's theory it is evident that on this view the Platyhelminthes are not true Coelomata. The evolution of the more complicated systems of genital organs amongst the Turbellaria out of the simpler arrangement in the Polyclada, has probably been the result of laying the eggs in numbers surrounded by a capsule. This led to a struggle amongst the eggs, resulting in the sacrifice of the smaller to the needs of the larger ova, and this to the production of weak ova to serve as food for the others, with the consequent differentiation of the ovary into germarium and vitellarium.

Class II.   TREMATODA.

The remaining two groups of Platyhelminths have taken to a parasitic mode of life and this has to a great extent influenced their organization. The term parasite is applied to an animal which lives at the expense of another without destroying its life. The Trematodes have lost the external ciliation of the skin, the

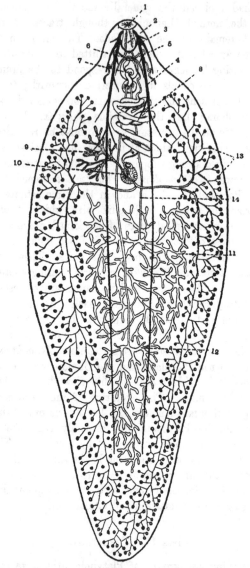

Fig. 336. Diagram of reproductive and nervous system of *Distoma hepaticum*
×about 8. From Leuckart.

1. Mouth.   2. Pharynx.   3. Nerve-ring.   4. Chief longitudinal nerve.
5. Beginning of alimentary canal.   6. Opening of penis.   7. Vesicula
seminalis.   8. Uterus.   9. Ovary.   10. Shell-gland.   11. Anterior
testis.   12. Posterior testis.   13. Yolk-glands.   14. Vas deferens.

ectoderm being everywhere covered with cuticle. In other features of their anatomy they present a great resemblance to the Triclade Turbellaria, and one family, the TEMNOCEPHALIDAE, may be described as intermediate between the two classes of Platyhelminths since its members still retain patches of ciliated skin. For the most part they live on or in the bodies of Vertebrates, attaching themselves either to the skin or to the alimentary canal or its outgrowths.

One of the most characteristic features of Trematoda is to be found in the suckers by which they adhere to their prey. Often indeed the lips of the mouth are thickened and muscular so as to constitute an oral sucker, but there is always a ventral adhesive disc provided with suckers or hooks or both. The mouth is situated at or near the anterior end of the body; it leads into an oral funnel opening into a muscular pharynx, which by alternate expansion and contraction pumps in the juices of the prey. Its action is thus different from that of the pharynx of Turbellaria, which, as we have seen, can act as a protrusible sucker. Behind the pharynx the alimentary canal divides into two parallel forks running back to the posterior end of the body, and beset with branches which in some cases may unite with one another across the middle line. It thus resembles what the alimentary canal of a Triclade would become were the mouth shifted to the anterior end of the body. The nervous system is remarkable for the fact that several trunks of equal size are given off from each side of the brain. The reproductive organs resemble those of a Rhabdocoele like *Mesostoma* ; thus the germarium is developed only on one ovary, of which it is a basal branch, and the testes each consist of a lobed organ directly continuous with the vas deferens. The main peculiarities are as follows : there is no spermatheca ; the spermatozoa from another individual enter either by a dorsal pore or two lateral pores, leading into a canal or canals which join the oviducts where they unite with one another. These ducts, totally unrepresented in Turbellaria, are called the "canals of Laurer." Further, the genital atrium is situated on the anterior part of the body in front of the ventral sucker. There is no uterus comparable with that of Turbellaria ; the so-called uterus being a long coiled tube composed of the conjoined oviducts (vitellarian ducts). Finally the testes are so large that there is not room for them side by side, but in order to stow them away one is situated behind the other.

Trematoda are divided into two Orders called respectively the

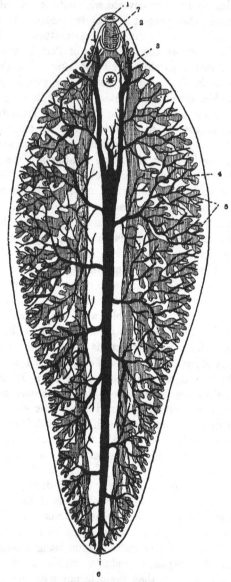

FIG. 337. Diagram of digestive and excretory system of *Distoma hepaticum*
×about 8. From Leuckart.

1. Mouth.   2. Pharynx.   3. Reproductive pore.   4. Branch of alimentary canal.   5. Branches of excretory system.   6. External
opening of excretory system.   7. Nerve-ring.

Monogenea and the *Digenea*. In the first named the egg gives rise to a larva which developes continuously into the adult; the main organ of adhesion is a disc situated at the posterior end of the body and armed with suckers or hooks, usually both, and there are two lateral vaginae from which the canals of Laurer lead inwards. The excretory system opens by two dorsal pores. One of the commonest of the Monogenea is *Polystomum integerrimum*, found in the bladder of the Frog. This animal has an adhesive disc bearing six suckers. The fertilized egg of *Polystomum* is discharged through the cloaca of the Frog into the water. After some time a larva hatches out which has a forked alimentary canal, but which is without genital organs and has no suckers, although the posterior adhesive disc is clearly differentiated. It is provided with a number of transverse bands of cilia by means of which it swims about until it finds a tadpole, to the skin of which it attaches itself. It creeps into the branchial chamber of its host and loses its cilia, and commences to develope genital organs and suckers. About the time of the tadpole's metamorphosis the Trematode wanders down the alimentary canal into the bladder. *Sphyranura osleri* is an allied form parasitic on the skin of the Urodele *Necturus*. Its posterior disc carries two large suckers and two hooks.

The *Digenea* give rise to larvae which become parasitic in some animal, where they give rise by gemmation to several generations of secondary larvae, which develope into adult forms only when they are swallowed by a second animal. The life-history therefore includes an alternation of generations and is only completed in two hosts. The Digenea further differ from the Monogenea in having as main adhesive organ a sucker situated on the anterior part of the ventral surface, in having only a single "canal of Laurer" which opens on the dorsal surface, and finally in the fact that the main trunks of the excretory system coalesce to form a single trunk which opens to the exterior by a median posterior pore. The Liver-fluke, *Distoma hepaticum*, is an example of the Digenea; it is parasitic in the liver and bile-ducts of the sheep, causing a wasting disease called sheep-rot. It gives rise to a larva consisting of a solid mass of cells, the outermost layer of which is ciliated. This larva cannot survive unless it reaches a pond-snail of the species *Limnaea truncatula*. In the pulmonary chamber of this animal it loses its cilia, enlarges and becomes hollow, forming a structure called the sporocyst, which sometimes divides into two

or more. Germ cells are budded off from the wall of the sporocyst
into the cavity. These by division form masses which develope
into secondary larvae called rediae, provided with a muscular
pharynx and a sac-like alimentary canal. These larvae have a pair
of blunt processes on the under side near the posterior end, by the
aid of which they move. They force their way out of the sporocyst
and enter the tissues of the snail, being found especially in the
liver. From the inner surface of their body-wall germ cells are
budded off which give rise to other rediae, which escape from the
parent by an opening near the anterior end. After a time the
rediae give rise to larvae of a third kind called cercariae.
These have suckers like the adult, and a forked alimentary canal
with a pharynx; they are provided with a tail stiffened by a rod of
gelatinous tissue recalling the Vertebrate notochord. By the aid of
the tail they work their way out of the snail and attach themselves
to blades of grass. The tail then falls off and they enclose themselves in a cyst of mucus, and remain there till they are eaten by a
sheep, from whose intestine they pass into the liver, where they
develope genital organs and become mature.

Class III. Cestoda.

The Cestoda are sharply distinguished from the two preceding
classes of Platyhelminthes by the total absence of an alimentary
canal. They are all internal parasites, living in the alimentary
canal of their hosts, and absorb the half-digested juices of their
hosts through all parts of their skin.

As in Trematoda, there is a well-marked cuticle which protects
the animal from the action of the digestive juice of the host.
The ectoderm has undergone an extraordinary modification. Its
cells have become long and filamentous, having long narrow necks,
the body of the cell with the nucleus being pushed downwards
into the subjacent tissues. These necks are however of very
various lengths, and so the ectoderm, although fundamentally a
single layer of cells, presents the appearance of a thick band of
many layers of nuclei. The ectoderm cells intermixed with longitudinal muscles form the cortical zone of the animal. Inside
this and surrounded by the circular muscles is the medullary
zone of parenchyma, in which the genital organs and the excretory
and nervous systems are embedded. Many of the cells of the
parenchyma secrete calcareous matter and form the so-called calcareous corpuscles.

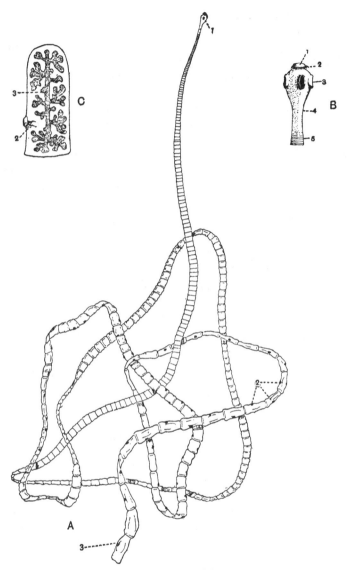

FIG. 338. *Taenia solium.* Slightly magnified.

A. Entire worm showing head and proglottides.   1. Sucker on head.
    2. Genital pores.   3. Ripe proglottis.
B. Head.   1. Rostellum.   2. Hooks.   3. Suckers.   4. Neck.   5. Com-
    mencement of strobilization.
C. Ripe proglottis broken off from worm.   2. Remains of vas deferens and
    oviduct.   3. Branched uterus crowded with eggs.

Each portion of the body of a Cestode which contains the reproductive organs is called a proglottis, and is budded off from the anterior portion which is called the head. The latter is provided with suckers, and generally, in addition, with a circle of hooks situated on a prominence called the rostellum (1, Fig. 338 B). With the head the Cestode adheres to its host; its hinder part or neck buds off proglottides in which new sets of reproductive organs are formed, and this process is repeated an indefinite number of times so that a chain of proglottides is formed. The oldest proglottis is thus the hindermost. This method of segmentation is called strobilization on account of its resemblance to the formation of the Ephyrae by a Hydra-tuba (p. 65); it differs from

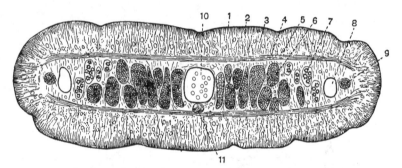

Fig. 339.   Transverse section through a mature proglottis of *Taenia* × about 12.

1. Cuticle.   2. Long-necked cells of ectoderm.   3. Longitudinal muscle fibres cut across.   4. Layer of circular muscles.   5. Split in the parenchyma which lodges a calcareous corpuscle.   6. Ovary.   7. Testis with masses of male germ-cells forming spermatozoa.   8. Longitudinal excretory canal.   9. Longitudinal nerve-cord.   10. Uterus.   11. Oviduct.

the segmentation of the Annelida, where the new segments arise not in the neck but in the tail.

The excretory system consists of a larger and a smaller trunk on each side, which unite in the last proglottis to end in a contractile vesicle opening by a median posterior pore. In each proglottis the trunks on each side are connected with their fellows on the opposite side by transverse canals. The nervous system consists of a ring in the head, whence two lateral trunks arise. There is no brain—the impressions the animal receives from the outside world being few and simple.

The reproductive organs have the same general structure as those of the Trematoda. The vitellarium is however often unpaired,

whereas there is usually a pair of germaria. There is a large
uterus (8, Fig. 340) which appears to be a lateral outgrowth
from the conjoined oviducts. From the circumstance that in
primitive Cestoda this uterus opens to the exterior independently
of the genital atrium it is believed that it corresponds to the canal
of Laurer in the Digenetic Trematoda. If this be so the functions
of the genital atrium and Laurer's canal have been exchanged in
Cestoda, for in Trematoda the atrium permits eggs to escape
whilst Laurer's canal admits spermatozoa from another individual,
in Cestoda these spermatozoa enter through the atrium whilst the
canal is enlarged to form a uterus. The testes are in Cestoda

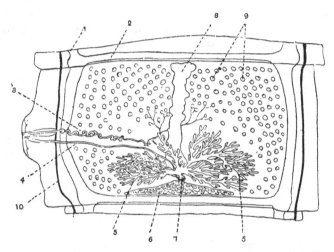

Fig. 340.   Diagram of a ripe proglottis of *Taenia solium*.   From Cholodkowsky
× about 10.

1. Longitudinal water-vascular canal.   2. Transverse water-vascular canal.
3. Vas deferens.   4. Vagina.   5. Ovary.   6. Yolk-gland.   7. Shell-
gland.   8. Uterus.   9. Testes.   10. Longitudinal nerve.

represented by a multitude of small rounded bodies (7, Fig. 339,
9, Fig. 340) from which excessively fine ducts are given off uniting
to form the single vas deferens ending in the penis. As the
proglottis gets older the eggs all pass into the uterus, which swells
enormously, displacing and destroying the other organs. In this
condition the proglottis drops off. This process is continually being
repeated so that a single parasite keeps on casting off portion after
portion, each charged with ova capable of developing. In this way

a Cestode firmly lodged in the alimentary canal of its host can
produce an almost indefinite number of eggs.

In the more primitive Cestoda the egg-cases, which as in Trematoda contain an ovum and a multitude of yolk-cells, escape from
the detached proglottis through the opening of the uterus into
water. In a short time a larva hatches out, consisting of an outer
layer of ciliated cells surrounding a solid internal mass which developes six chitinous hooks. After swimming in the water for a short
time the larva is swallowed by an aquatic animal, loses its outer
layer of cells, thus exposing the hooks and becoming what is now
termed an onchosphere. In the more modified Cestoda the uterus
has no external opening, and the eggs escape from the proglottis only
by decay. The remains of the proglottis, including the eggs, are
swallowed by some animals, and soon after an onchosphere hatches
out, no cilia being formed. In every case the onchosphere bores
through the alimentary canal of its host, and is carried by the
blood-stream to a suitable spot in the tissues where it fixes itself.
Once fixed, the larva increases very much in size and generally
becomes hollow so as to resemble a bladder. An infolding or
invagination of part of the outer layer now takes place, forming
a pouch, on the inner side of which the suckers and hooks of the
adult head make their appearance. The pouch is then turned
inside out and a well-marked head is thus formed. The larva is
now known as a bladder-worm or Cysticercoid and is found
parasitic in a great number of animals. A very common form is
Cysticercus pisiformis found in the coelom of the Rabbit attached
to the mesentery. *Coenurus cerebralis* is a dangerous Cysticercoid
in which the bladder can become as large as a plum and on which
not one but numerous heads are formed. It is found lodged in the
brain of the sheep and other domestic animals and causes the
disease known as gid or staggers. A still more dangerous form is
Echinococcus polymorphus, in which the bladder may become as
large as a man's head. This enormous vesicle buds off from its
inner surface secondary vesicles on each of which numerous heads
are formed. This parasite is found in the Pig, Sheep and even
Man, in the liver and other internal organs.

The Cysticercoid can complete its development only when
its first host is eaten by another animal. Then the bladder is
cast off whilst the head firmly attaches itself by its hooks and
suckers to the alimentary canal of its new host and commences to
bud off proglottides. Though the adult may attain an immense

length (as much as 20 feet) it is a much less dangerous parasite than the larva, and rarely produces worse symptoms than giddiness and a certain amount of abdominal pain. The adults are found chiefly in carnivorous animals, above all in the Dog, Wolf and allied species. Thus *Cysticercus pisiformis* becomes *Taenia serrata*, *Coenurus cerebralis* gives rise to *Taenia coenurus*, and *Echinococcus polymorphus* to *Taenia echinococcus*, all three infesting the alimentary canals of the Dog and its allies. The common *Taenia solium* found in the human intestine is developed from a Cysticercoid found in the muscles of the Pig.

Bothriocephalus latus, which may attain a length of 30 feet, is the largest Cestode found in Man. It belongs to a more primitive division of the class than *Taenia*, for it gives rise to a free-swimming ciliated larva, which is swallowed by the Pike or the Perch. In this it developes into a solid larva which' gives rise to the adult when the fish is eaten by man.

It is obvious that an animal which like a Digenetic Trematode or a Cestode depends for its survival on such a combination of lucky chances as that of transference from one particular species of animal to another animal of a definite kind must develope large powers of reproduction. In the Trematode this is chiefly manifested in the power of the larva to reproduce itself asexually, but in the Cestoda the power is in most cases only developed when the animal is in the adult condition. Considerable discussion has taken place as to whether the process of strobilization is, or is not to be regarded as a production of new individuals. When we recollect that the separated proglottides often retain life for some time after being cast out of their host, it would seem that there was much to be said in favour of regarding them as sexual individuals and the head as an asexual one. The most probable view on the whole is that of Lang, who suggests that in the ancestor of modern Cestoda the hinder part of the body which contained the genital organs was separated at maturity, as occurs in the case of some Polychaeta When the Cestode took to living in the alimentary canals of Vertebrata, the abundant food supply and favourable temperature stimulated the powers of regeneration so that the missing part was quickly reproduced, and by the hurrying on of this process of regeneration the process of strobilization was evolved, exactly as occurred in the Scyphistoma of *Aurelia*. It is interesting to notice that *Archigetes*, found mature in the coelom of the Oligochaet *Tubifex*, is the only Cestode which completes its development elsewhere than in a Vertebrate, and in this case only one proglottis is produced, which never separates from the head. To the posterior end of this proglottis an appendage is attached representing the bladder of the Cysticercoid, in which the six hooks of the onchosphere still remain embedded.

The Platyhelminthes are classified as follows:

Class I. TURBELLARIA.

Platyhelminthes with soft, usually leaf-like bodies and a ciliated ectoderm: rhabdites are often present: free-living.

Order I. Rhabdocoelida.

Turbellaria with a straight, rod-like alimentary canal and a protrusible pharynx.

Sub-order 1. Acoela.

The digestive system is represented by a mass of endoderm cells which contains no cavity: a short pharynx and single otocyst are present.

Sub-order 2. Alloiocoela.

The digestive canal is lobed: the testes are scattered.

Sub-order 3. Rhabdocoela.

The digestive canal is rod-shaped, the pharynx protrusible.

Order II. Dendrocoelida.

Turbellaria with a branched alimentary canal.

Sub-order 1. Triclada.

The alimentary canal has three main branches, one running forward and two running back. Male and female openings united.

Sub-order 2. Polyclada.

The alimentary canal has many branches sometimes anastomosing. Male and female openings as a rule distinct. Marine.

Class II.   Trematoda.

Parasitic Platyhelminthes with usually leaf-like, rarely cylin-drical bodies : no cilia on ectoderm : forked alimentary canal : ventral sucker or suckers and hooks often present : not segmented.

Order I.   Monogenea.

Trematodes with a sucker or suckers at the posterior end and in some cases another anteriorly.   Often external parasites or in the mouth, nose or branchial cavities.   Development direct.

Order II.   Digenea.

Trematodes with a sucker at the anterior end and another on the ventral surface or posteriorly.   Internal parasites and their organs of adhesion more weakly developed than in the Monogenea.   Development with an alternation of generations.

Class III.   Cestoda.

Parasitic Platyhelminthes with elongated and usually segmented body : no mouth or alimentary canal : suckers and hooks present but on the head only : the segments break off when ripe.

CHAPTER XXI.

PHYLUM NEMERTINEA.

THE Nemertines belong to that category of animals which would by most people be styled "worms"; that is to say they are long, soft-bodied animals, without limbs or appendages of any kind, which progress by undulatory movements of the body. Most Nemertines are marine, living under stones and amongst seaweed; a few are found in fresh-water and one or two species are terrestrial.

They swim in a graceful, undulating fashion, but they are much more sluggish in their movements than Annelida. It is a common and characteristic sight to see an isolated contraction passing slowly back along the body. Some species are minute, others, e.g. *Lineus marinus*, attain a length approaching 100 ft., and are perhaps the longest animals known.

The ectoderm consists of long, narrow, ciliated cells. In this they resemble the Turbellaria, but they differ from the latter animals profoundly in the structure of the alimentary canal. This organ is straight and unbranched; it begins in a mouth placed a short distance behind the anterior end of the body, and it ends in an anus placed at the extreme posterior end. For most of its course the alimentary canal is sacculated; that is, produced at the sides into a series of short broad pouches; otherwise it is not differentiated in any way.

The most characteristic organ of the Nemertine is the pro-boscis. This is a long tube lying above the alimentary canal, ending blindly behind, and opening to the exterior in front. The proboscis is a part of the outer skin invaginated, and when retracted it is surrounded by a closed space lined by a well-defined epithelium called the proboscis-sheath. This space contains a watery fluid, and its wall beneath the epithelium possesses powerful circular muscles. When these contract the proboscis is partially forced out of the sheath by being turned inside out and consequently protruding from the front end of the body at the same time. It can never be completely turned inside out, for certain cords traversing the proboscis-sheath restrain it. At the point in its wall which is at the anterior end when the process of turning inside out has reached its utmost limits, there is in the higher Nemertines a short lateral pouch in which a horny spike, the stylet, is secreted. Round the base of this open poison-glands, so it can be seen that the proboscis is an offensive organ for seizing prey. So far as we know all Nemertines are carnivorous. Amongst the lower Nemertines the stylet is not developed, nevertheless the proboscis can be employed to catch prey; it is quickly thrust out and coiled spirally round the victim and then retracted so as to push the prey into the mouth. This retraction is brought about by a muscular band which attaches the end of the proboscis to the sheath. The land Nemertines (*Geonemertes*) are said to travel by thrusting out the proboscis, attaching it to foreign objects and drawing up the body after it.

Underneath the ciliated ectoderm of Nemertines there

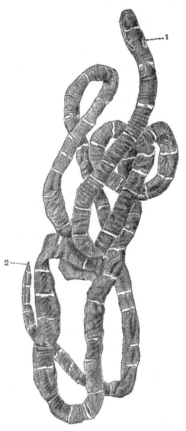

Fig. 341.　*Lineus geniculatus* × ¾.
1. Lateral slits on head.　　2. Anus.

is a series of powerful circular and longitudinal muscles : these layers are continued, but in the reverse order, on to the walls of the proboscis, since this is only invaginated skin. Under extreme irritation the extended proboscis is sometimes torn from the body—by the heightened pressure of the fluid in the sheath—

and in one instance such a disjoined proboscis was mistaken on account of its active contractions for a new species of Nemertine.

There is a well-defined central nervous system amongst Nemertines, consisting of a pair of ganglia lying at the front of and above the mouth—less frequently at the sides of it—and connected by a commissure which passes above the proboscis-sheath. The hinder parts of the ganglia are more or less distinctly separated as posterior lobes, and come into close relation with curious pouches of invaginated ectoderm, termed cephalic pits (Fig. 341), which seem in some cases at any rate to subserve the respiration of the nervous tissue, for the latter in some species contains haemoglobin. The ganglia are continued backwards into two powerful nerve-cords lying at the sides of the body. In the lower species these cords can be seen to lie just beneath the ectoderm and to be but thicker portions of a sheath of nerve-fibrils extending all round the body, and derived from the bases of the ectoderm cells. In the higher species the nerve-cords lie well within the muscles, and this sheath is not so evident.

Sense-organs in the form of eyes of simple structure often occur immediately over the region of the ganglia.

The interstices between the various organs inside the muscles are filled up with connective tissue, but there exist three longitudinal tubes with well-defined walls which are regarded as blood-vessels. Two of these tubes lie at the sides of the alimentary canal, and one is situated above it and below the proboscis-sheath. These vessels are connected anteriorly by arches and they unite with one another in both the head and tail.

The excretory organs have the form of a pair of branched tubes opening at the sides of the body not far behind the cephalic pits. Their branches end blindly and the terminations of these are closely applied to and even indent the wall of the lateral blood-vessel, while in each termination there is a tuft of cilia.

The Nemertinea are dioecious. The generative organs are exceedingly simple, consisting in both sexes of packets of cells situated along the sides of the body alternating with the pouches of the alimentary canal. No permanent genital ducts exist, but when the ova and spermatozoa are ripe they appear to make temporary ducts for themselves. The egg developes in many cases into a remarkable larva called a Pilidium. This is shaped something like a policeman's helmet with ear lappets. The edge of the helmet, including the lappets, is fringed with powerful cilia, and

there is besides a tuft of long
cilia at the apex. Underneath the
helmet is the opening of the mouth,
which leads into a sac-like gut devoid of an anus.

The adult is developed from the
Pilidium in an extraordinary way.
Four invaginations of ectoderm
appear on the under side of the
larva, at the sides of the alimentary
canal. These grow both upwards
and inwards until they completely
surround the canal. The inner walls
of the pockets form the ectoderm
of the adult. When the process is
complete the alimentary canal surrounded by the new ectoderm drops
out of the Pilidium and forms the
Nemertine.

Many species of Nemertine are
found on the British coast. As
examples we may mention *Lineus*
(Fig. 341), a long thin form without
stylets in the proboscis, and *Tetrastemma*, a short broad form with
four eyes and stylets in the proboscis.

The Nemertines, so far as our
present knowledge extends, form a
completely isolated group in the
animal kingdom.

From the circumstances that
some Rhabdocoel Turbellarians have
a protrusible organ in the anterior
part of the body, and that the excretory organs of Nemertines bear
a certain resemblance to those of
Platyhelminthes, it has been supposed that Nemertines and Platyhelminthes are allied. The totally
different character of the generative

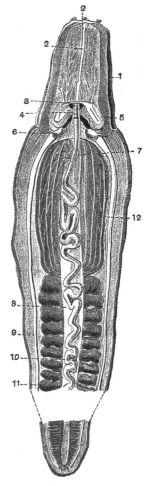

Fig. 342.   *Cerebratulus fuscus.*
Young transparent form × 7.
After Bürger.

1. Cephalic slits.   2. Opening
leading into the retracted proboscis.   3. Dorsal commissure
of nervous system.   4. Ventral
commissure. 5. Brain. 6. Posterior lobe of brain which comes
into connection with the cephalic
slit.   7. Mouth.   8. Proboscis
sheath.   9.   Lateral vessel.
10. Proboscis.   11. Pouches of
alimentary canal. 12. Stomach.

organs in the two groups and the presence of an anus in Nemertines tell strongly against this view, and indeed the resemblances will not stand minute investigation.

Of all the groups, however, about the nature of whose so-called "mesoderm" we are in doubt, the Nemertines have perhaps the greatest probability of turning out to be true Coelomata. The cavity of the proboscis-sheath, termed Rhynchocoelom by German authors, may very possibly be part of a true coelom; but we are not justified in assuming this until more exact observations have been made on its development.

The Nemertinea are divided into classes in accordance with the condition of the nervous system and the arrangement of the layers of muscles in the skin. These classes are as follow:

Class I. PROTONEMERTINI.

A nervous sheath underlies the entire ectoderm; the lateral nerve-cords, mere thickenings of this sheath, are situated outside the layers of muscles: no stylets in the proboscis.

Ex. *Carinella.*

Class II. MESONEMERTINI.

A nervous sheath underlies the entire ectoderm; the lateral nerve-cords are more deeply situated, lying between the outer circular and the inner longitudinal muscles: no stylets in the proboscis.

Ex. *Cephalothrix.*

Class III. HETERONEMERTINI.

A nervous sheath surrounds the body. Both it and the nerve-cords lie inside a special layer of longitudinal muscles, which are derivatives of the ectoderm cells, but outside all the other muscles: no stylets on the proboscis.

Ex. *Lineus, Cerebratulus.*

Class IV. METANEMERTINI.

No nervous sheath but definite peripheral nerves: the lateral nerve-cords lie within all the muscle layers: the proboscis is armed with stylets.

Ex. *Tetrastemma, Geonemertes, Malacobdella.*

Classes I—III are often united as the Class *ANOPLA*, characterised by the absence of stylets in the proboscis. Class IV is sometimes termed *ENOPLA* (i.e. armed) to contrast it with the remaining Nemertines or Anopla.

CHAPTER XXII.

PHYLUM ROTIFERA.

THE Rotifera are minute animals mostly confined to fresh water, a few only being found in the sea. Many of them swim about by means of a loop of cilia which encircles the front end of the body, but some are sessile. The motion of these cilia induced the early observers to think that the animal had wheels in front, whence the name (Lat., *rota*, a wheel; *fero*, to carry). There is a complete transparent alimentary tube, ending in an anus situated on the dorsal side in front of the hind end of the body, the latter forming a ventral projection called the foot. The first part of the tube is a stomodaeum, and lined like the rest of the skin with cuticle. This cuticle is thickened to form teeth which work on one another, the whole organ being called a mastax. The activity of the mastax enables one at once to distinguish a Rotifer from other minute animals.

Floscularia cornuta, often termed *F. appendiculata*, is a widely distributed species. It lives in ponds, ditches and pools, usually attached by its posterior end to some water-weed or other body and surrounded by a gelatinous tube into which the animal can retire. The length of the animal when extended is 0·2 to 0·3 mm. The body is covered by a thin cuticle, and may be divided into three regions, the head, the trunk and the foot. When the animal is extended the last-named is stretched out and smooth, but it is wrinkled when the animal withdraws into its tube. It terminates in a disc, on which opens the duct of two large glands which secrete an adhesive substance, by means of which the foot is anchored.

The anterior end of a Rotifer, on which the mouth opens, is called the disc. In *Floscularia* this is produced into five long tentacle-like processes fringed with stiff bristles which serve as a net to entangle small animals and plants. The cilia which produce

the current by means of which the prey is captured are situated
inside the ring of tentacles. They take the form of a horseshoe-
shaped band open dorsally, the mouth being situated in the centre

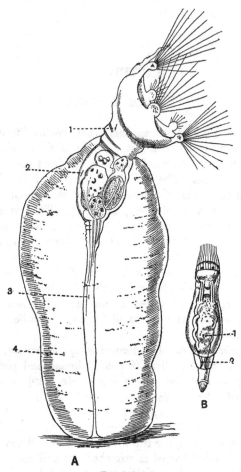

FIG. 343.

A. *Floscularia cornuta*. Female magnified. From Hudson.　　1. Head.
2. Trunk.　　3. Foot.　　4. Gelatinous tube in which the animal lives.

B. *F. campanulata*. Male magnified.　**1.** Vesicula seminalis.　2. Penis.

of the horse-shoe. This band of cilia is termed the velum: in
other Rotifera it is a complete circle but is folded back on itself in
such a way that a loop is produced which lies dorsal to the mouth.

This loop is called the trochus; it usually consists of more powerful cilia than the rest of the velum, which is termed the cingulum. The groove between the two bands in which the mouth lies is covered with fine cilia. In the genus *Rotifer* the trochus is composed of two almost circular lobes, which were described as wheels by the first naturalist who observed them, and this circumstance, as we have said, suggested the name given to the genus and afterwards bestowed on the whole phylum.

Within the cuticle which covers the body is the ectoderm. This takes the form of a protoplasmic layer with scattered nuclei, showing no cell-outlines. It surrounds a spacious body-cavity, probably a haemocoel, which contains a fluid with no amoebocytes; in this float the various organs of the body connected more or less closely together by connective tissue-cells. The muscles do not form a continuous layer underneath the ectoderm as is the case with the Platyhelminthes and the Nemertinea but consist of isolated bands, chiefly longitudinal, which retract the disc and the foot and bend the body in various ways, the recovery of its original shape being due largely to the elasticity of the cuticle.

The food, which consists of Protozoa and other small organisms, is swept into the mouth by the action of the surrounding cilia. The mouth leads through a wide vestibule into an oesophagus, which is ciliated and projects down into the so-called gizzard or mastax which, next

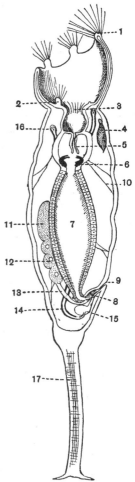

FIG. 344.   Diagram of *Floscularia*.

1. Circle of tentacles bearing bristles. 2. Velum. 3. Mouth leading to vestibule. 4. Brain. 5. Oesophagus hanging like a funnel into the crop. 6. Mastax with trophi. 7. Stomach. 8. Rectum. 9. Opening of cloaca. 10. Strands of muscles. 11. Yolk-gland part of ovary. 12. Ovarian part of ovary. 13. Oviduct. 14. Excretory duct opening into 15, urinary bladder. 16. A tag with a tuft of cilia. 17. Longitudinal and circular muscles in foot.

to the ciliary rings on the head, is the most characteristic organ of a Rotifer. The vestibule, oesophagus and mastax are all part of the stomodaeum. The cuticle lining the mastax is thickened so as to produce the trophi, which are hard, chitinous, chewing organs ; of these there are typically three, two mallei and an incus on which they strike. The incus, a Y-shaped piece, consists of two rami and a central piece or fulcrum. Each malleus is composed of a manubrium and an uncus or head, which may be toothed. The shape and arrangement of these hard parts is of great value in classification. In some species the mallei are absent altogether, and the two rami of the incus then work against one another like two lateral teeth. In the NOTOMMATIDAE the mallei can be protruded through the mouth and are used to cut into the cells of Algae on which the animals browse. In *Floscularia* and its allies there is a dilatation of the stomodaeum, called the crop, interposed between the mastax and the "oesophagus," and the latter hangs down into the crop just as a funnel might hang into a tumbler: the crop can be everted through the mouth.

After passing between, the jaws the food enters the stomach, which is lined with cilia ; here the food loses its original colour and becomes tinged with the brown secretion of the walls of the stomach. In most forms two salivary glands open into the gizzard and two gastric glands into the stomach, but these have not been clearly made out in *Floscularia*. The stomach is separated by a constriction from the ciliated rectum, and this ends in a non-ciliated proctodaeum into which the genital and excretory organs open. The alimentary canal of Rotifera, like that of the lower Vertebrata, thus terminates in a cloaca.

The excretory system consists of two longitudinal ducts consisting of columns of perforated cells, which bear a number of small tags hanging freely into the body-cavity (Fig. 345). Each tag consists of a cavity in which are several flagella, which show during life the peculiar flickering motion which is usually associated with the excretory system of Platyhelminths. In *Floscularia* four or five pairs of tags have been seen. The longitudinal ducts are usually connected by a transverse duct just under the disc, and they open as a rule into a capacious bladder which contracts at intervals and expels its contents into the cloaca and thus out of the body. It has been calculated that in some species this bladder expels a bulk of fluid equal to that of the animal about every ten minutes, and this fluid must be replaced by water which diffuses through the body-wall. This water doubtless brings with it oxygen and carries off carbonic acid which it has taken up from the tissues.

The principal part of the nervous system is a bilobed ganglion called the brain. This lies just under the disc on the dorsal side of the mastax; it bears two red eyes in *Floscularia*. In *Notommata* the brain is large, and on it more than one pair of eyes are situated. In the Bdelloida there is also a ganglion on the ventral side of the mastax, and a pair of circumoesophageal cords unite this with the brain. In *Floscularia*, as in Rotifera generally, there are three well-marked sense organs called antennae, consisting of prominences bearing stiff sense-hairs; one is situated in the mid-dorsal line, two are latero-ventral; these latter are in some genera fused with one another.

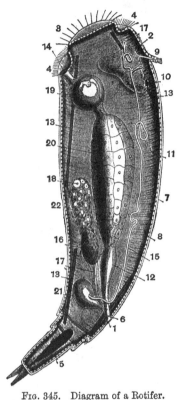

As in the Platyhelminthes, so in Rotifers, the ovary, which occupies a good deal of space in the body-cavity, usually consists of a vitellarium or yolk-gland and a germarium or true ovary. The latter lies between the former and the stomach; it is inconspicuous and is more or less hidden by the large cells of the vitellarium. Both glands may be paired. They are enclosed in a membrane continuous with the oviduct which opens into the cloaca behind the excretory duct.

The above description relates

Fig. 345.   Diagram of a Rotifer.

1. Anus.   2. Brain.   3. Trochus. 4. Cingulum.   5. Gland in foot. 6. Cloaca.   7. Cuticle.   8. Ectoderm. 9. Dorsal antenna. 10. Eye. 11. A ciliated "tag" of the excretory system.   12. Intestine. 13. Muscles.   14. Mouth.   15. Nephridial tube. 16. Ovum. 17. Oviduct.   18. Ovary.   19. Mastax. 20. Stomach.   21. Bladder. 22. Vitellarium.

chiefly to the female *Floscularia*, and in fact until 1874 the male was unknown. It is much smaller than the female (B, Fig. 343). As a rule the male Rotifer has a single circlet of cilia, a brain, eyes, excretory system and muscles all more or less reduced, but there is no mouth or alimentary canal.   The testis is large and the

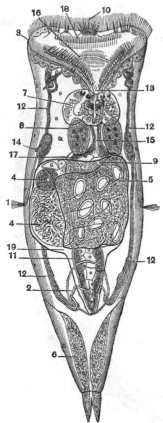

penis is introduced into the
cloaca of the female, or in some
cases is thrust through the wall
of the body, and then the eggs
are probably fertilized in the
ovary.

Floscularia lays two kinds of
eggs during the summer, both of
which are thought to develope
parthenogenetically. Both kinds
accumulate between the foot and
the tube in which the mother
lives. The larger eggs, which
average five to eight in number,
produce females; the smaller eggs,
whose origin seems to be determined by the temperature, may
amount to eighteen or twenty.
These produce the males. Towards the autumn the males
fertilize the females, and the
resulting eggs termed "winter-
eggs" are clothed with a thick
shell capable of withstanding
cold and drought. These live
through the winter and give rise
to females in the spring. A
similar alternation of summer
and winter eggs is met with in
certain Crustacea (Phyllopoda
and Ostracoda).

Fig. 346. *Hydatina senta*, ventral view.
After Plate. Magnified.

1. Lateral antenna.   2. Bladder.
3. Cingulum. 4. Eggs. 5. Vitellarium or yolk-gland.   6. Foot-
gland.   7. Gizzard.   8. Gastric
gland.   9. Germarium or ovary.
10. Lobes of "groove" bearing stiff
setae. 11. Intestine. 12. Excretory
tube. 13. Mouth. 14. Ciliated tag
of the excretory system. 15. Oesophagus.   16. Renal commissure
or transverse tube uniting kidneys
above mouth. 17. Stomach overlaid
by reproductive organs. 18. Trochus.   19. Uterus.

Rotifera are cosmopolitan, but
as a rule they inhabit fresh
water; about 700 species are
known, of which only one-tenth
live in the sea or in brackish
water. One species, *Synchaeta
baltica*, is pelagic and phosphorescent. A few are parasitic.

Hydatina senta is one of the commonest Rotifers, and is usually to be met with swimming about amongst the algae of green ponds. It possesses a shortly elongated, cylindrical body (Fig. 346). The disc bears a circular cingulum separated by a groove from the trochus; in this groove are five prominences bearing stiff setae. The posterior end of the body tapers and ends in a bifurcated foot on which a pair of glands open.

A great many of the Bdelloida live amongst the roots and leaves of mosses, etc., and these can survive being dried up for a long time, the body shrinking and sealing itself up in the cuticle. Apart from the species which possess this power Rotifera as a class are short-lived, and this is especially true of the males.

In certain respects, such as the nature of their excretory organs and of their ovaries, the Rotifers show some resemblance to the Platyhelminthes; but they differ from them profoundly in the alimentary canal. The velum, in its typical form consisting of trochus and cingulum, has been compared to the ciliated band which encircles the Trochophore larva. This band like the velum is seen on close inspection to consist of a prae-oral and a post-oral loop. The Trochophore larva is found in the life-history of the Polychaeta and some of the more primitive Mollusca, and it is believed by many to represent a common ancestral form. Should the comparison of the Rotifera with this larva be a just one we must regard the Rotifera as having been derived from the common stock of Annelida and Mollusca and to be therefore a very ancient group. There are however difficulties which arise when the comparison is carried into details, so for the present it is better to regard the Rotifera as a completely isolated phylum.

Leaving out of sight a few parasitic and aberrant forms the bulk of the Rotifera are classified as follows :

Class I. RHIZOTA.

Rotifera in which the foot is not retractile but forms a permanent organ of attachment. The animal lives in a tube.

Ex. *Floscularia.*

Class II. BDELLOIDA.

Rotifera which creep like a leech, using the foot as a sucker. They are provided with a protrusible adhesive proboscis on the dorsal surface, by which the anterior end of the animal may be made to adhere.

Ex. *Rotifer.*

Class III. PLOIMA.

Rotifera which swim freely, only occasionally attaching themselves by the forks of the bifurcated foot.

Ex. *Hydatina, Synchaeta, Notommata.*

Class IV. SCIRTOPODA.

Rotifera provided with hollow movable outgrowths **from** the body by means of which they leap.

Ex. *Pedalion.*

CHAPTER XXIII.

PHYLUM NEMATODA.

THE Nematoda include a very great number of species commonly termed Thread-Worms, or from the shape of their cross section Round-Worms. Certain species attain a great length, as long as a man, but more commonly they are small and insignificant and often microscopic. The general shape of the body is cylindrical, usually pointed slightly at each end; the surface is smooth with at most a few bristles, so that they easily insinuate themselves into the cracks in the damp earth, or between the tissues of the animals and plants in which for the most part they live.

As a rule, like most animals which pass their time in the dark, Nematodes are white or whitish-yellow in colour. The body is glistening and smooth, but not slimy. It is ensheathed in a thick cuticle, which is in some cases ringed. No locomotor organs exist, and the animals progress by wriggling, bending first to one side and then to the other. The cuticle, which is moulted about four times during the life of the animal, is secreted by an underlying layer called the ectoderm, in which no trace of cell limits can be detected, but nuclei are scattered in it. Along the middle dorsal and middle ventral line and along each side this layer is thickened and projects inward. The median ridges support two nerves, the lateral ridges two canals, which are believed to be excretory.

By these ectodermal thickenings the body-wall is divided into quadrants. In each quadrant there is a layer of longitudinal muscle-fibres of very peculiar appearance. In all muscle-fibres there is a patch of unmodified protoplasm surrounding the nucleus, the rest being modified into those fibrils which are the visible sign of a heightened contractile power. In the Nematoda, in which all metabolism is at a low ebb, only the outer-

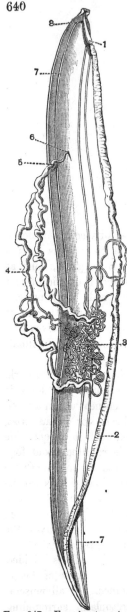

most layer of the muscle-cell is converted into fibrils, the great bulk consisting of unmodified protoplasm, which is often drawn out into an internal process running towards the dorsal or ventral ecto-dermal ridge. Muscle-cells of such a character are known only in Nematoda.

The body-wall, which has just been described, encloses a space which is traversed from end to end by the alim-entary canal. This space is full of a fluid in which amoebocytes float and it further lodges the reproductive system. It has no epithelial lining and appears to be a haemacoele. Besides this there is no circulatory system or heart.

The mouth is terminal and is usually surrounded by certain papillae or lobes, often three or six in number. It leads into an oesophagus, usually with a tri-angular cavity and thick muscular walls (Fig. 347). The oesophagus may be immediately followed by a second mus-cular bulb, called the pharynx, which sometimes has an armature of some bristles or spines. Both oesophagus and pharynx are parts of the stomo-daeum. Then follows the intestine, which is by far the largest part of the alimentary canal. The muscular oeso-phagus no doubt acts as a sucking organ, but there are no muscles and no cilia in the intestine. It is a simple tube formed of a single layer of cells, which both inside and out secrete a thin cuticle. Posteriorly it passes into a proctodaeum with muscular walls, and

Fig. 347. Female *Ascaris lumbricoides*, cut open along the median dorsal line to show the internal organs × 1.

1. The muscular oesophagus.   2. The intestine.   3. The ovary.   4. The uterus.   5. The vagina.   6. Its external opening.   7. The excretory canals.   8. Their opening.

this terminates in the anus situated a little in front of the end of the tail.

The nervous system consists of a ring round the oesophagus, which sends off in front six nerves to the mouth and its papillae, while behind it also gives off six nerves, the two more important of which run down the body in the above-mentioned median dorsal and ventral ectodermic ridges. Transverse commissures unite these main nerve-trunks at irregular intervals. With the exception of certain hairs and papillae to which a tactile sense has been ascribed, and in the free-living species certain eye-spots, no organs of sense are known in the Nematoda.

The excretory function is usually assigned to two long tubes which run along the lateral thickenings of the ectoderm. These tubes end blindly behind, but anteriorly the tubes approach one another ventrally and open by a common pore in the middle ventral line some little distance behind the mouth (Fig. 347). Each of these tubes is stated to consist of one immensely elongated hollow cell.

Nematoda are with few exceptions bisexual. The male is often smaller than the female and frequently has a curved tail. In both sexes the reproductive organs are tubular. In the male the organ consists of a long tapering tube much folded on itself opening into the proctodaeum close to the anus. In the uppermost and narrowest part of the tube there is a mass of protoplasm with nuclei ; lower down the mother-cells of the spermatozoa become separated, while the lowest part contains ripe spermatozoa. The names testis and vas deferens are given to the upper part and lower part respectively of this organ, but the whole is one continuous structure developed from a single cell in the embryo. In the female there are two similar tubes which unite to open in the mid-ventral line by an exceedingly short median piece termed the vagina. The vagina is situated about one-third the body-length from the head. In each tube it is usual to distinguish an upper ovary consisting of a mass of nucleated protoplasm, a middle oviduct where the bodies of the egg-cells have become separated from one another, and lastly a uterus where the eggs after fertilization are each provided with a shell. Each tube however, like the testis of the male, is developed in the embryo from a single cell.

In order to introduce the spermatozoa the male distends the vagina of the female by inserting two cuticular hairs, developed from the lining of the proctodaeum, called copulatory spicules.

It is a most peculiar characteristic of the Nematoda, which they share with the great group Arthropoda, that no cilia are found in any organ of their body. Even the spermatozoa have no flagella but move in an amoeboid manner. This absence of cilia, which are found in every other group of the animal kingdom except Arthropods and in many plants, has received hitherto no explanation. It is possibly correlated with the strong tendency of the protoplasm in both phyla to produce chitin. The eggs of Nematoda have a structure well adapted for histological investigation, and have been much utilized in researches on the behaviour of the nucleus before and during fertilization. As a rule the eggs are laid, but there are many species which produce their young fully formed.

Comparatively few species are free-living throughout their whole career, but these few are interesting. They are of small size and inhabit damp earth or mud, one family being marine. The mouth is often provided with movable spines and there are frequently eyes. These features suggest a relationship with the Chaetognatha, which, like the Nematoda, have a thick cuticle and only longitudinal muscles; and the idea receives support from the existence of a group of small marine "worms" called the CHAETOSOMATIDAE, which are probably also to be regarded as free-living Nematodes. These animals have two semicircles of movable bristles, situated one at each side of the mouth, and in addition a double ventral row of similar spines by means of which they creep about. Should this conjecture turn out to be well founded the Nematoda must be regarded as Coelomata, and it is interesting to speculate what has become of their coelomic sacs. Possibly, as in the Arthropoda, these have been reduced in size by the development of a haemocoele and are now represented only by the excretory and genital tubes; but until fuller details of the development both of Nematoda and Chaetognatha are known it would be unsafe to pursue these speculations further.

FIG. 348. *Ascaris lumbricoides*, cut open along the dorsal middle line × 1.

1. Oesophagus. 2. Intestine. 3. Testis. 4. Vas deferens. 5. Lateral excretory canals.

Taking the free-living forms as starting-point we can arrange the other families of Nematoda in an ascending scale of increasing parasitism, culminating in a form like *Trichina spiralis*, which is a perpetual parasite. This Nematode inhabits the intestine of its host (Pig, Man, &c.) where it lays its eggs. From these eggs larvae hatch out which bore through the walls of the intestine and get into the circulation, by which they are carried all over the body.

They encyst themselves in the muscles (Fig. 349) and do not develope further unless the flesh of their host is eaten by another animal, in whose intestine they become mature and then the cycle of development recommences. Their natural host is the Rat; the Pig is a secondary

FIG. 349. *Trichina spiralis*, encysted amongst muscular fibres. Highly magnified. After Leuckart.

host and being a gross feeder no doubt often devours rats and the remains of its own species, and thus the parasite is propagated. *Trichina* can however live perfectly well in Man, as the prevalence of the disease Trichinosis testifies. This disease is contracted by eating insufficiently cooked pork infested by the encysted larvae of *Trichina.* These become mature in the human intestine, and give rise to a second generation which cause severe and occasionally fatal symptons by boring through the intestinal wall.

Most Nematodes pass the earlier part of their existence in damp earth, during which they are known as Rhabditis larvae and bear a strong resemblance to some of the free-living forms. *Tylenchus tritici* forms the so-called Ear-cockles in wheat; these are brown galls replacing the wheat-grains and filled with encysted Nematodes. If the grains are beaten to the earth by rain the worms escape from the cysts and climb up the wheat stalks, where, after a generation which live in the flower, they again enter the grains. In *Sphaerularia bombi* the males and females live together in damp earth, but the fertilized female enters the body of a bee and here developes into a great sac filled with eggs. In *Syngamus trachealis* the eggs are laid in damp earth, and here develope into larvae, which are swallowed by poultry and develope in their windpipes into the sexual form, causing the disease called Gapes, which is often fatal. *Filaria sanguinis-hominis* lives in the human lymphatic glands; the embryos escape into the blood, whence they are

taken up by the Mosquito in whose body they develope. When the
mosquito bites they make their way again into the blood of man.
They are believed to be the cause of the strange tropical diseases
associated with the name Elephantiasis. But it would lead us too
far to enumerate all the modifications of the life-history produced
by parasitism. Suffice it to say that the Nematoda are perhaps the
most successful of all parasites; there is scarcely a phylum in the
animal kingdom which they do not attack. A smooth slippery
body which as a general rule causes little inconvenience to the host
and a low grade of metabolism requiring small supplies of oxygen
seem to have been the leading features in their success.

INDEX.

Names of genera are printed in italics. The figures in thick type refer to an illustration. In all cases the references are to the page.

Aard-vark, 549
Abdomen, 122, 152, 168, 172, 191
Abdominal, 405
Abdominal ganglion, 219
Abdominal pores, 364
Abdominal ribs, 474
Abducens nerve, 345
Abomasum, 572
Aboral pole, 71
Aboral sinus, 260
Acalephae, 66, 73
Acanthia, 189, 208
Acanthias, 411
Acanthobdella, 113
Acanthopteri, 409, 410, 415
Acarids, 198
Acarina, 198, 209
Acetabulum, 429, 467
Aciculum, 109
Acineta, 41
Acipenser. 396, 413
Acoela, 611, 624
Acopa, 330
Acridium, 186, 207
Acris, 453, 455
Acromion, 483
Actinometra, 283
Actinophrys, 26, 27, 40
Actinopterygii, 395, 413
Actinosphaerium, 26, 28, 40, 45
Actinozoa, 59, 73
Adambulacral ossicles, 256
Adder, 481
Adductor muscles, 226
Adhesive cells, 70
Adrenal body, 451
Aegithognathous, 517
Aeschna, 120, 121, 187
Aglossa, 452
Air-bladder, 392

Air-sacs, 511
Alary muscles, 175
Alauda, 519
Albatross, 518
Albumen-gland, 221
Alcedo, 519
Alces, 577, 601
Alcyonaria, 65, 73
Alcyonium, 59, 60, 61, 62
Alimentary canal, 86; of Arthropods, 133; of Echinoderms, 253, 272, 274; of *Hirudo*, 114; of Lamellibranchs, 230; of sea-urchin, 274; of *Sepia*, 241; of Vertebrates, 308, 319, 329, 347, 364, 375, 429
Alisphenoid, 395, 400
Allantoic bladder, 422
Allantois, 457, 545
Alligator, 459, 488, 491, 494
Alligator-turtles, 487
Alloiocoela, 611, 624 !
Allolobophora, 108, 116
Alpaca, 576
"Alternation of generations," in Coelenterata, 57; in Tunicata, 332
Alveoli, 272, 487
Alytes, 442
Amblystoma, 439, 443, 454
Amblystomatinae, 439
Ambulacral grooves, 250
Ambulacral ossicles, 255
Ambulacral plates, 268, 269
Amia, 396, 397, 413
Amiurus, 407, 414
Ammocoetes, 365
Amnion, 457
Amniota, 457
Amoeba, 13, 14, 40; in infusions, 17
Amoebocytes, 92, 98, 251
Amoebula, 26

Printed in the United States
By Bookmasters